EVOLUTION AND THE GENETICS OF POPULATIONS

A Treatise in Three Volumes

VOLUME 1

GENETIC AND BIOMETRIC FOUNDATIONS

Sewall Wright

THE UNIVERSITY OF CHICAGO PRESS
CHICAGO AND LONDON

To my wife Louise

Library of Congress Catalog Card Number 67–25533

THE UNIVERSITY OF CHICAGO PRESS, CHICAGO 60637

The University of Chicago Press, Ltd., London W.C.1

Published 1968

Printed in the United States of America

PREFACE

What was originally intended to be a single book has grown into three volumes. The subject is, nevertheless, much more circumscribed than the word evolution in the title implies. The central topic is population genetics: the deduction of the consequences for populations of the firmly established principles of the genetics of individuals, and the comparison of these deductions with observed properties of experimental and natural populations. The most important applications are to animal and plant breeding, to eugenics, and to interpretation of a phase of evolution.

It is convenient to distinguish four grand phases. First is the origin of life, about which rational speculation has become possible only in the last few years with the discovery of the chemical basis of genes and their action. Second is the origin of cells with perfected mechanisms for precise transmission of arrays of genes, for conjugation, and for orderly recombination of maternal and paternal elements. This was an enormously important phase and no doubt depended on the working out of increasingly precise genetic mechanisms in populations of cells, but population genetics, apart from peripheral speculations, takes the genetic properties of the perfected cell as a postulate. These properties give a reasonably firm base for interpretation of the third grand phase, the one primarily considered in this treatise, the transformation and splitting of species. These processes lead to the origin of new genera and higher categories, the fourth phase. The general inability of species to cross, however, precludes the operation of principles of population genetics, except through a succession of speciations. The factors involved in guiding the long-term course of evolution will only be touched on here.

Before going into the mathematical theory, it is desirable to consider its postulates and to consider which genetic processes are most generally important to avoid spinning out mathematics for its own sake. Considerable space has been devoted in volume 1 to reviewing genetic principles from this standpoint, including a review of the types of reproductive cycles in which mitosis, conjugation, and meiosis are combined in diverse ways. Another fundamental topic is the relation of genes to characters. Population genetics has relatively little concern with immediate gene action, but it is deeply concerned with the prevailing relations between genes and the complex characters with which the organism encounters its environment. Ultimately it is the relation of genes to selective value as a character that is most important. Too frequently, evolution is treated as a mere succession of favorable mutations rather than as a continual remolding of interaction systems in which the component genes cannot be treated as favorable or unfavorable in themselves.

At the other end of the bridge between the genetics of individuals and that of populations, the first question is how best to describe the characteristics of populations. Considerable attention is devoted to classification of patterns of variability, to statistical descriptions of such patterns, and to tests of significance of differences. Beyond description, the first step in interpretation depends on the mathematical theories of compound distributions and compound variables. The description and interpretation of systems of correlated variables is also of great importance. Path analysis is presented at length because of its usefulness in this respect, but also because it will be used extensively in volume 2 in deducing the consequences of systems of mating. Genetics and biometry are brought together near the end of volume 1 in the discussion of the genetic basis of quantitative variability.

The second volume will be concerned with mathematical theories of change in gene frequency as the elementary process in the transformation of populations. The effects of the directed pressures of recurrent mutation, of selection of various sorts, and of immigration, and the undirected effects of various random processes will be considered as they manifest themselves in populations of diverse structures.

It would seem desirable to illustrate each principle with a concrete example, but this leads to gross oversimplification. In nature, all sorts of processes are inevitably occurring simultaneously, and even in the best-planned experiments this phenomenon is largely unavoidable. For this reason, discussion of experimental evidence on inbreeding and selection is deferred to volume 3, where it is taken up along with the genetics of natural populations and with the general conclusions on the bearing of population genetics on evolutionary theory.

The point of view and the aspects emphasized necessarily reflect my experience. As a graduate student (1912–15), I was interested primarily in the interaction effects of genes, but was exposed to the earlier ideas on population genetics while assisting W. E. Castle in selection experiments and through frequent contacts with E. M. East.

I became more deeply involved during the next ten years, which I spent in the Animal Husbandry Division of the U.S. Department of Agriculture conducting experiments on the effects of inbreeding and crossbreeding on guinea pigs and trying to relate principles of population genetics to problems of livestock improvement. I wish to acknowledge here my great debt to G. N. Collins of the U.S. Bureau of Plant Industry for his warm encouragement and his criticisms during this period when I was developing my basic ideas.

My research program in the Zoology Department of the University of Chicago (1926–54) was about equally divided between continued studies of factor interaction in the guinea pig and theoretical studies of population genetics, two fields somewhat more related, for reasons implied above, than is immediately apparent. I owe much during this period to general discussions with my colleagues B. H. Willier and G. K. K. Link. From W. C. Allee, A. E. Emerson, and T. Park I learned most of what I know about the ecologic aspects of evolution. Among those outside the University of Chicago, I am indebted to J. L. Lush for our frequent discussions on the population genetics of livestock and to Th. Dobzhansky for the opportunity to collaborate with him in applying theory to studies of natural populations. I also owe much to the British investigators in population genetics whom I met during a year spent at the University of Edinburgh on a Fulbright Professorship.

After retiring from the University of Chicago, I was able to continue theoretical studies at the University of Wisconsin with encouragement especially from M. R. Irwin, R. A. Brink, and J. F. Crow, for which I am grateful. I am grateful to the National Science Foundation for a grant which has made this work possible.

CONTENTS

CHAPTER 1

Introduction

The term evolution means literally a rolling out. In addition to uses that are close to this literal meaning, it appears in discussions of change in things as diverse as stars, mountain ranges, the common law, and, as will be the case here, kinds of living organisms. It is not used to indicate all kinds of change, however. There is ceaseless change on the surface of the ocean but this is not an evolutionary process. Evolution always involves to some extent the opposite idea of persistence. It always refers, in short, to processes of cumulative change.

The idea of organic evolution seems hardly to have occurred at all to the ancient Greeks. Anaximander taught that "living creatures arose from the moist element as it was evaporated by the sun" and that "man was like another animal, namely a fish in the beginning," but this only very remotely foreshadows our present concept (cf. Nordenskiöld 1928). Greek thought oscillated between Parmenides' concept of eternal absolute being and Heraclitus' concept of ceaseless flux, without arriving at a philosophy of evolution. Aristotle indeed arrived (by observation of chick embryos) at a concept of development that was evolutionary in the broad sense, but only from the descriptive standpoint: his doctrine of epigenesis, the step-by-step elaboration of complexity from initial simplicity, contrasted with the doctrine of preformation of Hippocrates. This process, however, was in his view merely the more or less perfect realization of an eternal "form" rather than the development of a unique being.

Aristotle laid the foundation for a natural classification of some 500 kinds of animals by a clear perception of the difference between homology and analogy. He recognized a primary cleavage between the animals characterized by (red) blood and those he considered bloodless. The four great "genera" of the former were the viviparous quadrupeds (with which man and the cetaceans were recognized to be closely related), the birds, the oviparous quadrupeds (with which he came to associate the snakes), and the fishes (which he sometimes split into two groups—the oviparous bony fishes and the ovoviviparous cartilaginous selachians). The four great "genera" of the bloodless animals were the cephalopods, the higher crustacea, the insects (plus spiders, myriapods, annelids and other worms), and the shellfish (including not only molluscs with shells but also barnacles, sea urchins, and sometimes tunicates). He also knew of sea anemones, sponges, and other forms he considered to be intermediates between plants and animals. This hierarchic system might perhaps have suggested an evolutionary history had not permanence of "form" been so fundamental in his philosophy, and had he not also believed firmly that such permanent "forms" as those of all shellfish, many insects, and even eels might be realized abruptly in appropriate material by spontaneous generation.

There was no advance toward our present concept of evolution for nearly two millennia. The separate creation of all types of plants and animals and of man was a universally accepted tenet of religion in Europe, associated with ready acceptance of spontaneous generation. Even in the early seventeenth century, the chemist Van Helmont seriously set forth a recipe for producing mice from fermenting grain and dirty rags.

There could be no firm theory of evolution until the question of spontaneous generation was settled in the negative at least for all but the beginning of life. In 1680 Redi made an experimental test of the prevailing belief that maggots are generated spontaneously in decaying meat. He found that maggots did not appear in meat in vessels covered with gauze but that they did emerge from flies' eggs laid on the gauze. Redi was puzzled by the origin of internal parasitic worms and by the grubs in plant galls. The latter was soon correctly explained by Malpighi and others, but it was not until the nineteenth century that von Siebold cleared up the matter of the intestinal worms. Leeuwenhoek in the late seventeenth century maintained that the microorganisms he had discovered were produced only by their own kinds, but Needham made experiments about 1750 that seemed to prove the contrary. Spallanzani soon refuted Needham by more carefully conducted experiments. Nevertheless, it was not until the middle of the nineteenth century that Pasteur fully convinced the scientific world that there is no spontaneous generation of bacteria.

Organisms that produce offspring of a widely different type, which had often previously been described as belonging in a different higher category—colonial and solitary salps, polyp and medusa, fluke and redia—were found in the nineteenth century (the first by Chamisso, the others by Steenstrup) to be ones in which there was a wholly orderly alternation of generations. These instances could be interpreted as developmental phenomena not essentially different from such well-known metamorphoses within single life cycles as those of maggot into fly, caterpillar into butterfly, and tadpole into frog.

From ancient times, the existence of diverse races of men and breeds of domestic animals gave evidence of persistent minor changes in type within species. Belief in the transmission to the offspring of special characters acquired by the parents from their individual experience has also existed since antiquity, along with the belief in the persistence since creation of the essential types. These concepts opened the way for the speculative minds of eighteenth-century France to question whether such transformations might not have gone further than had been supposed. By the middle of the eighteenth century the idea of evolution was very much in the air.

Maupertuis (1698–1759) seems to have been the first to propose a general theory of evolution (cf. Glass 1959). This was based on a conception of heredity as an assemblage of numerous separate particles, which he derived from a study of a history of a four-generation human family in which polydactyly was inherited in what can be recognized now as the mode of a typical dominant gene. He noted that it could be transmitted to some of the offspring either by an affected mother or an affected father, to the confusion of both the ovist and animalculist schools of preformation. He suggested that the particles received from parents are distributed to appropriate parts of the body, from which representatives are reassembled for transmission to the offspring.

He supposed that changes in heredity might occur either from changes in these particles, brought about by climate or nutrition, or by irregularities in their distribution. He recognized the importance of natural selection of a crude sort. He also understood the importance of isolation in the origin of species. It may be seen that he anticipated to some extent all the major theories of evolution of the next century-and-a-half, except vitalism, which he rejected; and he insisted on the necessity of experiment in evaluating their roles. Maupertuis made little impression, however, on the biologists of his time or later, and even his initial great reputation as a physical scientist, based on his part in the first determination of the flattening of the earth at the poles and on his discovery of the principle of least action, suffered a precipitous decline from the grossly unjust, but devastating, ridicule of Voltaire.

Buffon (1707–86) is often credited with being the first pioneer of the modern doctrine of evolution, but according to Lovejoy (1959) it is questionable whether he did more to advance or to retard evolutionary thought. At first Buffon considered species to be merely arbitrary manmade groupings of more or less similar individuals, but in most of his later writings he adhered firmly to the doctrine of the separate creation of species, which he now believed could be sharply defined by the incapacity to produce fertile progeny from crosses, if crossing was at all possible. He did, however, support a limited evolution within species on the basis of the inheritance of effects of climate and nutrition. Still later, he came to recognize that the infertility of hybrids was a matter of degree, and he was willing to speculate on the possibility of a more extensive evolution within the genus or family than he had earlier.

A belief in the rigorous delimitation of species, even though associated with a belief in their permanence, was probably a necessary step toward a sound evolutionary theory. The attempt to define species rigorously goes back to John Ray (1627 ?–1705); and we have noted Buffon's contribution to the criteria. This concept reached a first culmination in the work of Linnaeus (1707–78). His invention of the binomial system of nomenclature and his attempt to apply generic and specific names to all known animals and plants firmly established the species concept and stimulated a multitude of collectors and museum men to the task of cataloging creation. Linnaeus listed 4,236 species of animals. This has increased to an estimated million, and the end is not yet in sight. Although the belief by Linnaeus' followers that the species is a sharply discrete entity possibly delayed somewhat the acceptance of organic evolution, it should be noted that Linnaeus himself in his later work found so much difficulty in defining species that he, like Buffon, questioned whether it might not be the genera, rather than the species, that were separately created.

In England, Erasmus Darwin (1731–1802), grandfather of Charles Darwin, developed a general theory of evolution on the basis of the inheritance of acquired characters. It was highly speculative and had little effect on biological thought.

The leading evolutionist at the turn of the century was Lamarck (1744–1829), whose name has come to be most frequently associated with the theory of evolution by the inheritance of acquired characters. This, however, was not the real core of his theory, which was that a tendency to progress from inert matter to a simple form of life, and thence up a single ladder of life toward man, was of the very essence of existence. This progress, he held, tended to be diverted along one line or another by the intractability of the environment. Here he stressed the inheritance of specially acquired characters, not as impositions from without, as supposed by Buffon, but as the result of active adaptation. Defeat and degeneration were also possible. In his later writings, however, he recognized major branchings of the line of ascent.

Cuvier (1769–1832), the leading comparative anatomist of the time and the founder of paleontology, threw the weight of his great influence against these evolutionary speculations. The fossil mammals that he described seemed too different from living forms to be ancestral. He developed the idea of a succession of eras each initiated by divine creation and terminated by a worldwide catastrophe, the last of which was the biblical flood.

Cuvier's associate Geoffroy Saint-Hilaire (1772–1844), on the other hand, accepted evolution, which he based on the occurrence of gross abnormalities that were known to appear in many organisms, including man. Although these are rarely viable, he maintained that occasional viable deviants could have provided the material for evolution by abrupt steps. The consensus of the zoologists of the time, however, was that this idea, as well as Lamarck's, had been completely discredited by the vigorous criticisms of Cuvier.

The matter of geologic time was a crucial one. Archbishop Ussher in the 1660's had calculated from the genealogies in Genesis that creation had occurred in 4004 B.C. As it was clear from well-documented history that there had been no appreciable evolution in the second half of the period since 4004 B.C., the evolutionists obviously required an enormously greater age of the earth for their theory. Even a scientist who did not accept evolution, but was trying to explain the cause of the structure of the earth's crust by events in a permissible past time, had to strain his integrity severely to account for thick formations high in the mountains that were largely composed of objects that looked extraordinarily like fossilized shells. Buffon insisted on a minimum of 75,000 years since creation for his theory about the earth's crust. Cuvier was able to accept the Noachian flood and its chronology, but he required a drastic reinterpretation of the biblical account of creation to accommodate the long succession of geologic periods, each with its characteristic forms of life and its terminal catastrophe. His views were widely accepted as the best possible reconciliation of theology and science and, in particular, were preferred to the uniformitarian interpretation presented by Hutton in 1788 under which "all past changes have been brought about by the slow agency of existing causes."

With the publication in 1830 of Lyell's *Principles of Geology*, however, the uniformitarian concept seemed inescapably demonstrated and came to be generally accepted, with its implication that the earth was millions of years old. The argument against evolution because of gross inadequacy of time was considerably mitigated. Nevertheless, Lyell did not extend his uniformitarian principles to organisms. He vigorously maintained that each species had been separately created and that the widely different faunas of successive geologic periods required separate acts of creation at widely scattered intervals.

Lyell's influence thus reinforced that of Cuvier in 1830 in delaying the acceptance of evolution. For nearly thirty years no biologist of high standing publicly professed belief in the theory. The gap was partly filled by others. A popular writer, Robert Chambers, called attention in 1844, forcefully if not always accurately, to the plain implications of Lyell's geology in conjunction with Cuvier's comparative anatomy and especially of Richard

Owen's concept of homology between species, exhibited by corresponding organs often with different functions, or with no function in some species. Chambers also called attention to the evidence from embryology. His book was immensely popular and probably did much to prepare the general public. So did the writings of Herbert Spencer in the 1850's, which brought out the incongruity, when applied in detail, of Agassiz' idea that the system of homologies on which the hierarchic natural classification was based reflected the thought of the creator.

This brings us to Charles Darwin (1809–82) whose *Origin of Species* (1859) finally precipitated the great revolution in thought brought about by the general acceptance of the theory. Darwin himself had become convinced of evolution as early as 1837, to a large extent because of his observations as a naturalist on the voyage of the "Beagle" to South America. He was impressed by the restriction, in the main, to South America of large groups of mammals—including both living and extinct forms (armadillos, sloths, anteaters, certain groups of rodents and of monkeys)—and the absence of important old world groups, with no apparent climatic reason. He also was impressed by the occurrence of multiple species of endemic genera on the Galápagos Islands, which was easily accounted for by common descent from a few immigrants but incongruous under separate creation. After years of accumulation of data on all aspects of the problem, including his theory of the cause of evolution (natural selection), he was brought to publication by receipt of a manuscript from A. R. Wallace, who had arrived at the same theory as a result of similar observations on the distributions of animals in the Malay region. The papers were presented simultaneously. The massive accumulation of data on all lines of evidence—classification, comparative anatomy, embryology, geographic distribution, and paleontology—and a plausible theory of causation that was wholly in the spirit of science and that covered a much wider range of possible evolutionary changes than did the inheritance of acquired characters, rapidly carried conviction to the biologic world and the literate public, where all previous presentations of evidence and theory had failed. The vigorous support and dialectical ability of T. H. Huxley played no mean role in this result.

The theory of natural selection was, of course, not wholly new. Maupertuis had stressed the inevitable elimination of all accidental variants that were unable to function adequately, and this had not escaped the attention of others. No one, however, had brought out so clearly the creative effect of selection operating on minute random variations in viability and reproductive capacity over a long period. Darwin and Wallace saw that among such variations there was an appreciable chance that some would be slightly superior to the prevailing type and would increase in numbers by natural selection, that these would ultimately establish a firm base from which a second small advance might be brought about in the same way, and so on, until a complicated adaptive change had emerged that would be utterly inconceivable as the result of chance at a single step. The failure to perceive this had been the fatal defect in Geoffroy Saint-Hilaire's theory.

The theory of natural selection rested on logical deductions from certain very general observations. These included the observation that variability can be found in all species that are carefully observed and that at least some of this variability is hereditary. The assumption of an appreciable chance that occasional slight accidental deviants would be superior could not be supported from direct observation of wild species, but it obtained support from the success of artificial selection in livestock, to which Darwin devoted much attention. Malthus' principle that all populations tend to increase in geometric ratio gave plenty of scope for natural selection.

There were, however, weaknesses and difficulties. Darwin recognized that the whole subject would remain in an uncertain state as long as the principles of heredity and variation remained obscure. He devoted great effort to remedying this situation by experimenting with flowering plants and pigeons, but he did not arrive at basic principles. Unfortunately, he never became aware of the simple rules of heredity that Gregor Mendel discovered in his experiments with the garden pea and reported on in 1866.

Darwin—and his critics—recognized other difficulties in the supposed discreteness of species and in accounting for the sterility of hybrids, for the origin of complex characters that seem useless until the whole complex is complete, for the origin of qualitative novelties, and for the numerous species differences that seemed of no value. He met these objections with varying degrees of plausibility.

The difficulty that troubled him most was the supposed swamping effect on new variations of random mating within the population. An engineer, Fleeming Jenkin, showed mathematically from the then generally accepted principle of blending heredity that the variability would be halved in each generation, except for new variability. Only variants with an enormous selective advantage would have a chance of being established by natural selection, contrary to Darwin's postulate that only occasional minute variations could reasonably be expected to be favorable. Darwin realized the force of this objection and in his later writings, he shifted more and more toward directed instead of accidental variation and thus to the inheritance of acquired characters, while still contending that natural selection was also important.

Along with the universal acceptance by biologists of evolution as a fact, there came to be increasing dissatisfaction, during the latter part of the nineteenth century, with natural selection as the master theory of causation. An objective account of the chaotic state of thought on this subject may be obtained from Vernon Kellogg's *Darwinism Today*, published in 1907 seven years after the rediscovery of Mendelian heredity but before this had a serious impact on the theory of evolution.

There was a bewildering array of theories proposed as supplements or alternatives to natural selection. Moritz Wagner in 1868 insisted on the all-importance of migration and spatial isolation in the actual splitting of a species from its parent species. He attributed the divergence in characters to the impact of different environments and to the incorporation of these changes into heredity, as supposed by Buffon. Isolation had, of course, been responsible for the peculiarities of geographic distribution that had especially impressed Darwin, but he had not recognized fully its importance in the actual origin of species. Wagner's emphasis on isolation is a useful supplement to Darwin's original theory, if divergence is explained by selection in relation to somewhat different conditions of life instead of by imposition of the effect of these directly on heredity.

Gulick (1872, 1888, 1905) agreed with Wagner in the importance of spatial isolation, but urged the likelihood of purely random changes in heredity in bringing about gradual divergence. His theory was stimulated by his observation of the apparently "nonutilitarian" character of the difference among the land snails of the endemic family *Achatinellidae* in the various, ecologically very similar, mountain valleys of the Hawaiian Islands. To some extent this recalls the role Maupertuis attributed to chance. This aspect of isolation can also be a useful supplement to natural selection in suggesting a possible explanation of species differences that are difficult to explain exclusively by selection.

Nägeli (1884) postulated a perfecting principle as an essential property of life that tends to bring about evolution automatically, largely irrespective of the environment. This was very close to Lamarck's view.

Cope (1887) developed a theory (archaestheticism) in which a primitive consciousness was the essence of all existence and thus preceded life and evolution. He stressed active opportunistic adaptation to the environment rather than automatic unfolding of trends. He assumed that the degrees of activity of the various growth centers in the body, determined by use or disuse, and the reciprocal inhibitions among such centers, determine the course of evolution of form, since in his view the developmental processes of the successive generations largely constitute the stream of heredity. The "élan vital" of Bergson was a wholly opportunistic vital principle, perpetually branching in the process of searching out possibilities of expansion against the resistance of inert matter.

More recently the paleontologist Osborn (1934), after considering and rejecting all other theories (natural selection, inheritance of acquired character, orthogenesis, mutation) as more than accessory in the course of evolution of such groups as the Titanotheres and the Proboscidea, concluded that the main cause is a perfecting principle he called aristogenesis.

The "emergent evolution" of Lloyd Morgan resembles these vitalistic doctrines in putting the essential principle beyond the scope of scientific analysis. It differs in holding that life, and later, consciousness, as well as certain morphological characters, "emerge" abruptly at certain stages in the course of evolution. The presentation of such views as scientific (rather than metaphysical) theories misconstrues the nature of science. It is, of course, obvious that the only direct knowledge anyone possesses is of his own stream of consciousness, but this is not verifiable by anyone else. The scope of science is that of verifiable knowledge and is thus restricted to the public, objective aspects of phenomena. The building up of the structure of science is immediately brought to a stop by postulating any property that is more than a convenient name for some verifiable phenomenon.

The doctrine of orthogenesis was advanced by Eimer toward the end of the last century to account for the preferential occurrence of variation in particular directions, rather than at random, which he believed he had observed in the study of color patterns in various kinds of organisms (butterflies, lizards, birds). He vigorously disclaimed any intention of proposing a vitalistic factor. What he maintained was that the organization already attained by evolution restricted the possible lines of variation, and that direction was given to variation by long persistence of the same environmental factors and the inheritance of this effect. Probably no one would question the first part of this, the only question being whether Darwin or anyone else in referring to random variation has ever meant more than variation in those directions which can occur accidentally on the basis of what the organism already is. Eimer's theory is thus essentially the same as Buffon's.

The term orthogenesis came to be used, especially by paleontologists, for a supposed organic inertia that could cause evolution along a certain line to overshoot the mark and lead to extinction. Orthogenesis has been invoked for the evolution of such characters as progressively greater total body size, more complicated folds in shells of ammonites, more or larger spines, longer horns or tusks, and so forth. This does not necessarily imply a vitalistic principle, but it does attribute a remarkable property to heredity. It should be added that the term is sometimes used merely descriptively for any long-time trend in evolution, in which case it is compatible with natural selection as well as with the inheritance of environmental effects.

Kölliker (1864) revived Geoffroy Saint-Hilaire's theory that evolution proceeds by large steps, under the name of heterogenesis. In its more extreme forms, for example in the hypothesis that the first bird came from a reptile's egg, this postulates such an enormous improbability by any natural means that it amounts to a special form of separate creation. The mutation theory of de Vries (1903) was more modest and had the great merit that it was the first of the post-Darwinian theories to be based on extensive observation of successive generations. He based his theory on researches with an American species of evening primrose, *Oenothera Lamarckiana*, that he found growing wild in the Netherlands, evidently from plants that had escaped from cultivation. This species was associated with a number of forms that differed in the multiple minor respects characteristic of species of the same genus, but the circumstances indicated common origin in the locality. Experiment showed that typical *O. Lamarckiana* continually produced small proportions of these forms, and that they either bred true (*O. gigas*) or lacked good pollen and segregated sharply into *O. Lamarckiana* and the new type, on fertilization with *O. Lamarckiana* pollen. The latter group were obviously not true species in the Linnaean sense, but the complexity of the differences that segregated together was at least suggestive of the origin of species by single "mutations".

De Vries carried out breeding experiments with many species of plants other than *O. Lamarckiana*. In the course of these he rediscovered (in 1900) Mendel's forgotten principles of the inheritance of simple character differences. These were also rediscovered in the same year by Correns and by Tschermak, with verification soon after by Bateson and others who were engaged in breeding experiments. It was immediately recognized that Mendelian heredity was a phenomenon that should throw new light on the theory of evolution.

About a decade before the rediscovery of Mendel's principles of heredity came the initiation of the statistical study of resemblances among related individuals by Galton (1889) and later by Karl Pearson (1894). The viewpoint of this "biometric" school came into violent conflict with that of Mendelian genetics after the rediscovery of Mendelian heredity in 1900. Population genetics arose from the attempts to reconcile these viewpoints.

CHAPTER 2

Modes of Inheritance and Reproductive Cycles

It is desirable at this point to carry our historical survey back near the beginning of the last century to the development of the cell theory by Lamarck, Dujardin, and Robert Brown and to its formulation as one of the basic principles of biology by the botanist Schleiden and the zoologist Schwann about 1840. This is the doctrine that all higher animals and plants are composed of essentially similar microscopic units that are also similar to certain whole organisms of minute size (protozoa, protophyta), and that these cells, each a jellylike droplet bounded by a membrane and containing numerous internal organelles (especially a nucleus), is in a real sense a complete organism carrying on its life processes, largely by and for itself. It was soon established that cells arise only by fission of preexistent cells.

The development of methods of differential staining, in the 1860's, led to the recognition of the rodlike chromosomes as entities present in a characteristic number in the nuclei of cells of each species. These were found to divide and to be apportioned equally to the daughter cells at ordinary cell division in a process known as mitosis; they formed pairs in the cells that were to give rise to reproductive cells (spores in plants, eggs and sperms in animals) which by segregating in the process of meiosis led to single representation in the reproductive cells, capable of restoring the double number on fertilization.

The precision of the processes by which persistence was insured through cell division and fertilization soon suggested that the chromosomes were the vehicles of heredity. Weismann (1892) drew the conclusion that whatever happened in cells that were not in the direct line of germ cells was a mere by-product of heredity, not involved in the stream of heredity itself. For the first time, he raised the question of whether there was any inheritance of acquired characters at all. He performed experiments with mice that indicated that at least one class of acquired characters, losses by mutilation, had no hereditary effect. Weismann became the founder of a school of evolutionists who were more Darwinian than Darwin in making natural selection the sole cause of adaptive advance.

Chromosomal Heredity

The possibility for more than empirical statistical treatment of the genetics of populations rests on the generalization that most hereditary differences follow certain simple rules that exactly parallel the observed mode of transmission of the chromosomes and their components in whatever form is studied. Table 2.1 shows the more important aspects of this parallelism in the

TABLE 2.1. Parallel principles in genetics and cytology

Genetics	Cytology
A. Usual rules (Mendel 1866)	
1. Persistence of genetic units (genes)	1. Persistence of chromosomes, including details of their fine structure
2. Single set of units transmitted by the germ cells	2. Single (haploid) set of chromosomes transmitted by gametes
3. Persistence of double set of units in individuals	3. Double set (zygoid) in fertilized egg and somatic cells
4. Segregation of alleles in the formation of germ cells	4. Pairing and disjunction in meiosis of homologous parts of homologous chromosomes
5. Random assortment of most nonalleles	5. Random assortment of nonhomologous chromosomes
B. Principles of linkage	
6. Partial linkage of some nonalleles	6. Limited number of chiasmata in meiosis at which exchanges occur between paired homologs
7. Crossing over in 4-strand stage	7. Chiasmata in 4-strand stage
8. Characteristic number of linkage systems	8. Agreement of number of chromosomes in haploids with number of linkage systems
9. Linear order of loci in linkage systems	9. Linear pattern of chromosomes
10. In many cases (especially in animals) one sex regularly heterogametic in one linkage system in which all genes exhibit sex linkage	10. Genetically heterogametic sex with one unpaired chromosome (XO) or unequal pair (XY)

ordinary course of events in the higher plants and animals that were studied first (e.g., garden pea, sweet pea, maize, snapdragon, the laboratory rodents, the domestic animals, and especially the fly, *Drosophila melanogaster*).

The first five genetic phenomena in Table 2.1 were discovered by Mendel (1866) in his studies of the garden pea before there was any knowledge of the corresponding cytologic phenomena. The chromosomes were first clearly recognized and their behavior in cell division throughout the reproductive cycle was worked out later in the nineteenth century by Flemming, O. Hertwig, Van Beneden, Strasburger, Boveri, Wilson, and others without any knowledge of Mendelian heredity. The rediscovery of Mendelian heredity in 1900 independently by de Vries, Correns, and Tschermak was soon followed by recognition of the exact parallelism of the phenomena demonstrated by the very different techniques of the breeder and the microscopist (Sutton 1903).

The next group of parallels was established during the next two decades primarily by Morgan and his associates, Sturtevant, Muller, and Bridges, with the fly *Drosophila melanogaster*, and has gradually been verified in many other groups. Even more impressive was the parallelism between numerous aberrant genetic phenomena and corresponding aberrant chromosome behavior (trisomy by Gates (1907) and tetraploidy by Lutz (1907) in *Oenothera*, nondisjunction by Bridges (1916) in *Drosophila* and in *Datura* by Blakeslee (1928). In several cases, specific chromosome aberrations were deduced from genetic exceptions long before a technique was available for observing the fine structures of chromosomes with sufficient adequacy for cytologic verification. Ultimately complete parallelism was established for such phenomena as translocation, inversion, duplication, and deficiency of chromosome regions.

The discovery that all types of mutation, including chromosome aberrations and the corresponding genetic phenomena, could be induced wholesale by X rays was made by Muller in 1928 (working with *Drosophila*) and by Stadler in 1928 (working with barley and corn). Painter's discovery (1933) that a detailed correspondence could be established between particular bands among some 5,000 visible in the giant chromosomes of the salivary glands of *Drosophila* and particular gene mutations finally made possible an extraordinarily complete demonstration of parallelism. These rules are among our basic postulates in the deductive aspects of population genetics.

Linkage and Crossing Over

It is not practicable to go into a detailed account of linkage phenomena here. The most important facts for our purpose are the gametic ratios from the two kinds of double heterozygotes in the ordinary case of a diploid zygote. The recombination fraction is designated c.

Genotype	AB	ab	Ab	aB
AB/ab	$\frac{1}{2}(1 - c)$	$\frac{1}{2}(1 - c)$	$\frac{1}{2}c$	$\frac{1}{2}c$
Ab/aB	$\frac{1}{2}c$	$\frac{1}{2}c$	$\frac{1}{2}(1 - c)$	$\frac{1}{2}(1 - c)$

The recombination values for genes in the same chromosome may differ greatly in oogenesis and spermatogenesis as in *Drosophila*, in which there is no crossing over under ordinary conditions in spermatogenesis, but all values up to virtual independence in oogenesis; or as in *Bombyx* in which the situation is the reverse. In many cases, including seed plants, birds, and mammals as far as studied, the differences are slight.

Where more than two pairs of alleles are involved in the same linkage system, the loci can be arranged in a unique order by putting those with the lowest amounts of recombination next to each other. In the case of three loci for which all recombination percentages are small, the largest percentage tends to be the exact sum of the other two, but with larger ones it falls off from this because of double crossing over. A crossover in one region, however, interferes with crossing over in adjacent regions, so that the percentage of double crossing over tends to be less (at short distances) than expected from the product of the total probabilities for the two component regions. This interference is measured (inversely) by the coefficient of coincidence, which is the ratio of observed double crossing over to the above product. At fairly long distances this ratio approaches unity.

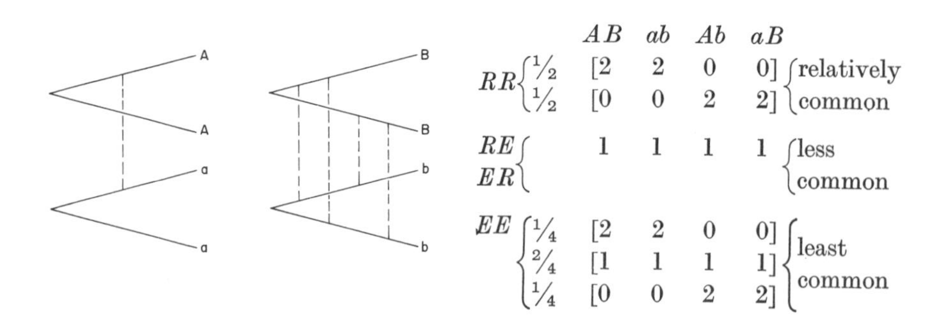

Fig. 2.1

The above mode of treatment is purely empirical. For a full insight into what happens it is necessary to turn to organisms in which all of the products of the reduction division can be recovered. It seems probable from all clearly worked out cases of this sort, e.g., *Neurospora* (Lindegren 1933), *Saccharomyces* (Winge and Laustsen 1937), *Chlamydomonas* (Pascher (1918), *Sphaerocarpos* (Allen 1930), and for special cases in such organisms as *Drosophila* (Bridges 1916), and *Habrobracon* (Whiting and Gilmore 1932) in which two of the products can be recovered, that crossing over always occurs in a four-strand stage. The typical pattern of assortment of two loci in different chromosomes is indicated in Figure 2.1, according to whether segregation is reductional (R) (no or an even number of chiasmata between centromere and locus) or equational (E) (odd number of chiasmata).

The pattern for two loci with coupling in the same chromosome (AB/ab) is as follows (Fig. 2.2), according to whether there is no chiasma (0), one (1), or two (2) between them.

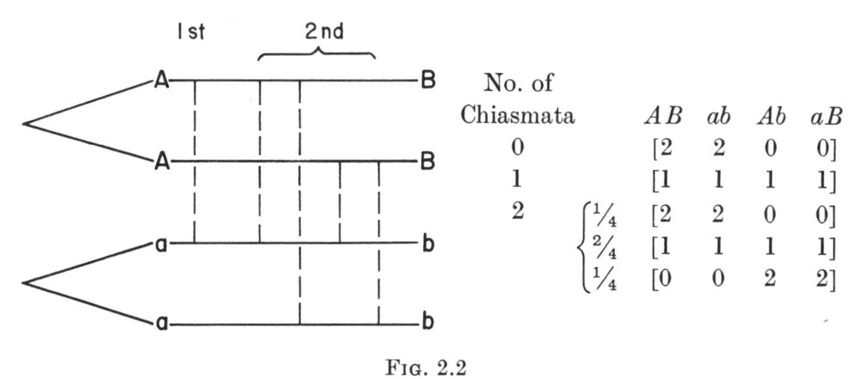

No. of Chiasmata		AB	ab	Ab	aB
0		[2	2	0	0]
1		[1	1	1	1]
2	1/4	[2	2	0	0]
	2/4	[1	1	1	1]
	1/4	[0	0	2	2]

Fig. 2.2

A rational theory of interference must obviously take account of the situation with respect to all four strands. It appears that exchanges do not occur between sister strands (Weinstein 1936). In certain cases evidence indicates that two-, three-, and four-strand double crossovers occur in the ratio 1 : 2 : 1 as indicated above, but this is not true in all cases (Whitehouse 1942). There is no decisive evidence yet on the actual mechanism of exchange.

The frequency of crossing over between two loci is not a constant quantity. It was shown by Bridges (1927) that in *Drosophila* it varies with the age of the mother in an irregular way. Plough (1917) found a marked effect of temperature in the neighborhood of the centromeres (increase with increased temperature). There is inhibition in the neighborhood of the ends of rearrangements. Inversion brings about the complete absence of detectable single crossovers within the inverted region in *Drosophila* oogenesis by relegating these to the polar bodies because of mechanical difficulties (Sturtevant and Beadle 1936). Inhibition of crossing over in one chromosome has the somewhat surprising effect of increasing it in others.

Plough's experiments demonstrated an 8-to-10-day lag in *Drosophila* in the effect of abrupt temperature changes on crossing over. This was found to approximate the interval between early prophase and laying and indicated that crossing over could not occur earlier than pachytene.

The recombination pattern is much more complicated in polysomic chromosomes than in disomic ones. In tetrasomics for example there are three possible kinds of heterozygotes with respect to a pair of alleles: *AAAa*, *AAaa*, and *Aaaa*. Cytologically, it appears that synapsis occurs only between two of the four zygotene strands at any given level (Darlington 1929). If the locus is so close to the centromere that the possibility of intervening chiasmata may be ignored, segregation may be treated as reductional between chromosomes (segregation patterns from $A_1A_2A_3A_4$:$A_1A_2 - A_3A_4$, $A_1A_3 - A_2A_4$ and $A_1A_4 - A_2A_3$ equally frequent). At the other extreme is the case in which the locus is so remote from the centromere that it may be assumed that there is an even chance whether two of the alleles in sister chromatids go to the same or opposite poles. Random association may be assumed, giving a probability of $1/7$ that the two genes in a gamete are duplicates (Haldane 1930). There is thus a probability of $1/28$ in this case that a particular gene in one chromosome of a tetraploid will be present in both chromosomes of a gamete. The probability for genes less remote from the centromere is thus between 0 and $1/28$.

Reproductive Cycles

The single set of chromosomes in gametes and the double set in individuals referred to in entries 2 and 3 of Table 2.1 are those characteristic of the higher plants and animals to which most early genetic studies were confined. It was well known that different cytologic relations were characteristic of other groups. As such cases came to be studied genetically, it was found that the parallelism of chromosome behavior and the principles of heredity continued to hold.

The customary classification of modes of reproduction as sexual or asexual is decidedly unsatisfactory in genetics. Reproduction is indeed often associated with the means of combining two lines of heredity, and this is often associated with sexual differentiation (femaleness: specialization for nutrition of offspring; maleness: specialization for bringing about union). On the other hand, there may be reproduction without syngamy and without sexual differentiation (as by budding), or without syngamy but by sexually differentiated females (parthenogenesis, apomixis); two lines of heredity may be brought together without sexual differentiation (conjugation in ciliates, fusion of fungus hyphae). Sexual differentiation may be restricted to gametes (some algae), or carried back to oogonium and antheridium (monoecious bryophytes), to male and female gametophytes (dioecious bryophytes), to spores and the organs that produce them of "asexually" reproducing sporophytes, to different flowers (as in maize), or to different sporophytes (dioecious seed plants). In animals, sexual differentiation may be restricted to gametes and the parts of the ovotestis that produce them. It may go back to the gonads of hermaphrodites or to separate male and female individuals.

Sexual differentiation is analogous to such devices as mating types in ciliates or in fungi (in which there are mating types in addition to sexual differentiation of oogonia and spermatia) or as self-incompatibility in many flowering plants. As important as such means of furthering biparental heredity (and care of the young in the case of sex) are in population genetics, they should not take precedence over the nuclear and corresponding genetic phenomena on which population genetics must be based deductively.

In dealing with modes of reproduction, it is convenient to consider primarily those based on the nuclear phenomena of the typical uninucleate cell in which these processes are remarkably standardized among organisms ranging from protophyta and protozoa to the highest plants and animals. The three basic nuclear phenomena in uninucleate cells are (1) equational mitosis, in which duplication and disjunction of all chromosomes leads to the production of daughter cells of the same genetic constitution as the parent cell; (2) the fusion of cells (syngamy by conjugation or fertilization), followed typically by nuclear fusion but in some fungi by persistence of a dicaryon; and (3) reduction or meiosis, in which pairing of homologous chromosomes and a single duplication, usually accompanied by exchanges, is followed by two divisions that sort out diverse haploid sets of chromosomes to four cells. Most of our discussion of population genetics will be based on these rather than on the processes in bacteria and viruses which were probably prior to such standardization in evolution. Certain phenomena in particular groups of higher organisms, however, require separate consideration.

Reproduction in which only equational mitoses intervene between generations should involve no nuclear or genetic change (except by mutational accident). The array of descendents, resulting wholly from such reproduction, constitutes a clone. Such reproduction may thus be termed clonogenic.

Reproduction from a cell that is the result of fusion of two cells, derived from different parents, gives biparental heredity whether the cellular fusion is followed by nuclear fusion or not. Even if the cells that unite come from the same individual (self-fertilization), different heredities may be brought together if the parent is heterozygous.

The process of meiosis sorts out one of what may be an enormous number of possible kinds of single sets of chromosomes from the double set of the nucleus in which it occurs. It leads to genetic diversity.

The reproductive cycle typically involves two phases, a zygoid (usually diploid) phase resulting from the fusion of two haploid (usually monoploid) gametes, and a haploid gamete-producing phase, that traces to a cell, derived by meiosis from a cell of the zygoid phase. Either phase may, however, be reduced to a single cell, a zygote that undergoes reduction on the one hand, or a haploid gamete produced directly by reduction on the other. More or less extensive clonogenic reproduction may intervene in either phase.

It is implied above that the haploid phase may actually carry more than a single set of chromosomes in whole or part and the zygoid phase more than two. The absolute number of representatives of a chromosome are customarily indicated by the terms monosomic, disomic, trisomic, tetrasomic, and so on, and of whole sets by monoploid, diploid, triploid, tetraploid.

If both the haploid and zygoid phases exist as more than single cells, they are usually very different morphologically and thus usually differ greatly in exposure to selective factors. These differences do not depend on the absolute number of chromosomes but apparently are related to the mode of reproduction. If clonogenic reproduction intervenes in either phase, there may be a great morphological difference between individuals that reproduce in this way and ones whose reproductive cells undergo meiosis, although in other cases the same individual may reproduce in both ways.

I. Cycle of syngamy and reduction
    A. Organism haploid. Union of gametes followed by reduction (many flagellates, most chlorophyceae).
    B. Alternation of haploid gametophyte and zygoid sporophyte.
        1. Phases morphologically similar (some chlorophyceae, most rhodophyceae and phaeophyceae).
        2. Phases dissimilar.
            a. Major phase haploid (bryophytes).
            b. Major phase zygoid.
                (1) Free but diminutive gametophyte or prothallium (most pteridophytes).
                (2) Female gametophyte remains in ovary, male gametophyte is a pollen tube (spermatophytes). In angiosperms the female gametophyte (embryo sac) typically contains only eight nuclei, the male gametophyte only three—a tube nucleus and two generative nuclei—one of which fertilizes the egg cell, while the other

unites with two other nuclei of the embryo sac to give rise to the triploid endosperm.

C. Organism zygoid.
   1. Unicellular organisms in which biparental heredity is not associated directly with reproduction. Meiosis followed by loss of all but one product, equational division of this, and either union of the products (autogamy) or reciprocal exchange of one of them with another individual (conjugation). Ciliates such as *Paramecium*.
   2. Multicellular organisms. Haploid gametes produced directly by meiosis. Some algae, most metazoa.

D. Zygoid females produce haploid eggs by meiosis. Haploid males produce sperms of their own genotype (many rotifers, hymenoptera, and scattered cases among other groups of insects).

II. Clonogeny (any number of sets of chromosomes)
   1. Exclusive clonogeny. Cyanophyceae and many bacteria as far as known. Sporadic species or races of higher plants that multiply only vegetatively or by apomixis, sporadic species of animals with ameiotic parthenogenesis and no males as far as known. The discovery of a variety of modes of biparental heredity in certain bacteria and viruses make it probable that exclusive clonogeny is always the result of recent evolutionary loss of mechanisms for syngamy.
   2. Predominant clonogeny. All unicellular plants and animals in which syngamy and reduction occur at all; some higher plants, colonial animals, sometimes with morphological alternation of generations as in hydromedusae. Animals that reproduce by ameiotic parthenogenesis except under unfavorable conditions (most rotifers, cladocera, aphids, and others).
   3. Clonogeny for relatively few generations. Reproduction by budding may be intercalated in either the gametophyte or sporophyte generation in many higher plants; fission in turbellaria, annelids, etc.; ameiotic parthenogenesis; paedogenesis in trematodes; in the fly, *Miastor*, etc.; polyembryony in hymenoptera and in armadillos.
   4. Sporadic polyembryony; identical twins in man, cattle, and others.

Special Cases

There are many special phenomena not included in the foregoing table. In the basidiomycetes, for instance, the haploid basidiospores produced in the fruiting bodies by meiosis give rise to haploid mycelia. Biparental heredity is brought about by fusion of compatible hyphae. These may belong to the same mycelium, with presumably the same sort of nucleus in some species, but in more cases there must be a difference at a mating type locus and in still more, in two such loci. This gives rise to a mycelium in which each cell contains descendants of both nuclei (a dicaryon); the fruiting bodies (sporocarps) are formed in this, and in certain cells (basidia) the two nuclei at last fuse and undergo meiosis to form in each case four haploid basidiospores, completing the cycle. In some plants and many animals, a sex-determining mechanism, one sex XX, the other XY or XO, brings about special modes of inheritance with respect to one chromosome, sex linkage, which requires special consideration. In the fly *Sciara*, males as well as females are zygoid, but combined cytologic and genetic analysis indicates that all paternal chromosome are lost in a monopolar reduction division so that transmission over two or more generations is like that in category I-D (Metz 1938).

Biparental heredity and recombination is achieved in certain bacteria and viruses by processes other than the standardized cycle of syngamy and reduction of higher organisms. Genetic units may be transmitted by molecules of DNA (transforming principle) from dead pneumococcus cells to living ones of a different genotype (Avery *et al.* 1944). Recombination occurs sporadically in *Escherichia coli* (Lederberg and Tatum 1946). Blocks of genes can be carried by bacteriophage from a strain of *Salmonella* that is lysed to one in which the phage is tolerated, in which it takes a place in the DNA of the bacterium (Zinder and Lederberg 1952). Among sufficiently related bacteriophages a mixed infection in a bacterium results in individuals of mixed heredity as a result of synthetic processes that seem to involve a succession of individuals in the population of prophages in the host (Visconti and Delbrück 1953). Population genetics takes an especially complicated form in organisms in which syncytia can occur. This is the case in many fungi, in which hyphae of different strains may fuse to form heterocaryons containing mixed populations of nuclei. There is the possibility of evolution of complex systems of haploid nuclei from such mixtures or from mixtures arising by mutation (for example, *Neurospora*) (Lindegren 1942). Pontecorvo and Roper (1952) find that fusion of different nuclei (caryogamy) occurs as a rare event in heterocaryons of *Aspergillus* species. Crossing over occurs with moderate frequency in the mitotic divisions of such nuclei with segregation and recombination as consequences. Haploidization of these zygoid nuclei also occurs with moderate frequency. A reproductive cycle of a more conventional sort occurs in both *Neurospora crassa* and *Aspergillus* by union of haploid gametes, followed by formation of reduced ascospores. There is also clonogenic reproduction from conidia containing one or more nuclei. As occasional caryogamy, somatic crossing over and haploidization occur in species in which the conventional cycle is unknown; it follows that the absence of such a cycle does not preclude evolution by recombination in these forms.

□ □

CHAPTER 3

The Gene

The results of the evolutionary process in a species are presumably registered largely in patterns of organization of its chromosomes. Following Sturtevant (1913) geneticists have described these patterns in terms of linear successions of genes. Population genetics, in its deductive aspects, has rested largely on the concept of the gene as an entity that has a certain frequency in the population under consideration. In recent years, however, the gene concept has come under considerable criticism. The sense in which we shall use the term thus requires consideration.

Mendel used the word "Merkmal" for his dominant and recessive alternatives. Bateson translated this into "character." Under a widely adopted hypothesis of Bateson and Punnett, the dominant alternative was looked upon as due to the presence of a "unit character," and the recessive merely as the absence of this. It was, however, soon established that hereditary differences in characters tend to split up into many independent pairs of alternatives, and that the effects of these are often difficult to identify as characters. It came to be recognized that unit (genetic) factor is a better term than unit character, and a little later Johannsen's term "gene" was welcomed as a distinctive one for the basic unit of heredity.

Meanwhile, Cuénot (1904), on analyzing the heredity of coat color of the mouse, had clearly demonstrated a set of multiple alternatives as well as several independent pairs, and it came to be recognized at least by 1915 that multiple sets were not uncommon. It was natural, at first, to extend the presence and absence hypotheses to multiple alleles by postulating differences in the quantity of the gene. It had, however, been recognized that unitary genetic differences might lead to multiple character differences, and it soon became clear that the pleiotropic effects of multiple alleles do not, in general, fall into the same order.

Some ambiguity has remained in the frequent use of the term "gene" to mean both the common substrate of alternative units and a particular one of these. The term "locus" will be used here for this substrate, "gene" will be restricted to a particular representative of a locus considered by itself, and "allele," a contraction of Bateson's term "allelomorph," will be used for a particular representative in relation to others. "Allelic" will be used in the sense of Bateson's (1909) definition of "allelomorphic": "alternative to each other in the constitution of the gametes." Allelic is thus used to describe a genetic relation, not a physiologic one as has been done in recent years by some authors. Alleles may be complementary as well as noncomplementary and in extreme cases may seem to have no more in common in their differences from type than random nonalleles, although one can usually trace a thread of physiological similarity of some sort through an allelic series. From studies of loci with effects for which there are very delicate means of recognition (antigen, self-incompatibility) it appears that there may be hundreds of alleles at a locus.

Conceptions of the locus as a physical entity were, of course, guided by cytologic observations. The early demonstration of the persistence of individualized chromomeres and their pairing at synapsis (Wenrich 1916), Belling's demonstration (1928) of as many as 2,000 such particles in favorable plant material, culminating in the demonstration of some 5,000 identifiable bands in the salivary chromosomes of *Drosophila*, and the location of particular loci in or near such bands by Painter (1934) and Bridges (1935) led to general acceptance of the particulate nature of loci. Morgan's (1922) and Muller's (1929) estimates of average size of loci as in the millions in terms of molecular weight seemed to give adequate scope for patterns capable of mutating at an enormous number of different sites within a single locus, and thus an adequate basis for the complexity of observed allelic series.

It will be convenient to quote from a review article by Wright (1941*c*) as a basis for consideration of the changes brought about by more recent studies.

The breeder in referring to a gene means merely a member of a set of alternative conditions in the hereditary material. Whether the condition in a particular case is one of addition, of loss of material or of rearrangement, he does not ordinarily know. In cytogenetics and physiological genetics, however, it becomes desirable to be able to refer to the actual material at a certain locus as it is when affected by a particular allele. The term gene has come to be used also in this sense.

The question whether the genes in this sense are discrete entities or merely regions in a continuum is an old one. Breeding data merely prove that a certain difference behaves as unitary in the tests that have been applied, but there is no assurance that such a difference may not turn out to be composite by other tests. From this point of view, genes are merely regions of the chromosomes within which crossing over or other breakage has so far not been observed to occur.

There are a number of possibilities both with respect to the physical and physiological discreteness of the genes. 1) The genes may be physical units, separated from each other by nongenic material within which exchange at crossing over or breakage under x-ray treatment is likely to occur. 2) The genes may be physical units in immediate contact with each other, the breaks occurring only between genes. 3) The genic material of a chromosome may be a continuum capable of breakage at any point.

If the gene is a physical unit, it may act as a physiological unit, but it is also possible that it may be composite physiologically or that a group of adjacent genes may act together differently than if widely separated. If the genic material is a continuum, there may be more or less widely spaced centers of physiological activity separated by relatively inert material or there may be overlapping regions that act as physiological units.

At that time geneticists were beginning to find, somewhat to their surprise, that there had been two viewpoints toward the gene all along. In one view, the genes were looked upon as structural units, regions in the chromosomes within which, as noted above, crossing over or other breakage had so far not been observed to occur. Others had always looked upon the gene primarily as a physiological unit. Goldschmidt (1946), who was of the latter school, felt that the whole gene concept had been destroyed when it was discovered that chromosome rearrangements in which the nearest breakage point is at a considerable distance from a locus often produced position effects simulating those of mutations in the locus in question. These observations were not at all disturbing to those who thought of the gene as merely a unit of heredity with physiological relations with other such units. Sturtevant (1925), the discoverer of position effect, had taken this latter view.

The idea that genes control the properties of cells and hence the characters of organisms by determining the enzymes that are present traces to the earliest years of the century (Cuénot 1904; Garrod 1902, 1908). The simplest hypothesis for such determination is that the pattern of the gene is somehow reflected in that of protein molecules synthesized under the influence of the gene or its primary product (Wright 1927). The extensive demonstration of one-to-one relations between genes and elementary metabolic processes in microorganisms that was initiated by the investigation of Beadle and Tatum in the mold, *Neurospora*, supported the concept of the gene as a pattern in the chromosome that has this template property, irrespective of whether it is a component of a larger block that does not suffer breakage or, on the other hand, is capable of occasional breakage internally. They gave massive support to the one gene one enzyme hypothesis.

The most revolutionary recent advances in genetics have been on the chemical side. It has been known since Miescher's work in the 1870's that chromosomes are largely composed of protein and nucleic acid, but until recently, nucleic acid seemed ruled out as the site of specificity by its supposedly monotonous tetranucleotide structure. Most geneticists up to the middle 1940's assumed that there was little choice but to suppose that specificity resided in the pattern of successive amino acids. Then came the discovery that the pneumococcus-transforming principle, derived from dead organisms of one type and capable of conveying the specificity of this type to living organisms of another type, was probably pure DNA (deoxyribonucleic acid) (Avery, MacLeod, and McCarty 1944). The biochemists, moreover, came to recognize that there was after all a virtually unlimited basis for differences in specificity in the sequence of four sorts of nucleotides in chains of indefinitely great length. The demonstration of constancy of DNA per chromosome set (Boivin, Vendrely, and Vendrely 1948, Mirsky and Ris 1949, Swift 1950) added to the confidence in its genetic significance. There followed Chargaff's demonstration (1950) of the equality or near equality in the numbers of guanine and cytosine nucleotides and of adenine and thymine nucleotides in the DNA extracted from a great variety of organisms, and the Watson–Crick model of DNA (1953) based on X-ray diffraction studies which gave significance to the above equalities. In this model, DNA usually consists of two complementary and oppositely directed but intertwined helices, each with a phosphate-sugar backbone, carrying at each level a purine or pyrimidine group with hydrogen bonding to its complement on the other strand (Chargaff's pairs). Watson and Crick noted that separation of these strands under certain cell conditions, and synthesis by each of its complement, provided a mechanism by which any specific pattern would tend to be maintained indefinitely, and one in which any change in this pattern, accidental or induced, would be followed by an equally persistent multiplication of the new pattern. It is not necessary for our purpose to go into the recent studies in which specific DNA molecules have been multiplied in the test tube (after priming in the presence of an enzyme [Kornberg 1962], or to speculate on the problem of the untwisting of the two strands).

Free bacteriophage consists of a double, or in some cases single, strand of DNA and little else within a protein capsule (with a tail for attachment). This capsule does not enter the host cell. The genetic material of other viruses (e.g., polio, the plant viruses) seems to be similar except that DNA may be replaced by RNA (ribonucleic acid, in which ribose takes the place of deoxyribose and uracyl of thymine). The genetic material of bacteria seems to be a little more complicated.

There seem to be complications, however, in the threadlike prophase chromosomes of higher organisms. These are some hundred times the diameter (20 Å) of a Watson–Crick molecule of DNA. The electron microscope reveals a hierarchy of numerous fibrils, those at each level twisting about each other to form ones twice as big (Kaufmann and McDonald 1957, Ris

1957). It appears that there must either be a great many parallel ultimate strands or an enormous amount of folding, both of which raise difficulties for a theory of crossing over.

Since 1940, there have been notable advances in the analysis of fine structure from purely genetic evidence (foreshadowed by the theory of step allelomorphism of Dubinin [1929] and Serebrovsky [1930]). Oliver (1940) demonstrated rare crossing over between supposed alleles at the lozenge locus of *Drosophila melanogaster*, and Lewis (1942) demonstrated position effects following rare crossing over between star and asteroid, which are probably located in the components of a double salivary band. These initiated a period during which many supposedly single loci in the same species were successfully split. Similar results have been obtained in several other organisms, notably *Zea mays*, *Neurospora*, *Aspergillus*, and *Saccharomyces*.

Interpretation took several directions. In some cases, subdivision of a system of supposed alleles with multiple effects seemed to yield components with single effects and thus led to the hypothesis that heredity is made up, after all, of ultimate particles that behave as units both structurally and physiologically. Some proponents of this view have indeed not hesitated to divide seemingly complex loci into simple components on the basis of effects alone. Unfortunately, thorough analyses of the effects in cases in which supposed loci have actually been divided by crossing over have not indicated any clear-cut separation into physiologically simple units. Maps based on complementarity do not fully agree with those based on crossing over (Giles 1965; Green 1965).

Another interpretation is directed toward complete abolition of the classical structural gene: that disruption by crossing over or by other means may occur between any two nucleotide pairs of the DNA molecule. Pontecorvo and Roper (1956), for example, have pointed out that the amount of recombination (5×10^{-4}) between the outer two of three demonstrated components of the white superlocus of *Drosophila melanogaster* is 62 times as great as the smaller amount between the two adjacent ones (8×10^{-6}) and that a similar calculation from four separable mutations in a superlocus of *Aspergillus* gives a ratio of 150 ($1.8 \times 10^{-3}/1.2 \times 10^{-5}$). These suggested the possibility of unlimited divisibility.

Demerec (1957) and his associates working with *Salmonella* have found successions of physiological units corresponding to several successive steps in the synthesis of such substances as histidine, cystine-methionine, and tryptophane. This contrasts with the usual wide scattering in the chromosomes of the loci controlling such metabolic steps in *Neurospora* and *Aspergillus*, although sporadic cases of clustering are now known in these and other organisms. Each of the physiological units is capable of breakage (and inactivation) at many points and thus seems to be composed of many structural genes.

The process of subdivision of blocks concerned with a single physiological process has gone farthest in studies of phage by Benzer (1957), Stresinger and Franklin (1957), and Edgar and Epstein (1965). Benzer introduced the term "cistron" for a physiological unit (interpreted to be such by the criterion of a difference between cis- and trans-recombinants of internal mutations) and "recon" for a block within which crossing over has not been detected (i.e., a classical structural gene). He also uses the term "muton" for a unit of mutation. We have noted that the classical gene was assumed to contain a vast number of discrete sites at which mutation might occur. Benzer's analyses of a very large number of mutations of phage T4 of the colon bacillus, all characterized by inability to multiply in one or more host strains, indicated that crossing over may occur between any two nucleotides (probability about 10^{-4}). The structural gene (recon) thus seems to be reduced to a single nucleotide, although a physiological unit (cistron) may have more than 400 nucleotides (4% crossing over in its range, which is more than 20 times as large in crossover units as the complex loci of *Drosophila* or *Aspergillus* referred to above). The total length of the single phage chromosome was estimated as at least 200% on the same scale.

As noted, the maps based on complementation do not always correspond to ones based on crossing over. There may be apparent complementation in trans-recombinants between mutant sites that are located between ones that show more or less failure of complementation, and thus seem to belong to the same cistron. This ambiguity in the definition of the cistron, coupled with the reduction of the classical gene (recon) and the mutational unit (muton) to a single nucleotide, makes it seem best to follow Demerec in using the term "gene locus" for a unit of the DNA that determines a single polypeptide molecule, even though divisible by rare crossing over.

There are still complications, however. Successions of cooperating loci (operons) have been found by Jacob and Monod (1961) in bacteria that are called into action or suppressed in unison by a terminal member (the "operator"), itself controlled by a repressor. Thus the whole block behaves to some extent as a physiological unit at a higher level than the gene locus.

Interpretation of cases of apparent rare crossing over between two mutations in which only one of the products of meiosis is recovered has been complicated by recognition of a class of mutations, "gene conversions" (Lindegren 1953, M. B. Mitchell 1955), that occur only in heterozygotes but not by ordinary crossing over, although they are likely to occur in the neighborhood of a crossover. This phenomenon was foreshadowed by a study of a "mutable" locus in *Drosophila virilis* by Demerec (1928) and by

Winkler's (1930) conversion theory, designed to account for abnormal tetrad ratios in fungi and mosses (a phenomenon for which, however, there are other possible explanations in most cases [Winge and Roberts 1954]). In the well-established cases of gene conversion, the products of meiosis show the expected 2:2 segregation at all heterozygous loci on both sides of the one in question. At this itself, one allele has been replaced in one of the products by the other allele (3:1) or has been modified (2:1:1). The rate is definitely higher than that of mutation in homozygotes.

There are some general reasons for suspecting the existence of preferential breakage points in *Drosophila*. Let us assume that the chromosome consists of a succession of physiological units, each involving some 1,000 nucleotide pairs. If the breakage points of chromosome rearrangements occur at random between nucleotide pairs, there would be only one chance in a thousand that a break would not disrupt a physiological unit and only one in a million that there would be no disruption at either end. Yet 40% of 332 X-ray induced random translocations observed by Patterson (1934) and his associates were viable as homozygotes and over 90% of these were fertile. There is no excessive mortality before eclosion in such experiments (cf. Muller and Settles (1927) that might indicate loss of crossovers within physiological units.

Similarly, nearly every crossover should cause at least isoallelic mutation if one supposes that the population carries many isoalleles in each physiological unit, at more or less equal frequencies. Such mutations may indeed occur, but an evolution toward a condition that would insure greater stability of the more successful patterns would seem more likely.

In addition to such general arguments, we have the intensive study of Muller (1956) on rearrangements in a restricted region at the left end of the X chromosome of *Drosophila melanogaster*, which indicated a very limited number of breakage points. The Greens' study (1956) of 18 lozenge mutations in the same species indicated, through very extensive tests, that these fell into just three groups, so that crossing over occurred between but not within them. There are other studies with similar results.

We should, however, note the possibility that physiologically significant patterns in the DNA may be separated by long inert stretches in the same material. The results of Patterson and his associates can be accounted for if there is twice as much inert as active material, even if breakage occurs at random.

But let us return to the possibility of an evolutionary mechanism for protecting useful sequences of nucleotides from disruption (cf. Wright 1959*a*). The identification of Bellings' ultimate chromomeres with genes has been questioned on the ground that the former have been observed to be merely sites of incipient coiling in leptotene. It is, however, generally agreed that they always occur at certain definite places. This is certainly true for the highly specific banded structure of the salivary chromosomes of *Drosophila* (thick and thin bands, double and single bands, and so forth). Precocious coiling of definite regions suggests intrastrand synapsis of very short tandem duplications, each of which should yield a tight double loop. This was demonstrated in a special case in maize by Laughnan (1955). Perhaps such tandem duplications have been much more common in the course of evolution than even Bridges (1935) or Metz (1947) supposed and constitute relatively stable structural genes by largely restricting breakage to unduplicated intervening regions. It is to be expected that such duplicants would become differentiated in the course of time and thus often permit mutation at one step.

It has long been evident that inversions of various lengths are common in the course of evolution. The result of tendencies toward repeated tandem duplication of very small length, toward differentiation of these duplicants, and toward the occurrence of inversions would be subdivision of the chromosome into rather small regions of more or less similar material, bounded by unconformities. Such regions would behave like supergenes consisting of several related loci, the mutations of which would behave as pseudoalleles, separable by rare crossing over in accordance with the amount of intervening unduplicated material. Although the modes of origins of physiologically significant patterns of nucleotides and of structural genes are independent in this view, one would expect natural selection to bring about a tendency toward coincidence and to restrict unduplicated material to regions of no physiological significance in which disruption would have no physiological consequences, even if not at exactly the same level in different strands of the chromatid.

Ordinary crossing over is assumed here to occur only in inert regions of the chromonemata (perhaps many stranded) of homologs, twisted about each other in such a way that exchange at one level interferes with other exchanges for long distances. Studies of fine structure suggest that there may be a different sort of crossing over within loci in which double crossing over tends to occur at very short distances (negative interference). Perhaps very short double exchanges may occur very rarely between parallel strands of looped duplicants within a locus, giving rise to mutations if there has been differentiation of the duplicants, or giving rise to conversions in heterozygotes if between strands in homologous loci.

Heterochromatin

Heitz (1928) described parts of the chromosomes (or whole chromosomes) that are out of phase with the bulk of the chromatin in their staining reactions as heterochromatin (as opposed to euchromatin). In *Drosophila* such material occupies disproportionately little space in the enormously enlarged salivary chromosomes and does not show the orderly banded pattern. It includes very few demonstrable genes but tends to exert an irregular inhibitory influence on euchromatic loci if brought into their vicinity by a chromosome rearrangement (Schultz 1936). Extra Y chromosomes, although tending to inhibit the effects of certain genes (Cooper 1956), reduce the inhibitory effect of other heterochromatin (Gowen and Gay 1934). Minor effects on quantitative characters have been described (Mather 1944; Goldschmidt, Hannah, and Piternick 1951). This is also the case with the supernumerary chromosomes of many species, which vary in number and are relatively dispensible (Beermann 1957). A tendency toward nonhomologous pairing is a noteworthy characteristic of heterochromatin.

Constituents of the genome of maize, known from their easy transposability without recognizable chromosome rearrangement, their inhibitory effect on adjacent loci, and in some cases their tendency to cause chromosome breakage at their site (McClintock 1952, Brink and Nilan 1952) have been interpreted as probably heterochromatic. Heterochromatin has been described in some cases as less easily disrupted than euchromatin (cf. review, Hannah 1951).

There are various interpretations of the differences between hetero- and euchromatin. The suggestion of Caspersson (1941) developed further by Pontecorvo (1944) that heterochromatin may consist of many replications of a few genes, the products of which are needed in bulk, is an attractive one. This could account for all of the properties noted above. Large masses of the same active material might be expected to become low points in concentration gradients of raw materials and high points in concentration gradients of products, especially at low temperature (Sturtevant 1925; Offermann 1935). The inhibitory effect on adjacent unadjusted genes and of extra Y chromosomes on some distant ones, the greater inhibitory effect of extra Y chromosomes on other heterochromatin and hence on the inhibitory effects of the latter may be examples. Irregularities of various sorts (the variegation of V-type position effects, heteropycnosis) may also result from great differences in concentration of abundant products. There is, however, another interpretation that has had many advocates, that heterochromatin differs from euchromatin in a structural pattern in such a way as to interfere with gene function.

Mutation

The ordinary objective of science is to make formulations that are of predictive value. Unique and unpredictable events are the negation of this objective. Mutations constitute the raw material of population genetics, and it is possible that almost every mutation is a unique and unpredictable event. This difficulty is, of course, not restricted to population genetics. In absolute terms, every event has unique aspects. The first task of any branch of science is classification—the assembling of more or less similar phenomena into categories. In the case of mutation, there is a gradation from kinds of events that occur only rarely in the history of a species, but then are of major importance, to ones with individually small and closely similar effects that recur frequently enough as a class to be treated statistically.

In the former category are the mutations that involve whole sets of chromosomes. Doubling of the entire diploid set of chromosomes in an individual of a species results in an autotetraploid. Triploidy results from union of a gamete from a tetraploid, or from a gamete in which diploidization has occurred, with a normal gamete. Triploids are sterile or nearly so and can form a population only by clonogeny. Tetraploids can also persist only in an independent population, but this may be panmictic within itself. Allelic genes in such a population exhibit tetrasomic instead of disomic Mendelian heredity which thus requires consideration. Character changes are usually inconspicuous, the most conspicuous being associated with increased cell size. There is usually reduced fertility, partly from irregular meiosis but to a large extent from disturbed physiology. There is the possibility of an ultimate evolutionary return to diploidy in behavior, with a doubled number of pairs, by differentiation of homologs accompanied by preferential pairing. There are probably species in nature that have arisen in this way, but the trend of opinion is that polyploids that have arisen from hybrids are much more important.

At the opposite extreme are the allotetraploids or amphidiploids, formed by doubling the chromosome set of a hybrid between two species so remote that there are no chromosomes sufficiently homologous to pair in meiosis. Such a hybrid is sterile but may establish a clone. Doubling its chromosomes may lead to completely normal meiotic pairing and fertility, as was predicted by Winge (1917) and soon verified experimentally in many cases. Sterile triploids are produced in crosses with both parental species. Since an amphidiploid also has a novel array of characters, it has the attributes of a new species from the moment of origin.

Between these types are tetraploids from hybrids between forms that are closely enough related to permit considerable pairing of chromosomes or segments of chromosomes. Stebbins (1950) has called these segmental allotetraploids. They resemble amphidiploids in showing a distinctly new phenotype (typically intermediate between the parents) and stability in some respects, but they differ in also showing considerable segregation of parental characters. Thus they have the potential for evolving into diverse stable types, sometimes close enough to one of the parental species to suggest origin as autotetraploids.

An allotetraploid may give rise to an autoallooctoploid by a second doubling. A sterile triploid from a cross between an allotetraploid and one of its parent species may give rise to an autoallohexaploid or, if crossed with a third species, to an allohexaploid (as in the common wheats: Kihara 1959). These are the more important classes of polyploids but others are known (Stebbins 1950).

The crucial event, chromosome duplication unaccompanied by cell division, occurs spontaneously with moderate frequency under certain conditions (low temperature at the right time, regeneration from a cut surface, and so forth). It can be induced with high frequency by treatment with certain drugs, notably colchicin (Blakeslee and Avery 1937).

Polyploidy has been very common in the evolution of species of higher plants but seems to have occurred much less frequently among animals. Hybridization followed by backcrossing may lead, by a process known as introgression, to incorporation of one or more chromosomes or chromosome segments into a local population that otherwise is the same as one of the parental species (Anderson 1949).

The addition or loss of single chromosomes by nondisjunction (trisomics, monosomics) results in simultaneous change in many characters. These effects segregate within the population in Mendelian fashion, although there are usually viability differences that seriously distort the ratio. The homozygotes (tetrasomics, nullosomics) are usually inviable. Alleles carried by an extra chromosome show trisomic ratios instead of the usual disomic ones.

Intrachromosomal duplications and deficiencies are common types of mutation. Both may be treated as gene mutations, with certain qualifications. Thus Bar eye in *Drosophila melanogaster* may be treated in population genetics as merely a semidominant sex-linked mutation, but one that undergoes further mutation (to double bar or to type) with unusually high frequencies (about 10^{-3} per generation instead of the usual 10^{-5} or less). It was found independently by Bridges (1936) and by Muller (1936) to involve duplication of at least half a dozen bands in the salivary chromosome. Sturtevant (1925) showed the high mutability to be due to unequal crossing over. The dominant mutations of the notch type in the same species were found to be lethal when homozygous and to be due to deletions of varying length while always involving a certain locus (Mohr 1923). An interesting feature of such deficiencies is their inability to complement the effects of recessive mutation in the region that has been lost (pseudodominance).

Other chromosome aberrations (inversion, translocation, transposition) consist of rearrangements without any immediate addition or loss of material. They may, however, be associated with position effects because of disruption of physiological relations between genes at opposite sides of a rearrangement point.

Inversions interfere greatly with crossing over in the region covered and to some extent beyond. Single and most double crossovers result in dicentric chromosomes and acentric fragments, which are lost. The only crossovers within an inversion that are transmitted in oogenesis in *Drosophila* are two strand doubles (Sturtevant and Beadle 1936), and as no crossovers occur in *Drosophila* males, an inversion creates a more or less extensive chromosome region in which all mutations behave as if allelic, and the inversion may be treated as if a Mendelian superlocus. Multiple systems of overlapping inversions are found in many species of *Drosophila* and can be treated in population genetics as sets of multiple alleles (Dobzhansky and Sturtevant 1938).

A reciprocal translocation between two chromosomes leads to linkage of these and the formation of gametes with duplications and deficiencies as well as normal and balanced mutant gametes. In plants the gametophyte generation acts like a sieve to remove unbalanced haploid combinations. Thus with respect to the sporophyte, the translocation can be treated as if a Mendelian mutation, allelic to normal (Belling 1927).

		Sporophyte			Gametophyte				
Normal	$++$	$\frac{AB}{AB}$	$\frac{CD}{CD}$	$+$	AB	CD			
Semisterile heterozygote	$+M$	$\frac{AB}{AD}$	$\frac{-\ BC}{-\ DC}$	$+$ M	AB AD	CD BC	AB AD	BC lost DC lost	
New normal	MM	$\frac{AD}{AD}$	$\frac{BC}{BC}$	M	AD	BC			

After correlating certain types of genetic change with demonstrable chromosome aberrations, geneticists were left with the class of apparent point mutations. As the criteria are essentially negative, it is not surprising that this is being gradually reduced by the recognition of special phenomena. We have already referred to gene conversion. Another is the class of highly mutable genes. Before considering these, however, it will be well to take up the great residual class of mutations that recur spontaneously only at very low frequencies and irrespective of other genes as far as is known, but at relatively high frequencies when treated with certain agents. The characteristic rates of overall occurrence of point mutations are of fundamental importance in population genetics.

Stadler (1930) made a study of seven loci in maize, chosen merely because they all affected the endosperm and would thus be recognized in kernels. The rates of occurrence were highly diverse (500, 100, 11, 2.4, 2.2, 1.2, less than 1 per million kernels). Muller *et al.* (1949) found the average rate in 9 specific loci in oogonia in *Drosophila melanogaster* to be about 10^{-5}. Estimates based on the total number of sex-linked lethal mutations were of the same order. Studies in spermatocytes and oocytes indicate a considerably lower rate. Russell's studies (1962) of seven specific loci in the mouse indicated a rate of about 10^{-5} in spermatogonia. Haldane (1935) estimated that hemophilia arises by mutation in human X-chromosomes at a rate of 2×10^{-5}. Similar estimates have been made for several other rare family traits. Those for autosomal dominants or sex-linked recessives range from 10^{-4} to 4×10^{-6} (Neel and Schull 1954), with median at about 2×10^{-5}. Estimates for recessives are, on the whole, similar but more liable to misinterpretation because of selection for or against heterozygotes.

Lewis (1948) obtained a surprising result for the self-incompatibility locus in *Oenothera organensis*, a species restricted to the Organ Mountains of New Mexico and probably containing less than 500 individuals in nature, although some 45 alleles have been found in these. A style heterozygous for any allele prevents adequate growth of the pollen tube that carries it. He found no mutation of the natural type in 10^{-8} pollen grains. Mutations of another type, not found in nature, occurred with a rate of 10^{-6}.

Rates in bacteria are often very low per cell division (10^{-8} or less) (Luria and Delbrück [1943], Demerec [1955]). Novick and Szilard (1952), however, in very accurate determinations in their chemostat, found the rate per locus per hour to be independent of sixfold differences in rate of division.

Various biases may enter into some of these figures. Mutations with very slight effects (isoalleles) are likely not to be counted. On the other hand, special loci may be a selected lot since they are known to be ones at which mutations have already occurred. Thus in man only mutant conditions that occur frequently enough to provide adequate material can be studied. In *Drosophila melanogaster* mutations are known in about 10% as many loci as there are salivary bands. Perhaps many loci include undifferentiated duplicants which protect against recognizable single point mutations.

The rate of occurrence of mutations is affected by temperature with coefficients in *Drosophila* of about 2.5 per generation or 5 per unit of time (Muller 1928, Timofeeff-Ressovsky 1934). Very much higher rates occur on exposure of the larvae to high or low temperatures, which are fatal to many (Plough and Ives 1935).

Ionizations and activations due to X-rays, radioactive isotopes, and so forth induce gene mutation and chromosome breakage at rates that rise linearly with dosage (because due to single hits) and thus without any threshold. Rearrangements that depend on two remote breaks depend on the accidental occurrence of two inducing events within a period too small to permit healing of the first break; thus they tend to increase as the square of the dosage, usually qualified by differential viability if observation is based on progeny classes instead of on the chromosomes themselves (Sax 1941). Small inversion and deficiencies due to such agents tend, however, to increase linearly with dosage, indicating that they are caused ordinarily by a single event (single cluster of ionizations or ramifying chemical consequences of a single release of energy). With protons and neutrons, which leave large dense tracks of ionization, even rearrangements that depend on two remote breaks behave as though they were caused by a single hit in the cell. Only a minute fraction of spontaneous mutations in *Drosophila* (less than 0.1%) depend on cosmic rays or natural terrestrial ionization (Muller and Mott-Smith 1930; Timofeeff-Ressovsky 1934). The proportion is undoubtedly much greater, however, in longer-lived organisms. Activation by ultraviolet tends to cause less drastic mutations than those due to X-rays, but it can also produce chromosome rearrangements.

A major breakthrough in mutation theory came with the demonstration by Auerbach (1949) that certain chemicals, notably the mustard gases, were potent sources of all sorts of mutations. As with the other agents, there seemed to be little specificity in the kinds of mutations produced. Induction by pretreatment of media by irradiation (Stone *et al.* 1947) or by chemical treatment (Wyss *et al.* 1947) are examples of influences on mutation rates that indicate that chains of chemical processes intervene between such an event as ionization and the occurrence of a mutation.

Recent attempts at chemical induction of mutations have been guided by knowledge of the chemistry of DNA. Thus Gieres and Mundy (1958) found that nitrous acid, expected to break phosphate bands, is a very potent mutagen. Moreover, treatment with analogs of the natural bases of DNA has led to highly specific replacements of either one of the complementary pairs of nucleotides by the other (Freese, 1963).

Certain recently recognized types of mutation must be distinguished from point mutations. At present, they seem to be special cases of only secondary importance in population genetics, but this situation may change radically with increased knowledge.

Highly mutable loci were first studied intensively by Emerson (1914) in variegated pericarp in maize. This category has turned out to depend, at least in some cases, on the remarkable class of easily transposable loci in maize discovered by McClintock (1951, 1956). She discovered an element Dissociation (*Ds*) that moves about frequently from place to place in the genome in certain strains. Its presence is revealed in some cases by inhibition of the effect of an adjacent gene (for example, white instead of colored aleurone) and by frequent apparent mutation to the dominant gene in streaks in the kernels from cells in which it has moved to a new place. *Ds* also tends to bring about chromosome breakage at its site, followed by loss of the distal region and thus apparent mutations from dominant to recessive in heterozygotes for this region. A necessary condition for transposition and thus for either apparent dominant mutation of adjacent genes, or breakage and apparent recessive mutation of distal genes, was found to be the presence of another transposable element Activator (*Ac*) elsewhere in the genome. This element, however, also tends to inhibit adjacent genes. Brink and Nilan (1952), on studying pericarp variegation of maize (P^{vv}), found that the occasional mutations to self red (P^{rr}) that affected whole kernels or groups of kernels (instead of merely streaks on a kernel) were apt to be associated with an adjacent area of similar size of exceptionally pale variegated seeds. The mutation was interpreted to be due to removal of the modulator, *Mp*, from its site near gene *P* to another site that, on assortment to a sister cell, reduces the transposability of *Mp* in the old site adjacent to *P*, giving pale variegation in the area derived from this cell. On introduction of *Ds* into the strain, it appeared that *Mp* behaved like *Ac* and is probably the same.

An earlier example of determination of high mutability is the mutation of the ordinarily stable white aleurone of maize (due to *a*) to red (*A*) in streaks on the kernels and anthers (where demonstrable genetically) on introduction of an element *Dt*, associated with a heterochromatic knob in a different chromosome (Rhoades 1938). The element *Dt* has been shown to be transposable, but it has not been shown that mutation of *a* is due to removal of an adjacent element.

The relation of a number of cases of high mutability of genes described by Demerec (1928, 1929) in *Drosophila virilis* to these cases is not clear, but control of mutability of distant loci is a feature in common. A notable feature of mutable miniature wing was the occurrence of alleles in which mutations to the dominant normal condition were restricted to the soma (wings), although the original unstable allele mutated in both soma and germ line. We have referred to the restriction of mutation of one of Demerec's mutable genes (reddish) to heterozygotes (e.g., reddish-yellow mutating to the dominant type) as resembling conversion, except for its greater than usual frequency. McClintock has noted that many apparently stable gene mutations may be due, not to internal change at the locus, but to proximity to an inhibitory element that is not subject to transposition because of absence of the necessary activator.

The phenomenon of paramutation, described by Brink (1958), is the most unorthodox type of genetic change to date. Gene R^r (anthocyanin in aleurone and plant tissues) is a well-known, fairly stable gene of maize. In the highly inbred strain with which Brink worked, the endosperm genotype R^rrr has a certain grade of strong mottling. Other alleles R^{st} and R^{mb} give mottled patterns even when homozygous (stippling and marbling respectively.) When pollen from $R^{st}R^r$ or $R^{mb}R^r$ is used on colorless $r^r r^r$ or $r^g r^g$, the segregants that carry R^r come out very much lighter than expected, in all kernels. This change, R to R', turns out to be transmitted on pollinating recessives again, although in some cases at least there is some reversion. On inbreeding the modified R' plants, it appears that there is considerable reversion of $R'rr$. Experiments with trisomics ($R^{st}R^r r$), in which it is believed that only two of the chromosomes pair in meiosis at any given level, indicate that the effect does not occur at meiosis. The effect is not transmissible through r gametes and thus does not behave like a cytoplasmic component. R^{st} mutates fairly frequently to self red R^{sc}, but these do not behave like R^r. They are not subject to paramutation by R^{st} and most, but not all, of them are still capable of inducing paramutation in standard R^r. Finally modified R' itself seems to have acquired a feeble capacity to induce paramutation in standard R. The most noteworthy result is the induction in 100% of the paramutable genes of change in physiological effectiveness that persists to later generations, even though this is subject to gradual decay.

Irregularities in the Mechanism of Heredity

This chapter began with the assumption, based on extensive evidence, that the chromosome mechanisms of most organisms are such that there is 1:1 segregation of alleles. Significant deviations from equality in backcrosses are, indeed, common but it would seem safe to assume that, in general, the mere replacement of one gene by an allele has no effect on segregation and that such deviations must thus be due to selective advantage of one allele. This, however, need not be expected to apply in a case in which there is a structural difference between the segregating entities and asymmetry in meiosis. The latter usually occurs in oogenesis both in plants and animals. We may refer to textbooks of cytology for numerous examples (e.g., Swanson 1957, chap. 9).

Cases of unequal segregation in the formation of male gametes are more disconcerting than in oogenesis. In flies of the genus *Sciara* (Metz 1938) the combined genetic and cytologic evidence indicates that all paternal chromosomes are cast out in a monopolar spindle in spermatogenesis, although necessarily maternal in the preceding generations. The result is that all genes of all chromosomes are transmitted as if sex linked, except that the genes in the paternal chromosomes produce their characteristic effects in the males that receive them.

In several species of *Drosophila*, a so-called sex ratio gene has been found in high frequencies in nature in the X chromosome (Gershenson 1928, Sturtevant and Dobzhansky 1936). In this case unequal segregation leads to sex ratios in excess of 90 females to 10 males from males that possess this gene.

Sandler, Hiraizumi, and Sandler (1959) described a case in *Drosophila melanogaster* in which males which were heterozygous for an entity, Segregation-distorter (*SD*), found in nature, transmitted the condition to more than 95% of their offspring, as indicated by linked genes. This entity was located in or near the centromeric heterochromatin of chromosome II. The phenomenon did not occur if synapsis in the region was prevented; from this and other evidence, it seemed to depend on a direct effect of *SD* on the corresponding region of the normal chromosome (in spermatogenesis only), which rendered sperms nonfunctional with the latter. It was later found that normal chromosomes differed widely in sensitivity to this effect, and also that some *SD* segregants from chromosomes in which recombination had occurred gave much less extreme deviations from normal segregation than at first. A remarkable result, suggestive of Brink's paramutation, was that the properties of both *SD* and *SD*+ tended to be modified permanently by heterozygosis with insensitive *SD*+ and conversely. Crossing over in the immediate vicinity of *SD* was found to be greatly reduced, suggesting that it was a structural modification of some sort but too small to be cytologically evident.

One very extensively studied case of divergence from normal segregation in a mammal, not due to obvious selection, is that of the *t*-locus in the mouse (Dunn 1957). The mutation *T* (brachyury) results in a short tail when heterozygous (t^+/T) and is lethal when homozygous. A great many alleles have been found, widely distributed in nature, of which the heterozygotes t^+/t^x are normal, the heterozygotes T/t^x tailless, and the homozygotes t^x/t^x lethal. It is remarkable that the heterozygotes between lethal *t*-alleles t^x/t^y are usually normal. The point of interest in the present connection is that male heterozygotes (but not female) may produce highly aberrant ratios. With certain of the alleles) 95% of the sperms carry t^x to only 5% normal. In other cases, there is approach to equality or even deficiency of the *t*-allele. It seems probable that the locus is compound in some sense and not unlikely that there is some structural aberration, but it has not been demonstrated that the cause of its behavior is disturbed meiosis. There is, indeed, evidence for physiological selection in a demonstration of change in the segregation ratio on changing the relation of ovulation and coitus (Braden 1958).

Sandler and Novitski (1957) proposed the term meiotic drive for cases in which heterozygotes produce two kinds of gametes in unequal frequencies for cytologic reasons. The phenomenon is one that would obviously be disastrous for adaptive evolution if at all widespread, and it is probable, therefore, that natural selection tends to prevent its occurrence. As far as we know, it is, indeed, decidedly uncommon. Equal segregation will, in general, be assumed in developing the principles of population genetics, but the consequences of departures from equality must also be considered.

CHAPTER 4

Nongenic Heredity

The definite rules of Mendelian heredity give the firmest basis that we have for statistical deductions about populations. These, however, need to be supplemented by consideration of the less well understood modes of inheritance that trace to antonomous entities other than the chromosomal genes. It was, indeed, a common early view that the latter have to do only with superficial characters and that the fundamental distinctions between taxa are cytoplasmic (Loeb 1916).

We can only study the heredity of differences. As the complex interactions among the effects of factors came to be understood, many that first seemed non-Mendelian were later shown to have a genic basis. At present, an overwhelming proportion of analyzable differences have proved to be Mendelian in a broad sense, and most of the residual, unanalyzable variability follows statistical rules that point definitely to a Mendelian interpretation. There are, however, a considerable number of carefully studied differences that are definitely not Mendelian in any sense.

Microorganisms

Heredity is much less distinct from self-regulatory physiological processes in unicellular organisms than in multicellular ones. Each daughter cell after fission has half the substance of the parent in contrast with the enormous reduction in mass that intervenes between parent and offspring in sexually reproducing higher organisms.

The persistence of effects of injury or of acclimatization through hundreds of fissions of protozoa was attributed by Jollos (1934*a*) to what he called *Dauermodifikationen*, with the implication that they were more related to current physiology than to genetic mechanisms. When he wrote, the roles of micro- and macro-nucleus and of autogamy were not well understood, and the possibility that the adaptive changes may have involved gene recombination or even mutation and selection does not seem excluded. Clearly cytoplasmic, however, were the considerable delays in the expression of genes after crossing, described by Jennings (1940) and his students in *Paramecium*. De Garis (1935) showed that after crosses between strains differing in size by as much as twenty-fold in one case, each exconjugant line at first reflected the size of the strain from which it came. In the course of 22 to 36 fissions without autogamy, these lines came to the same size, which, although usually intermediate, might be larger or smaller than either parent according to the common zygote nucleus. Genetic differences in rate of fission behaved similarly (Sonneborn and Lynch 1934). This "cytoplasmic lag" implies some autonomy of the cytoplasm before reconstitution under genic influence.

A different sort of cytoplasmic lag has been found for certain other characters of *Paramecium aurelia*, of which antigenic specificity, studied by Sonneborn (1948) and Beale (1952), may be taken as representative. The antigens in question were demonstrated by the induction in rabbits of antisera capable of immobilizing *Paramecia* that carried the corresponding antigen. It was found that there were several loci with two or more alleles, each capable of determining a particular antigen but only in a cytoplasm that was in an appropriate state. Each cytoplasmic state was found to permit only the alleles of a particular locus to function. Each state was favored by certain environmental conditions, such as a particular temperature, but showed considerable persistence after change to unfavorable conditions. A clone might retain the same state for fifty or more generations under antagonistic conditions before changing rather abruptly to the state appropriate to these conditions (revealed by shift to the antigens controlled by another locus).

After crossing homozygous clones, differing in two loci (for example, strain 90, genotype $d^{90}d^{90}g^{90}g^{90}$, and strain 60, genotype $d^{60}d^{60}g^{60}g^{60}$, in cytoplasmic states D and G respectively, one exconjugant clone persisted for a long time in state D and produced only the antigens determined by d^{90} and d^{60} while the other persisted for a long time in state G and produced only the antigens determined by g^{90} and g^{60} although both have the same genotype, $d^{90}d^{60}g^{90}g^{60}$ (Beale 1952). Beale found random assortment among the homozygous clones after autogamy in a 1:1:1:1 ratio. These principles also applied to a third locus, s, and the appropriate cytoplasmic state, S, in these experiments.

The significant point in the present connection is that the cytoplasm has a permanent repertoire of alternative states, each of which exhibits a limited sort of "heredity" in its expression. The situation suggests the existence of alternative self-regulatory states of the cell as a whole or, at least, of some extensive system in it such as the plasma membrane rather than autonomous "plasmagenes" in Winkler's terminology (cf. Wright 1945).

Many unicellular organisms, however, have a complex array of organelles which divide at fission with a regularity that suggests necessary genetic continuity. These are the centrioles, the blepharoplasts, and kinetosomes at the bases of flagellae and cilia, and the various types of plastids. The mitochondria, present in all living cells, also appear to divide and to be apportioned at cell division.

In certain cases, organelles have been lost irreversibly without killing the cell or preventing its multiplication, at least for a considerable time (reviewed by Ephrussi 1953). Thus treatment of trypanosomes with acriflavine has been shown to cause failure of the kinetoplast to divide and to result in a permanently akinetoplastic strain (Werbitzki 1910). Again, the rate of division of the chloroplasts of *Euglena wesnili* has been reduced by various means, resulting in irreversible loss. In this case, however, the strain could not be maintained indefinitely (Lwoff 1950).

One of the most thoroughly analyzed cases of extragenic heredity is that of a small type of cell found regularly in yeast cultures (Ephrussi 1953). It is incapable of respiration but retains the capacity to ferment sugar. A whole series of respiratory enzymes is absent. The frequency in typical cultures was 1 to 2%, shown to be the result of a balance between a "mutation" rate of about 0.002 and a selective disadvantage under aerobic conditions. The rate could be accounted for by random assortment at cell division of some ten autonomously dividing particles and occasional failure of a daughter cell to receive any. As in the trypanosomes, rates of irreversible loss approaching 100% could be induced by acriflavine and related substances.

Crosses between the normal and small cells gave rise to clones that all behaved like normal ones, as expected from random apportionment of cytoplasmic particles, although marker genes showed the usual 2:2 segregation in each ascus. Interestingly, Mendelian mutation occurred with similar effects. In these mutant small cells, however, the significant particles were evidently not lost but were merely not functioning in the presence of the recessive gene, since crosses with the nongenic small type gave rise to 100% normal diploids followed by 2:2 segregation of the genic types.

Another very instructive case is that of the "killer" character in certain races of *Paramecium aurelia*, studied exhaustively by Sonneborn (1943) and his associates. Certain strains give off one or another toxic substance of a class known as paramecin, which is lethal after a time to strains that do not produce it. After conjugation, each line retains its own character unless conjugation has been sufficiently prolonged to permit mixing of cytoplasm, in which case both lines become killers. This property has been shown to depend on Feulgen-staining "kappa" particles, typically several hundred in number, which multiply more or less independently of the organism. As in

Euglena, it is possible to slow down the multiplication of those kappa particles, relatively, to division of the whole cells and ultimately to eliminate them from a line which thereupon becomes sensitive (Preer 1950). Kappa is subject to mutations (in its own constitution) which modify the quality of paramecin. It can be interpreted as a symbiont, but it is not wholly independent of the genotype of its host. A gene *K* must be present to enable it to persist. Kappa is absent in strains of genotype *kk* and soon disappears in segregants of this genotype.

Another very favorable system for study has been found by Sager (1965) in *Chlamydomonas* (a haploid flagellate). She has been able to induce some thirty kinds of "nonchromosomal" mutations by sublethal concentrations of streptomycin where chromosomal mutagens failed. These largely resembled chromosomal mutations in the variety of traits affected, but differed radically in mode of transmission. Pairing is between mating types, mt^+ and mt^-, dependent on Mendelian genes. The reduction division of the zygote gives regular 2 : 2 segregation of these and other Mendelian alleles, but all four typically have the nonchromosomal gene of the mt^+ parent. In a small fraction of the cases, however (much increased by growth at high temperature), the nonchromosomal genes of both parents were transmitted to all four offspring and segregated to one or the other pure type in the course of several divisions. The important point was established that different pairs of alternatives segregate independently. Moreover, in some cases new allelic mutants occasionally appeared, probably by intragenic recombination. The results suggest a very considerable importance of nonchromosomal genes in the evolution of *Chlamydomonas*.

Reciprocal Differences in Crosses of Higher Organisms

Cell size and content are usually extremely different in the uniting gametes of higher organisms except for the nuclei. The usual similarity of the progeny of reciprocal crosses between strains or species suggests nuclear heredity of the differences, although there may possibly be autonomous submicroscopic particles that are transmitted equally by the gametes. Most investigations of possible extragenic heredity have started from cases in which reciprocal crosses have given consistently different progenies. Most of those have turned out to be genic with some asymmetry in gene content of gametes or with a delay of a generation in phenotypic expression.

Thus reciprocal crosses may differ because of differences in the number of representatives of one or more of the chromosomes in the uniting gametes. A common special case arises from the sex-determining mechanism. If one sex is XX and the other XY or XO, the offspring of the homogametic sex should be alike in reciprocal crosses, but those of the heterogametic sex may be different.

Many cases of reciprocal difference in F_1 have been found to lead to results in later generations that are exactly what would be expected if the characters of all of the offspring in a brood reflected the maternal genotype instead of their own. This might come about because of a maternal physiological influence either on the egg or on later stages, as in the seeds of flowering plants and the embryos and fetuses of mammals. It may also come about if gene products act in the oocyte with effects that carry through to the reduced egg and to the zygote, which from the standpoint of the physiology of development may be considered as a single organism, developing under three successive heredities. Special researches are necessary to discriminate between those possibilities.

Juvenile stages often differ in color in broods from reciprocal crosses with Mendelian segregation of color of the F_3 broods (*Bombyx mori*, Toyama 1913, Tanaka 1914); *Gammarus chevreuxi*, Sexton and Pantin 1927; *Ephestia kühniella*, Caspari 1933). In the last case, transplantation of normal tissues (with gene A) to recessive females (with aa) enabled the latter, although mated with aa males, to produce offspring whose color indicated diffusion of a soluble pigment due to A into the egg.

Fused (fu/fu) is a recessive character of *Drosophila melanogaster* that permits viability if from an fu^+/fu mother but not if from an fu/fu mother, although the latter can produce viable offspring if her eggs are fertilized by fu^+ sperm. In this case, Clancy and Beadle (1937) found that transplantation of fu/fu ovaries to a normal host did not permit development of fu/fu offspring, indicating that fu^+ must act either in the oocyte or in the zygote to permit viability.

In the snail *Limnaea peregra*, dextral coiling is dominant over sinistral in broods, according to the maternal genotype (Boycott and Diver 1923, Sturtevant 1923, Boycott *et al.* 1930). Direction of coiling cannot be a direct consequence of maternal phenotype (which need not agree with maternal genotype). The occasional occurrence of mixed broods in which the sinistral gene can be demonstrated to have mutated indicates that the character is due to gene action in the oocyte (Diver and Andersson-Kottö 1938).

Such late morphological characters as polychaeta in *Drosophila funebris* (Timofeeff-Ressovsky 1935), number of lumbar vertebrae in mice (Russell and Green 1943), and fused tail in mice (Reed 1937) show a lag of a generation in expression of Mendelian genes and suggest oocyte effects rather than maternal physiological effects. But when one notes that mere age in female guinea pigs makes a great difference with respect to presence or absence of the little toe and in the amount of white in the spotting pattern of her progeny, it must be recognized that discrimination can only be based on direct experiment. After allowing for these Mendelian interpretations of reciprocal differences, there remains a considerable number of well-studied cases in which none of them apply.

Extragenic Heredity in Higher Plants

There are both hermaphroditic and exclusively female plants in certain species of plants (*Satureia hortensis* and in *Cirsium oleraceum*). The former breed true, the latter, pollinated by hermaphrodites, produce only females (Correns 1908, 1937). Moreover, *C. oleraceum* females produced only females if pollinated by *C. canum*, a species consisting exclusively of hermaphrodites, and continued to do so if the hybrids were backcrossed generation after generation to *C. canum* until indistinguishable from the latter in all other respects. Apparently this persistent femaleness implies exclusive transmission down the straight female line and thus presumably in cytoplasm. East noted, however, that if the hermaphrodites are *aa* and the females *Aa*, and it is supposed that gene *A* not only results in femaleness but also somehow prevents the occurrence or functioning of egg nuclei with *a*, these results would be accounted for. A considerable number of cases are now known in which segregation is disturbed other than by lethality.

This possibility seems to have been eliminated in a case of male sterility in maize studied by Rhoades (1933). This also followed the straight female line. Rhoades introduced it into heterozygotes with markers in each of the ten chromosomes. There were no disturbances in the segregation of these, indicating that there was no gene for male sterility in any of them. The male steriles occasionally produced some good pollen but this did not, in general, transmit the normal condition. There were, however, a few cases in which the male sterile plants produced normal mutants. On the other hand, Rhoades (1950) obtained wholesale mutation from normal to cytoplasmic male sterility as an effect of homozygosis in a certain gene iojap (*ij*).

There have been a considerable number of other cases of male sterility that behave as if transmitted by the cytoplasm. One of these exhibits a different sort of relation to a gene than that referred to above. A cross between two varieties of flax, tall and recumbent (Gairdner 1929) gave normal plants in both reciprocal crosses, but male steriles appeared among those that traced in the female line to the procumbent variety (*P* cytoplasm) in a way that demonstrated dependence on segregation of recessive genotype *mm*. No male sterility appeared in any cross that traced in the female line to the tall variety (*T* cytoplasm). It is assumed that the tall variety was (*T*)*mm* and the procumbent $(P)MM$. The results could be interpreted as being caused by failure of normal functioning of mm in P cytoplasm. There was no indication of induced mutation of the cytoplasm in this case.

Another type of behavior was observed by Blakeslee (1921) in a variant of *Datura stramonium* having abnormal leaf shape, absence of spines, and low vigor, as well as partial male sterility. This "quercina" variant was usually transmitted to all descendants in the female line. There was, however, occasional transmission by the good pollen, which was sometimes present. This variant was found to spread after grafting and was thus interpreted as due to a virus, although it was not transmissible by innoculation.

Nongenic Chlorophyl Defects

The great majority of the cases of nongenic heredity that have been studied in higher plants have to do with chlorophyl defects. This by no means implies that the chloroplasts are wholly autonomous in their properties. Mendelian differences are in fact much more frequent than non-Mendelian ones in species that have been studied intensively. In maize, for example, Emerson, Beadle, and Fraser in 1935 listed more than 90 genes with effects on chlorophyl, some concerned with uniform reduction (virescent, luteus, white, etc.), some with developmental patterns that cut across cell lineages (for example, zebra banding) and some with cell lineage streaks. These authors referred to only two nongenic chlorophyl defects. Even among the nongenic cases in higher plants, there is reason to believe that many, perhaps most, are due to effects of other cytoplasmic constituents on the plastids, rather than to autonomy of the plastids themselves.

Most of the nongenic chlorophyl defects that have been studied have been of the nature of variegations (*status albomaculatus* of Correns) but a few have been of uniform pale color. The usual mode of behavior is illustrated by the first described case of nongenic heredity, variegation in *Mirabilis jalapa* (Correns 1908, 1909). Most of the leaves exhibit a coarse to fine variegation but some leaves or whole branches are yellowish white, others pure green. Flowers on green branches, however pollinated, produce only the normal green variety and flowers on white branches, however pollinated, produce only white seedlings, which soon die. Flowers on variegated branches produce a mixture. Anderson (1923) and Demerec (1927) found this sort of behavior in the two phenotypically different variegated strains of maize. On planting the seeds according to their positions on the ear, they obtained a coarse mosaic of different degrees of variegation among the seedlings as well as groups of wholly green and wholly white seedlings. Correns and Wettstein (1937) listed 50 other species in which this type of variegation had been studied; and this number would undoubtedly be much greater now.

The simplest explanation is obviously that of the more or less random assortment of a mixture of normal and defective plastids at each cell division until homogeneity is reached, one way or the other. Unfortunately, Correns was unable to find any evidence for such mixtures within cells. He concluded that there was diseased cytoplasm, held in check in the meristem but later proceeding either to irreversible damage to all plastids or to complete cure. Correns' observations have been confirmed by others. Gregory (1915), on the other hand, found apparent mixture in variegated *Primula sinenses*, but his interpretation of his observations has been questioned. Recently, Michaelis (1959) has studied numerous matroclinous variegations induced in *Epilobium hirsutum* by radioactive isotypes. He found three groups: one with cells containing a mixture of normal and degenerate plastids and a strong correlation between neighboring cells; another without obvious mixture, perhaps because of less contrast between the types of plastids; and finally, one in which intermingled cells showed either wholly normal or wholly degenerate plastids. He showed in the first class that the segregation pattern would be accounted for by random assortment of 10 to 20 units, agreeing well with the observed number of plastids (average 12) in meristematic cells; but in the last class, the pattern would require segregation among hundreds of units (possibly microsomes) and simultaneous action on all plastids of diffusing injurious substances, with the result dependent on whether these substances were above or below a threshold. This interpretation is, however, still moot (Wettstein and Eriksson 1965, Hagemann 1965).

The situation found by Andersson-Kottö (1930) in the fern *Scolopendrium vulgare* is interesting. The 64 spores from any sporangium on a variegated plant all gave rise to pale green gametophytes or all to normal green ones. There was no variegation in either kind of gametophyte. Union of gametes from green gametophytes produced only green sporophytes, but if either or both gametes came from a pale gametophyte, the sporophyte was variegated with abrupt transition between green and white. Thus in this case transmission was not restricted to the female line. The pattern clearly did not represent directly any sort of assortment of green and white entities.

We may note here, parenthetically, that in certain species of seed plants (Viola and Pisum, de Haan 1933), occasional transmission by the pollen has been described. In one of the first described cases, that of *Pelargonium zonale*, a periclinal chimera that breeds as if pure white, both reciprocal crosses with pure green produced extremely coarse variegation (sectorial chimeras that occasionally gave rise to periclinal ones). Baur (1909) interpreted this as a transmission of plastids by both ovule and pollen.

Winge (1919) described a case of variegation in *Humulus japonicus* that was phenotypically of the usual type and was transmitted only in the female line in crosses with normal green, but was radically different in its behavior in that there was no segregation into true-breeding green or white. Andersson-Kottö (1923) found a similar situation in the variegated fern *Adiantum cuneatum*. All sporangia produced a mixture of dark-green and rapidly dying pale-green gametophytes. In this case, the gametophytes became variegated with abrupt transitions from dark to pale green or the reverse in the leading cell of each wedge. All sporophytes also were variegated with abrupt transitions. Alternative physiological states of the cytoplasm, with some tendency to persistence but capability of changing abruptly in either direction, seems more plausible than any sort of assortment of cytoplasmic components. Ikeno (1930) described a case in *Capsicum* in which normal green bred true, but variegation was transmitted to all offspring by either ovule or pollen. This is interpreted as a virus infection.

It was noted above that there are cases in which there is Mendelian segregation of normal and variegated. Sô (1921), in connection with such a case in barley, suggested that the gene for variegation acts by inducing a plastid mutation (cf. Imai 1928). Rhoades (1950) has given the most complete demonstration of this sort of case using a white-streaked recessive mutation, iojap (*ij*), of maize. Green by iojap produced only green, but the reciprocal produced some streaked and even white (lethal) offspring. On backcrossing white-streaked heterozygotes from the latter cross to normal, ij^+/ij^+, some progenies were obtained that were 100% streaked or even white, although as indicated by the segregation of a linked marker, 50% of these must have been ij^+/ij^+. Clearly the plastid defect for which ij/ij was a condition had become autonomous. We have referred to the induction of nongenic male sterility by the same gene. This occurred in the same experiments but there was no correlation in incidence, indicating that the effects were in different constituents of the cytoplasm.

A similar phenomenon has been described by Woods and DuBuy (1951) in catnip (*Napeta cataria*). A recessive, *mm*, is associated with a high mutation rate in the mitochondria causing multiple visible differences in plastids and pale green color, transmitted after crosses largely in the maternal line.

Crosses between Species and Subspecies

There has been a considerable number of cases in which persistent matroclinous heredity has been observed with respect to characters other than chlorophyl defect or male sterility. Most of these have occurred in crosses between species or at least geographic races. The most extensive have been those in mosses, studied by Wettstein (1924, 1928) and his students. Differences in size, form, and color of the sporophyte were strongly matroclinous in reciprocal crosses of *Funaria hygrometrica* × *F. mediterranea*, and persisted in backcrosses to the paternal species. This was also true of leaf characters of diploid hybrid gametophytes obtained by regeneration from sporophytes. There was extreme matrocliny in the difficult generic crosses *Physcometrium piriforma* × *F. hygrometrica* through six generations of backcrossing to the paternal species. Wettstein was impressed by the contrast between apparently exclusive genic heredity in crosses between varieties of the same species and increasing importance of the "plasmone" with increased taxonomic difference between the parents.

Sirks (1931) studied many strains of two subspecies, major and minor, of *Vicia faba*. Only genic heredity was found in crosses within the two subspecies, but in the many crosses between them, there were systematic reciprocal differences in quantitative characters that persisted in F_2. Stem length was greater but fruit length shorter in minor cytoplasm. These crosses were not all actually matroclinous, however, because of important genic differences among strains of the same subspecies in those same respects that in some cases resulted in superficial patrocliny. There was also an interesting complication in that a chromosome marked by six pairs of alleles showed no crossing over in the F_1 hybrids with minor plasm; and no homozygotes with respect to genes from the major parent appeared in F_2 in association with 25% seed abortion. There was normal crossing over and segregation in the reciprocal cross (major cytoplasm). This leads to consideration of other cases in which it is not a matter of matroclinous heredity of a character transmitted by the cytoplasm but of some sort of incompatibility between genome and plasmone.

Most of the species of *Oenothera* are multichromosomal structural hybrids in which large blocks of chromosomes segregate as units. Thus *O. Lamarckiana* with a ring of 12 chromosomes and 1 free pair at metaphase behaves as a hybrid between a "gaudens" and a "velans" complex, which largely breeds true because of failure of both homozygotes. *O. Hookeri* on the other hand, with seven pairs of chromosomes, shows normal meiosis.

Renner (1925) has shown that *Lamarckiana* plastids do not function well with either a velans/*Hookeri* or homozygous *Hookeri* genome, while all other combinations of plastids and genomes derivable from crosses between these species function normally. Thus while *Hookeri* × *Lamarckiana* gives only dark green F_1 and later descendants, the reciprocal cross gives both dark green gaudens/*Hookeri* hybrids and pale green velans/*Hookeri*. It is interesting that the latter have spots and occasional branches that are dark green, attributed by Renner to *Hookeri* plastids received in small numbers with the pollen. The selfed gaudens/*Hookeri* hybrids produced dark hybrids of their own constitution and also pale green *Hookeri*/*Hookeri* segregants. The latter did not show dark spots, which is interpreted as due to the low probability of receiving *Hookeri* plastids in the pollen in this case. Similar results have been obtained in several other *Oenothera* crosses. A very important point is the reversion to full color of plastids that have been inhibited for a generation by an incompatible genome, on returning to association with a compatible one, in contrast with the effects of genotype *ij*/*ij* in maize.

In experiments with *O. Berteriana* (genome *B*/1) and *O. odorata* (*v*/I) by Schwemmle (1938), there was lethality of zygotes in which *odorata* plastids were associated with *B*/1, 1/I, *vv*, or II, and pale green color when *odorata* plastids were associated with *B*/I or 1/*v*. *Berteriana* plastids gave normal plants with all these genomes.

There were also differences in various morphological characters in combinations of the same genome with different cytoplasms (in the broad sense), including differences in pollen size, shape of leaves, size of petals, and length of hypanthium. By breeding from flowers on dark green branches (with paternal plastids) in predominantly pale plants, it was possible to obtain combinations of plastids, general cytoplasm, and genomes from different sources. As far as chlorophyl was concerned, it was the combination of plastids and genomes that turned out to be significant. The plastid-genome combination, irrespective of the straight female line was also most significant for the lethal effects referred to above and for certain aspects of morphology (leaf shape), but the combination of general cytoplasm received in the straight female line and genome was most significant for petal size and hypanthium length.

Schwemmle (in contrast with Renner) found gradual recovery of plastid color in the incompatible combinations referred to above after four or five generations of selfing. His evidence indicated that this and certain other changes were to be attributed to change in the genome rather than in the plastids. There were, however, possibilities of crossing over and selection so that this does not necessarily imply induced mutation (cf. Caspari 1948).

There are several other plant genera in which the cytoplasm from one species and genomes tracing in part or wholly to another show more or less incompatibility (reviewed by Correns and Wettstein 1937). The most intensively studied cases have been of certain *Epilobium* species, carried farthest by Michaelis and his associates (1933). The hybrid *E. hirsutum* × *E. luteum* shows considerable disturbance of development in F_1 and complete lethality in the third backcross to luteum. The reciprocal hybrids are more normal but there is much pollen abortion. Repeated backcrossing to hirsutum pollen (always from the same clone) gave an appearance of segregation in the early generations but gave apparently pure hirsutum after 8 crosses ($1h^8$), and this continued to $1h^{24}$, except for the persistence of almost complete pollen sterility in most lines. The experiment brought out the remarkable persistence of the properties of luteum cytoplasm.

Nevertheless, there was marked improvement in pollen fertility in some lines. Michaelis is inclined to attribute this to transfer of small amounts of cytoplasm with the pollen and subsequent reassortment of components of the mixed cytoplasm. In another cross, *E. hirsutum* × *E. parviflorum*, alterations occurred among branches of the same plant, directed toward elimination of developmental disturbances that Michaelis again attributes to differential assortment of cytoplasmic entities and to competition and selection within individuals.

Crosses among a great many local races of *hirsutum* demonstrated that there might be as great, persistent cytoplasmic differences as in species crosses. Cytoplasmic differences were found between strains growing in nature only 0.5 to 3 km. apart. These observations are in contrast with Wettstein's generalization.

There has been a considerable number of cases of merogones, seeds from plants of one species that have apparently developed with a genome derived exclusively from the pollen of another species. Cytoplasmic influence is claimed in some cases, but in others no such influence was apparent—e.g., pure *Euchlaena mexicana* from a seed of *Tripsacum dactyloides*, pollinated by *E. mexicana* (Collins and Kempton 1916). Nevertheless, our most important conclusion is perhaps that the cytoplasm of a species may acquire highly persistent properties that make it incompatible with the genome of a related species.

Nongenic Heredity in Metazoa

Extraordinarily few cases of nongenic heredity have been clearly demonstrated in metazoa. Goldschmidt (1934*a*, *b*) has interpreted the results of certain crosses between geographic races of the gypsy moth, *Lymantria dispar*, on the basis of such heredity, but there is an alternative interpretation because of the sex-determining mechanism, XY in females and XX in males.

European females × males of certain Japanese races produced intersex females but normal males. Backcrossing these intersex females to Japanese males repeated this result, and half the females in F_2 were intersexes. The reciprocal cross gave only normals of both sexes but some intersex males appeared on backcrossing to European males and also in F_2. Goldschmidt oscillated between attributing those results to cytoplasmic heredity and to a prematuration effect of the Y chromosome on all oocytes. An ingenious experiment (Goldschmidt 1934*b*) in which the transmission of the Y chromosome was separated from the line of cytoplasmic transmission by use of males that were extreme intersex females genetically, seemed to settle the issue in favor of cytoplasmic heredity, but Winge (1937) showed that they could also be accounted for by assuming zygotic determination of sex by X and Y chromosomes of different potencies in different races.

Goldschmidt also found persistent matrocliny in reciprocal crosses with respect to the extent of larval pigmentation. Here again there is no decisive discrimination between cytoplasmic heredity and the Y chromosome.

Another frequently cited case of apparent matroclinous heredity is one described by Kühn (1927) in the wasp, *Habrobracon juglandis*. Dark and light strains were derived by selection. Both reciprocal crosses gave progenies a little darker than the light strain but with a slight matroclinous difference. This difference was undoubtedly maintained in the sons (haploid), in the daughters from backcrosses to the light strain, and in their sons. Both strains were, however, extremely variable and it is not clear that they had become isogenic. The data as published do not give adequate evidence that the original reciprocal F_1's were necessarily identical in genetic modifiers.

Jollos (1934) found evidence that exposure of larvae of *Drosophila melanogaster* to high temperature caused matroclinous *Dauermodifikationen* that persisted several generations (as well as gene mutations). Plough and Ives (1935) confirmed this to some extent in a very large and carefully controlled experiment, but the apparent cytoplasmic effect was much smaller. Heat treatment of females (or of both sexes) produced $0.63 \pm 0.03\%$ sporadic irregularities in development in the second to sixth generations in the straight female line, in contrast with only $0.21 \pm 0.02\%$ in control lines and lines in which only the males had been treated. It appears that near lethal treatment with heat had produced persistent cytoplasmic damage which lowered the threshold for irregular development.

L'Héritier and Teissier (1947) found a strain of *Drosophila melanogaster* in which susceptibility to CO_2 was usually transmitted in the straight female line, but to some extent by males. None of the chromosomes were involved. L'Héritier and de Scieux (1947) found that either injection of hemolymph, or of tissue grafts, from susceptible to resistant females led to the production by the latter of susceptible offspring. A strain of *D. simulans* was made susceptible. The conclusion was that the pertinent entity was a virus.

Murray and Little (1935) found strongly matroclinous, reciprocal differences that persisted to later generations in crosses between strains of mice with high and low incidence, respectively, of spontaneous mammary tumors. It was shown later (Bittner 1937) by exchanging litters at birth that the line of transmission was through the milk, rather than through the ova, and that a virus was involved in this case.

Recently, however, Kitzmiller and Laven (1960) have presented a clear case of nongenic heredity in mosquitoes with no indication of a virus. It had been found that crosses between many local races of *Culex pipiens* produce nonviable embryos in one direction, although normally viable ones in the other direction. In other cases, both reciprocal crosses give nonviable embryos (or both viable). There was no significant intergrading among these classes, although they cut across the morphologically recognized subspecies. Laven (1959) has studied intensively crosses between two German races, one from Hamburg (*Ha*) and one from Oggelshausen (*Og*). *Og*♀ × *Ha*♂ gave inviable embryos, *Ha*♀ × *Og*♂ gave normal offspring. This continued through 60 generations of backcrossing to *Og*♂. Backcross males in which all three pairs of chromosomes were from *Og*, as shown by markers, mated with *Og*♀ gave inviable embryos, indicating transmission of an *Ha* factor in the cytoplasm down the straight female line.

There are, of course, a great many cases among animals in which both reciprocal crosses give sterile or inviable progeny, which prevents any analysis of the mechanism. It is certain that chromosomal and genic differences play a role in isolating species, but it is possible that cytoplasmic differentiation may also be a major factor in this respect.

CHAPTER 5

Gene and Character

The unit characters of the early geneticists probably seemed to most other biologists to reflect merely another revival of Hippocrates' ancient doctrine of preformation, discredited by all who—like Aristotle, Harvey, Wolff, von Baer, and later embryologists—had seriously attempted to trace the step-by-step elaboration of complexity in the course of development. Most of the early geneticists did little to disavow this interpretation. They were too busy in working out the laws of transmission of these units and the patterns of organization in the germ cells to have time for consideration of physiological implications. In pursuing this task, they were interested in characters mainly as markers for genes. They were interested, for the most part, only in "good" genes consistently associated with easily classifiable characters. The catch words used for convenience in naming genes often seemed to others to imply that they thought of the organism as a mosaic of unit characters.

The extent to which even the earliest geneticists actually were preformationists can easily be exaggerated. Interaction effects such as those responsible for the familiar modifications of the 9:3:3:1 F_2 ratio (9:3:4, 9:7, etc.) were recognized very early (Cuénot 1904, Bateson 1909) and interpreted as due to epigenetic chains of reactions.

It is now generally accepted that each gene acts by controlling the synthesis of a correspondingly specific protein, which in turn often functions as an enzyme in controlling cellular processes. The occurrence of only four kinds of nucleotides in the DNA molecules of which the genes are composed, but twenty kinds of amino acids in the proteins, requires a code of at least three nucleotides to each amino acid. The first suggestion was that the sequence of such groups of nucleotides directly assembles a corresponding sequence of amino acids by some sort of steric relation. It soon became evident, however, that the process was less direct. One of the paired DNA molecules of an active gene synthesizes the complementary RNA molecule ("transcription") in the presence of an enzyme, RNA polymerase, and of certain other conditions. Three different classes of RNA molecules are produced in this way by three different classes of genes, and these are required in each protein synthesis. The specificity of the DNA is ultimately "translated" into that of the protein molecule by means of large, short-lived molecules of "messenger" RNA (mRNA). Each of these becomes associated in the cytoplasm of the cell with a ribosome, a small granule of which a second sort of RNA (ribosomal or rRNA) is a major constituent. This seems to be nonspecific with respect to protein synthesis. One of these granules is believed to move along the mRNA molecule as the site at which the amino acids, corresponding to the pattern of the mRNA molecule, are added one by one to a forming polypeptide chain. This process requires the cooperation of a limited number of kinds of small RNA molecules, "transfer" or tRNA. Each of the latter becomes bonded to a particular kind of amino acid by means of a specific enzyme. The complete compositions of several of these tRNA molecules have been determined. The first, a carrier of alanine (Holley *et al.* 1965), was found to consist of just 77 nucleotides in contrast with thousands in a typical mRNA molecule. A particular group of three of these nucleotides is free to form bonds with a complementary group of three on the mRNA molecule. This occurs with the group that comes next on the ribosome to a group to which bonding has already occurred. In this way, the amino acid carried by the tRNA molecule is brought into juxtaposition with the end of the growing polypeptide chain to which it becomes joined by a peptide bond. It is believed that the chain is terminated by action of particular tRNA molecules (cf. Watson 1965).

This translation process must depend on a code that relates groups of three mRNA nucleotides to specific amino acids by way of complementary "anticodons" in the tRNA molecules. The first such relation was established by the use of synthetic polyuridylic acid which, acting like mRNA, synthesized polyphenylalanine in cell-free systems extracted from colon bacilli and supplied with all twenty of the amino acids (Nirenberg and Mathaei, 1961). In a surprisingly short time, this and other groups had established a code for all amino acids. With a few exceptions, of which two may be related to chain termination, each of the 64 possible sequences of three nucleotides is known to take up a particular amino acid. Since there are only twenty nucleic acids, several different groups of three nucleotides take up the same amino acid ("degeneracy" of the code).

The First Grand Phase of Evolution

The evidence to date indicates that the same code applies to organisms as different as the colon bacillus, tobacco, and man and hence probably to all organisms. The association of the anticodon at one end of the folded tRNA molecule with an amino acid at the other end seems to be arbitrary from the chemical standpoint. Although it may not be wholly arbitrary from that of natural selection, there is a strong indication that all existent organisms trace to a single form in which a particular code had become established. The similarities in the basic metabolic processes of all existent organisms also point toward common origin.

There are obvious difficulties in the evolution of a system in which DNA duplication, RNA synthesis, the bonding of tRNA with amino acid, and protein synthesis in the ribosomes all depend on specific polypeptides, which require the whole system for their synthesis. At present we can only speculate on the first grand phase of evolution in which this circular mechanism became established. It presumably required an enormous period of natural selection among imperfectly functioning systems before it reached its present state.

The processes of establishing an efficient metabolic system and orderly processes of equational cell division and of reduction following cell fusion may be considered a second phase. This treatise, however, is concerned only with subsequent phases, based on cells that have evolved such orderly processes of gene duplication, of translation of gene specificity into protein specificity, of mitosis, and of meiosis after at least occasional syngamy that a mathematical theory of the statistical consequences in populations is feasible.

Genes and Selective Values

Natural selection within a population necessarily operates directly on individuals and thus on genotypes as wholes rather than on allelic genes. The most significant "character" in this connection is by definition the "selective value" of the genotype. This, however, is an abstraction that is practically impossible to measure. It becomes necessary to focus on less comprehensive characters such as viability and fecundity or, still more narrowly, on specific internal adaptations such as the various aspects of metabolism, morphology of internal organs, homeostatic mechanisms, or on specific adaptations to the external world such as instincts, intelligence, size and form appropriate to a particular niche, strength, speed, weapons, armor, concealing coloration. As analysis is carried down toward the immediate effects of gene replacements, the relation to selective value tends to become more remote and contingent.

Gene and Character at Various Levels

It is desirable to go somewhat farther into the relations between gene replacements and characters at various levels. The most direct, as already discussed, is that between gene mutation and protein molecule. Considerable information has been obtained from cases in which the exact composition of protein molecules has been worked out. There are cases of replacement of a single amino acid by a gene mutation to which a direct selective value can be assigned. Sickle-cell anemia in man is due to a type of hemoglobin that differs from the normal type in this way (Ingram 1956). Genetically, this anemia depends on a gene with partial dominance over its normal allele, the heterozygotes being only slightly anemic while the homozygotes are severely affected. It is rare in most parts of the world, as expected from its disadvantageous character, but it is very common in parts of Africa in correlation with the prevalence of malaria to which the heterozygotes show relatively high resistance (Allison 1955). The effect of this gene in this case is of negative or positive selective value (the latter only up to a certain gene frequency), according to the environmental conditions. There are a considerable number of other mutant types of hemoglobin in man that are similarly related to selective value.

Differences among the hemoglobins of different species presumably trace to similar gene mutations. It by no means follows that each of these became established directly by selection. Selective advantage in the species in question may well have depended on association of the properties of the new hemoglobin with changes in many other aspects of blood or tissue physiology, and thus on a much more complex "character" than mere change in the hemoglobin by itself.

Elementary Metabolic Processes

The study of metabolic processes in such favorable organisms as *Neurospora*, *Aspergillus*, *Escherichia*, and *Salmonella* has been characterized by the demonstration of long branching chains of reactions in which each link depends on preceding ones and is controlled by a particular locus. In experiments with *Neurospora*, mutations were induced in various ways and were discovered when single spore cultures failed to grow on media containing the minimum constituents for growth of the type strain, in contrast with normal growth on a liberally supplemented medium. By progressive cutting down of the supplement, it was often found that addition of some one simple substance (or one of a group of closely related substances) was all that was needed for full growth (Beadle and Tatum 1941). In one case (see Fig. 5.1), arginine is needed for growth of one mutant type, either citrulline or arginine will do for another, and any one of the three—ornithine, citrulline, or arginine—suffices for a third (Srb and Horowitz 1944). The analysis of this case has since been carried much farther. Although some genes have been shown to affect a metabolic process only indirectly, others have been shown to deter-

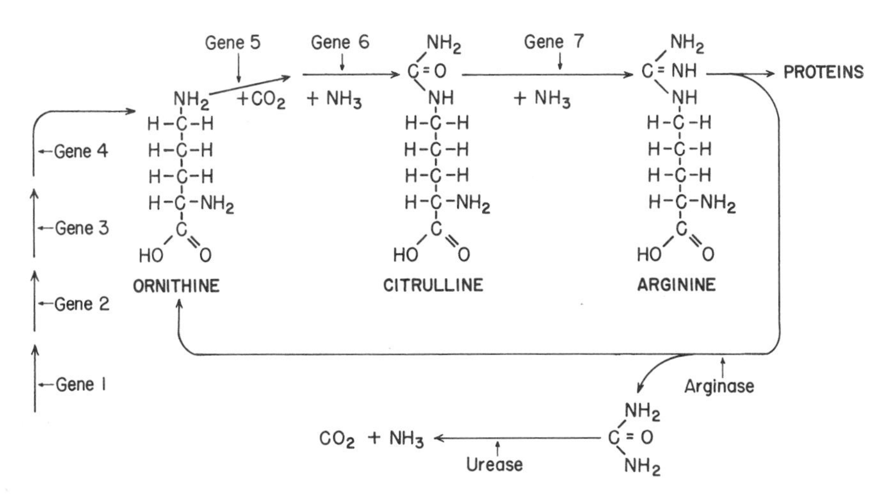

FIG. 5.1. Genes related to steps in the arginine cycle (after Srb and Horowitz, 1944).

mine enzyme differences so that the effect is only one step beyond the immediate gene product.

The Developmental Process

A given array of genes and cytoplasmic constituents of the oocyte, supplemented later by those introduced by the sperm, determine chains of metabolic processes that result in a kind of protoplasm with particular properties, including for instance, response to surface forces by the formation of a special sort of semipermeable membrane; and response to trivial asymmetries in the play of external stimuli by polarization along a certain axis (anterior–posterior), and by establishment of dorsoventral differentiation and, perhaps, of a plane of right-left symmetry. In the course of cleavage, different previously repressed genes are called into play or provide the substrates for action of other genes. This process may, in principle, lead to the step-by-step elaboration of a pattern to any degree of complexity.

It is not necessary here to go into the processes of formation of germ layers, into the inductions and the processes of differential growth by which the morphologic pattern is formed, or into the stages of histologic differentiation that often seem to delimit sharply the places in which particular genes may be called into play and bring about the apparent differences in cell heredities revealed on transplantation. Such processes, in the nervous system, presumably give the basis for genetically determined aspects of behavior, and hence even to genetically determined extraorganic structure. At each level, physiology is registered in structure which gives a firm basis for physiology or behavior at the next higher level. There is no fundamental difference in kind between the relation of genes to differences in such structures as the webs of different species of spiders and the differences in the arrangement of their eyes. Both are indirect (Fig. 5.2).

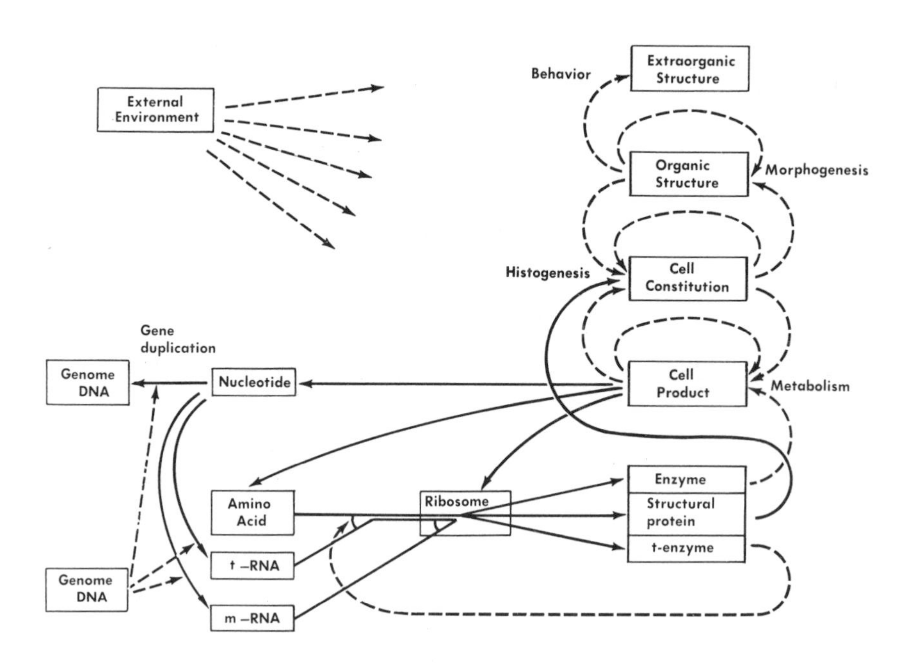

FIG. 5.2. Diagram of relations of genome and external environment to structure at successive levels of organization. (Redrawn from Fig. 21, Wright 1963.)

It was formerly common for biologists, other than geneticists, to attribute the development of characters partly to physiological processes and partly to heredity, as if heredity could operate by some sort of sympathetic magic, independently of physiological channels. The attitude of physiological genetics has been that characters are *completely* determined by physiological processes, but that the genes are (with minor qualifications) the ultimate internal physiological agents.

If, instead of tracing forward from primary gene action (Fig. 5.2) we consider everything that may affect a particular process at a particular time in development, these fall naturally into four categories: (1) local gene action, (2) the chain of past events in the line of cells in question, (3) correlative influences from adjacent cells and from other parts of the body, and (4) external environmental differences. Since the second and third may be analyzed, step by step, on this fourfold basis, and local gene action must be evoked by products of previous events, the ultimate factors are the array of hereditary entities in the egg and sperm and the succession of external influences. Both the prospective and retrospective points of view are indicated in Figure 5.3 (Wright 1934*a*, 1963).

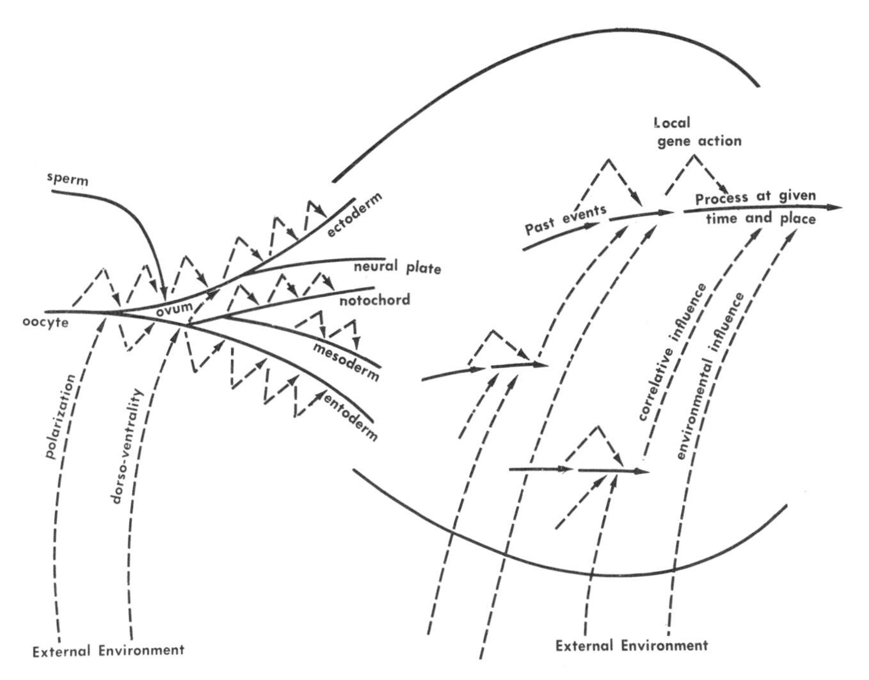

FIG. 5.3. Diagram of relations of genetic and environmental factors to the development of a chordate. (Redrawn from Fig. 1, Wright 1934*a*.)

Generalizations

There are a number of broad generalizations that follow from this netlike relationship between genome and complex characters. These are all fairly obvious but it may be well to state them explicitly.

1) The variations of most characters are affected by a great many loci (the multiple factor hypothesis).

2) In general, each gene replacement has effects on many characters (the principle of universal pleiotropy).

3) Each of the innumerable possible alleles at any locus has a unique array of differential effects on taking account of pleiotropy (uniqueness of alleles).

4) The dominance relation of two alleles is not an attribute of them but of the whole genome and of the environment. Dominance may differ for each pleiotropic effect and is in general easily modifiable (relativity of dominance).

5) The effects of multiple loci on a character in general involve much nonadditive interaction (universality of interaction effects).

6) Both ontogenetic and phylogenetic homology depend on calling into play similar chains of gene-controlled reactions under similar developmental conditions (homology).

7) The contributions of measurable characters to overall selective value usually involve interaction effects of the most extreme sort because of the usually intermediate position of the optimum grade, a situation that implies the existence of innumerable different selective peaks (multiple selective peaks).

These will be considered in succession for the most part, except that it will be convenient to interpolate discussion of the genetics of a number of interaction systems as wholes between generalizations 5 and 6 for purposes of illustration.

The Multiple Factor Hypothesis

A number of systems in which characters have been shown to be affected by allelic differences at multiple known loci will be discussed later. The evidence for the multiple factor hypothesis in unanalyzable, quantitative variability will be deferred to chapter 15, after the discussion of biometric methods.

Universal Pleiotropy

Pleiotropy has a broader meaning in population genetics than in physiological genetics. Grüneberg (1938) drew a distinction between "genuine" pleiotropy in which two effects trace to two different primary actions of the gene and "spurious" pleiotropy in which they trace to the same primary mode of action in different parts of the body, or else one effect traces to the other. If each gene has only one primary effect, the synthesis of a specific kind of polypeptide, genuine pleiotropy does not exist, unless one treats this polypeptide as a composite in which different components may be responsible for different effects. Grüneberg was no doubt right in holding that most cases of apparent pleiotropy are of one or other of the secondary sorts that he called spurious. From the standpoint of population genetics, these distinctions are, however, unimportant. There is pleiotropy if two effects contribute differently to selective value.

There can obviously be no genuine pleiotropy, in the physiological sense, of the mutation for sickle-cell anemia, since only one amino acid is replaced in the product of the locus, but the results of the disadvantageous anemia and the advantageous resistance to malaria contribute differently to selective value and may be considered as genuine pleiotropic effects from the standpoint of population genetics. Similarly the white coat color and pink eyes of albinism, caused by a single mutation, are no doubt cases of spurious pleiotropy from the physiological standpoint, but they might contribute very differently to selective value, under arctic conditions for example. Even effects that seem exactly the same in different parts of the body may contribute differently to selective value in some cases, for instance, in connection with a pattern of concealing coloration.

From the condensed descriptions of the mutants of *Drosophila melanogaster* (Bridges and Brehme 1944) it is at once apparent that a large proportion have conspicuous effects on seemingly unrelated characters such as bristle and wing form, aristae and bristles, wings and legs, legs and body color, eyes and bristles, and so on. A great many, with characteristics that are not obviously at a disadvantage, are strikingly low in viability or fecundity or both. Moreover, nearly all mutations that have been tested in competition with wild type are rapidly outbred, even though no deficiency in viability or fecundity is apparent in ordinary experiments. Pleiotropy seems, indeed, to be virtually universal in the *Drosophila* mutations that are sufficiently conspicuous for use in Mendelian experiments.

In slower-breeding animals, such as mammals and birds, there is less evidence for minor effects on viability or fecundity of mutations. There are, however, many cases of serious defects in these respects associated with the character change for which the mutations are named, as well as many cases of effects on two or more conspicuous characters. Specific examples will be given in other connections. The available evidence indicates that pleiotropy is virtually universal.

Uniqueness of Multiple Alleles

For some time after multiple alleles were discovered, it was a common belief that they differed merely quantitatively, since they usually seemed to have effects that differed only in degree. This became untenable, however, when it was found that parallelism was often lacking among pleiotropic effects (Wright 1916). It is obvious now that the difference in the sequence of nucleotides by which alleles differ is a qualitative one and that differences in the sequence of amino acids in the polypeptide molecules that are their primary products are also qualitative.

It is sometimes convenient in population genetics to assign numerical values to members of an allelic series in considering their effects on a single measurable character. This, however, is probably never more than a rough approximation. The order of effect even on such a character may be different among combinations with genes at other loci, and in dealing with such a complex character as selective value, the essential uniqueness of each allele must never be forgotten.

The most striking phenotypic evidence of this uniqueness of alleles is probably to be found among loci that determine the presence of various sorts of antigens. This uniqueness was apparent in the first case, that of ABO isoagglutinogens of human erythrocytes, shown by Bernstein (1924) to depend on three alleles I^A, I^B, and i. Gene I^A whether in I^AI^A, I^AI^B, or I^Ai, determines the presence of an agglutinogen that responds to an antibody present in the serum of individuals lacking it; the same statement can be made of I^B. Gene i was at first supposed to produce no agglutinogen, but the situation has become more complicated since. Alleles have been distinguished, such as I^{A_1} and I^{A_2}, that seem qualitatively alike but are different in the strength of their responses. Actually both I^{A_1} and I^{A_2} determine antigens that respond to anti-A_2, while I^{A_1} determines an antigen that behaves as if it carries an additional component, which is never found by itself. The R series of antigens of human erythrocytes for a time seemed to be composed of three pairs of alternatives, C,c; D,d; and E,e, all but d demonstrable by specific reagents, and it was suggested that these were really three linked loci (Race 1944) in spite of very extreme deviations from randomness in the frequencies of combinations in populations. There is a similar situation at the M,N; S,s locus in man. Wiener and Wexler (1952) and Owen (1959) have pointed out that the systematic identification of positive reactions to antisera with single specific entities breaks down completely in cases in which the chemical constitution of groups of antigens are actually known (as in experiments by Landsteiner and van der Scheer).

The most complex case yet described is the B locus of cattle with more than 250 alleles, which are distinguished by the pattern of responses to more than 30 different antisera. It is convenient to name the alleles by their positive responses but asymmetrical relations like that of I^{A_1} and I^{A_2} in man abound. A more complex situation is found among the B, G, and K "factors" of many of the alleles. Many of them react as if they include both a B and a G factor, some include B but not G, others G but not B, and still others, neither. All four combinations are common and suggest the occurrence of two completely linked loci. On the other hand, the factor *K*, demonstrable by a specific anti-*K* serum, while common in association with *BG*, has never been found with the other three combinations (Stormont 1959). It is most appropriate to look on the *B* complex as due to a single locus with an indefinitely large number of unique possible alleles rather than as a block of 30 or so linked genes, even though the locus may on very rare occasions be broken in some way. Many other similar but less extensive systems are known in cattle and other animals.

The self-incompatibility loci of many plants are somewhat similar. At least 45 alleles have been described in *Oenothera organensis* (Emerson 1939), a species in which the entire population was probably only about five hundred. Williams' (1947) studies indicated several hundred in a particular population of red clover. The number in the whole species must be immense (Lewis 1949). The criterion for specificity in this case is inhibition of pollen-tube growth on a style in which one of the two alleles is the same.

Dominance

The phenomenon of dominance does not exist where the observed characters are the immediate products of the two alleles formed independently, without significant competition. This situation (called codominance) is found with allelic hemoglobins and other proteins. Codominance is also the rule for the apparent factors of the complex specificities of antigens that, as noted, must somehow reflect specificities of the genes but not as directly as in the preceding case. Not surprisingly, it is a rule that has exceptions (Stormont 1959, Owen 1959). In the case of the self-incompatibility alleles of many plants, each allele (in the style) functions independently of the other in diploids, but there are complications in tetraploids that indicate interactions between the alleles of the diploid pollen grains (Lewis 1949).

Apparent codominance has also been found in some allelic series determining characters that are presumably more remote from primary gene action than those above. There are a number of notable examples among pattern-determining loci.

The Asiatic ladybird beetle, *Harmonia axyridis*, has 12 or more different true breeding color patterns in the elytra. These include yellow with or without small black spots, and black with yellow spots differing in number, size, and location. These have been found by Hoshino (1940) and Tan (1946) to depend largely on a single allelic series. The heterozygotes combine the black areas of the homozygotes of both alleles, as if each were called into play locally, independently of the other. There is epistasis if the pattern of black of one allele completely covers that of the other, but otherwise both patterns show more or less completely. Thus there is dominance of the melanic character, not of the genes. The allele for self-yellow seems to be recessive to all the others under certain conditions, but under environmental conditions that bring out small black spots, these may show in heterozygotes with other alleles.

There is a similar situation in nature in the fish, *Platypoecilus maculatus* (Gordon and Gordon 1950). In this case each of seven alleles determines the presence of black in a restricted area near the base of the tail, while there is one (the commonest) that determines no such pattern. The last behaves as a universal recessive, while heterozygotes among the others show black wherever one or both of their homozygotes show it.

The case is more complicated in the numerous different color patterns found in grouse locusts (*Tettigidae*). In *Paratettix texanus*, Nabours (1930) found 21 basic patterns on pronotum and posterior femora due to a single allelic series (no crossovers in very extensive data). A pattern of varied shades of gray behaved as a universal recessive, whereas localized areas of chalky white, of pale yellow, or of various shades of mahogany red, determined by different alleles, exhibited almost complete dominance over gray. There was usually intermediacy in combinations of the other colors.

Another such pattern factor was, however, closely linked and a second, loosely linked. In other species, e.g., *Apotettix eurycephalus*, similar pattern factors behaved as if in several closely linked loci rather than nearly all in one, suggesting that the apparent alleles are due to variations in a complex locus. As noted earlier, however, the term locus has ceased to have the absolute meaning it once had. Other more or less similar cases can be cited.

Interactions between the chains of processes initiated by genes at the same and other loci are, however, the rule with most characters and give rise to the phenomenon of dominance. Degree of dominance and alterations of it are necessarily of importance in population genetics.

In spite of the enormous number of experiments in which the state of dominance of alleles has been reported, it is difficult to obtain an accurate idea of what the situation actually is because there has been little careful study of degrees of dominance under varied genetic backgrounds. As a practical consideration in planning and reporting experiments in which the exact degree is of no concern, the terms "recessive" and "dominant" have usually been applied merely to indicate which homozygote is more easily distinguishable from the heterozygote. This is apparent in Mendel's paper and in most of the early studies with *Drosophila*. The number of genes I have studied in guinea pigs has been small enough to permit close attention to the varying degree of dominance (and the interaction effects) in all combinations that seemed likely to be of interest (Wright 1960*c*, *d*).

Guinea pigs have been crossed rather extensively with their wild ancestor, the Peruvian cavy, *Cavia cutleri* (Castle 1916*a*, Wright 1916) as well as with the Brazilian cavy, *C. rufescens* (Detlefsen 1914). These were homallelic as far as indicated by appearance and by tests in the following color factors (discussed later): *S*, *Si*, *Gr*, *E*, *A*, *C*, *F*, *P*, *B*, except that in some or all of the Brazilian cavies *A* was replaced by an allele (A^r), demonstrated later to be intermediate in effect between *A* and *a* of the guinea pig. There is considerable reason to attribute gene *dm*, another color factor, to *C. cutleri*, but there is no information on the pair *Mp*, *mp*. The type genes of these species concerning hair direction are *r*, *M*, *st*, and probably *re*. Another type gene is *px*, instead of *Px*, the latter being a mutation that tends to restore the pentadactyl foot in *Pxpx* but produces a lethal monster in *PxPx*. At this point, it will be convenient to list these according to their usual dominance relations, without going further into their effects (Table 5.1). $E/e^p, e$ indicates the complete dominance of *E* over both e^p and *e*; $S|s$ indicates that heterozygotes are at least statistically intermediate between the homozygotes, although they may overlap one or both of the latter individually; and $r\backslash R$ indicates that the type gene *r* is completely recessive to the mutant *R*.

TABLE 5.1 Usual dominance relations of pairs of alleles of the guinea pig. $E/e^p, e$ implies dominance of *E* over e^p and *e*; $S|s$ implies intermediacy; $r\backslash R$ implies recessiveness of *r*.

		Loci	Non-type alleles
DIFFERENCE RECOGNIZABLE IN WILD TYPE BACKGROUND			
Wild type dominant	$E/e^p, e; A/A^r, a; C/c^k, c^d, c^r, c^a; P/p; B/b$	5	(10)
Heterozygote usually intermediate	$S\|s; Si\|si; Gr\|gr; F\|f; re\|Re; px\|Px$	6	(6)
Wild type recessive	$r\backslash R; st\backslash St$	2	(2)
DIFFERENCE NOT RECOGNIZABLE IN WILD TYPE BACKGROUND			
Heterozygote usually intermediate	$Dm\|dm; Mp\|mp; M\|m$	3	(3)
RELATIONS BETWEEN NON-TYPE ALLELES			
Wild type dominant	A^r/a		(1)
Heterozygote usually intermediate	$e^p\|e; c^k\|c^d, c^r, c^a; c^d\|c^r, c^a; c^r\|c^a$		(7)

The wild type allele is completely dominant over lower alleles (10 in number) in only 5 of the 13 loci in which the differences are recognizable on a wild-type background. There are 6 loci at which the heterozygotes are typically intermediate and 2 at which the wild-type allele is completely recessive. There are 3 loci in which the difference would not be recognizable in a wild-type background (with the possible exception of *Dm*, *dm*), and in all of these the heterozygote tends to be intermediate. In seven of the eight cases of heterozygotes between nontype alleles, the heterozygotes tend to be intermediate and the one exception involves the gene A^r, introduced into the guinea pig from *Cavia rufescens*, in which it seemed to be usually, if not always, the type allele.

There are important qualifications. Genes *P* and *C* show overdominance in combinations *E* (or e^p) *Cbb* and *E* (or e^p) *Pbb*, respectively. *Stst* is intermediate in the presence of *Rmm*. All of the sixteen heterozygotes, described as usually intermediate, overlap one or both of the homozygotes and dominance is easily shifted by selection. In some cases, one of the homozygotes may be completely dominant in certain combinations.

The earliest hypothesis on dominance seems to have been the presence and absence theory of Bateson (1909), developed when only pairs of alleles were known. He supposed that one representative of a gene, as a physical entity, is ordinarily enough to give the character in full and that the recessive character is due to absence of this entity. This hypothesis broke down with the discovery of series of multiple alleles in which each has a specific effect, and the demonstration of reverse mutation.

In a hypothesis sketched in 1929 and elaborated later (Wright 1934*a*, 1941*a*, *c*) I postulated a chain of reactions between the primary gene product (assumed to be an enzyme) and the observed character. It was assumed that the gene controlled agent (A) at a given step, formed at rate x, reacts with a substrate S formed at the rate s, to produce a product P at the rate p. With an agent of low activity the rate of formation of the product may be expected to be proportional to x giving semidominance as far as this step is concerned, but with a very active agent that tends to exhaust S, it will be limited by the rate at which substrate is supplied over the given period. If one representative of the gene in question uses up the substrate as rapidly as supplied, the gene will be completely dominant over an inactivated allele.

The simplest hypothesis of this sort is that in which the amount of product is determined sharply by whichever of the interacting substances A or S is in defect. This may be called the bottleneck hypothesis (Fig. 5.4*A*):

$$p = x \quad \text{if } x \leq s$$
$$p = s \quad \text{if } s \leq x.$$

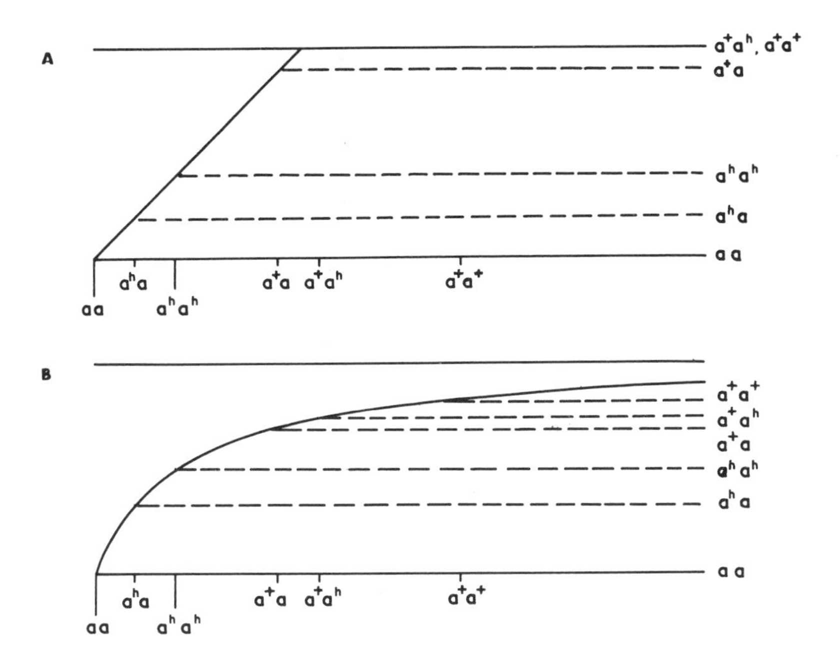

FIG. 5.4. Relation of gene dosage (abscissas) to phenotype (ordinate) on the bottleneck hypothesis (above: part *A*) and on the partitioning hypothesis (below: part *B*). (Redrawn from Fig. 7, Wright 1934*a*.)

Another simple hypothesis is that the rate of formation of the product depends on competition between the action of constructive and destructive agents with rate constants c and d, respectively, and thus apportionment according to their strengths. This may be called the partitioning or multiplicative hypothesis:

$$p = \left(\frac{c}{c+d}\right)s, \quad \text{or} \quad p = \left(\frac{x}{x+d}\right)s \quad \text{if } x = c.$$

This is a hyperbola that rises from zero with an initial slope s/d toward s its asymptote (Fig. 4*B*). The rates become absolute amounts when multiplied by the duration of the reaction.

More generally, the gene-controlled agent A and the substrate S may be supposed to react with rate constant c to give the product (Fig. 5.5), but to be diverted into nonproductive channels with rate constants b and d, respectively, during a period of flux equilibrium.

BIMOLECULAR REACTIONS

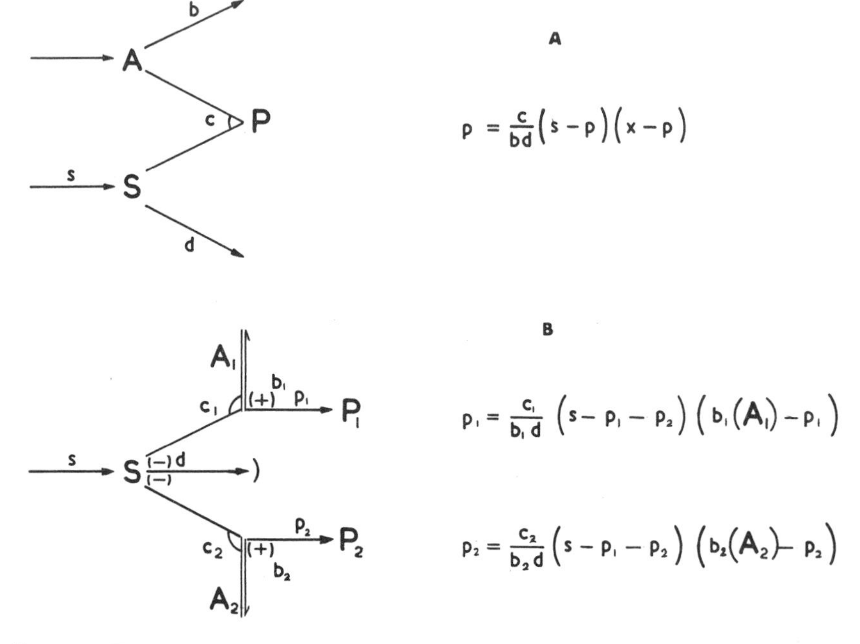

FIG. 5.5. Relation of rates of production of product (p) to gene dosage (x), rate of formation of substrate (s), and the reaction rates b, c, and d indicated in the diagram under the intermediate (bimolecular) hypothesis (above: part A). A more complicated system is indicated below (part B) in which enzymes A_1 and A_2, determined by alleles in a diploid, compete for substrate S. (Redrawn from Fig. 5, Wright 1934*a*.)

$$x - cAS - bA = 0$$
$$s - cAS - dS = 0$$
$$p = k(s - p)(x - p), \qquad k = c/bd.$$

This is a hyperbola that rises from zero with initial slope $sk/(sk + 1)$, but falls off as it approaches the asymptote s. The initial tangent intercepts the asymptote where $x_I = (sk + 1)/k$. The ratio p_I/s at this point is

$$[(sk + 1) - \sqrt{(sk + 1)}]/sk.$$

As k is made larger, there is approach to the bottleneck hypothesis at which $p_I/s = 1$. As k approaches 0, there is approach to $p = [x'/(x' + d)]s$, $x' = cx/b$, as under the partitioning hypothesis, under which $p_I/s = 0.50$.

Various values of k and p_I/s are shown below.

	Partitioning			Intermediate			Bottleneck
k	0	0.083	1.78	8	48	9800	
p_I/s	0.50	0.51	0.625	0.75	0.875	0.99	1

The situation is a little more complicated for a first step in which the gene product is an enzyme (Fig. 5.5*B*).

There may be other agents than that tracing to the gene in question that can transform the same substrate to contribute to the product. To allow for this, it is merely necessary to add a constant to x in all of the above formulas. With such a supplement, complete inactivation of the gene does not lead to complete absence of the character. The dominance relations are essentially similar.

The gene may act destructively on the substrate. In this case, it is not a resultant of the product of the reaction formed at rate p that is observed, but a resultant of what is left of the substrate after removal of the product. Letting y represent the grade of the character, $y \propto p$ in the preceding cases, but $y \propto (s - p)$ in the case considered here. The value of y is thus measured by the interval between the appropriate one of the curves described above and the upper limit s. The dominance relations in terms of gene activity are the same as with a positive character.

The product of the reaction, whether formed at rate p or $s - p$, may react with a second substrate, positively or negatively. The degree of dominance after the first step need not carry through the second step without increase, but some increasing dominance along the chain is most probable. As already noted, if complete dominance is achieved at any step, it cannot be undone at any later step, assuming that specificity differences among the alleles do not carry through.

Under the theory up to this point there cannot be less than exact semidominance of a character due to an active gene product over partial or complete inactivation, but there may be any degree up to complete dominance. In a series of alleles, it is possible under the bottleneck hypothesis that all may exhibit semidominance with each other, but it is more likely that the higher ones will pass through near dominance to complete dominance of the highest. Under the other hypotheses, degree of dominance rises systematically with increasing activity, but theoretically never quite reaches completeness with the highest.

Many multiple allelic series conform approximately to this description. Thus in the series E, e^p, e found in black, tortoise-shell, and yellow guinea pigs, respectively, Chase (1939) found the following percentages of black (within the colored areas in the presence of ss). In this case, however the "substrate" is available space (limit 100%), rather than a substance.

	ee	e^pe	e^pe^p	Ee	Ee^p	EE
With $S-$	0	68	84	100	100	100
With ss	0	61	79	100	100	100

Intensity of color in guinea pigs is due primarily to the c-series of alleles. The lowest allele, c^a, gives albinism in c^ac^a. The percentages of the amount in genotype $ECPFB$ (intense black) are as follows (Wright 1960*d*).

c^ac^a	c^dc^a	c^dc^d	c^rc^a	c^rc^d	c^rc^r	c^kc^a	c^kc^d	c^kc^r	c^kc^k	$C-$
0	40	66	44	76	87	77	90	96	93	100

There are irregularities in this case. The greater dominance of c^d over c^a than of c^r over c^a is undoubtedly significant. The apparent overdominance of c^kc^r is less certainly significant.

An allele due to complete inactivation of the type allele has been conveniently termed an "amorph"; one with partial inactivation a "hypomorph"; and one with greater activity, a "hypermorph" (Muller, 1932). This is consistent with the uniqueness of each mutation in its chemical constitution. If a mutation made possible a new sort of effect, Muller termed it a "neomorph." If this implies reaction of its product with a substrate on which that of the type allele had no effect, the latter would be an amorph with respect to this particular effect. Such completely dominant mutations as R (rough fur) and St (star rosette on forehead) in the guinea pig may well be of this type.

In any of the schemes discussed so far, a very active primary gene product is more likely than one with slight activity to exhaust the available amount of substrate and lead to complete dominance of strong effects over inactivation of any degree. Conversely, a very slight primary effect on metabolic processes is more likely to be represented at all subsequent steps by agents that do not deplete their substrates much and thus lead to incomplete dominance of the ultimate effect. The actual prevalence of dominance among minor factors is an outstanding problem in population genetics, and there are many possible complications of this simple theory, related to the theory of gene interaction.

Gene Interaction

Interaction effects are least likely where the observed character is closely related to primary gene action as it is in the case of allelic differences in protein composition. This seems to be borne out by the evidence.

The serologic responses of antigens are not so close, but are also largely independent of the rest of the genome. Interactions among these are, however, by no means unknown. Irwin (1947) and his associates have demonstrated that certain antigens that are homozygous in the parents are absent in the hybrids of many species of doves and certain hybrid ducks, and that such hybrids exhibit "hybrid" antigens not present in either parent. Fox (1949) has demonstrated interactions among the antigenic effects of mutations induced in *Drosophila melanogaster* by X-rays.

Interaction effects necessarily occur with respect to the ultimate products of chains of metabolic processes in which each step is controlled by a different locus. This carries with it the implication that interaction effects are universal in the more complex characters that trace to such processes. There are usually also effects on dominance.

The product at one of the steps in such a chain of processes may be subject to destruction up to a certain level, by substances tracing to other genes. A gene that would otherwise exhibit incomplete dominance over an inactivated allele may actually be recessive to the latter because the heterozygote falls below this threshold (Wright 1927, 1929). In the following case, genotype *ff*, which greatly reduces the activity of the product of genes c^k and c^d in yellow (*ee*), causes these genes to become recessive (at birth) to inactivation of the *c*-locus (albinism, c^ac^a). Six months later c^kc^k (and c^kc^d and c^dc^d) have also fallen below the threshold. The figures below are percentages of the amount of yellow at birth, and in later pelages.

	At birth				Later pelage			
	c^ac^a	c^kc^a	c^kc^k	$C-$	c^ac^a	c^kc^a	c^kc^k	$C-$
eeFF	0	18	37	100	0	13	31	70
eeFf	0	14	31	83	0	8	28	53
eeff	0	0	4	34	0	0	0	14

Supplementary agents that react with the same substrate as the product of the gene under consideration may vary because of changes in another locus. This can be represented by substituting $(x_1 + x_2)$ for x in the formulas. The gene *F*, which is of major importance in a step in the production of yellow pigment, is a feeble substitute for P in a corresponding step in the production of black (or brown) pigment. The amounts below are percentages of the amount in intense black ($ECPB$).

	$ppff$	$ppFf$	$ppFF$	Pff	PFf	PFF
ECB	1	12	20	100	100	100
Ee^de^dB	0	3	8	66	66	66

If both factors are sufficiently active to exhaust the substrate, the F_2 ratio is 15:1 instead of 12:1:2:1, as above.

Another sort of interaction is that in which there are two or more levels of substrate because of allelic genes back of it. If the reaction of genes at a second locus with these substrates is of the partitioning type, the products with the lower substrate are in constant ratio to those with the upper one.

If

$$p_1 = \left(\frac{x_1}{x_1 + d_1}\right)s_1 = s_2,$$

then

$$y = \left(\frac{x_2}{x_2 + d_2}\right)s_2 = \left(\frac{x_1}{x_1 + d}\right)\left(\frac{x_2}{x_2 + d}\right)s_1.$$

The F_2 ratio is thus of the 9:3:3:1 type, assuming that there is near dominance of the upper allele in both cases and that neither of the lower alleles represents complete inactivation. If there is complete inactivation in either case, this reduces to 9:3:4, and if there is complete inactivation in both, it becomes 9:7.

If the reaction with the substrates at two levels (second step) is of the bottleneck type, the situation is more complicated. The change or lack of change in dominance under various conditions may be seen from Figures 5.6*A* and 5.6*B*. There can be no F_2 ratio of 9:3:3:1 since one or the other of the lower alleles must be a limiting factor for the other reaction. There can be ratios of 9:3:4 reducing to 9:7 with complete inactivation at one or both loci. The lower substrate, if not zero, acts as a ceiling that tends to impose dominance in the second reaction. The results are intermediate under the intermediate type of reaction with the substrate.

In the case below, bb, which replaces sepia (B) by the slightly different color, brown, also lowers the ceiling 50% or more, with respect to the effects of the c-series. The product of c^d (like that of c^a and c^r) suffers from thermolability up to birth (Wolff, 1955). The recovery in later pelages (if B is present) more than offsets the general tendency toward reduction of intensity manifested by C (and c^k). The greatly lowered ceiling, with bb in later pelages, almost obliterates all differences in the c-series except that due to inactiva-

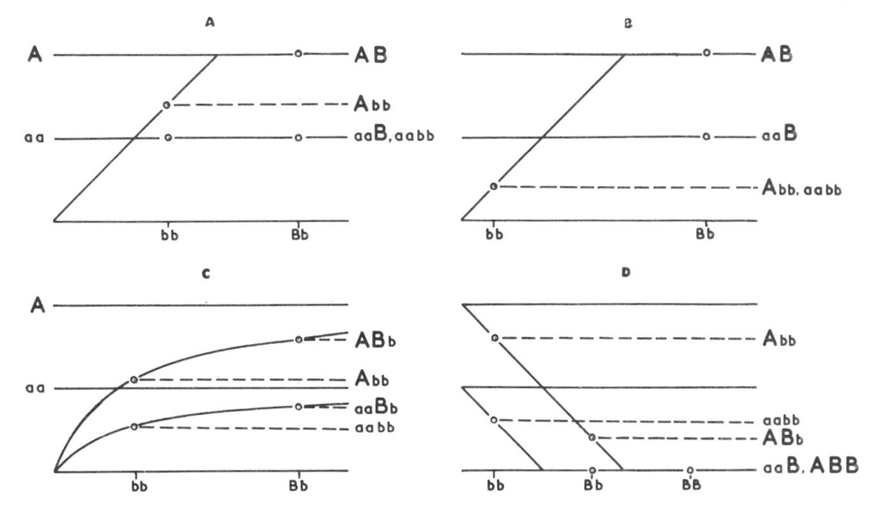

FIG. 5.6. Some two factor relations between genotype and phenotype. Two levels of substrate are determined by *aa* and *A*. Action of *bb*, *Bb* is constructive according to the bottleneck hypothesis in upper row (parts *A* and *B*); according to the partitioning hypothesis in the lower left (part *C*), and destructive according to the bottleneck hypothesis in the lower right (part *D*).

tion (c^a), and c^d becomes completely dominant over c^a. This contrasts with the complete recessiveness of c^d to c^a in *eeff*.

	At birth				Later pelage			
	c^ac^a	c^dc^a	c^dc^d	C	c^ac^a	c^dc^a	c^dc^d	C
EPB	0	46	66	100	0	55	72	93
EPbb	0	33	42	50	0+	35	35	38

Although replacement of *B* by *bb* imposes a lowered ceiling on the action of the *c*-series and thus a tendency toward dominance of all alleles over albinism, the replacement of *P* by *pp* tends to reduce action of the *c*-series more according to the multiplicative (partitioning) pattern. There is little or no ceiling effect.

	At birth				Later pelage			
	c^ac^a	c^dc^a	c^dc^d	C	c^ac^a	c^dc^a	c^dc^d	C
EppB	0	4	8	20	0	2	3	8
Eppbb	0	5	10	15	0	3	6	10

The F_2 ratio in the case of heterozygotes Cc^aPp is 9:3:4 with respect to coat color and 9:7 with respect to eye color, which is pink with *pp* as well as with c^ac^a.

The primary products of alleles at a locus may be able to react with a number of somewhat different substrates. This introduces no complication if the effects are proportional. If the different substrates lead to pleiotropic effects these will be parallel.

The specificity differences among alleles make it more probable, however, that their efficiencies with different substrates will be different; consequently, pleiotropic effects will not be parallel. This is clearly true for the *c*-series. The sepias (*E*) below are in terms of 100 for intense black (*EPB*), and the yellows (*ee*) are in terms of 100 for intense yellow (*eeFF*). The colors are too different to use the same scale of quantity. The heterozygotes of lower alleles are intermediate.

	At birth					Later pelage				
	c^ac^a	c^rc^r	c^dc^d	c^kc^k	C	c^ac^a	c^rc^r	c^dc^d	c^kc^k	C
eeFF	0	0	39	37	100	0	0	44	31	70
eeff	0	0	4	4	34	0	0	1	0+	14
EPB	0	87	66	93	100	0+	92	72	81	93
EppFFB	0	3	8	15	20	0	4	3	7	8

Gene c^r seems to be completely unable to take part in the production of phaeomelanin (yellow). The heterozygotes c^dc^r and c^kc^r (in *eeFF*) are indistinguishable at any age from c^dc^a and c^kc^a. The c^d heterozygotes produce 17%, rising to 23%, whereas those with c^k produce 18%, fading to 13%. Gene c^d is more efficient than c^k after its greater thermolability up to birth ceases to operate.

These differences may create complicated competitive processes. There seems to be no competition in phaeomelanic pigment cells (*ee*) (in the absence of sootiness factors, not considered here), but there is clearly competition in the eumelanic pigment cells in the hair follicles. The inverse relation between the efficiencies of c^d and c^r in producing phaeomelanin and eumelanin in the coat in *EP* is probably due to this, since c^d produces much more eumelanin than does c^r in the eyes where no phaeomelanin is formed, even in pink-eyed intense yellows (*eeCppF*). With either *eeP* or *EP*, eyes are black with c^dc^d, c^dc^r, c^dc^a, but dark red with c^rc^r and light red c^rc^a, pink with c^ac^a (Wright 1916). The greater dominance of c^d over c^a than of c^r over c^a in *EPB*, in spite of the larger amounts of pigment in the latter case, can be attributed to the competitive effects that reduce the efficiency of the c^d-product. The fact that c^dc^dPF and c^dc^dPff both have the same grade of sepia indicates that the reduced efficiency of c^d, in spite of difference in capacity to produce yellow in eec^dc^dF and eec^dc^dff, is prior to action of F and thus presumably of P (at the same step) and is due to competition of eumelanic and phaeomelanic precursors (in excess) for c^d-product. There should be no such competition for C-product, the dominance of which indicates that it is in excess. That there is some capacity to produce yellow in eumelanoblasts is shown by the nearly pure yellow of *ECppffmpmp*. The amount is, however, much less than with *ee* (7% instead of 34%). Although the c^r-eumelanic product is extraordinarily efficient in conjunction with P, it is of low efficiency with ppF.

A second competitive process is indicated by the apparently complete absence of yellow in *ECppFmpmp*. This contrasts with approach to yellow in many cases in $Ec^dc^dppFmpmp$. In the competition of C-eumelanic and C-phaeomelanic products for F, the former dominates, whereas in that of c^d-eumelanic and c^d-phaeomelanic products neither wholly dominates at this step. It appears that the reduction of eumelanic pigment with *Epp* by genes of the type of *mp* must follow this competition.

E, in the presence of *ppff*, acts largely as an inhibitor of yellow. Letting $eeCF = 100$:

	Ec^dc^d	eec^dc^d	EC	eeC
ppff	0	4	7	34

In this case, the F_2 ratio from $EeCc^d$ (or $EeCc^k$) is 3:9:1:3, or practically 3:10:3, but becomes 3:9:4 from $EeCc^r$ or $EeCc^a$ and becomes 3:13 for Eec^dc^r or Eec^dc^a.

The estimates of the effects of *Mp*, *mp*, specific modifiers of *Epp* are as follows, with $ECPF = 100$.

	mpmp	*Mpmp*	*MpMp*
ECppFF	13	19	25
ECppFf	8	14	20
ECppff	0	1	1

There was evidence for a segregation of a second modifier of *ECppFF* that in combination with *mpmp* gives a color indistinguishable from that of Ec^rc^rppFF (about 4 instead of 13 on the above scale), but this was lost before it was clearly demonstrated. These very pale sepias showed no trace of yellow.

Action of alleles on different substrates permits each allele to be dominant over those lower in the series as with A, A^r and a (light agouti, dark agouti, and black). It also permits more or less dominant inhibitory alleles in the same series with recessive inactivating ones, of which examples are known (e.g., *C* locus in maize).

The most remarkable interaction effect is in browns, *Ebb*. The three loci *C*, *P*, *F*, the type alleles of which contribute to higher intensity of brown in combinations up to a certain point, tend to reduce intensity beyond this in a way known as dinginess. In low grades, scattered hairs on cheeks and nape are diluted except at the tip. In the highest grades all hairs are diluted, except at the extreme tip, over most of the coat.

	ppff	*ppFF*	*Ppff*	*PpFF*	*PPff*	*PPFF*
ECbb	1	15	53	50	40	30
Ec^kc^kbb	0	12	49	49	49	49

The optimal combination seems to be *CPpff*. Since Cc^aPpFF was slightly more intense than *CCPpFF*, it is probable that the optimum is actually Cc^aPpff with *CCPPFF* some 40% lighter (but extremely variable).

The product of *CPF* must be assumed to have pleiotropic effects. Primarily, it acts on a substrate controlled by *bb* to produce something, *x*, that is necessary in the production of brown, but if in excess of the *CPpff*-product, this excess reacts with a different substrate to produce something that tends to destroy *x* almost quantitatively. The fact that the tips of the hairs are ordinarily intense suggests that there is so much of the *bb* substrate there that no excess of *CPF* product exists even in *CCPPFF*.

The *CPF* product also determines the intensity of sepia pigment, but in my colony *CCPPFF* was always intense black in the presence of *EB*, with no trace of dilution. Since, however, the amount of pigment and presumably of substrate was twice or more as great with *B* as with *bb*, little or no excess *CPF* product is to be expected. Ibsen and Goertzen (1951) observed a slight dilution of intense black *ECCPPFFB* in the presence of homozygotes of a semidominant gene *W* (whitish) that was present in their colony and that so enhances the dinginess of browns of genotype *ECCPPFFbb* that much of the hair below the tip is pure white, something never observed in my colony. Dinginess is affected by many other factors, ones that affect the upper limit for eumelanin, $\sum$ (*leu*), age, presence of androgens (Wolff 1954) (color fading in males but not females), white spotting (*ss*), and other modifiers $\sum$ (*Db*).

Finally, there is a type of dilution that affects both eumelanin and phaeomelanin and seems to depend on the vitality of the pigment cell. With *sisi* (silver) there is a sprinkling of white hairs in the coat at least on the belly that is capable of extension by selection to such an extent that the animal is pure white except on head and feet.

Si is completely dominant in low grades, but there may be much ventral silvering in *Sisi* where *sisi* approaches white. There seems to be a slight dilution of color in *sisiC*. With lower *c*-alleles there is some dilution with *Sisi* and more with *sisi*. This is most conspicuous in c^dc^a and c^rc^a and is more extreme in yellows than in sepias or browns. A second pair of alleles *Dm*, *dm* produces no silvering and has little or no effect if *C* is present; but with lower *c*-alleles, *Dmdm* and *dmdm* dilute color very similarly to *Sisi* and *sisi* and act cumulatively with these.

The genotype *sisidmdm* is pure white, except for occasional pale spots on the cheeks, and shows reduced eye color. The animals are anemic (with 25% fewer red blood cells) and have a high mortality after birth (chap. 9). The males are completely sterile because of absence of spermatogonia in testes that are only one-fourth normal size. The females are also often sterile, and if not, of low productivity.

Apparently these genes affect the viability of certain types of cells, including the melanoblasts; and the depressed condition of the latter, in hair follicles in which they survive at all, affects deleteriously the processes determined by the lower *c*-alleles, but not those by *ppF*, *ff*, or *bb*. In the table below, the number of plus factors (*Si*, *Dm*) is indicated by number. The effects on action of c^k and c^r parallel those on c^d. With *E*, 100 refers to intense black; with *ee*, 100 refers to intense yellow.

	No. of plus factors, *Si* and *Dm*				
	0	1	2	3	4
EC	0	90	100	100	100
Ec^dc^d	0	38	54	64	74
Ec^dc^a	0	22	29	40	49
eeC	0	64	80	100	100
eec^dc^d	0	22	30	40	49
eec^dc^a	0	5	12	19	25

The F_2 ratio from *SisiDmdm* in the presence of *EECC* is essentially 15 black to 1 white, apart from minor variations in intensity. With lower *c*-alleles there are distributions of intensities of either sepias (*E*) or yellows (*ee*) that can be interpreted in the ratio 1:4:6:4:1.

Melanin Pigmentation

At this point, it is desirable to give a more integrated survey of the interaction effects of the genes that affect melanin pigmentation in the guinea pig. The color of the skin, coat, and eyes depends on the presence of pigment granules (sepia, brown, or yellow) or their absence (white). The granules are produced in special cells, melanocytes, located in the hair follicles, basal layer of the epidermis, choroid coat of the eye and certain other tissues, and in retinal cells. The melanocytes migrate from the neural crest. Absence of pigment may be due to failure of the neural crest cells to reach the normal site, to death in the site, or (in albinos, c^ac^a) to failure of melanogenesis without death of the cell.

Melanin pigment has been shown to be derived from tyrosine by way of dihydroxyphenylalanine (dopa). Dopa oxidase activity of various genotypes of the guinea pig has been studied in frozen sections (Kröning 1930*b*, W. L. Russell 1939), in colorless extracts from skin (Ginsburg 1944); tyrosinase activity as well as dopa oxidase activity has been studied in rates of oxygen consumption and also in pigmentation in homogenates of fetal skin (Foster 1956). Foster's studies indicated a profound difference between high activity in cells with genes *EaaCP* (potentially black or brown) and low activity in ones with *eeCPF* (potentially yellow) or in the subterminal phase of the agouti pattern (potentially yellow) of *EACPF*. This difference was more or less obscured in the other studies, probably because of saturation. No activity was found with any lower *c*-alleles (*C* absent) in Foster's rate studies, or with c^rc^r including blacks, c^rc^a or c^ac^a by any technique, but a weak dopa reaction was found with c^k and c^d compounds in the frozen sections. Replacement of *P* by *pp* reduced activity in eumelanic cells (pale sepia or pale brown), and replacement of *F* by *ff* reduced that in phaeomelanic cells (dilute yellow) and in the pale brownish yellows *ECppff* as compared with *ECppFF*, but less in both cases than might be expected from the reduction in coat color. On the other hand, the reactions were actually somewhat stronger in browns (*EbbCP*), especially in dingy browns, than in the corresponding blacks, perhaps because of less utilization and loss of tyrosinase. Genes *E*, *C*, *P*, and *F* seem clearly to be concerned positively with tyrosinase activity, and the *c*-alleles are apparently those that are directly concerned with specificity. Not surprisingly, there is no activity in frozen sections or extracts from white areas of spotted animals.

The estimates of relative quantities of pigment were based primarily on visual grades from comparisons of animals with standard squares of skin, chosen so that each grade was barely distinguishable from the preceding (sepia and brown, 1 to 21; yellow, 1 to 13). A grade was assigned each animal at birth and often later. The corresponding amounts of pigment relative to intense black or intense yellow were estimated by extraction from weighed samples of hair and colorimetry (E. S. Russell 1939, Heidenthal 1940, Wright and Braddock 1949, Wright 1949*a*). Later, reflection meter readings (R) were taken of the coats in many genotypes using amber, green, and blue filters. Indexes closely paralleling the visual grades were obtained by the equation $I_x = 10(\overline{\log R_w} - \overline{\log R_x})$, where $\log R_w$ is the average logarithm of the readings of white and R_x in the reading in question. A final intensity index, I, was obtained from the sum of the indexes from the three filters. Relative quantities of pigment were derived from suitable regression equations on the visual grades (chap. 12).

Eye colors were described merely visually. The black eye with *CPBSiDm* is diluted slightly by *sisidmdm* and by c^rc^r (dark red) and much more by c^rc^a (light red). The dilution of brown eyes (*bb*) acts cumulatively with these. Pigment is almost eliminated in the pink eyes of *pp* and wholly absent in albinos (c^ac^a).

The primary differentiation in quality is between cells that produce eumelanin (sepia, brown) and phaeomelanin (yellow). The most important genes are the three alleles E, e^p, and e. EE, Ee^p and Ee are typically self-eumelanic; ee is typically self-phaeomelanic and e^pe^p and e^pe show a mosaic of these colors. These cannot, however, be supposed to determine the actual differentiation but merely to give a predisposition. Thus with EA, each hair follicle produces the same sort of yellow as with *ee* during a phase of the growth cycle of each hair, in spite of the presence of E. The result is the agouti pattern of the wild cavies, in which each hair has a subterminal yellow band in otherwise eumelanic hairs on the back and a terminal and longer band on the belly. With A^r the yellow bands are narrower on the back and subterminal on the belly. In the parallel case in the mouse (Silvers 1958), the site of the primary action has been shown, by migration of melanocytes into grafts, to be in the epidermal cells rather than in the eumelanocytes themselves. This is also known to be the case with the striped pattern of cats (Toldt 1912) and of swine (Hickl 1913), in both of which the pattern is visible in stripes of thick and thin epidermis before pigmentation has occurred. The orderly pattern of the Brown Leghorn fowl is also under epidermal control (Wang 1943). On the other hand, pigment cells in animals with pure yellow at birth because of *ee*, may become eumelanocytes later (sootiness) under the influence of unanalyzed genes and low temperature.

This brings us to the genes involved in intensity of the various colors, acting within the pigment cells themselves. In Figure 5.7 a quality index (Q) from the ratio of the index with the blue filter to the total index is plotted against the total index for various genotypes. The great difference between the yellows at the top and the eumelanic colors below is brought out. Among the latter, genotype *ECppff* approaches yellow; pale browns (*Eppbb*) are considerably yellower than pale sepias (*EppB*), and within each of these classes, those with c^d are yellowest and those with c^r least. The dark browns

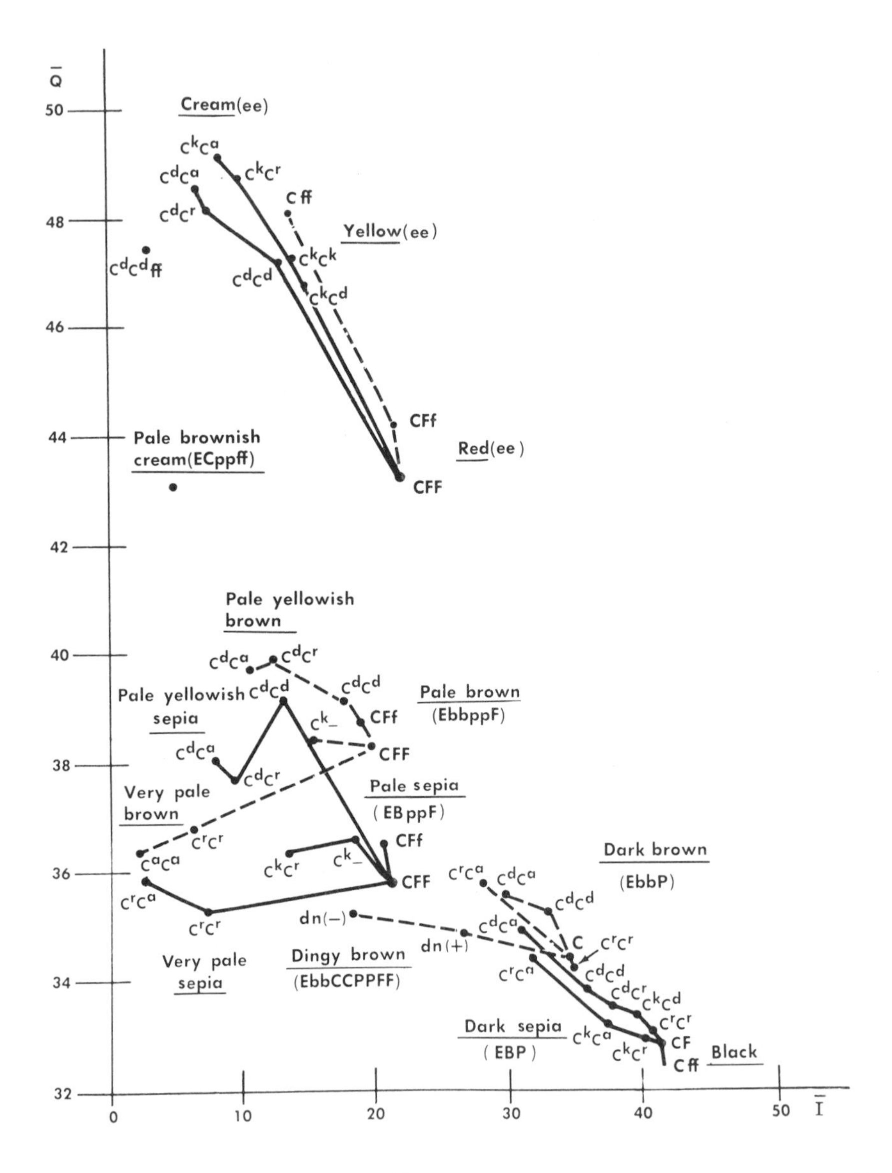

FIG. 5.7. Mean index of quality ($\bar{Q}$) plotted against mean index of intensity ($\bar{I}$) for various genotypes of the guinea pig concerned with coat color. (Redrawn from Fig. 26, Wright 1963 [from Fig. 6, 1959*c*].)

(*EPbb*) are only very slightly yellower than the others; these differences reflect the competition between the eumelanic and phaeomelanic processes that occur within the eumelanocytes as discussed earlier. The great range of intensity of dingy browns of genotype *EbbCCPPFF* is indicated by *dn*(−) and dn(+).

Figure 5.8 gives the estimates at birth for the major genotypes (with segregation of minor ones averaged). Figure 5.9 gives the effects of combinations of *Si*, *si* and *Dm*, *dm* and the *c*-combinations in *EPB* sepias and *eeFF* yellows. Figure 5.10 gives the values as affected by temperature and age, at birth and about six months later (increase with c^a, c^r, and c^d, decrease with c^k and *C*). In these, intense black (*ECPB*) above is 100 and intense yellow (*eeCFF*) below is 100.

The interpretations of these interactions along the lines of the discussion in the previous section are assembled in Figure 5.11. In some cases, factors known genes (modifiers, $\sum$ followed by an indicator of the gene effect modified) and developmental and environmental factors that affect a particular character, are merely bracketed. As far as practicable, however, postulated reactions and interactions are indicated by arrows or by lines joined by an arc to indicate a joint reaction. Destructive effects, responsible for a threshold, are indicated by a joint reaction followed by a reversed arc. An arc opening to the right, where arrows separate, indicates alternative processes. A + or − sign after a factor indicates whether it increases or decreases the process in question. Three character complexes that introduce white into the coat are shown at the left side. One of these, silvering due to gene *si*, its supplementation by gene *dm* to produce dark red-eyed whites *sisidmdm*, and the interaction effects with the *c*-alleles have already been discussed.

Grizzling, due to gene *gr*, produces a pattern that differs from silvering in that it is not present at birth and only gradually produces a sprinkling of white hairs on whatever color is determined by the rest of the genome. As with silvering, the heterozygotes overlap both homozygotes, and the amounts of white in both heterozygotes and homozygotes can easily be shifted by selection.

The third sort of pattern of color and white, spotting, usually centers in the gene *s*, but low grades can occur in *SS* with a sufficient array of "modifiers." The pattern ranges from self-color except for a little white on feet, nose, or belly, to self-white with eyes of whatever color is characteristic of the rest of the genome. In intermediate grades, there are irregularly placed areas of color—large on the back; small, but most regularly present, on the head. Colored spots are often separated either wholly or partially by narrow white streaks. The median amount of white in inbred strains with *ss* range from 10% to 98% (chaps. 6 and 11), depending on the array of modifiers, but each

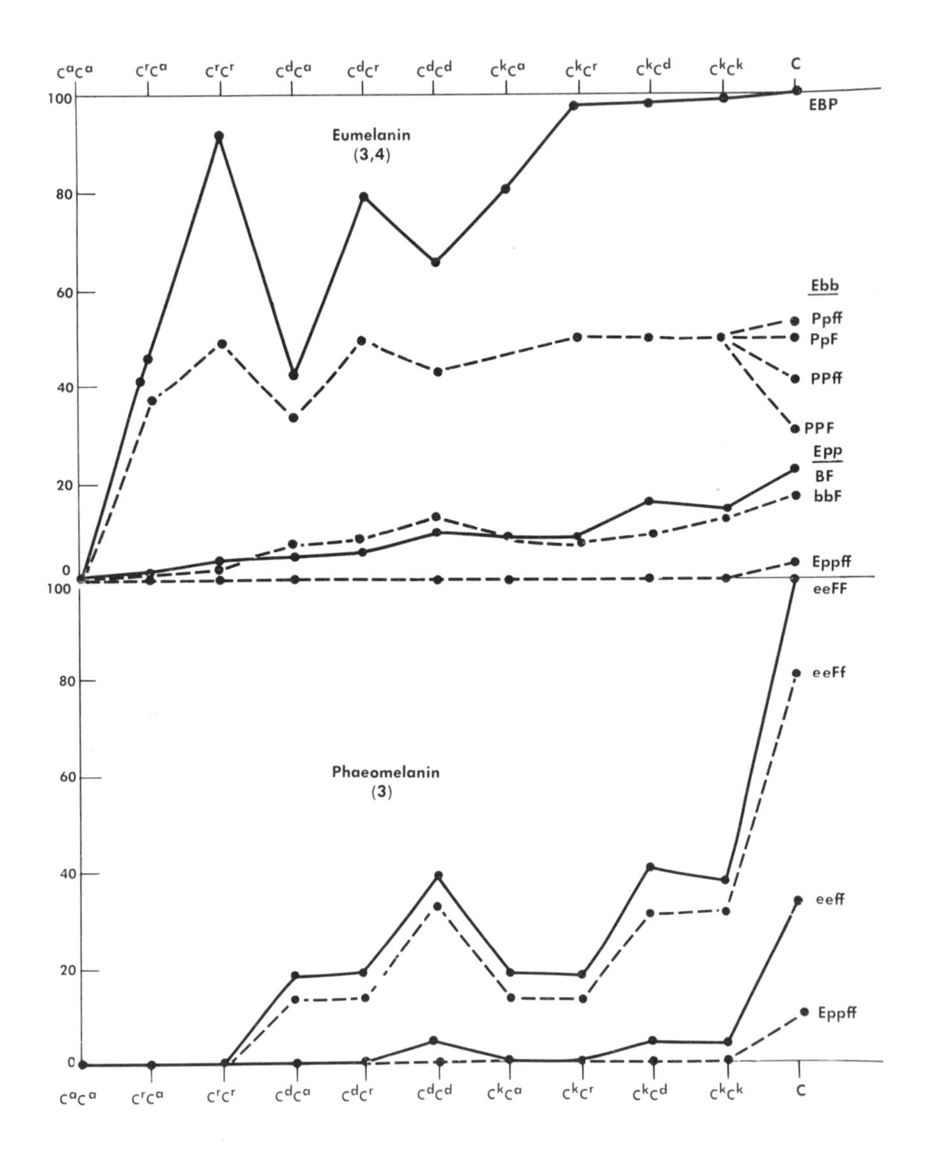

FIG. 5.8. Relative amounts of pigment (eumelanin above and phaeomelanin below) in coats of guinea pigs of various genotypes. (Redrawn from Fig. 23, Wright 1963.)

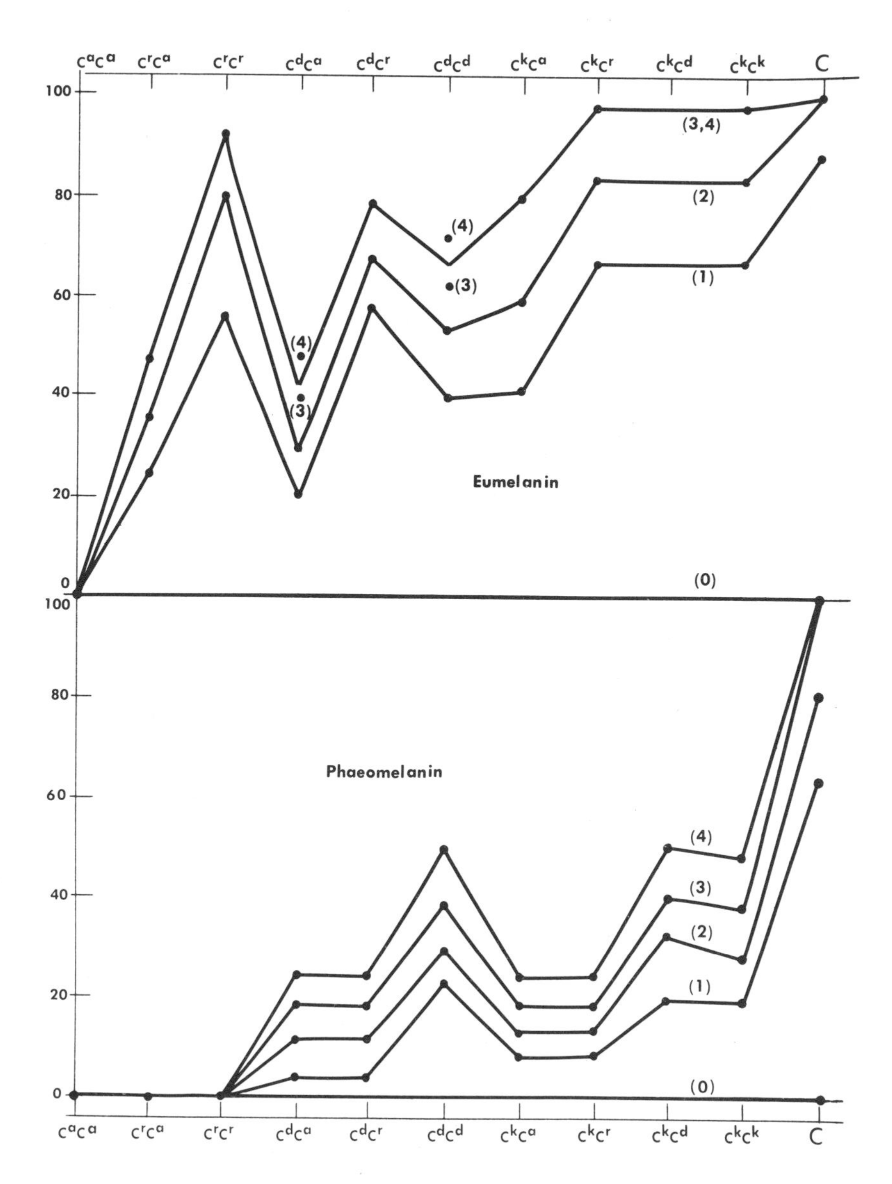

FIG. 5.9. Relative amounts of eumelanin (above) and of phaeomelanin (below) in coats of guinea pigs of genotypes *EPP* and *eeFF* respectively, in association with *c*-compounds and with 0 to 4 plus factors at the loci *Si*, *si*; *Dm*, *dm*. (Redrawn from Fig. 24, Wright 1963 [data from 1959*d*].)

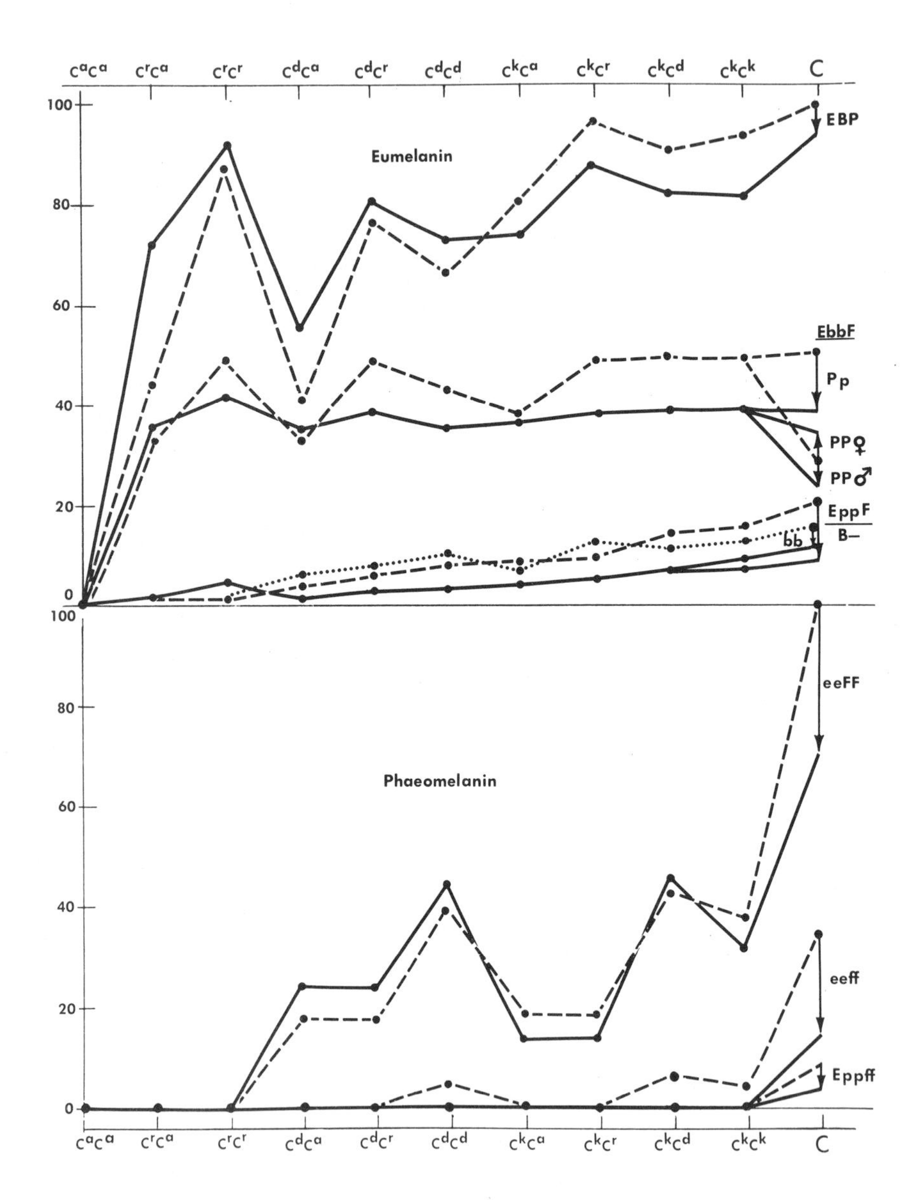

FIG. 5.10. Relative amounts of eumelanin (above) and of phaeomelanin (below) in coats of guinea pigs of various genotypes at birth (broken or dotted lines) and at about six months of age (solid lines). (Redrawn from Fig. 25, Wright 1963.)

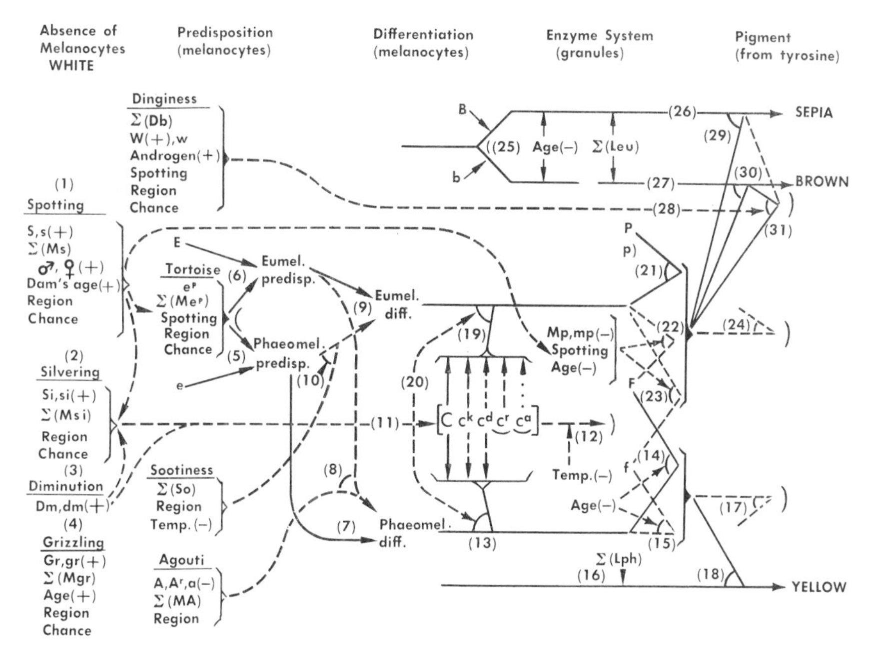

Fig. 5.11. Factor interactions in the determination of coat color of the guinea pig. (Redrawn from Fig. 22, Wright 1963.)

varies enormously because of accidents in development. There is no correlation between parent and offspring and, indeed, no correlation between presence of color or white at points one third of the length of the animal apart on the back (chap. 14). The median amount of white in females is a little greater than in males (chap. 9) and, in offspring of each sex, increases as the mothers become older.

Again the heterozygotes overlap both homozygotes and all three genotypes —*SS*, *Ss*, and *ss*—show correlated changes with changes in the array of modifiers. The dominance relations are interesting. As with E, e^p, e, dominance depends on available space (limits, 0% and 100%).

Table 5.2 gives comparisons of *SS*, *Ss*, and *ss* with three different background heredities (Wright 1928, Wright and Chase 1936). Strain *A* was homallelic *SS* by many tests, yet 12% had small amounts of white. Strain *B* was wholly *ss*. It covered the entire range except probably for self-color, and thus overlapped strain *A* phenotypically. The crossbreds, *Ss* with 24% self-color and up to 95% white, overlapped both homozygotes broadly, but were largely of low grade. The backcross $(AB)B$ was so nearly the sum of 50% strain *B* and 50% F_1 that no important difference in minor factors is indicated.

TABLE 5.2. Grades of piebald (parts of white in 20) in *SS*, *Ss*, and *ss* in three different genetic backgrounds. Strains *A* and *B* (eight factor dominant and recessive, respectively) were unrelated, but the results of the backcross (*AB*) *B* indicate similarity in piebald modifiers.

STRAIN	GENOTYPE	PERCENTAGE FREQUENCIES									No.
		0	x–2	3–5	6–8	9–11	12–14	15–17	18–20	White	
A	*SS*	88	12								354
A × *B*	*Ss*	24	69	3	1	1	1	1			838
B	*ss*	0	6	9	11	16	20	17	21		851
(*AB*) *B*	$\frac{1}{2}Ss$, $\frac{1}{2}ss$	10	38	12	9	7	9	9	6		437
$\frac{7}{8}D$	*SS*	100									
	Ss	96	4								76
	ss	0	70	21	7	1	1	1			145
$\frac{15}{16}(13)$	*SS*	50	50								34
	Ss	0	31	30	21	5	9	4			90
	ss						3	25	59	13	270

From data of Wright and Chase (1936).

One of the whitest of the spotted strains (no. 2, median 94% white) was crossed with the self-colored strain D (SS) and produced 91% self, 9% low-grade spotted. Even after two backcrosses to D, 3 of the 305 offspring had traces of white indicating about 4% among these that were still Ss. A few heterozygotes were found by test crosses and interbred, and ss was extracted. These produced no self-colored offspring but were largely of low grade, very different from strain 2 from which they derived s. In this case, S is completely dominant except for the rare occurrence of traces of white in Ss.

The above tests of $\frac{7}{8}$ blood D were made by crossing with the whitest of all of the inbred strains (no. 13 with median 98% white). This strain differed little genetically from strain 2, however, as shown by many crosses carried to F_2. The heterozygous offspring (Ss) were 19% self-colored and 81% low-grade spotted, and thus much like $F_1(A \times B)$. Repeated backcrossings were made between individuals with less than 50% white (presumably Ss) and strain 13. These continued to give almost the same bimodal curve from $\frac{3}{4}$ (13) to $\frac{127}{128}$ (13). The upper half (90 animals) among those $\frac{15}{16}$ or more of strain 13 were only slightly less white than pure 13 and are combined with 180 from matings among them ($ss \times ss$). The lower half from these later backcross generations included no selfs and many with 50% white or more. SS was recovered from matings between low grade individuals of the late backcross generation ($Ss \times Ss$) but only constituted 12.5% of the total, indicating that about 50% had traces of white.

The parallel variation of SS, Ss, and ss is apparent. It is probable that selection in heterogeneous populations such as A and B would be effective, whether restricted to SS (as in A), to ss (as in B), or to Ss (as in AB) in producing almost complete dominance of S.

One of the most interesting features of the spotting pattern is the amount of pleiotropy. Although the colored areas reflect in the main the color expected from the rest of the genome, there may be differences in quality or intensity, distributed in a pattern related to that of color and white.

The most striking effect is on the tortoise-shell patterns due to e^pe^p or e^pe. These are changed from a sprinkling of yellow hairs on a eumelanic ground to the tricolor pattern of eumelanic spots on a yellow ground, both somewhat mixed and both on a white ground. Spots of the same or different colors are often separated partially or completely by white streaks, but in the case of two eumelanic spots, there may be separation by a yellow streak or a streak yellow at one end, white at the other. Thus the relation of eumelanic spots to yellow is similar to that of both to white.

Similarly, spotting tends to break up the orderly pattern of silvering ($sisi$) into spots of strong and weak silvering often separated by white streaks. It also tends to break up the relatively uniform dinginess of browns

(*ECCPPFFbb*) and less frequently the uniform pale sepia (of *EBppF*) and pale brown (of *EppFbb*) into large blocks of different intensity, often separated partially or wholly by white streaks.

Other genotypes are affected very rarely, if at all, but apparent somatic mutations in which a spot of the recessive color appears on a heterozygote (*Pp*, *Bb*, *Ff*) have been observed in 0.03% of the animals of the colony in similar relation to the spotting pattern in the other cases. A similar situation has been found in other mammals (Castle 1929, Dunn 1934).

The spotting pattern seems to be due to differentiation in the skin that tends to interfere with the migration of melanoblasts from the neural crest (Wagener 1959, Schumann 1960). The white regions seem to be due to the interference, and the pleiotropic effects directly to modifying influences from the spotted pattern of differentiation of the skin.

In summary, the system of factors that affect melanin pigmentation of the guinea pig includes a few essential loci in which the type allele (*E*, *A*, *C*, *P*, *B*) is completely or (*F*) incompletely dominant. Another group of type genes (*S*, *Si*, *Gr*) are completely dominant, or nearly so, over their mutants as long as the array of modifiers is such that these mutants have little effect, but they are incompletely dominant in associations with effective modifiers. Other loci (*Dm*, *dm*; *Mp*, *mp*; *W*, *w*) make little or no difference except in combination with mutants of the preceding groups and exhibit semidominance. Beyond these demonstrated sets of alleles there is much quantitative variability, partly genetic, partly environmental, and partly from developmental accident. At least ten different kinds of variability of this sort are distinguishable. Crosses between genetically determined extremes give intermediates with no analyzable segregation in F_2. The simplest interpretation is that these depend on multiple loci with intermediate heterozygotes (Wright 1960*c*).

Study of effects in combination indicates a complicated network of interacting processes with numerous pleiotropic effects. There is no reason to suppose that a similar analysis of any character as complicated as melanin pigmentation would reveal a simpler genetic system. The inadequacy of any evolutionary theory that treats genes as if they had constant effects, favorable or unfavorable, irrespective of the rest of the genome, seems clear. It is desirable, however, to consider briefly studies of color characters of other organisms, and in more detail, studies of a number of different sorts of characters.

Colors of Other Organisms

The interaction systems responsible for the colors of other higher vertebrates that have been studied are similar in complexity. E. S. Russell (1949) has made quantitative studies of the differences among many genotypes of the mouse in color, size, shape, numbers, and clumping tendency of pigment granules in the medulla and cortex of hairs at different levels along the axis. The primary conclusion that emerged was that replacement of any particular gene by an allele tends to bring about changes in several or all of the above

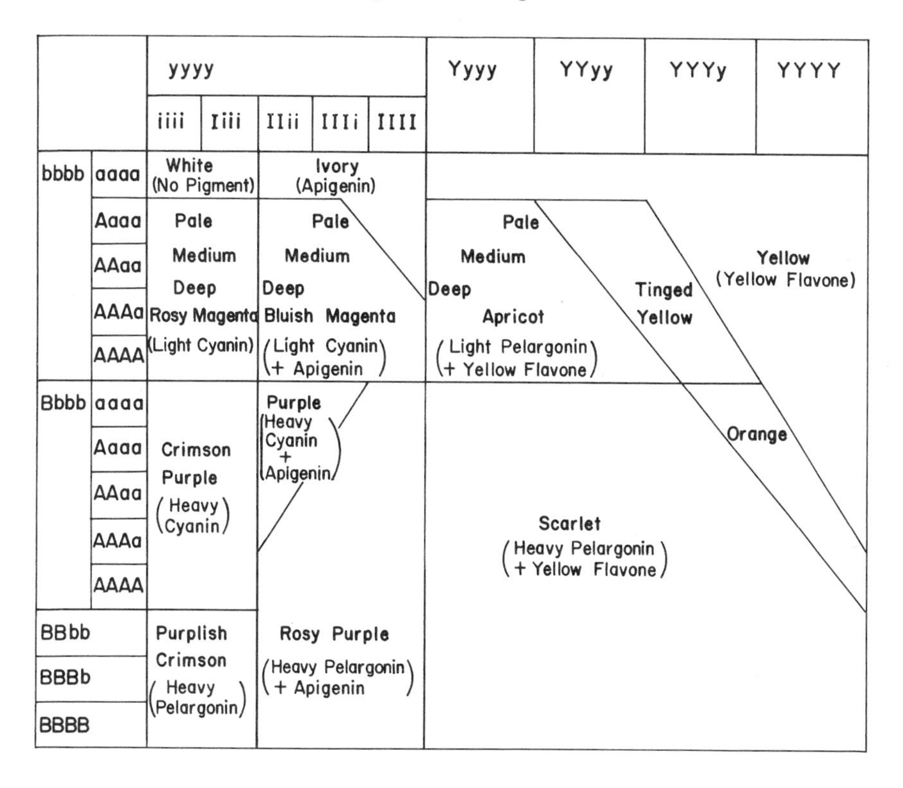

FIG. 5.12. Relation of genotypes of *Dahlia variabilis*, a functional tetraploid, to flower colors.

respects and to do this to different extents in different genetic backgrounds. This pleiotropy is probably secondary. A simple primary change in tyrosinase, by replacement at the *c*-locus, may distribute its effect over changes in intensity of color, in size, in shape of granules, and in their number and distribution in the hair.

The intensive studies of the interactions among the effects of the eye color genes of *Drosophila melanogaster* (Bridges and Morgan 1923, Beadle and Ephrussi 1936) lead to similar conclusions with respect to the complexity of the pattern of interactions. This is also true of numerous studies of the anthocyanins and anthoxanthins responsible for flower colors (Bateson 1909, Wheldale 1916, Lawrence and Scott-Moncrieff 1935). The last reference presents points of special interest because it deals with a tetraploid, *Dahlia variabilis*, in which dosage effects could be studied to a greater extent than in diploids. Figure 5.12 is intended to give some idea of the complexity. There are thresholds, dosage effects, changes of dominance, cooperative interactions and competitive ones. Whatever the primary effects of the genes, at the level of flower color, replacement at a locus exerts a pull one way or another on a complex network of reactions.

Hair Direction

It is desirable to compare the variations of morphological characters with those of color and pattern in regard to the nature and complexity of interaction effects. Hair direction (which is related to the pattern of epidermal ridges in the skin) is a pattern in which variation can be classified relatively easily since it is essentially only two instead of three dimensional.

In wild cavies (*rr MM rere stst*) the hairs are directed away from the snout on the body (with minor qualifications) and toward the toes on the legs (smooth fur). On introducing the dominant gene *R* into a genotype that is otherwise like that of the wild species, there is in general merely reversal of hair direction on the feet, especially on the hind feet (grade E), but occasionally there is also a slight crest along the back (Wright 1916, 1949*b*).

Those with *RMm* are highly variable but most of these have either a strong dorsal crest (grade D) or a single pair of rosettes halfway along the back (grade C). Some of those called grade C had only a single dorsal rosette. This was almost always (96%) on the right side, a curious example of regular asymmetry. A few are called grade E and a considerable number in certain strains were called grade B, or better CA, with no rosettes on head or belly (as in grade C) but with two pairs of dorsal rosettes (as in grade A).

With *Rmm*, there are typically two strong pairs of dorsal rosettes, anterior and posterior, radiation from the ears, a strong forehead rosette, eye, cheek, and groin rosettes and feathering along the midline of the belly (grade A). The number of dorsal rosettes, however, varied in both directions. It might be increased (grade A++) or it might be reduced to two (AC). In cases of asymmetry (A+) rosettes again tended to occur more on the right side than on the left.

Genes *M*, *m* are specific modifiers of the effect of *R*, with no detectable effects in *rr*. Other specific modifiers were clearly present. As indicated, high-grade *RMm* (grade CA) differed markedly from low-grade *Rmm* (grade AC). Apparently the number of dorsal rosettes was especially subject to modifiers, $[\sum (MRd)]$, in both these genotypes. The effects of these modifiers were, however, confounded by a very considerable amount of nongenetic variability. Thus the regression of offspring on mid-parent in matings of *Rmm* × *Rmm*, including an excess of the extremes AC and A+ +, was only 0.48, indicating that only about half of the variability was genetic. Three generations of selection did not suffice to fix grade AC.

As indicated above, eye rosettes ordinarily appear only in roughs of high grade (Rough A: *Rmm*). Strong eye rosettes were, however, recorded in a few individuals that were otherwise of grade E, genotype *RMM*, with no record of the character in their ancestry (other than in grade A). Analysis of the descendants of one of these indicated that the anomalous eye rosettes depended on a gene *Re* with some degree of dominance. Penetrance (but not always expressivity) was high in homozygotes, *ReRe*, but only about 50% in heterozygotes, *Rere*, in the presence of *RM*. In the absence of *R* it was only about 50% in homozygotes and very low, but not zero, in heterozygotes. There was often irregular asymmetry, including a few cases in which there was a strong rosette about one eye and none about the other. This indicates that developmental accident played a considerable role. There is thus some favorable interaction effect from *R* but much more irregularity in expression than with *R rere*.

Another aspect of the full rough pattern of *Rmm*, the forehead rosette, can also appear in the absence of *R*. All animals in the colony of this type traced to three obtained from outside. This character Star (*St*) behaved in a remarkably different way from Rough-eye (*Re*) in two respects. First it showed no intergradation with normal and behaved as a simple Mendelian dominant in eight successive backcrosses to a smooth furred line (*rr stst*) in which it had never been present. Second, instead of showing enhancement in the presence of *R*, there was a tendency toward reciprocal inhibition. In *rrSt* the forehead rosette is almost invariably (99.6%) single and very flat. In about 19% of *RMMSt*, it is replaced by a weak, double rosette. This is combined with the rosettes of grades D or E as expected, but the average grade is reduced. In *RMmSt* the reduction is still more conspicuous. Finally in *RmmSt*, in which an exceptionally strong forehead rosette might have been expected by summation of effects, there was actually only a slight irregularity of hair direction. The anterior dorsal rosettes of grade A were also weak and far to the side, leaving a broad smooth shield anteriorly on the back. This last feature refers to *RmmStSt*. In this case only, *St* ceased to be completely dominant and *RmmStst* showed a great variety of intermediate conditions of the dorsal rosettes. It is interesting that *RMM* and *RMm* tend to inhibit the effect of *St* in a part of the coat, the forehead, in which they have no visible effect in the absence of *St*, and that *St* in turn inhibits the effects of *RMm* and *Rmm* in a region, the anterior dorsum, in which it has no visible effect in the absence of *R*.

Gene *St* has an interesting pleiotropic effect in a strong tendency toward a white spot in front of the center of the forehead rosette. Bock (1950) showed that this does not require even heterozygosis in the spotting factor *s*, and that it tends to be inhibited slightly in roughs E, D, and C (*RMMSt*, *RMmSt*), but strongly in full roughs (grade A, *RmmSt*). The results indicate that the white forehead spot is a secondary consequence of the effect of *St* on the skin of the forehead that is manifested in the forehead rosette, since the neutralizing effect of *R* on the latter, especially in *Rmm*, also tends to prevent the white spot. Figure 5.13 gives an interpretation (Fig. 5.13).

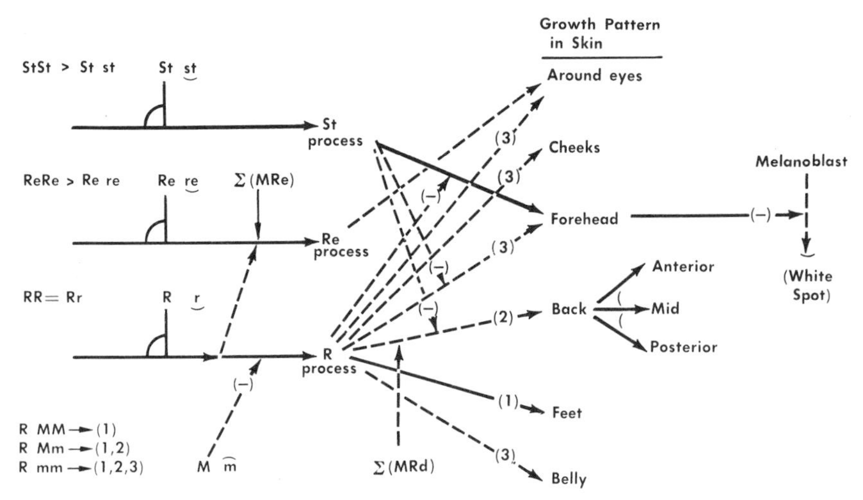

Fig. 5.13. Diagram of factor interactions in the determination of hair direction and of a pleiotropic effect on a white forehead spot in the guinea pig. (Redrawn from Fig. 27, Wright 1963.)

Another element of the full rough pattern of *Rmm*, the feathering on the belly, has also been observed independently of *R* in a certain inbred strain. It followed an irregular course of inheritance, neither a simple dominant nor a simple recessive. Similarly, dorsal irregularity of hair direction has occurred independently of *R* but so sporadically that mere developmental accident seemed indicated.

The most interesting aspect of the genetics of hair direction is its revelation of the extraordinary diversity in the relation of genes to characters and, in particular, in kinds of interactions that may be found even in a rather limited genetic system. The supplementation of the effects of a few major genes by extensive quantitative variability is again to be noted.

Atavistic Polydactyly

We turn now to a different sort of morphological character, the occurrence of atavistic digits (Wright 1934*c*, *d*). The guinea pig, like all the species of Caviidae, lacks the thumb, big toe, and little toe. A fourth toe, exactly like the little toe of related rodents is not, however, uncommon. Among 22 closely inbred strains, there were 11 that were invariably 3-toed on the hind feet (including strains 13 and 32), the little toe appeared sporadically in five (including strain 2, the major branch of which was, however, entirely free of the trait), and in six the incidence was fairly high (including strain 35). A strain *D* with 100% occurrence of well-developed little toes was produced by Castle (1906, 1911) by selection.

Crosses were made between strain *D* (4-toed) and the 3-toed strains 2, 13, and 32, and between *D* and strain 35 with 31% 4-toed in a branch descended from a single mating in the twelfth generation. The results from 2 × *D* and 32 × *D* gave complete dominance of 3-toed in F_1 (except for one individual in 26 in the latter), and possible 3:1 and 1:1 ratios in F_2 and the backcross to strain *D*, respectively, in both cases. These results suggest segregation of a single essential recessive factor for the little toe, but tests of the supposed segregants from (2 × *D*) × *D* gave results that completely exploded this hypothesis. The 3-toed segregants produced only 23% 3-toed in 186 instead of 50%, whereas the 4-toed segregants gave nearly the same result (16% 3-toed in 119 instead of none). The most plausible hypothesis seemed to be cumulative action of multiple factors on the underlying physiology, and two thresholds, one for development of the little toe, below which there is homeostatic control of the normal 3-toed foot, and a slightly higher threshold, or better ceiling, above which development of the little toe is controlled homeostatically (Fig. 5.14).

The cross 13 × *D* gave 67% 3-toed and 33% 4-toed in F_1. Tests of the two F_1 types gave results that did not differ significantly either in F_2 or in the backcross to *D*. There was no simulation whatever of one factor heredity, but the results fit well with interpretation of multiple factors and two thresholds. Finally in strain 35 with 31% 4-toed, there was no difference in the results of 3-toe × 3-toe and 4-toe × 4-toe within subbranches (although considerable differences among these were probably due to minor mutations [cf. chap. 12]). The variability was due primarily to an effect of the age of the mother (high percentage of 4-toed from immature mothers [chap. 6]) and secondarily to a seasonal (nutritional?) effect (excess 4-toed in winter

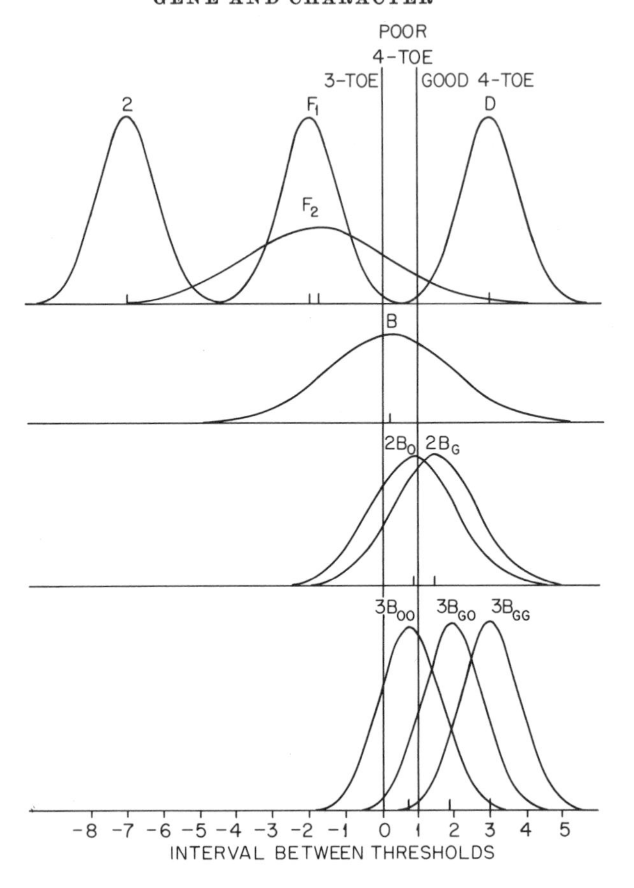

FIG. 5.14. Theoretical distributions of factors determining development of little toe on an underlying scale with thresholds for any little toe and for perfect development, in two inbred strains of guinea pigs, 2 and D; F_1, F_2, and first, second and third backcrosses to strain D. $2B_0$ and $2B_G$, refer to second backcrosses from first backcrosses without little toe and with good development, respectively. $3\ B_{00}$ refers to third backcrosses from second and first backcross animals lacking little toes. In $3B_{G0}$ the second backcross parent had a good little toe but its first backcross parent had none In $3B_{GG}$ both had good little toes. Areas beyond or between thresholds indicate frequencies. (From Fig. 1, Wright, 1960*e*, *J. Cell. Physiol.* 56:127.)

and early spring [chaps. 9 and 12]). Most of the young from 35 × D were 4-toed and in this case it was the percentage of 3-toed that increased in F_2. These F_2 results and those of backcross to 35 again fit the threshold hypothesis (Fig. 5.15). The numerical results will be considered later (chap. 15).

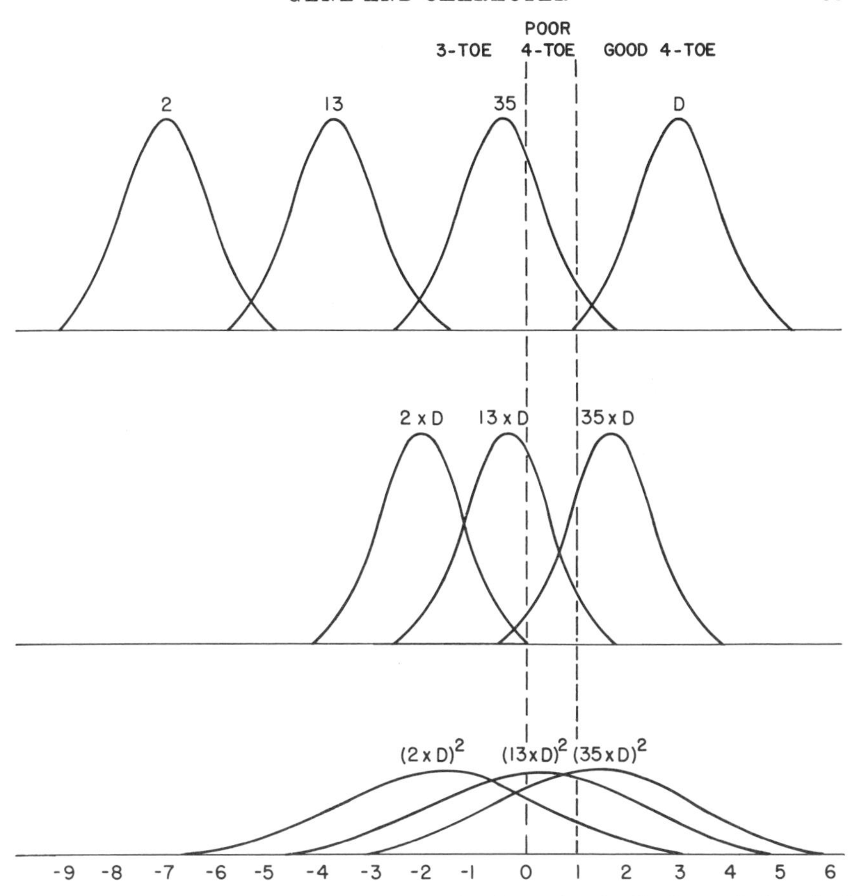

FIG. 5.15. Comparison of inbred strains of guinea pigs 2, 13, 35, and D and of F_1 and F_2 of crosses of D with the others on the same scale as in Figure 14. (Redrawn from Fig. 28, Wright 1963.)

Similar little toes have been restored in a wholly different way, in this case in association with thumbs and big toes resembling those of other rodents. A single mutant individual (in a 3-toed strain) showed these characters imperfectly developed. It was demonstrated that a dominant gene (*Px*) with variable penetrance was responsible. In the stock of origin, about 82% of *Pxpx* showed one or more of the atavistic digits. After certain crosses penetrance fell to 20% or even less (chap. 6) (Wright 1950).

Crosses with strain D (4-toed) revealed an interesting interaction effect. It was not surprising that penetrance of the little toe rose from 82% to 100% in F_1 by combining two heredities, both of which favor it. More interesting is the fact that among those with *Pxpx*, penetrance of the thumb rose from 74% to 100%, although thumbs had been wholly absent in strain *D*. Similarly penetrance of the big toe (in *Pxpx*) rose from 2% to 18% in F_1 and to 55% in the back cross to strain *D*, although it was wholly absent in *D*.

The effects of *Pxpx* were usually restricted to this restoration of atavistic digits, but there were a few other abnormalities that appeared occasionally: fusions in the sternum, ventral flexure of the feet, and, very rarely, a sixth digit. On the average, young with *Pxpx* weighed 7% more than *pxpx* littermates.

The homozygote *PxPx* turned out to be grossly abnormal. It was found by Scott (1937, 1938) that over 90% died at about the twenty-sixth day of gestation and were absorbed. These that reached birth (68 days) had short clubbed legs with broad paddle-shaped feet having 5 to 12 digits each (chap. 6). The hind legs were rotated 180° and lacked tibiae. There were microphthalmia and either hydrocephaly or a protruding brain. The animals were harelipped and grossly abnormal in their internal anatomies. The defect was manifest at about 18 days of gestation in hind-limb buds of double width and overgrown mid- and hind-brains.

There was thus extreme pleiotropy in *PxPx*. The usual localization of the effects in *Pxpx* to the feet appears to be a threshold matter. The most remarkable feature of this case is perhaps the production of a seemingly normal atavistic type of foot in the heterozygote but a monstrous type of foot, associated with monstrous development of most other parts of the body, in the homozygote.

Otocephaly

The extraordinary pleiotropy in the case of *PxPx* seems to require that the basic metabolic defect give rise independently to diverse consequences in different parts of the body. In another type of monster, otocephaly, that has been studied extensively (Wright and Eaton 1923, Wright and Wagner 1934, Wright 1934*b*), a graded series of abnormalities was restricted to the head, but ranged from mere shortness of the mandible to complete absence of jaws, nose, eyes, brain anterior to the medulla, and almost complete absence of the skull except for distorted ear capsules and a reduced occipital ring (aprosopus) (Figs. 5.16 and 5.17). In contrast with *PxPx*, all the defects could be traced to a single localized source—an acute inhibition, in some degree, of the prechordal mesoderm or organizer at a very early stage.

The defect is essentially the same as that most characteristically produced by a great variety of environmental agents in fish, amphibia, and birds, when applied as early as the undivided egg but not after the primitive streak stage.

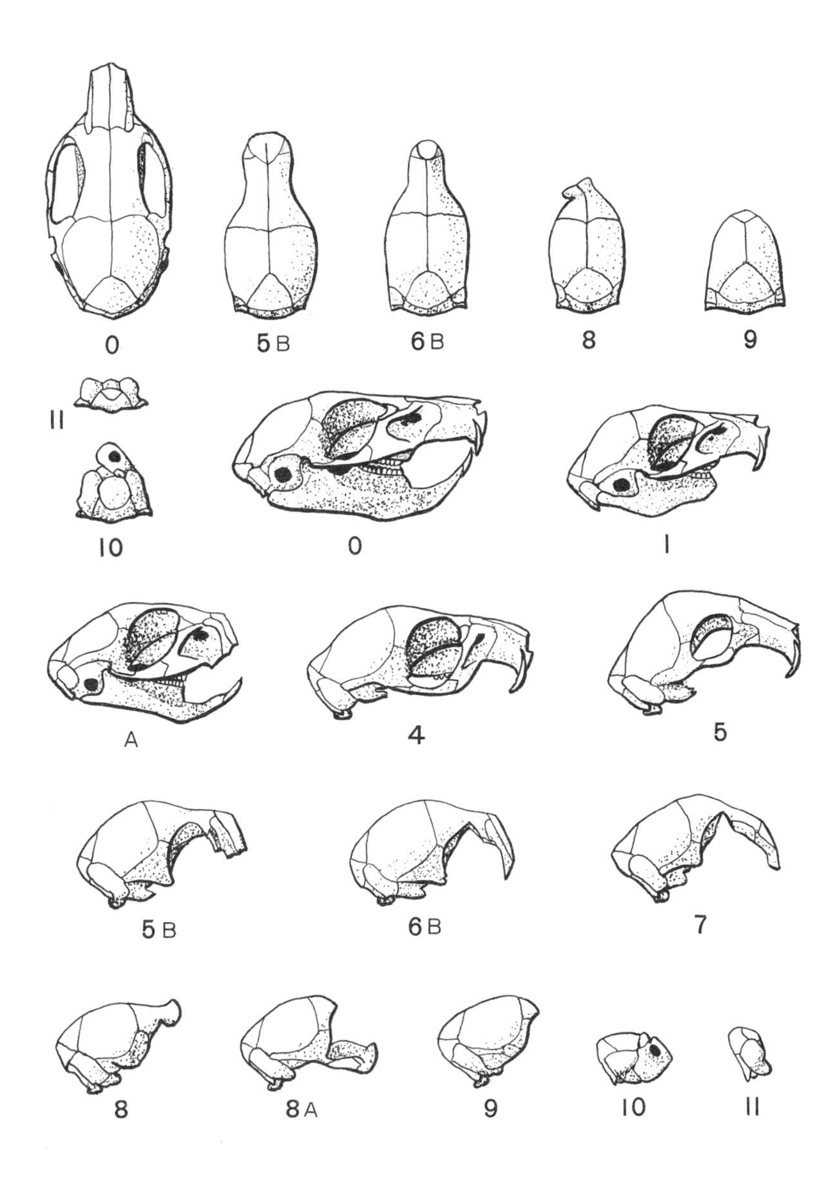

FIG. 5.16. Dorsal and lateral views of skulls of normal (*O*) and otocephalic guinea pigs of the grades indicated. (From Plate 1, Wright and Wagner, 1934, *Am. J. Anat.*, 54:442.)

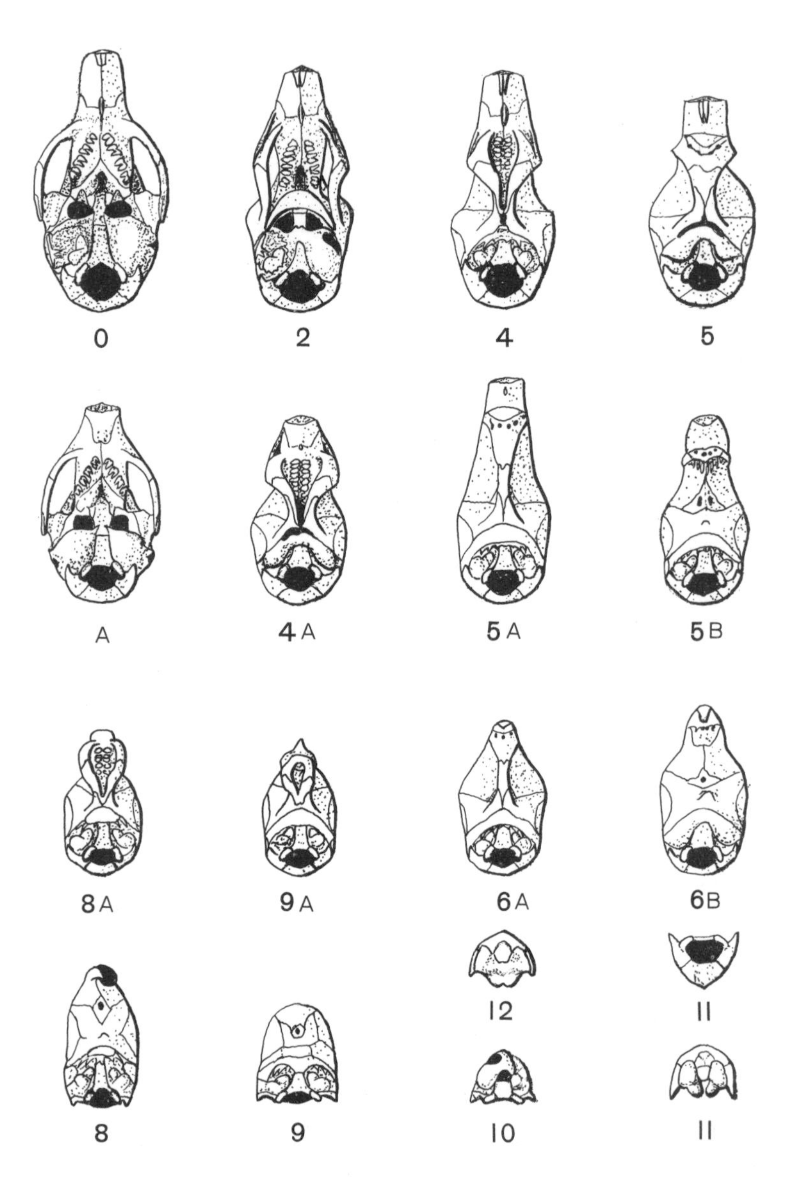

FIG. 5.17. Ventral views of normal (*O*) and otocephalic guinea pigs of the grades indicated. Occipital views are shown of grades 11 and 12. (From Plate 2, Wright and Wagner, 1934, *Am. J. Anat.* 54:443.)

The secondary effects of the genetic factors are probably no more specific and are essentially like those of the environmental ones.

Manifestation in the guinea pig was always a threshold phenomenon, dependent on conjunction of the genetic factors with hypothetical environmental ones, largely peculiar to each separate individual. The incidence was 0.06% in the general colony, but there were segregating, higher incidences in early branches of one inbred strain (no. 13) that led to fixation of an incidence of 5.5% in one line from the thirteenth generation of brother–sister mating. This persisted in all subsequent branches for more than 25 generations, except for one branch that started from a dominant mutation in the nineteenth generation of brother–sister mating, which raised the incidence to 18%, and was maintained for 15 years by selection from high-producing matings. A large subbranch relapsed to 5%. In all lines, abnormal females were twice as numerous as abnormal males (chap. 6).

A considerable number of other abnormalities—anotia, microphthalmia, ventral flexure of foreleg, hydrocephaly—showed a similar pattern as far as studied. There was considerable tendency to concurrence of defects in individuals or offspring from the same mating, indicating again that the secondary effects of the genes were as nonspecific as the environmental factors, and that the specificity was rather in the points of weakness in the developmental process at the time of concurrence of the gene-controlled and environmental effects (Wright 1960*e*).

Morphological Characters of Other Species

Green's (1954) study of differences among inbred strains of mice with respect to number of presacral vertebrae (25, 26, or 27) indicated a situation somewhat similar to that described for multifactorial polydactyly in the guinea pig. Maternal heredity played an important role in this case, revealed by matroclinous differences between reciprocal crosses, demonstrated by ovarian transplantation to depend on uterine environment.

Grüneberg especially (1951, 1952) has made systematic studies of numerous anomalies that show different frequencies in different inbred strains, with results in crosses that indicated what he called "quasicontinuous" variability. An interesting case involved the absence of third molars, especially the lower ones, in 12% of an inbred strain, and invariable presence in another. No genetic differences could be found between normals and defectives in the former, and there was again evidence of environmental influence in a strong tendency to concurrence in litters. This trait was lost altogether in reciprocal F_1's and F_2's and in backcrosses. Measurement of the size of the third molars in the two strains showed, however, that although there was little variation in the normal strain (coefficient of variability 2.3%), the other showed a smaller average size and much greater variability (6.8%). There was clear evidence of association of small size with absence within litters. Again the normal condition behaved as dominant in reciprocal crosses, and little evidence of segregation of size was apparent in F_2. The average fell, however, in backcrosses to the abnormal strain, especially if the mother was from the latter. There were clearly multiple genetic and nongenetic factors: a threshold below which the tooth failed to develop at all and above which it was extremely susceptible to deleterious factors, and a higher region on the underlying physiological scale, not quite a ceiling, at which this extreme susceptibility ceased and variability was merely characteristic of ordinary quantitative variability (chaps. 6 and 11).

If we turn to *Drosophila*, we find a very large number of morphological mutants, including ones that are unlocalized except for tissue (e.g., spineless, forked bristles, and so forth), but also more or less localized ones (e.g. scute, Dichaete). As noted earlier there is a great deal of pleiotropy.

There are mutations of irregular penetrance, and also sporadic abnormalities, characterized by different frequencies among strains and susceptibility to environmental conditions (e.g., abnormal abdomen). The relative frequencies of such cases seems to be less than in the mammals that have been studied, but this is probably merely because of the greater concentration by the *Drosophila* workers on "good" genes for studying linkage systems and mutation rates rather than characters for their own sake as in the mammals.

There is an important class of genes in *Drosophila*, however, that are comparable in irregular penetrance and susceptibility to environmental difference to those studied in mammals. These are the homeotic mutations in which an organ tends to appear in a wrong part of the body, perhaps replacing a very different organ; e.g., aristae by legs (aristapedia), mesothorax (with wings) by metathorax (with halteres) (tetraptera), the reverse (tetraltera), oral lobes by antenna-like or leglike structures (proboscipedia), growth of palpus from eye (kidney and lobe alleles), extra joints on the legs (four-joint and others), sex-combs on other than the first legs of males, and so forth. We will return to these later.

The situation is similar in other groups of organisms, plants as well as animals. But again it is much easier to find lists of "good" genes, suitable for linkage studies, than to get an unbiased picture of all of the hereditary deviations that affect particular characters.

Homology

The tracing of evolutionary lines of descent is based largely on the concept of homology. This may be defined as a similarity between organs of two species in structure, physiology, development, and relation to other parts that is closer than can be accounted for by function or coincidence, in contrast with analogy in which the similarities in all respects are no greater than can be so accounted for. Most attempts to trace phylogenies are expressed in terms that seem to imply preformation. The morphologist tends to treat a certain organ as the homolog, in an absolute sense, of a more or less similar part of another species, excluded thereby from being the homolog of any other part. Each part is treated as if it had an independently evolving heredity. In this viewpoint there seems less difficulty over evolutionary changes in the proportions of such a part (e.g., of each named mammalian tooth or bone, Osborn's [1925] "allometrons") than in parallel evolution of similar parts (different molars, different vertebrae), or in the convergence of nonhomologous parts (membrane and cartilage bones), or in the appearance of new organs (Osborn's rectigradations or aristogenes), even though closely similar to others that have long been present (new dental cusps or even whole teeth, mesonephros and later metanephros in addition to pronephros).

Morphologists also, however, use the term homology in connection with replications of similar structures within the same organism. Corresponding parts on the right and left side of an organism are usually almost perfect mirror images. Segmented organisms (annelids, anthropods, vertebrates) exhibit serial homologies. There are also the similarities among nonsegmental replicated organs such as hairs, ommatidia, leaves, flowers. It is important to appreciate the relation of genes to both phylogenetic and ontogenetic homology.

To the geneticist, confident that all genes are, as a rule, distributed equally to all nuclei, the simplest situation is that in which genes determine a metabolic difference manifested alike in all parts of the body. The metabolic mutations of *Neurospora* are an example in a simple organism. In a more complex one, the demonstration by Gregory and Castle (1931) that two- or threefold differences in the sizes of strains of rabbits depend on corresponding differences in rates of growth and division of cells from the 8–16 cell stage of the embryo on, indicates universal gene action.

The situation is very different with Snell's (1929) recessive dwarf mouse demonstrated by Smith and MacDowell (1930) to depend on defect of the anterior pituitary. The gene is presumably present in all cells of the body but is interpreted as having a primary effect only in this one tissue.

Most cases in complex organisms are intermediate. The sites of primary effect, as far as can be determined, are seldom as universally distributed as those that determine the size differences of rabbit strains, but also are seldom as narrowly localized as that underlying the restricted growth of the dwarf mice.

The genes that predispose toward otocephaly or anotia in the guinea pig must produce their primary effects very early, well before histogenesis. There are others that produce effects that become manifest considerably later but nevertheless are responsible for general metabolic disturbances not restricted to any one type of cell. Thus *PxPx* in the polydactyl monster of the guinea pig results in simultaneous delay in differentiation and correlated overgrowth of such diverse tissues as those of the hind brain and the limb buds. Probably in most cases, however, the primary effects are restricted to particular differentiating tissues (blood cells, melanoblasts, epidermis, cartilage) as Grüneberg (1947) especially has emphasized.

Whether concerned with a metabolic process that can occur in all kinds of cells or in only a particular kind, however, there is often a more or less orderly restriction of demonstrable effect to cells in a particular region of the body. Such regions may be distinguished by the impact of various sorts of the correlative factors, especially metabolic gradients (Child, 1911). In a broad sense, localization of the effect may be considered a threshold phenomenon within the developing organisms.

The genetics of hair direction furnishes simple examples of the relation of gene action to homology. Starting from the homogeneous pattern of the smooth-furred wild cavy, *r r MM stst*, with all hair directed away from the snout and toward the toes, the replacement of *r* by *R* leads to the same result—the appearance in both hind feet of independent centers from which the hair radiates. Here, as in almost all other cases of morphological expression of gene action, there is homology of right and left sides that rests merely in the calling into play of the same gene or genes under almost identical developmental conditions.

Examination of the feet shows that the hair actually radiates from partings along each side of each toe connected across the top by a part to form a pattern like a letter H. Thus there is replicative homology in the pattern of the separate toes. Moreover, with favorable modifiers an almost identical pattern appears on the foretoes illustrating serial homology of fore and hind feet as well as an additional illustration of the replicative homology among toes.

Even in genotype *RMM* there may be a swerving of the hair streams on the back to produce a slight crest along the midline (Rough D). This becomes the rule in genotype *RMm* and is often associated with well-defined rosettes on each side (Rough C) halfway along the back. Bilateral homology is again illustrated, but the only homology with the parts most characteristically affected by *RRMM* is tissue homology.

The developmental conditions are not, however, quite the same on right and left sides, and apparently the threshold for formation of rosettes is somewhat lower on the right side than on the left. Such regular asymmetry can hardly be traced to an external cause. Directly or indirectly it must be traced to the basic asymmetry of the mammalian embryo revealed in the regularly asymmetrical arrangement of the viscera. The penultimate cause must apparently be sought in an effect on steric asymmetry of polarized molecules in the developing egg and the ultimate cause in steric asymmetry of certain key genes, presumably not *R* or *M*.

The shift from prevailing localization of a single pair of rosettes, halfway along the length of the back in the presence of *RMm* to two pairs, one anterior and one posterior with *Rmm*, indicates a principle of apportionment of space rather than that of localization by morphological features, illustrated in the reversals on the toes. The radiation from the eyes and ears, and, less clearly, on forehead and cheeks, illustrate again localization by morphological features. This also applies at least in part to the reversals on the belly. All of these point toward the general principle that replicative homology depends merely on the same genes coming into expression wherever the developmental conditions are sufficiently similar.

The atavistic return of digits—lost in the evolution of the whole family Caviidae—in certain genotypes raises the question of homology with respect to a character more typically considered in discussions of the subject than pattern of skin growth and hair direction. In polydactyly, it is clear that there are no genes that determine digits in the preformist sense. They, together with certain nongenetic factors, merely favor or oppose manifestation of developmental potentialities that are latent in all cavies. We must suppose that all the genes under which any part of the ancestral pentadactyl foot developed were so deeply involved in the development of the other parts of the foot and of the organism as a whole that most of the genetic systems concerned with the thumb, big toes, and little toes are necessarily still present. The evolutionary loss of these digits was presumably due to some simple superimposed inhibitory process that stopped the formation of lobes on the limb buds prematurely. Any genetic or environmental effect that inhibits this inhibitory process releases the process by which these digits had been shaped in the remote ancestors.

Multifactorial polydactyly may conceivably involve ancestral genes that have been carried at low frequencies throughout the history of the Caviidae or have been restored by reverse mutation, but this is hardly likely for *Px*. Yet if *Pxpx* acts merely by inhibiting a relatively simple inhibition acquired in evolution, the thumbs, big toes, and little toes that develop in this genotype under the released ancestral heredity may be considered essentially homologous to these digits in mammals that have never lost them, in spite of the monstrous characteristics of the foot in *PxPx*.

The homeotic mutations of *Drosophila* can be interpreted similarly. It must be supposed that the developmental conditions are not very different in the rudiments of aristae and legs, for example. A gene (aristapedia) that shifts those in the aristae toward those in the legs may provide the basis for calling into action step by step the whole chain of reactions that give rise to a leg. Moreover, genes that bring about modified segmentation of the tarsus (such as four joint) but that normally find no basis for action in the aristae find such a basis if the arista is replaced by a leg (Waddington 1943). Again, conditions are not so different in the different legs but that a structure, the sex combs, that normally appear only on the first legs of males may be enlarged on the first leg and brought into expression on the other legs by a single gene replacement, without necessarily implying that sex combs were ever characteristic of these legs in the ancestry (Slifer 1942).

Intermediate Optima and Multiple Selective Peaks

Cases in which there is reciprocal reversal of effect among combinations of loci are very important in the theory of evolution. The combination effects of *R*, *r*; *St*, *st* on the forehead rosette in guinea pigs (*mm* present) give an example. Both *Rstst* and *rrSt* (differing by two replacements) have strong forehead rosettes, whereas *RSt* is almost smooth and *rrstst* quite so.

Interaction effects of this type are undoubtedly not very common among ordinary characters. They must, however, be almost the rule in the character that is of first importance in population genetics. This is selective value. The reasons are partially indicated in the following:

Selection whether in mortality, mating, or fecundity, applies to the organism as a whole. A gene which is more favorable than its allelomorph in one combination may be less favorable in another. Even in the case of cumulative effects, there is generally an optimum grade of development of the character and a given plus gene will be favorably selected in combinations below this optimum but selected against in combinations above the optimum (Wright 1931).

We may give as an illustration the series of guinea pig colors ranging from white to intense golden agouti in which replacements act, in the main, cumulatively. The optimum color in nature is undoubtedly an intermediate cream agouti, judging from the color of the wild Peruvian cavy (*Cavia cutleri*), the wild ancestor of the tame guinea pig. There are several ways by which more or less close simulation of this color can be produced by dilution of golden agouti (*EABCPFSiDm*), each differing from the others in two or more respects and thus at a different selective peak. The actual genotype of the wild cavy is none of these, being the same as that of the golden agouti except for multiple minor dilution factors, different from those known in the laboratory (with the probable exception of *dm*).

The prevalence of pleiotropy also tends to bring about multiple selective peaks. Genes that have favorable effects at all will, in general, also have one or more unfavorable effects, with the net effect depending on the total array of genes. Evolution depends on the fitting together of favorable complexes from genes that cannot be described in themselves as either favorable or unfavorable.

The simplest model for dealing with selective value as a character seems to be the cumulative action of many genes on an underlying character, but with selective value falling away from an intermediate optimum, with the multiple resulting selective peaks differentiated by second-order pleiotropic effects. In general, the existence of complex patterns of factor interactions must be taken as a major premise in any serious discussion of population genetics and evolution.

CHAPTER 6

Types of Biological Frequency Distributions

Introduction

Since population genetics necessarily deals with the statistical properties of populations, the whole mathematical theory of statistics is pertinent. This is a very large subject, too large for anything approaching exhaustive treatment here. The aspect that will be of primary concern is the mathematical description of patterns of variability and their interpretation in terms of underlying causes. The second-order statistics required for judging the degrees of reliability of parameters and for testing the significance of differences will be introduced for use in the discussion of concrete cases, but a full presentation of the theory and of many important practical considerations involved in analysis of data must be left to the textbooks devoted to these subjects.

The Problem of Describing a Population

A complete description of any biological population is a wholly impossible task. Even without attempting to go beyond phenotypes, it would involve a list of all of the individuals, accompanied by descriptions of all observable characters, not only as adults, but throughout development, and also of the successive environments to which each had been exposed. Successive generations would have to be recorded in this way and connected by pedigrees for even a superficial evaluation of the roles of heredity and environment in determining the various characters.

Even a very restricted description of this sort, based on only a sample of the population, would be too unwieldy to be of any immediate use. The information must somehow be condensed into a form that can be grasped by the mind. Adequate condensation is difficult because each individual is an integrated whole with a unique complex of interrelated characters and a unique history. No measurements of single characters or environmental conditions can be considered in isolation without losing much of their significance. Yet it is a practical necessity to choose what seem the most significant variables and start by calculating a few statistics in each case to bring out as adequately as possible how each varies. Statistics can then be extracted from joint distributions of all sorts to obtain some grasp of the interrelations.

The importance of the choice of variables may be illustrated by a simple example. Early studies of a cross between squashes with fruits of widely different sizes and shapes based on the two variables, length (L) and diameter (D), gave a fairly adequate description since cross sections are always roughly circular. The nature of the variability of F_2 populations from wide crosses indicated that multiple factors were involved. By using two different variables, total size (weight, $W \propto LD^2$), and a form index ($F = D/L$), Sinnott (1931, 1935) found, however, that the analysis of various F_2 populations revealed clear segregation of either one or two major loci with respect to form, although there was still multifactorial determination of size. From a purely mathematical standpoint, the two sets of variables are equally adequate since those of either set are mathematical functions of those of the other set. From the biological standpoint, the set in which one of the variables proved to be genetically simple was obviously more instructive.

It is probably usually true that measures of volume or weight, whether of the organism as a whole or of some one organ, associated with appropriate indexes of form are more instructive than linear measurements. On the other hand, indexes must be based on measurements and their use involves certain statistical pitfalls.

Types of Frequency Distributions: Dimensions

A primary classification of frequency distributions may be made according to the number of dimensions used in ordering the classes. In some cases there is no order that is significant. The number of individuals of different species captured in a locality is an example (except for frequency itself).

The present chapter will be concerned with variability in one dimension. Variability in two or more dimensions will be deferred to a later chapter, but it may be well to give an example here. Table 6.1 shows the joint distribution of stature and weight in 1,000 male students (18–25 years old) at Harvard in 1914–16 (Castle 1916*b*). It is obvious that a given weight has different implications for men of different heights, and thus that the one-dimensional variables are somewhat arbitrary abstractions from the complex realities.

Frequencies in Time and Space

The description of the numerical and structural properties of populations must be based on frequencies in time and space. The form of the frequency distribution of population size over a succession of years tends to be rectangular

TABLE 6.1. Joint distribution of height and weight of 1,000 Harvard students

WEIGHT IN KG.	HEIGHT IN CM.															TOTALS
	155–57	158–60	161–63	164–66	167–69	170–72	173–75	176–78	179–81	182–84	185–87	188–90	191–93	194–96	197–99	
44–46			1													1
47–49	1		3	1												5
50–52	1	2	1	6	4	6	2									22
53–55	1	4	8	15	12	8	7	2			1					58
56–58		1	4	10	15	19	20	11	3	2						85
59–61		1	5	8	22	43	25	21	11	4	2					142
62–64	1		2	8	9	31	39	29	21	10	2	2				154
65–67			1	2	10	21	25	39	30	18	4		1			151
68–70			1	1	9	6	30	27	32	16	13	2		1		138
71–73				2	4	5	18	20	12	18	15	4	2			100
74–76					1	4	11	15	6	7	9	6			1	60
77–79					1	2	2	8	5	7	4	4	1			34
80–82							4	6	3	4	6	2				25
83–85							2	1	2	3	2	2				12
86–88					2	1	2			2						7
89–91										1	1					2
92–94							1	1								2
95–97																
98–100																
101–3																
104–6								1			1					2
Totals	4	8	26	53	89	146	188	181	125	92	60	22	4	1	1	1,000

From Table 1, Castle, 1916*b*.

if there is no trend, trapezoidal if there is a linear trend, exponential if there is unrestricted growth, and sigmoid if limited growth. Cyclic variations in numbers are found in many species, especially in relation to the seasons, but sometimes with longer periods than a year. The frequency distribution of age classes is an important property of populations.

The most significant frequency distributions of individuals in space are, of course, usually two dimensional. The nature of a distribution may, however, be sufficiently indicated by transects. The simplest situation is that in which the distribution is rectangular, indicating uniform occupation of its territory. More often there are centers of high frequency with numbers tapering off toward zero in both directions along any transect until a second center of high frequency is approached.

Figure 6.1 shows the numbers of red-eyed and orange-eyed flies of the

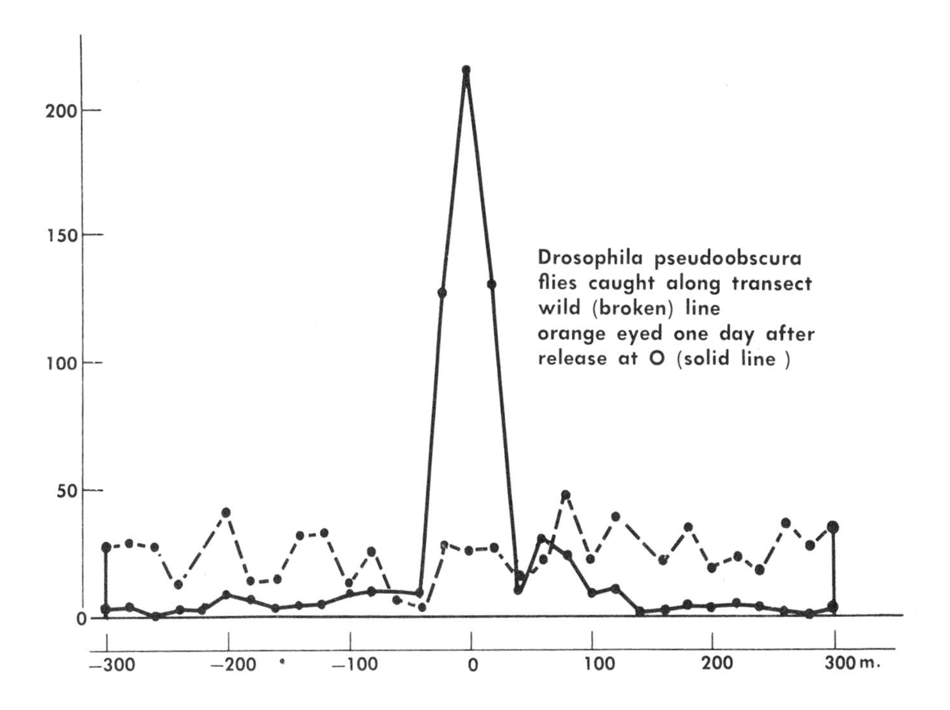

FIG. 6.1. Distribution along a line of traps of 635 orange-eyed *Drosophila pseudoobscura*, released at 0, at stations 20 m apart, after one day (solid line); in comparison with frequencies of 740 wild flies of the same species caught at the same time. (From data of Dobzhansky and Wright, 1943.)

species *Drosophila pseudoobscura*, caught in traps at 20-meter intervals along a transect on Mt. San Jacinto in California (Dobzhansky and Wright 1943). The fluctuations in the numbers of red-eyed flies (wild) largely reflect minor heterogeneities in the area (high frequencies near oaks and pines). The orange-eyed flies had been released the day before at the station at O so that the distribution reflects the pattern of dispersion. It is a good example of a highly leptokurtic distribution, one with a high narrow peak but wide extension of low frequencies about this.

Morphological and Physiological Frequency Distributions
Continuous Variability

The qualitative characteristics of populations are defined by statistics extracted from the frequency distributions of morphological and physiological characters of the individuals. We will consider first the principal forms taken by the distributions of continuous variables. In such cases, every individual has a unique measurement if a sufficiently fine scale is used, but these values are usually more or less densely clustered about one or more points on the scale (modes). A frequency distribution can be obtained by tabulating the cases that fall in successive equal intervals. These frequencies can be presented graphically by plotting them as the heights of rectangles erected from the appropriate intervals (histograms) or as ordinates at the midpoints of the intervals, connected at the top by lines (frequency polygons). The former represents the frequencies more accurately, but the latter is usually more convenient for comparison of different distributions on the same graph. The greater accuracy of the histogram is here of minor importance, since tests of differences depend in any case on arithmetic treatment of the sets of frequencies rather than on visual judgments.

The most common type of continuous distribution in reasonably homogeneous data is that in which the values are most densely clustered about a single mode and taper off more or less symmetrically in both directions to give a bell-shaped curve on smoothing the histogram. Quetelet (1835) recognized the close approach of the distribution of human statures to the normal probability curve, well known as that which random errors in physical measurements tend to approximate. This type of distribution is illustrated in Figure 6.2*A* for the height of Harvard students in Table 6.1.

Galton (1879) noted that appreciable positive skewness (longer tail in the positive direction) may be expected (if there is relatively great variability) because of the greater effect of a given factor when applied to a large base rather than to a small one. Positive skewness is illustrated by the weights of the same students (Fig. 6.2*B*).

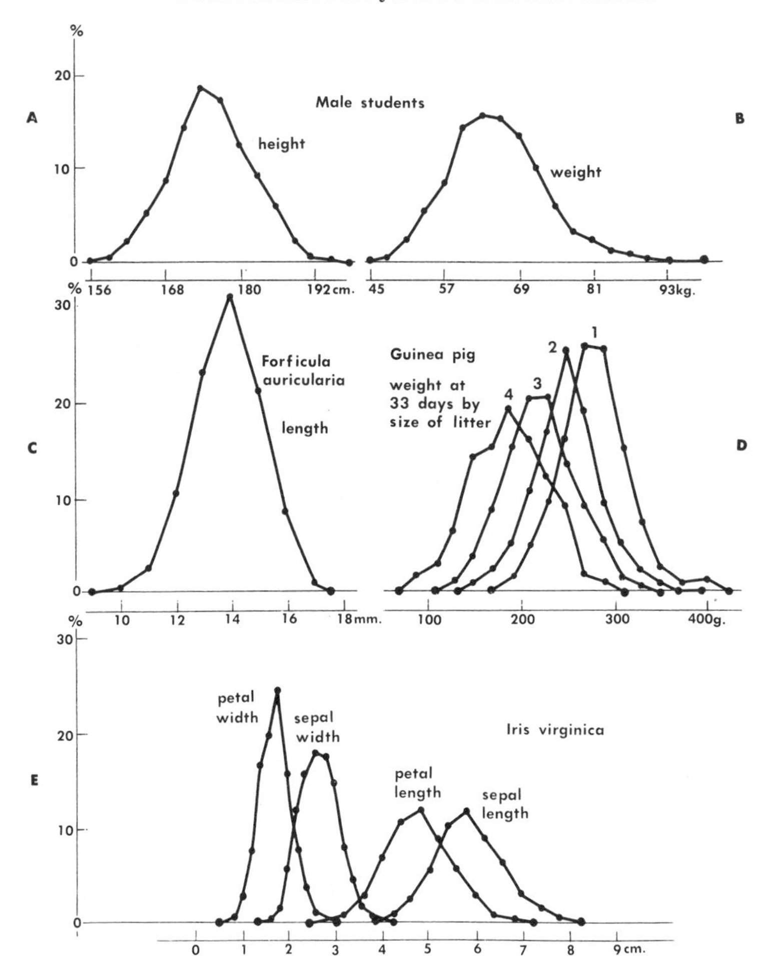

FIG. 6.2. Some unimodal distributions of sizes. Numbers of individuals. *A*, *B* 1,000 male students; *C*, 1,519 *Forficula*; *D*, 80, 382, 530, and 205 guinea pigs in litters of 4, 3, 2 and 1 respectively; *E*, 1,584 ± flowers of *Iris virginica*. (From data: *A* and *B*, Castle 1916*a*; *C*, Diakonov 1925; *D*, Wright and Eaton 1929; *E*, Anderson 1928.)

Not all such distributions show asymmetry, however. There is none in the lengths of earwigs (Diakonev 1925) in Figure 6.2*C*. No appreciable asymmetry appears in the 33-day weight of guinea pigs in litters of from one to four shown in Figure 6.2*D* (Wright and Eaton 1929). On the other hand, the distributions of length and width of petals and sepals in flowers of *Iris* *virginica* collected over its entire range by Edgar Anderson (1928) mostly show slight positive skewness (Fig. 6.2*E*). Whether skew or not, all of these distributions show the characteristic bell-shaped form.

This is also true of the distributions of a number of physiological characters in Figure 6.3. The distribution (Fig. 6.3*C*) of sucrose content in sugar beets (de Vries 1909) is interesting because it shows negative skewness. In this case a small proportion below 12% (probably less than 1%) were not reported. Such truncation is much more serious in the records of speed of American trotters (Fig. 6.3*E*), recorded only if the time for a mile was 2:30 or less. Records above 2:29 must be discounted. Considering only those that were 2:29 or less in the random sample of horses 5 years old or more taken from the register by Galton (1898), the mode was little, if any, below the truncation point in records for 1892–93 but it is evident that it was below in those for 1894–96. Truncation also presents a problem in dealing with the records for butterfat production of dairy cows (Fig. 6.3*F*). In the Register of Merit of the American Jersey Cattle Club, the requirement for cows less than 3 years was 250 lb. in 365 days. This rose linearly to 360 lb. at 5 years. Figure 6.3*F* shows distributions for one of the age classes in a tabulation made by Davidson (1928) of the records up to and including 1920. Figure 6.3*D* shows a distribution of lengths of diapause in gypsy moths (Goldschmidt 1933). Figure 6.3*B* shows the distribution of intelligence quotients among foster and own children in a study made by Burks (1928) from a California population restricted to persons of north European ancestry. The distribution (Fig. 6.3*A*) of another sort of physiological character is of basal metabolism of men and women (Harris and Benedict 1919). As this character would be expected to reflect body size to a large extent, it is desirable to deal with relative rates of some sort. It turned out that variability is reduced most by taking basal metabolism in ratio to an estimate of body surface, based on height and weight. The distributions of such indexes are the ones shown in Figure 6.3*A*. All these physiological characters show distributions of the same general type as the size characters.

As already noted, indexes are also often of more significance in analyzing morphological variability than absolute sizes of parts. Figure 6.4*C* shows the distributions of the ratio of the distance between the eyes (frons) to head width in two laboratory populations of house flies (*Musca domestica*) from Egypt designated as forms *cuthberti* and *vicina* (Peffley 1953). Figure 6.4*B* shows the distribution of the ratio of the length to the diameter in shells of the snail *Limnaea palustris* (Mozley 1935). Figure 6.4*B* shows the distribution of an index of relative head length in a population of rattlesnakes, *Crotalus viridis viridis*, collected while hibernating near Platteville, Colorado (Klauber 1938). An adequate form index requires some care in this instance

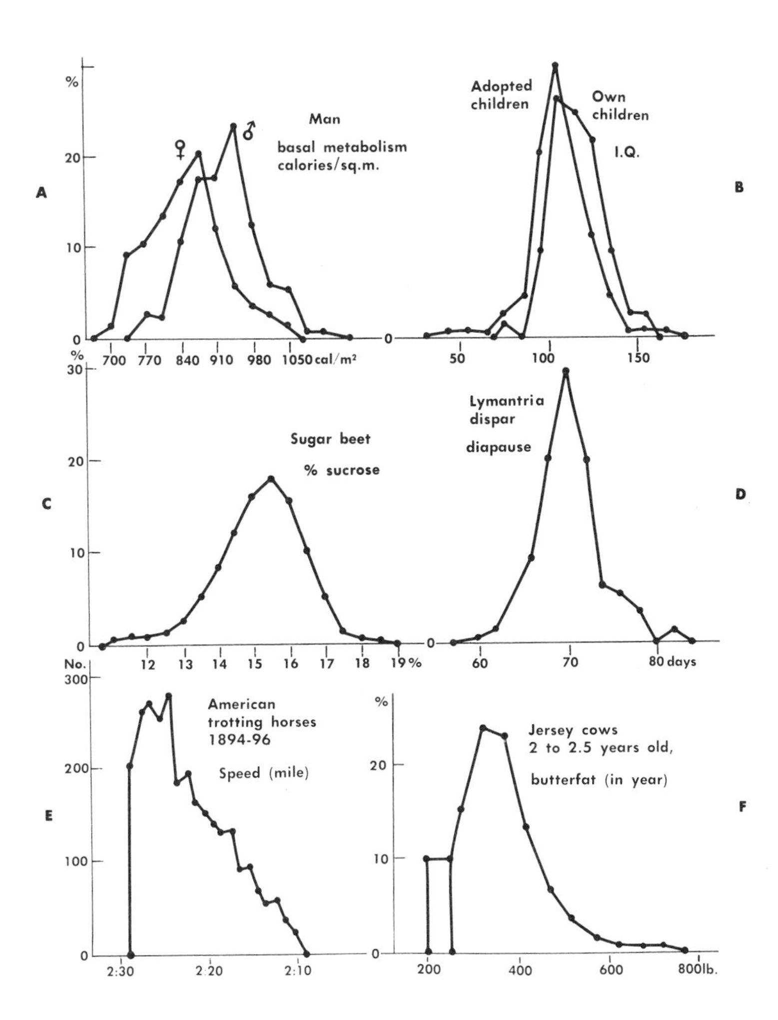

FIG. 6.3. Some unimodal distributions of physiological variables. Numbers: *A*, 103 women, 136 men; *B*, 214 adopted children, 105 own children; *C*, 42,997 sugar beets (about 1% truncation below 12% sucrose); *D*, 205 *Lymantria dispar*; *E*, 3,310 American trotting horses (about 27% truncation below 2:29 min for mile); *F*, 2,565 Jersey cows (about 10.5% truncation below 250 lbs per year). (From data: *A*, Harris and Benedict 1919; *B*, Burks 1928; *C*, de Vries 1909; *D*, Goldschmidt 1933; *E*, Galton 1898; *F*, Davidson 1928.)

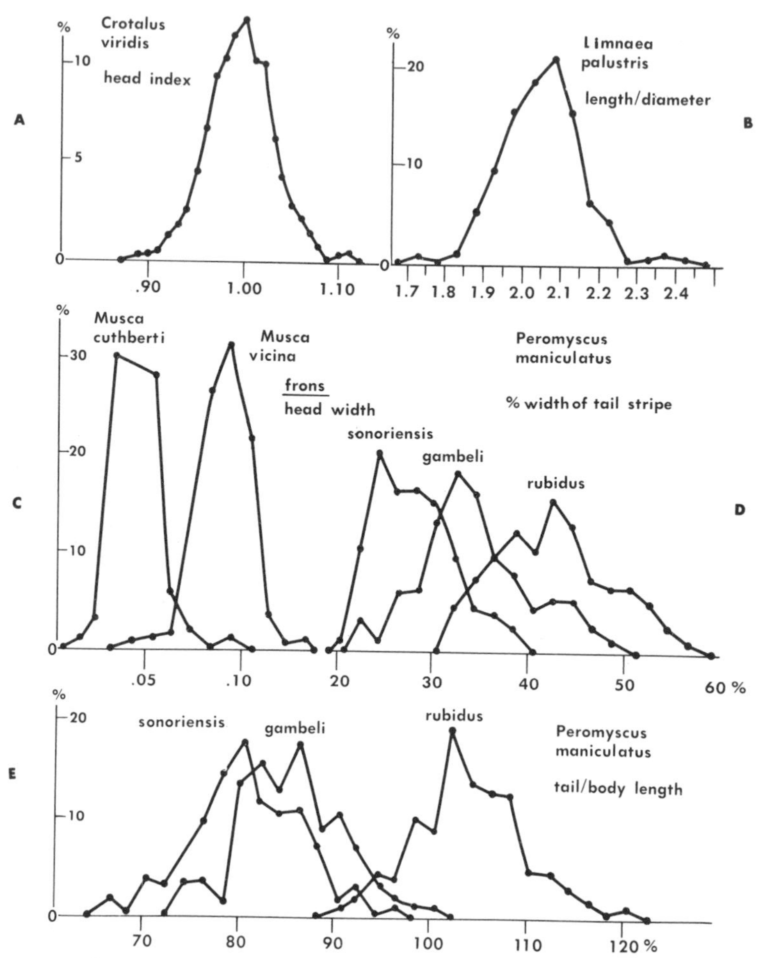

FIG. 6.4. Some unimodal distributions of indexes. Numbers: *A*, 833 *Crotalus viridis*; *B*, 271 *Limnaea palustris*; *C*, 139 *Musca cuthberti*; 283 *Musca vicina*; *D*, 133, 116, and 140 of subspecies *sonoriensis*, *gambeli*, and *rubidus*, respectively, of *Peromyscus maniculatus* (tail stripe); *E*, 138, 118, and 142 of the same subspecies (tail/body length). (From data: *A*, Klauber 1938; *B*, Mozley 1935; *C*, Peffley 1953; *D*, *E*, Sumner 1920.)

since mature animals continue to grow, and any tendency of a simple ratio to change with age (heterauxetic growth) becomes important. Klauber found that a good estimate of mean head length for a given total length (L) in centimeters was given by the equation $0.03553L + 6.968$. He took head length in ratio to this estimate to obtain the distribution of an independent form index.

Figure 6.4*E* shows the distribution of a form index, tail length in ratio to body length, in three wild populations of the deer mouse, *Peromyscus maniculatus*, collected in different localities in California by Sumner (1920). These represent three subspecies. These and many other form indexes again show distributions of the same general type as the other classes of characters.

Among the many characters he measured in populations of deer mice, Sumner also determined the percentage of the width of the dorsal stripe in the tail circumference. This gave a convenient index of the extent of pigmentation which differed greatly among the populations. Figure 6.4*D* shows the distributions for the same three populations as in Figure 6.4*E*. The conspicuous positive skewness of the Victorville population, with mode at only 25% in comparison with the relative symmetry of the Eureka distribution with mode at about 43%, is to be noted.

This brings us to consideration of the peculiarities of distributions of percentages which become marked as the necessary limits, 0% and 100%, are approached. Figure 6.5*A* shows the extreme asymmetry of the distribution of percentages of black required in association with red, yellow, and white to match the skin color of Jamaicans in a study of the genetics of skin color by Davenport (1913). Figure 6.5*C* gives the distribution of percentage of white in the coats of males of three strains of guinea pigs (Wright 1926*a*, Wright and Chase 1936). The representatives of strains 35 and 2 involved here are taken from descendants of single matings derived from many generations of brother-sister mating and were demonstrably almost isogenic with respect to spotting, in spite of the enormous variability owing largely to accidents in development. Strain *DS* derived its spotting factor (*s*) from strain 2 but was extracted after at least three generations of backcrossing to a closely inbred self-colored strain (*SS*). The extreme positive skewness of *DS* with median at 6.5% contrasts with the extreme negative skewness of strain 2 with median 93.2%, as well as with the moderate negative skewness of strain 35 with median at 62.5%. These illustrate the relation between skewness and mode that is characteristic of many percentage characters. No self-colored animals appeared in *DS* in spite of its low median, and only 3% self-white appeared in strain 2 in spite of its very high median. There were, however, 23.8% self-whites in the whitest of the strain, no. 13, showing that this limit can be fully reached in extreme cases. Moreover, the limit at 0% is easily reached in genotype *Ss* (which ranges from 100% self-colored to 100% spotted, according to the modifiers that are present) and is almost always fully reached in genotype *SS*. There is damping, but not complete damping of variability at the limits in this case, in contrast with the probable

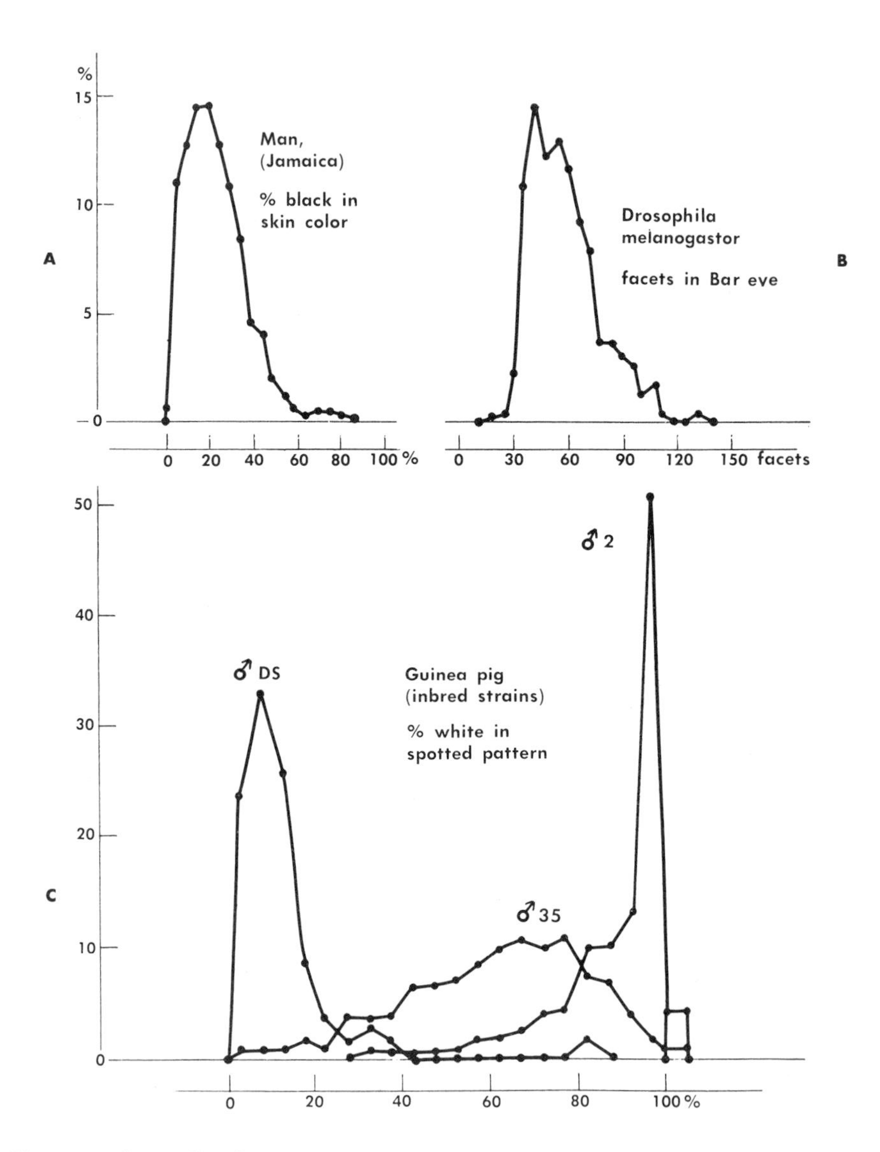

FIG. 6.5. Some distributions of percentages and a meristic distribution that resembles two of the former. Numbers: *A*, 1,086 men; *B*, 488 *Drosophila melanogaster*; *C*, 85 strain *DS*, 751 strain 35, and 820 strain 2, male guinea pigs. (From data: *A*, Davenport 1913; *B*, Zeleny 1922; *C*, Wright 1926*a* and Wright and Chase 1936.)

complete damping in human skin color. There may be other cases in which the limits merely truncate otherwise normal variability, or even ones in which variation is exaggerated as the limits are approached. Cases in which the limits can be fully reached bring us to the category of threshold distributions, which will be discussed later. Figure 6.5*B* illustrates a distribution (facets in a Bar-eyed strain of *Drosophila melanogaster* [Zeleny 1922]) superficially similar but really very different from these.

Bimodal and Multimodal Distributions

If measurements cluster significantly about two values instead of one, the first suggestion is that there is gross heterogeneity. It may turn out that a collection contains two species closely similar except in the respect considered. On recognition of this situation it is often possible to find some other criterion by which a complete separation can be made. Another common cause is heterogeneity with respect to discrete age classes. Figure 6.6*C* shows the distributions of the total lengths of the rattlesnakes *Crotalus viridis viridis* collected in a den near Platteville, Colorado (Klauber 1937). The cluster about the low mode consists of juveniles, hatched in the year in which they were collected. There is so little overlap with those that are over a year old that the distribution of juveniles can be separated with no important error. There is, however, an indication of bimodality in the distribution of adults, at least in the males. The lower mode depends largely on the yearlings, but there is evidently much overlapping with still older classes. Even in the females the distribution is highly platykurtic, indicating its composite nature.

No sharp line can be drawn between heterogeneous and homogeneous populations. The variability in a "homogeneous" population is due to the operation of multiple factors. A separation according to high or low values of any one of these would give two distributions with different modes. It is only when the differences due to one factor play an overwhelming role that heterogeneity is recognizable. This is brought out clearly in Figures 6.6*A* and 6.6*B*. These figures represent the distribution of lengths (L) and diameters (D) in an F_2 population of squashes, derived from a cross between disk shaped and spherical strains (Sinnott 1931). The disks (dominant) and spheres (recessive) could be distinguished without overlap by using the form index D/L. The clusters about the two principal modes in the distribution of lengths (L) represent the two segregating types with only a little overlap. In the diameter (D), there is considerable separation of the means of the dominants and recessives (as determined by the form index), but there is so much overlap that the total distribution is unimodal with little indication of

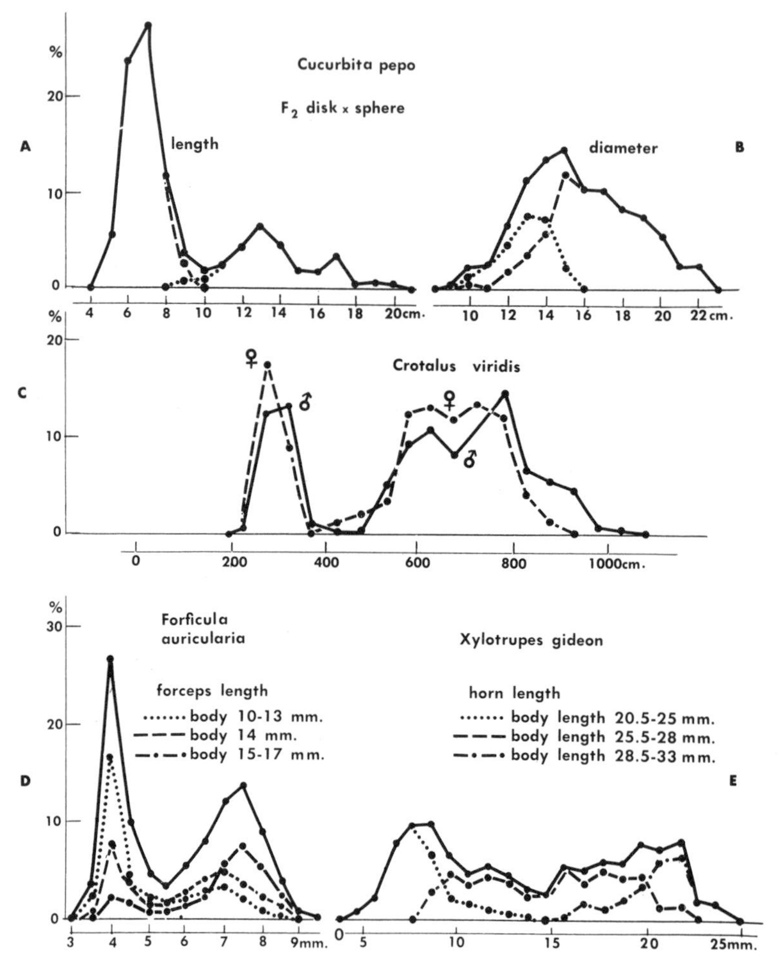

FIG. 6.6. Some bi- and multimodal distributions of sizes. Numbers: *A*, *B*, 205 *Cucurbita pepo*; *C*, 399 ♀, 459 ♂ *Crotalus viridis*; *D*, 1,519 *Forficula auricularia*; *E*, 316 *Xylotrupes gideon*. (From data: *A*, *B*, Sinnott 1931; *C*, Klauber 1937; *D*, Diakonov 1925; *E*, Bateson and Brindley 1892.)

heterogeneity. The segregating locus is here merely one of many individually minor causes of variability in diameter and the population may be considered to be essentially homogeneous in this respect.

Figures 6.6*D* and 6.6*E* illustrate an unusual type of bimodality. Male earwigs (*Forficula auricularia*) from the same colony ordinarily show marked bimodality with respect to forceps length (Bateson and Brindley 1892).

Figure 6.6*D* gives the total distribution for 1,519 individuals, collected over several years by Diakonov (1925) and analyzed more fully by Huxley (1927*a*). There was no bimodality in body length (Fig. 6.2*C*). Diakonov showed that there was no essential genetic difference between the types and no sharp environmental difference. The small-sized individuals (10–13 mm.) show both types but a big excess of ones with small forceps. In those of average body length (14 mm.), there are fewer with small forceps and more with long ones. The large individuals (15–17 mm.) show still fewer with small forceps and still more with large ones. The average forceps length within the two types rises only slightly with increase in body size. Diakonov concluded that the two types represented two different equilibrium points in development. Huxley suggested that the essential difference was a discontinuous difference in the length of the growth period after the onset of strongly heterauxetic growth of the forceps, perhaps dependent on whether there were one or two subsequent molts. He suggested that favorable conditions, reflected in large body size, favored the extra molt.

Bimodality in the length of the horns of the beetle, *Xylotrupes gideon*, also described by Bateson and Brindley (1892) and further analyzed by Huxley (1927*b*), seems to be somewhat similar but is more complicated (Fig. 6.6*E*). The small beetles tend to produce only short horns, the large ones only excessively long horns, and bimodality is exhibited only by the class of beetles of intermediate size. Moreover, the two modes are here much closer together than the modes for the small and large beetles. There seems to be little heterauxetic growth of the horns under unfavorable conditions, much under favorable conditions, but great lability in the amount under the intermediate condition as indicated by average body length.

It should, perhaps, be added that many apparently multimodal distributions are ones in which the irregularities of form are due merely to accidents of sampling where the number of cases is small.

Dichotomies

The simplest type of frequency distribution from the standpoint of immediate description is the dichotomy. A single parameter, the proportion in one of the two classes, gives all of the information available from the distribution on the nature of the population, except that specification of the size of the sample and degree of independence of the observations are needed in judging the reliability of this parameter.

There are, however, radically different sorts of dichotomies with respect to causation. The simplest is that in which the two classes reflect directly the presence of one or the other of two alternative ultimate factors. This is a point dichotomy. All cases of clear-cut Mendelian segregation of dominant and recessive alleles come here, and many cases of sharp dimorphism in natural populations are known to be caused by point dichotomy. Føyn and Gjøen (1954) have given a beautiful example in the serpulid worm, *Pomatoceros triqueta*, of which a collection from Oslofjord yielded 2,628 with brown branchial crown, 218 with blue scattered among the browns (and 2 with orange). Experiments showed that blue was a simple dominant over brown. The most common case of a point dichotomy is, of course, the presence of two sexes in all forms in which the decisive factor is chromosomal.

A very different sort of dichotomy is illustrated by the percentages of winged and wingless aphids within clones of *Macrosiphium solanifolii* studied by A. F. Shull (1932). In a given clone, the percentage was affected by many factors including the durations of periods of alternating light and darkness, or of temperature, or of nutrition, all in the late prenatal stage, and the winged or wingless characters of the mother (offspring tending to reversal). Thus the result depended on a complex of factors of which no one was decisive except under carefully controlled conditions. A remarkable feature was that different clones responded differently, as illustrated by the following comparison between clones A and B under the same conditions of exposure of the parents to light and with other conditions controlled in each period (Table 6.2). Clearly, there are interaction effects that are far from additive.

TABLE 6.2.

Clone	Continuous Light				Cycle 8 hr. Light 16 hr. darkness			
	Wingless	Winged	No.	% Winged	Wingless	Winged	No.	% Winged
A Aug.–Oct. 1928	505	4	509	0.8	103	265	368	72.0
Dec.–Mar. 1929	362	22	384	5.7	54	150	204	73.5
B Aug.–Oct. 1928	490	952	1,442	66.0	1,070	172	1,242	13.8
Dec.–Mar. 1929	177	233	410	56.8	166	37	203	18.2

From data of Shull (1932).

The important point here is that two sharply alternative types clearly depend on whether the physiological state at a critical time, resulting from a complex of interacting genetic and environmental factors, is on one side or the other of a certain threshold. This is referred to as a threshold dichotomy.

Many groups of organisms have no chromosomal mechanism of sex determination. In most the individuals are simultaneously hermaphroditic, but in some, sex depends on the stage of development—as in certain pulmonates—in which each individual develops first into a male and then transforms into a female (or in some cases the reverse). In other cases sex is determined by environment. In the Gephyrean *Bonellia* (Baltzer 1925) the larvae develop into large-sized females if isolated, but into minute males if exposed to secretions from females (or to certain chemicals). In these cases sex ratio depends on a threshold instead of a point dichotomy. The difference in type of dichotomy thus does not depend on the nature of the character but on whether there is or is not a pair of sharply alternative ultimate factors that is normally decisive.

With respect to underlying causes, a threshold dichotomy resembles an arbitrary dichotomy of a continuous variable such as that given by classifying the men in Figure 6.2*A* into two classes, "short" and "tall," according to whether they are below or above an arbitrary height. A point dichotomy is radically different. It may be surmised that in studying the interrelations among variables, the mathematical treatment of threshold dichotomies should be designed to give the same results as far as possible as if the underlying continuous distributions of causal complexes were available, while the treatment of point dichotomies should be that of two sharp alternatives.

Point Distributions

There are many cases in which natural populations include multiple alternative types, whether due to multiple alleles or to recombinations of two or more segregating loci. This takes us to multidimensional variability, unless the types all fall into a single order phenotypically. There are many series of multiple alleles known in the laboratory and some in nature having effects on a given character that can be arranged in such an order. The form of the distribution may be quite erratic since there may be no orderly relation between frequency and phenotypic effect. On the other hand, the random combination of the effects on a single measurement of multiple independent loci may give a frequency distribution that approaches continuous bimodal or unimodal distribution in form, according to whether or not there is one pair with outstanding differential effect.

Threshold Distributions

Many distributions have essentially only one type on one side of a threshold but continuous variability on the other. One has already been discussed:

distributions of percentages in which there is an important class at 0% or at 100%. The threshold here is one of mathematical necessity. We consider now ones in which it is physiological, the threshold being that point at which a self-regulatory (or homeostatic) process breaks down.

Otocephaly in the guinea pig, which has been discussed earlier, provides an example (Wright and Eaton 1923, Wright 1934*b*). Table 6.3 shows the number of these monsters in five groups in my colony. Group I with 61 of the monsters in about 110,000 is fairly typical of the usual incidence. These animals had no ancestry of a particular inbred strain, no. 13. Some 5,300 animals of mixed ancestry but with at least half blood from strain 13 (group II) produced 75 or 1.4%. Strain 13 itself (group III) produced 1.0% (34 in 3,507) exclusive of the descendants of a particular mating in the thirteenth generation of sib mating. Group IV includes all descended from this mating (13–13–1) except for a cluster derived from a mating in the nineteenth generation. All the evidence indicated that group IV, with 203 otocephalic in 3,689 young or 5.5%, was homozygous in all pertinent genes. All branches, whether from continued brother-sister mating or from crosses within the group, produced about the same percentages for more than 30 years, except for certain lines tracing continuously to mating 13–19–1 (group V), which produced 212 in 1,168 or 18.2%. This increase in percentage in group V is interpreted as owing to a dominant mutation, which persisted in certain lines but was lost abruptly in one (here included in group IV).

All these groups, with incidences varying from 0.06% to 18.2%, show essentially the same distribution of grades of defect, from mere shortening of the lower jaw (grade 1) to at least grade 10 with no jaws, single median ear opening and reduced external ears, no nose, no eyes, little or no forebrain, and only posterior parts of the skull.

We have treated this as an example of continuous variability beyond the threshold of normality. There is, however, indication of a second major threshold at the point at which the mandible is completely lost (grade 4 from which grade 3 differs only in having two adjacent ear openings instead of one). Grade 4 is much the most numerous (43%) indicating that it covers a wider range on the scale of factors. Only in a relatively small proportion of the cases, does defect go beyond this to bring about a succession of reductions of the maxillary (grade 5) and of the brain and associated parts (grades 6 to 12). There seems to be continuity in this reduction, but the external signs show a succession of thresholds (Figs. 5.16, 5.17).

The situation is complicated by the occurrence of defects of the premaxillary and maxillary, which occur to some extent out of order. Thus premaxillary defect without mandibular defect occurred in 3 cases (grade A), but in general such defect occurred only with complete loss of the mandible

TABLE 6.3. Frequencies of grades of otocephaly in groups of guinea pigs.

	A	1	2	3	4	5	6	7	8	9	10	11	12	?	oto	Total	% oto	6–12 / A–12
I		6	12	5	18	2	2	1	4	2	3			6	61	110,000	0.06	21.8
II	1	10	12	14	29	5	1	1	1		1				75	5,300	1.42	5.3
III		3	4	5	14	1	1	1		2	2			1	34	3,507	0.97	18.2
IV		20	39	32	86	4	6	1	1	4	7			3	203	3,689	5.50	9.5
V	2	16	23	34	102	7	8	2	5	2	4	3	1	3	212	1,168	18.15	12.0
♂	1	17	32	27	83	4	8	4	4	5	7		1	5	198	59,084	0.34	15.0
♀	2	35	55	58	154	15	10	2	6	5	10	2		7	361	59,084	0.61	9.9
?		3	3	5	12				1			1		1	26			8.0
Total	3	55	90	90	249	19	18	6	11	10	17	3	1	13	585	123,664	0.47	11.5
%	0.5	9.6	15.7	15.7	43.5	3.3	3.2	1.0	1.9	1.8	3.0	0.5	0.2					

From data of Wright (1934*b*).

I, no ancestry from strain 13
II, crossbreds involving strain 13
III, strain 13, excluding descendants of 13–13–1
IV, descendants of 13–13–1, except high cluster from 13–19–1
V, high cluster from 13–19–1

and usually with more or less associated maxillary defect (grade 5). These defects were usually complete in grade 6 or at least 7, but in two cases reduced maxillaries carrying molar teeth were present below a cyclopean eye (8A, 9A). There were other minor irregularities. Such deviations from a single order are typical of threshold distributions with respect to a complex developmental defect.

In one respect, this series is definitely atypical, however. Ordinarily the degree of "expressivity" in those that are defective at all increases with the "penetrance," using Vogt's convenient terms (cf. Stern 1960). In the present case, the percentage of high grade monsters (grades 6 to 12) is significantly greater in group I, with minimum penetrance. Moreover, this percentage is actually, if not significantly, greater in males than in females, although penetrance is only about half as great in the former as in the latter. The underlying cause of all the defects is believed to be partial inhibition of the prochordal mesoderm. It appears that if this is sufficiently severe to inhibit completely the portion of the anterior neural crests responsible for the mandibular mesectoderm, extension to the anterior medullary plate (of which the preceding is a prerequisite) depends on largely independent developmental factors. The threshold type of distribution also applies to several other abnormalities of the guinea pig such as microphthalmia-anophthalmia, ventral flexure of the forefeet, and axial abnormality (Wright 1960*e*).

Grüneberg and his associates have studied some 30 skeletal anomalies of the mouse that have widely varying incidences among inbred strains and in the same strain under different rations. In general, the average degree of expression of a trait increased with the percentage of occurrence (Grüneberg 1952, Deol 1958, Deol and Truslove 1957).

Waddington (1955) produced strains of *Drosophila melanogaster* in which the crossveins between the fourth and fifth longitudinal vein tended to be deficient, by selecting from a wild stock those individuals that responded most strongly to heat treatment as pupae. There were no such defects in the untreated wild stock but the selected strains came to develop them without treatment. Strains with relative slight defect (L_2) and strong defect (L_5) were produced by further selection.

L_2 males show a distribution of the threshold type with 42% normal and frequencies of defect that decrease with increasing grade of defect. The distribution of females is similar except that the frequency of normals is much lower (16%). In L_5, normals are completely absent in both males and females and most of the flies have less than one fourth of the crossvein left. The results of crosses indicated multifactorial differences (Waddington 1955).

These could be considered as continuous distributions on a scale of percentage of occurrence of the crossvein. It is, however, possible for the crossvein to be represented in excess through occurrence of extra portions of vein attached to it. Waddington produced strains E_2 and E_5 with different average amounts of excess by selection from wild strains in which excess occurred spontaneously as a rare abnormality.

The results are somewhat similar to those for defect, but in this case, we are clearly dealing with continuous variability beyond the threshold for abnormality of a character, in which there is developmental "canalization" of the normal condition, in Waddington's terminology. Crosses between the strains with excess and defect showed that these involved independent regulatory systems (Waddington 1955).

Continuous Variability above a Breaking Point

Grüneberg (1951) has described a case that seems to be opposite to those described above (Fig. 6.7*A*). About 12% of the third lower molars were absent in one inbred strain of mice (*CBA*) and those present were on the average small and highly variable in contrast with the invariable presence of relatively large and only slightly varying third lower molars in another strain (*C57Bl*). That absence was causally related to small size was indicated, among other things, by the highly significant association of absence on one side with especially small size on the other. In this case the self-regulatory process seems to have been impaired by the genotype of strain *CBA*, leading to great variability. If the complex of developmental factors is sufficiently unfavorable, there is a threshold at about three-fourths normal size, at which the tooth fails to develop at all.

A breaking point of this sort is almost the rule for factors that affect the viability of the organism as a whole. There is considerable resistance to arrays of such factors, as indicated by only moderate increases in the mortality percentage up to a certain point in their accumulation, but beyond this the percentage jumps to nearly or wholly 100%. This may be illustrated by the extensive studies of Dobzhansky and his associates (e.g., Dobzhansky and Queal 1938) on the viability of homozygotes with respect to random chromosomes from wild populations of various species of *Drosophila*. The testing of each chromosome requires three generations. Figure 6.7*B* shows the results of tests of second chromosomes (usually involving 200–300 flies per F_3 test) of *D. willistoni* collected in Brazil by Pavan *et al.* (1951).

Most of the homozygotes fall into a roughly normal distribution centering at about 90% of the control (heterozygous) viability with relatively few in the range between 10% to 60%. There were, however, many that were either completely lethal or nearly so. Many studies have shown that this pattern is characteristic of chromosomes of species of *Drosophila*.

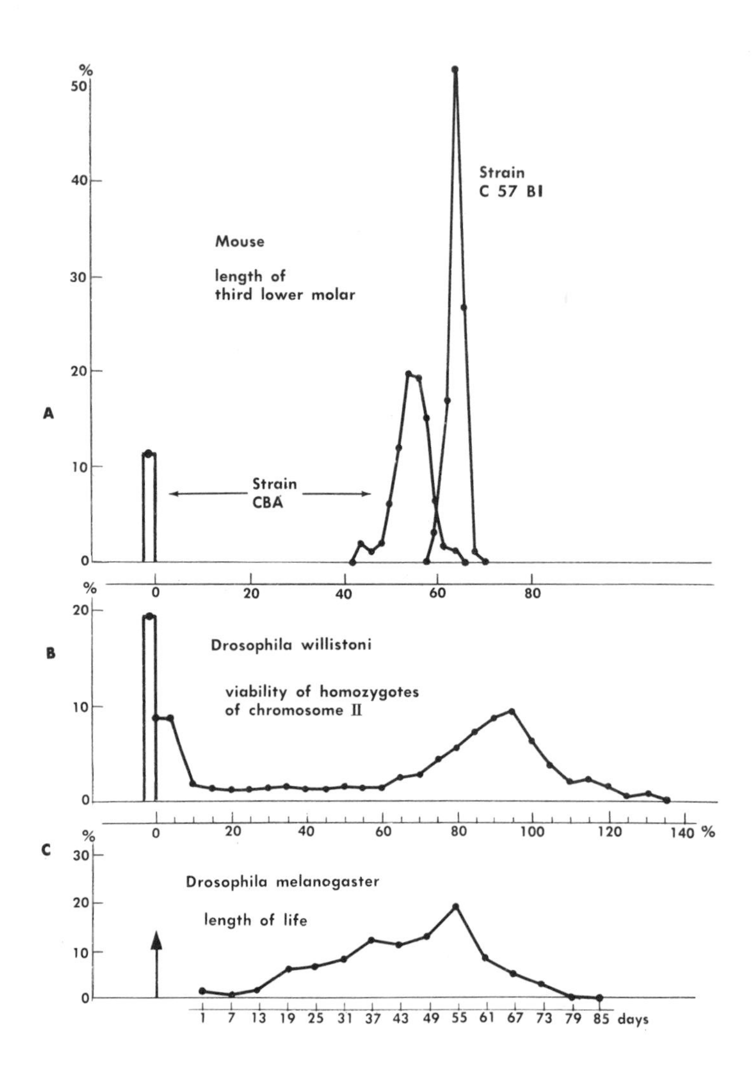

FIG. 6.7. Some distributions (size, viability) that are unimodal except for a subthreshold class. Numbers: *A*, 205 *CBA* mice; 100 *C57Bl* mice (no subthreshold class); *B*, 2,004 *Drosophila willistoni*; *C*, 2,822 *Drosophila melanogaster*. (From data: *A*, Grüneberg 1951; *B*, Pavan *et al.* 1951; *C*, Pearl 1928.)

The situation is usually somewhat similar in age-mortality curves such as that of wild *Drosophila melanogaster* shown in Figure 6.7*C* (Pearl 1928). The mortality before emergence is not shown.

Meristic Variability

Most organisms show replications of similar parts. The number of these parts may be the same in all but a few "abnormal" individuals, or it may vary widely both within and among populations. These variations have been shown in many cases to be due to both genetic and environmental factors. Since each unit is usually fully developed, if present at all, and the array of factors may vary continuously (at least if environment plays any role), it must be supposed that the continuity of the developmental effect is interrupted by a succession of thresholds. Frequency distributions of meristic characters are especially convenient in studies of variability, not only because counting is usually easier than measuring, but because the number of parts is more likely to be independent of general size than a measurement.

An illustration of the transition from a simple threshold character to a meristic one is the number of digits on the feet of guinea pigs, which has already been discussed. The guinea pig, like all species of Caviidae, normally has four toes on each forefoot (no pollex) and three toes on each hind foot (no hallux and no fifth digit). It was also noted, however, that an apparently atavistic little toe occurs sporadically in many strains, with considerable frequency in some and that a strain (*D*) was developed by Castle (1906) in which all individuals had this perfectly developed.

Table 6.4 shows the distributions on left and right hindfeet in an inbred strain, no. 35, which was characterized by a high frequency of the little toe but complete absence of thumb and big toe. As the correlation between the feet is imperfect it is clear that the two sides are acted upon by somewhat different complexes of developmental factors. Two thresholds are recognized: that of any development of the little toe on a foot and that of perfect development. The marginal totals show the distributions for each foot separately. Somewhat higher percentages cross both thresholds on the left side than on the right side.

Because of the existence of developmental factors that act separately on the two sides, there are no sharp thresholds for the animals as wholes. Nevertheless, the factors that act on the animal as a whole are so much the most important, even within an isogenic line, that the convenience of using a single grade for an animal outweighs whatever inaccuracy there may be. It is convenient to recognize three classes: animals with only three toes on both hind feet (00). An intermediate class (*Int*) and animals with well-developed

TABLE 6.4. Percentages with little toe absent (0), imperfect (*Int*), and well developed (*G*), on right and left feet of 2,332 guinea pigs of strain 35.

		RIGHT FOOT			
		0	*Int*	*G*	Total
	0	67.2	5.2	1.6	74.0
LEFT	*Int*	7.9	5.3	2.0	15.2
FOOT	*G*	2.7	3.1	5.0	10.8
	Total	77.8	13.6	8.6	100.0

From data of Wright (1934*c*).

little toes on both hind feet (*GG*). Table 6.5 shows the distributions of animals according to the age of the mother. It brings out the orderly decrease in penetrance as the maturity of the mother increases. In this case, the changes in penetrance (*Int* + *GG*)/(Total) are paralleled by changes in degree of expression (*GG*)/(*Int* + *GG*).

As noted in chapter 5, a mutation (*Px*) occurred in a normal strain (*I*), which tended to restore not only the little toe but also the thumb and big toe.

TABLE 6.5 Percentages with little toes both absent, intermediate (*Int*), or both good (*GG*), according to age of mother among guinea pigs of strain 35.

Age of Dam	00	*Int*	*GG*	Total	$\frac{Int + GG}{\text{Total}}$	$\frac{GG}{4T}$
3–6	42.4	43.1	14.5	408	57.6	25.1
6–9	59.2	35.4	5.4	463	40.8	13.2
9–15	71.2	25.3	3.5	722	28.8	12.0
15–45	81.9	16.9	1.2	741	18.1	6.7
Total	67.2	27.8	5.0	2,334	32.8	15.4

From data of Wright (1934*c*).

The homozygotes were grossly abnormal and usually died and were absorbed early in development, but occasionally they did not die until birth. Since many *Pxpx* show no extra digits or no other than little toes, they cannot always be distinguished from *pxpx* in segregating progenies. In progenies that include monsters that came to birth, it can safely be assumed that one-third of the other young, on the average, are *pxpx*, two-thirds, *Pxpx*. Table 6.6 shows the percentage distributions separately for thumbs, little toes, and big toes in four groups of matings that produced *PxPx*, after subtracting one-third (none with extra toes) in the first three cases, in which more than one-third were of this type, and subtracting all normals (7) and all with little toes but no other extra digits (14) in the case of group *ID* which had 50% or more ancestry of strain *D*. The other groups had no ancestry of either strains *D* or 35. The percentages on right and left feet did not differ significantly in any of these cases and are here averaged. As with *pxpx* of strain 35, three classes are distinguished with respect to each kind of extra digit: none (0), intermediate (*Int*) and good (*G*). No well-developed thumbs were observed in the first group, whereas the little toe was always present in the last group (*ID*). Otherwise, all the distributions for these digits have two thresholds; one for any development and one for perfect development of the digit in question. The effectiveness of the genes for the little toe, fixed in strain *D*, in increasing the incidence of thumb and big toe in the presence of *Pxpx*, although these digits are never found in *D*, is brought out.

Again, it would be convenient to assign a single index to each animal as a whole, but the imperfect correlations among the six extra digits raise difficulties. There are 729 possible combinations of the grads 0, *Int*, and *G* for the

TABLE 6.6. Percentages with thumbs, little toes, and big toes in each case with both absent (0), both good (*G*), or on intermediate condition in guinea pigs of genotype *Pxpx* in four groups of different ancestry.

	THUMB			LITTLE TOE			BIG TOE		
	0	*Int*	*G*	0	*Int*	*G*	0	*Int*	*G*
I-13, Ro	65.9	34.1	0	27.2	63.2	9.6	100.0	0	0
A[1]	36.5	54.1	9.4	47.3	48.5	4.2	100.0	0	0
Misc.	40.3	37.9	21.8	34.4	49.0	16.6	92.7	4.5	2.8
ID	1.1	22.0	76.9	0	0	100.0	89.1	6.5	4.4

From data of Wright (1935*a*).

separate digits. Actually not all of these occur, since the big toe was not found in these data except in the presence of both thumbs and both little toes (not necessarily perfectly developed). Nevertheless, there are a great many combinations. Moreover, it is clear that there are significant differences among strains in the order of the thresholds for thumbs and little toes. Table 6.7 shows a highly condensed classification of the animals as wholes into those with no extra digits (0/00), those with one or both little toes but no others (0/*T*0), those with one or both thumbs but no others (*T*/00), those with thumbs and little toes both represented but no big toes (*T*/*T*0) and those with all three of the kinds of digits represented (*T*/*TT*). The 20 animals of the last class included 10 with all 6 digits, but in only 2 of these were all 6 well developed. The Table brings out the difference between I-13, Ro with 44% (0/*T*0) 0% (*T*/00) and group A[1] with only 8% (0/*T*0) but 23% (*T*/00). It may be added that one animal in *ID* in which both thumbs and both little toes were present and well developed, but both big toes were absent, also had a rudimentary sixth toe on its right forefoot.

In the homozygotes *PxPx*, the broad paddle-shaped feet have large numbers of digits. As there was some branching of metacarpals and metatarsals, counts were made of the numbers of separate nails on each foot. The average numbers were about the same on all feet (8.75, 8.47, 8.54, and 8.47 on left and right forefeet and on left and right hindfeet, respectively). The percentages of feet with numbers from 5 to 12 are shown in Table 6.8 for the same groups of matings as for *Pxpx* in Table 6.7. These distributions are of approximately normal type. The modifiers of strain *D* tend to increase the number of digits in *PxPx* as well as in *Pxpx* and *pxpx*. The total distribution for feet is shown in Figure 6.8*A*. Table 6.9 shows the distribution of total

TABLE 6.7. Guinea pigs of same four groups as in Table 6.6, classified according to presence of one or both thumbs (above), one or both little toes (below left), or one or both big toes (below right). All of genotype *Pxpx*.

	$\frac{0}{00}$	$\frac{0}{T0}$	$\frac{T}{00}$	$\frac{T}{T0}$	$\frac{T}{TT}$	No.
I-13, Ro	17.5	44.4	0	38.1	0	63
A[1]	23.0	8.0	23.0	46.0	0	87
Misc.	18.5	18.5	11.3	41.9	9.7	124
ID	0	0	0	82.6	17.4	46

From data of Wright (1935*a*).

TABLE 6.8. Number of nails per foot in guinea pigs of genotype *PxPx* of the same four groups as in Tables 6.6 and 6.7.

	5	6	7	8	9	10	11	12	No. Feet
I-13, Ro			12	27	17	2	1		59
A[1]	1	5	14	30	14	5			69
Misc.		2	13	20	23	21	8		87
ID				3	8	10	5	3	29
Total	1	7	39	80	62	38	14	3	244

number of nails (range 28 to 44) in 48 animals in which counts could be made on all four feet.

The pectoral and pelvic fins of fishes are homologous to the limbs of higher vertebrates, but the pattern of rays is a meristic character of which the elements are much less individualized than the digits of higher vertebrates. The dorsal and anal fins are organs of the same general sort in fishes and show the typical approximately normal distribution (of number of fin rays) of a meristic character (*Menidia menidia notata*, Hubbs and Ramey 1946) (Fig. 6.8*D*).

The vertebral column is composed of somewhat individualized elements in mammals and birds and of relatively unindividualized ones in lower vertebrates. Green (1954) has studied the genetics of differences in the number of presacral vertebrae among several inbred strains of mice. The typical number of vertebrae in strain *P* is 25, but there are occasionally 26, as well as intermediates in which a vertebra is double on one side but not the other. Thus the distribution may be interpreted as indicating a threshold for any development of an extra vertebra and a second one for its full development, analogous to the distribution of extra toes in guinea pigs of strain 35. Strain *DBA*/26 shows a similar distribution except that more transgress the

TABLE 6.9. Total number of nails on all four feet of 48 guinea pigs of genotype *PxPx* on which all four feet were uneaten.

28	29	30	31	32	33	34	35	36	37	38	39	40	41	42	43	44	Total
3	2	4	3	7	6	2	5	6	2	1	2	2	2	–	–	1	48

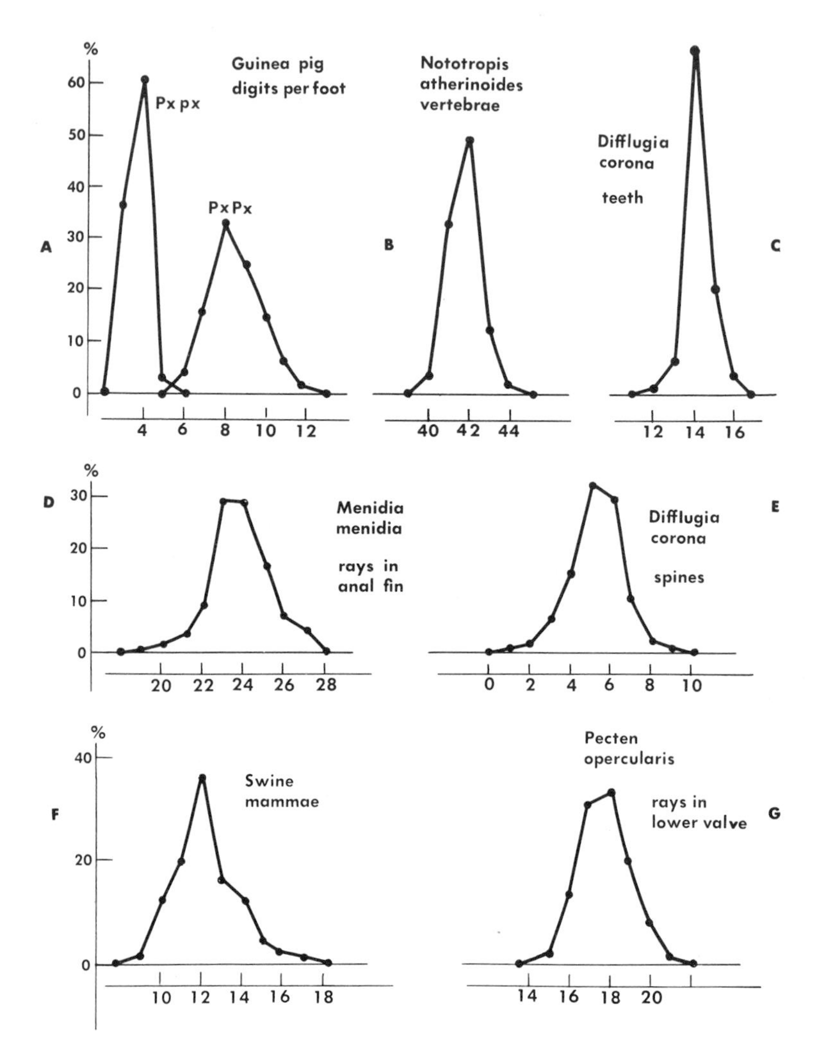

FIG. 6.8. Some meristic distributions (animals). Numbers: *A*, 542 *Pxpx*, 244 *PxPx* guinea pigs; *B*, 1,088 *Nototropis atherinoides*; *C*, 452 *Difflugia corona* (teeth); *D*, 245 *Menidia menidia*; *E*, 4,645 *Difflugia corona* (spines); *F*, 5,970 swine; *G*, 508 *Pecten opercularis*. (From data: *A*, Wright 1935*a*; *B*, Hubbs 1922; *C*, Jennings 1916; *D*, Hubbs and Ramey 1946; *E*, Jennings 1916; *F*, Parker and Bullard 1913.)

thresholds. In strain *C57Bl/10*, the norm has become 26, but small percentages show only 25 or the intermediate condition. In strain *NB* the normal is also 26 but in this case a very small percentage transgress the threshold for incomplete development of an additional vertebra. This trend goes much farther in *BALB/c* and *SEC/2*. In the latter, 27 has become on the whole the norm for number of vertebrae but a considerable percentage has only 26 or an intermediate, and a few only 25. The cross between the extreme strains (*P* with norm at 25 and *SEC/2* with norm at 27) gave F_1 with a norm of 26 and only a small percentage falling below, and F_2 with norm also at 26 but a little variability in both directions.

Figure 6.8*B* shows the variation in the number of vertebrae in the fish *Nototropes atherinoides*. Hubbs (1922) has shown that the number is subject both to genetic differences (between different local populations) and to temperature at a critical period in development. The distribution is again of the near normal type of many meristic characters. Other examples are given in Figure 6.8, which will be discussed later.

There is also, however, a large class of meristic characters that show distributions that are far from normal. Figure 6.9*A* shows the distribution of the number of tentaculocysts in a jelly fish *Ephyra* (Browne 1895). The number is eight in more than 78% of the individuals; yet there is a wide range of variability, 5 to 14. Evidently the developmental process is ordinary highly standardized or canalized but if regulation breaks down, there is no great barrier between one meristic step and the next in either direction.

De Vries (1910) has studied a similar situation in the number of ray florets in *Chrysanthemum segatum*. Figure 6.9*C* shows a count of 1,000 flowers made by Ludwig in Thuringia. More than half had 13 rays but there was a range from 7 to 21. This pattern was somewhat exaggerated in the progeny of selfed plants with 13 rays. Other strains of the same species, however, showed an equally strong standardization at 21 rays. Standardization was not so great, however, as to prevent rapid progress by selection.

In many cases, there is variation in only one direction from the "normal" number, the threshold pattern of variability already discussed. Thus in a tabulation of a wild population of *Ranunculus bulbosus* in the Netherlands, de Vries found that about 91% had 5 petals but the remainder ranged up to 14 (Fig. 6.9*D*). After two generations of selection, however, less than 10% had 5 petals and the mode had advanced to 9. There may also be unidirectional variability below the normal number. Among 1,145 flowers of *Weigelia amabilia*, he found that 77.5% had 5 petals, but 17.1% had 4 and 5.3% 3.

The individualized macrochaetae on the head and thorax of *Drosophila melanogaster* show distributions of the threshold type where there is any variability at all. MacDowell (1917) and Payne (1918*a*, *b*) found that the

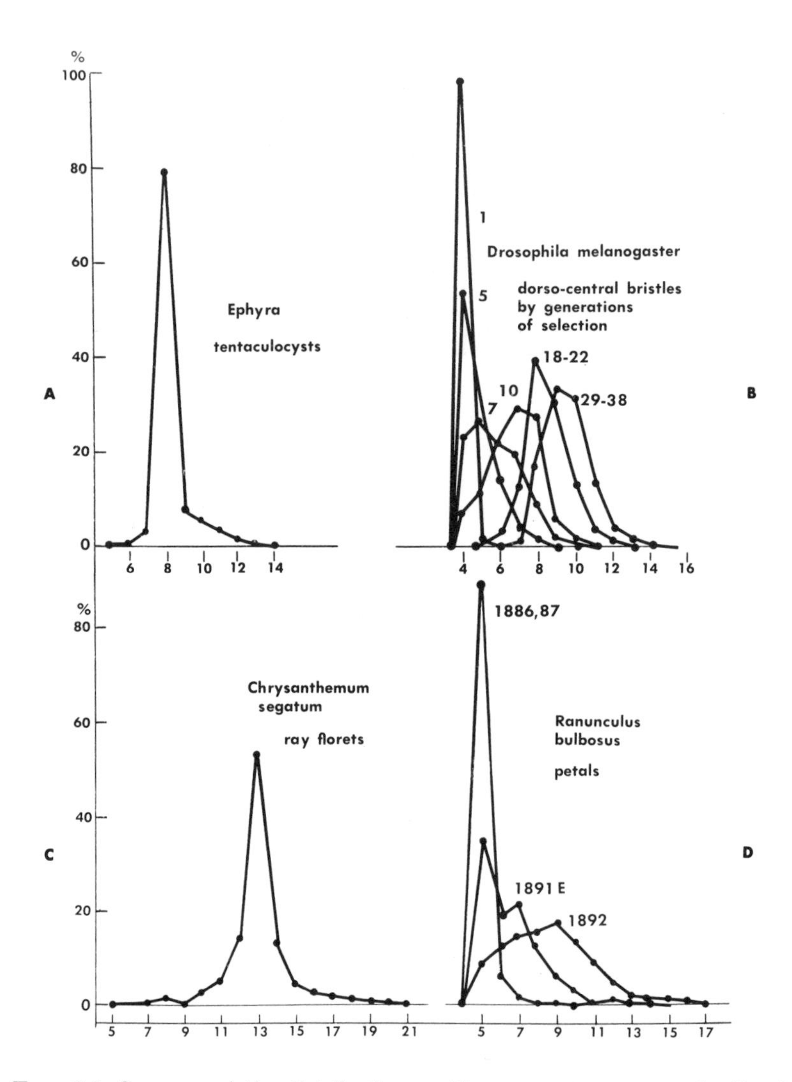

FIG. 6.9. Some meristic distributions with strong to no standardization of particular numbers. Numbers: *A*, 1,116 *Ephyra; B*, 120 (1), 1,663 (5), 986 (7), 1,544 (10), 2,860 (18–22), and 3,138 (29–38) *Drosophila melanogaster*; *C*, 1,000 *Chrysanthemum segatum*; *D*, *ca.* 700 (1886–87), 128 (1891 *E*), 4,405 (1892) *Ranunculus bulbosus*. (*A*, Browne 1895; *B*, Payne 1918*b*; *C*, *D*, *E*, de Vries 1910.)

occurrence of sporadic dorsocentral bristles in excess of the normal four gave an effective basis for selection. As selection proceeds the distribution passes from the threshold type to near normal (Fig. 6.9*B*). Payne started from 1 in 613 with an extra bristle and selected for 38 generations.

The microchaetae on the ventral abdominal segments have also been the objects of selection experiments, which will be discussed later. These unindividualized bristles show near normal distributions in wild populations.

The number of mammae on swine, studied by Parker and Bullard (1913), show an interesting pattern of distribution intermediate between the threshold and near normal types (Fig. 6.8*F*). There is almost a threshold at ten (or better at five on a side, since there is a rather strong correlation between sides). Less than 0.1% had four on a side; however, the mode is not five but six on a side.

The number of scales on various parts of the body constitute favorable meristic characters for study of variability in reptiles and fishes. Figure 6.10*E* shows the remarkable multimodal distribution of number of scale rows of a subspecies (*annectans*) of the snake *Pituophis caterops* as given by Klauber (1946). The reason for the multiple modes is a very strong tendency toward right-left symmetry, much greater than for mammae in swine. The modes are at odd numbers because of the occurrence of a median row. Considering only odd numbers the distribution is near normal with mode 33. The number of scales on the fourth toe of the lizard *Cnemidophorus t. tessalatus* (Klauber 1941, Fig. 6.10*F*) varies from 17 to 24 with a mode at 20 that is so high it suggests some standardization.

In a very extensive study of the distributions of scale numbers in rattlesnakes, Klauber (1941) finds considerable standardization in those of certain regions, typical normal variability in others. Thus there appears to be a little more standardization of the supralabials (Fig. 6.10*C*) than of the infralabials (Fig. 6.10*D*), both on the basis of more than 4,800 individuals. The subcaudals (Fig. 6.10*B*) show little indication of standardization. We will consider later how such vague impressions can be tested (as they were by Klauber).

Figure 6.10*A* shows the percentage distributions of number of scales in a subspecies of shiner *Boleosoma nigrum*, as given by Hubbs and Ramey (1946). The number of scales is large and variability is great. The distribution seems to be of the same positive skewed type characteristic of quantitative variation with a large coefficient of variability.

Figure 6.5*B* shows the distribution of facet numbers in an unselected bar-eyed strain of *Drosophila melanogaster* (Zeleny 1922). Again, we are dealing with large numbers of unindividualized elements and the forms of the distributions are of the usual positively skewed type of quantitative variability.

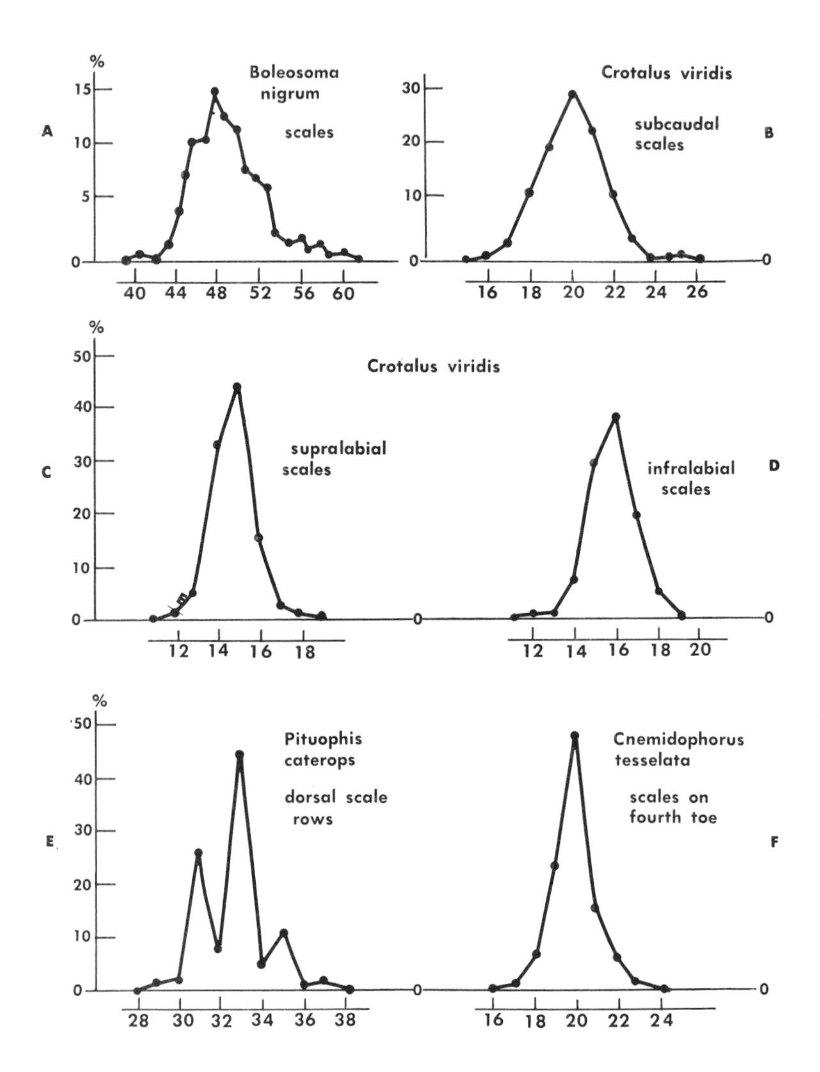

FIG. 6.10. Some meristic distributions. Numbers: *A*, 397 *Boleosoma nigrum*; *B*, 331 *Crotalus viridis* (subcaudals); *C*, 4,887 *Crotalus viridis* (supralabials); *D*, 4,827 *Crotalus viridis* (infralabials); *E* 533 *Pituophis caterops*; *F*, 1,424 *Cnemidophorus tessilata*. (From data: *A*, Hubbs and Ramey 1946; *B, C, D, F*, Klauber 1941; *E*, Klauber 1946.)

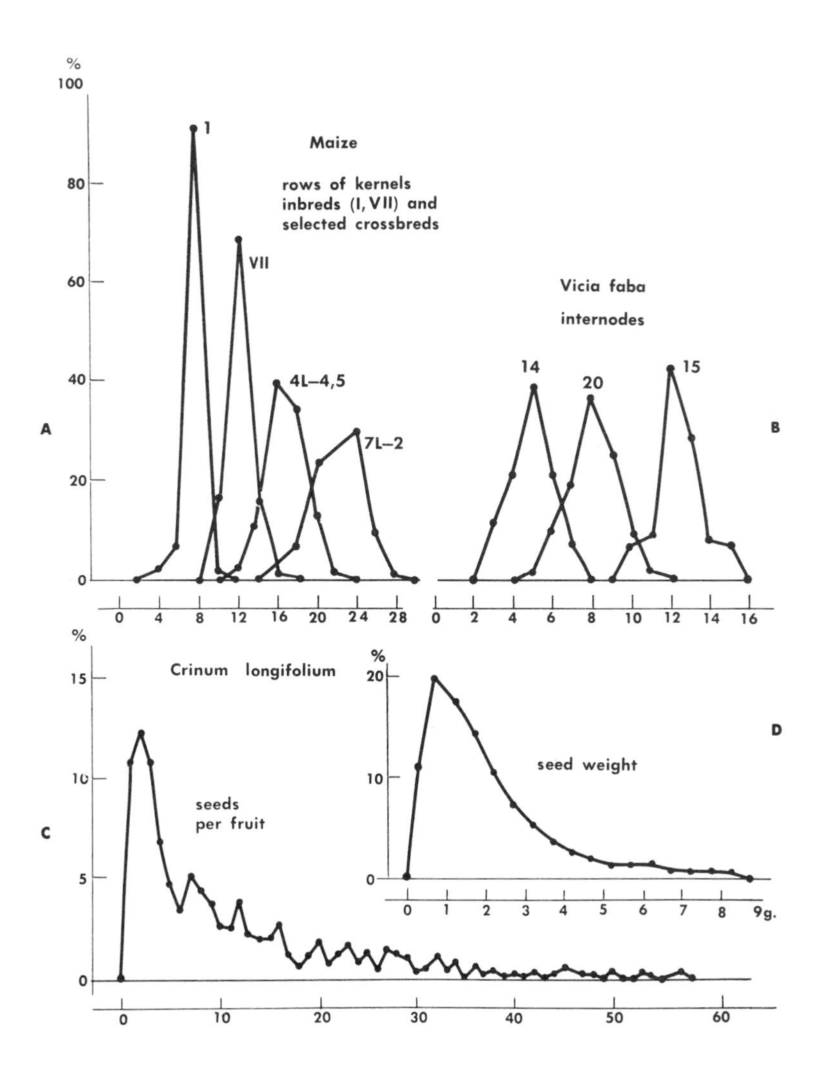

FIG. 6.11. Some meristic distributions and a related continuous distribution (plants). Numbers: *A*, 408 (I), 905 (VII), 241 (4L-4.5) 257 (7L-2) maize; *B*, 323 (14), 248 (20), 243 (15) *Vicia faba*; *C*, 1,000 (seeds per fruit); *D*, 2,000 (seed weight) *Crinum longifolium*. (From data: *A*, Emerson and Smith 1950; *B*, Sirks 1932; *C*, *D*, Harris 1912.)

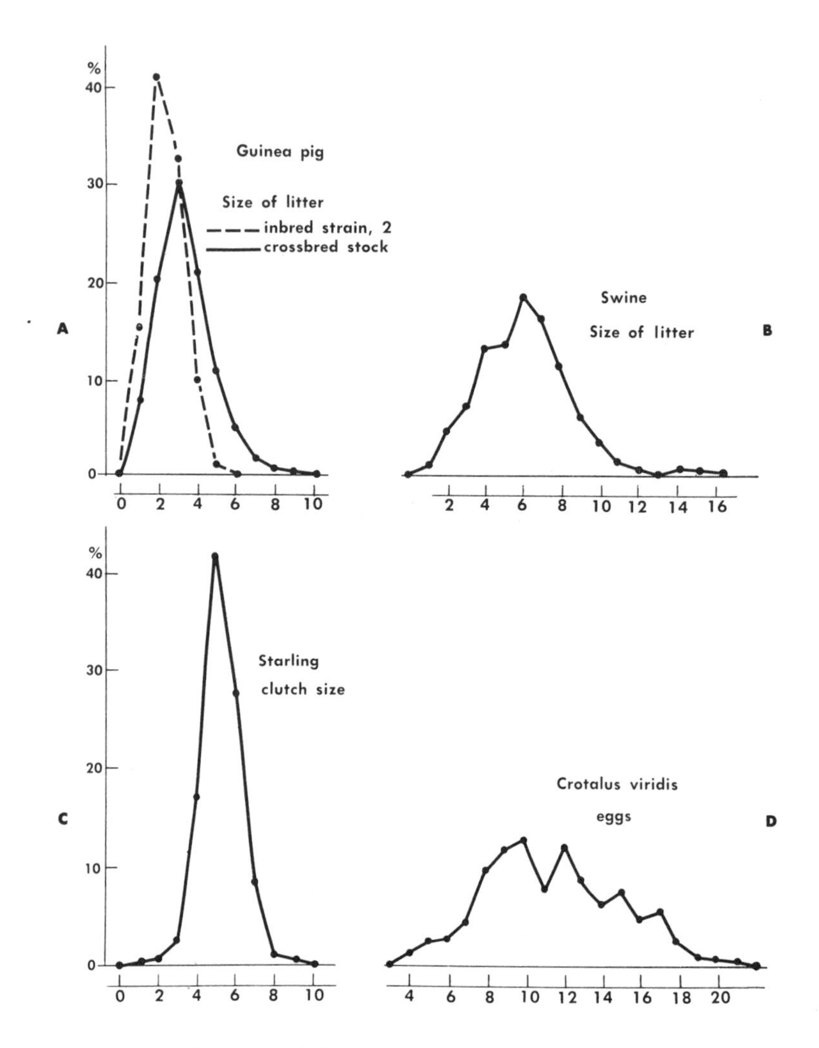

FIG. 6.12. Distributions of litter or clutch sizes. Numbers: *A*, 1,067 (2), 1,277 crossbred guinea pig (about 2% truncation of both below 1); *B*, 1,000 swine; *C*, 1,592 starling; *D*, 294 *Crotalus viridis*. (From data: *A*, Wright and Eaton 1929; *B*, Parker and Bullard 1913; *C*, Lack 1948; *D*, Klauber 1936.)

The shells of *Difflugia corona* usually have irregular spines, formed by pseudopodia in the fission product that does not retain the old shell. They are not individualized structures, and the clone illustrated in Figure 6.8*E* (data of Jennings 1916) shows a fairly normal distribution. Figure 6.8*C* shows the distribution in one of the clones of the number of "teeth" formed at fission about the aperture.

The distribution of number of rays on the lower valves of a population of scallops, *Pecten opercularis* (Fig. 6.8*G*), reported by Davenport (1904) illustrate near normal variability in another animal phylum.

There are some peculiarities in the distribution of number of rows on ears of maize. The double row is really the unit. Figure 6.11*A* presents a number of distributions drawn from the very extensive studies of Emerson and Smith (1950). These include the largest of the 8-rowed, and 12-rowed inbred lines, two strains from plus selection on the basis of four of the 12-rowed strains, and a strain from plus selection on the basis of seven 12-rowed strains. The amount of variability increases with the mean but there is little or no skewness, although some curtailment of variability is apparent at the lower ends of the distributions.

Figure 6.11*B* shows the distributions of number of internodes in three lines of *Vicia faba* as reported by Sirks (1932). In two of them the distribution is almost triangular. Figures 6.11*C* and 6.11*D* show extraordinary positive skewness of seeds per fruit and seed weight in *Crinum longifolium* (Harris 1912).

Finally, we give a number of distributions of size of brood. Figure 6.12*A* shows the distribution of size of litter in an inbred and in a crossbred strain of guinea pigs. The last shows some positive skewness, the other seems more nearly symmetrical and approaches normality, except that there is necessarily interruption of the lower tail below one. The underlying factor complexes may well vary continuously, but their effects cannot be recorded at the lower end of the range. The situation is somewhat like that in such artificially truncated distributions as those of mile records of trotting horses and butterfat production by Jersey cows. The distributions of size of litter in swine (Parker and Bullard 1913; Fig. 6.12*B*) of clutch size in starlings (Lack 1948; Fig. 6.12*C*), and of number of eggs in rattlesnakes *Crotalus viridis* (Klauber 1936, 1941; Fig. 6.12*D*), are all of the near normal type, with slight suggestions of deviations.

CHAPTER 7

Statistical Description of Frequency Distributions

Frequencies as Probabilities

It is often convenient to represent a frequency array as the sum of terms, each consisting of a symbol designating the class (A_i) associated with the proportional frequency represented by f_{A_i}, or merely f_i, if not ambiguous.

$$(7.1) \qquad [f_1A_1 + f_2A_2 + \cdots + f_KA_K] = \sum_{i=1} (f_iA_i), \qquad \sum_{i=1}^{K} f_i = 1.$$

The frequencies may be considered approximations of probabilities (P_i) pertaining to classes in a hypothetical infinite population.

Similarly, the joint frequency distribution of two variables may be represented by $\sum_j \sum_i [f_{A_iB_j}(A_iB_j)]$ or merely $\sum \sum [f_{ij}A_iB_j]$ in which (A_iB_j) is the designation of the joint class and the frequency is $f_{A_iB_j}$ or merely f_{ij} if it is understood that the variables are to be taken in the order indicated in the class symbol. The frequencies may again be considered approximations of probabilities.

The probability of occurrence of a certain class (A_i) of one variable in association with a specified class (B_j) of another is a conditional probability represented by:

$$(7.2) \qquad P_{A_i|B_j} = P_{A_iB_j}/P_{B_j}.$$

Similarly

$$P_{B_j|A_i} = P_{A_iB_j}/P_{A_i}.$$

Thus the joint probability may be written in terms of conditional probabilities as

$$(7.3) \qquad P_{A_iB_j} = P_{B_j}P_{A_i|B_j} = P_{A_i}P_{B_j|A_i}.$$

Variables are independent, or combined at random, if all the conditional probabilities are the same as the total probability ($P_{A_i|B_j} = P_{A_i}$ for all i and j) (which implies that $P_{B_j|A_i} = P_{B_j}$ for all i and j).

$$(7.4) \qquad P_{A_iB_j} = P_{A_i}P_{B_j} \quad \text{if } A \text{ and } B \text{ are independent.}$$

The array of joint probabilities can be written as the product of the separate arrays, if there is independence, noting that a "product" of class terms means merely association.

$$\sum_j \sum_i (P_{A_i B_j} A_i B_j) = \sum_i (P_{A_i} A_i) \sum_j (P_{B_j} B_j) \tag{7.5}$$

if A and B are independent. This product principle can obviously be extended to any number of independent variables.

In the great majority of cases, probabilities can only be estimated from observed frequencies, as indicated above. There are certain cases, however, in which estimates can be made before any frequencies have actually been observed. These have to do with alternative elementary events that may confidently be assumed to be equally probable. The probability of a compound event is the proportion of all of the equally probable elementary events that are included in it. Historically, the theory of probability was based largely on experiments with well-shaken coins or dice or well-shuffled cards, in all of which the alternatives are geometrically and materially so much alike, barring fraud, that any residual differences, including the distinguishing marks, may be judged to play no role in the sort of process by which the event is decided. The greater the knowledge (including possibilities of fraud) on which a judgment of equal probabilities of elementary events is based, the more reliable the a priori probability. If actual frequencies deviate significantly from a priori expectation based on dice of which there is only rather superficial knowledge, it may be that careful measurement of faces, determination of center of gravity, etc., may suggest a reason. An a priori probability, supplemented by a specific explanation of deviations from expectation, gives a deeper understanding than a mere empirical frequency distribution by itself. Moreover, accurate measurements may give a basis for a revised a priori probability that can then be subjected to more extensive empirical tests. The scientific process consists to a large extent of an indefinitely extended alternation of a priori inferences and empirical tests.

Population genetics rests in part on phenomena that yield a priori probabilities approximating in reliability those cited above. Experiments with many kinds of organisms indicate that the distinction between alleles ordinarily has no differential effect on the segregation of homologous chromosomes in the reduction division, even where this is asymmetrical as in oogenesis. Moreover, the maternal or paternal origin of nonhomologous chromosomes ordinarily has no detectable effect on their assortment. There are exceptions to these principles, but they hold generally enough for the development of a first-order theory, even though this may need to be supplemented in some cases. The assortment of genes in the same chromosome can only be determined empirically but even in this case, the principle of the linear order of the genes and the incompletely understood principles of interference give some basis for deductions. Union of gametes ordinarily seems to be at random with respect to segregating alleles. There is more probability of disturbance of expected ratios when it comes to segregating individuals, but a hypothesis of differential viability at one stage or another can be tested. Population genetics is, in short, a field especially well suited for the alternation of a priori deduction and empirical testing.

Accordingly, we shall have frequent occasion for representation of arrays of alleles in the gametes of heterozygotes by such an expression as $[\frac{1}{2}A + \frac{1}{2}a]$ and for such expression $[\frac{1}{2}A + \frac{1}{2}a]^2$ for the array of F_2 zygotes with respect to one locus, and $[\frac{1}{2}A + \frac{1}{2}a]^2[\frac{1}{2}B + \frac{1}{2}b]^2$ with respect to two loci in different chromosomes.

The reliability of these a priori probabilities makes it possible to use effectively the principle of conditional probability in estimating the probability of causes (Bayes' rule). We note that P_{B_j} is equal to $\sum_i P_{A_i}P_{B_j|A_i}$ for all i's and that, therefore,

$$P_{A_i|B_j} = \frac{P_{A_i}P_{B_j|A_i}}{P_{B_j}} = \frac{(P_{A_i}P_{B_j|A_i})}{\sum_i (P_{A_i}P_{B_j|A_i})} \cdot \tag{7.6}$$

As an example, consider the case of dominant individuals, D–, produced by $Dd \times Dd$ and tested by mating to a recessive ($D– \times dd$). The initial array of probabilities for dominant individuals is $[\frac{1}{3}DD + \frac{2}{3}Dd]$. If a recessive appears this immediately shifts to certainty of Dd. If, however, there is a succession of dominant offspring and no recessives, the probability of DD gradually rises without ever quite reaching certainty. Let P_{m_1} be the probability that a test mating is $DD \times dd$ and P_{m_2} the probability that it is $Dd \times dd$. Let P_n be the probability of producing n dominants and no recessives.

m_i	Mating type	A priori yield P_{m_i}	$P_{n\|m_i}$	Product $P_{m_i}P_{n\|m_i}$	A posteriori probability $P_{m_i\|n}$
$m_1 =$	$DD \times dd$	$\frac{1}{3}$	1	$\frac{1}{3}$	$2^{n-1}/(2^{n-1} + 1)$
$m_2 =$	$Dd \times dd$	$\frac{2}{3}$	$(\frac{1}{2})^n$	$\frac{2}{3}(\frac{1}{2})^n$	$1/(2^{n-1} + 1)$
		1		$P_n = \frac{1}{3}\left[\frac{2^{n-1}+1}{2^{n-1}}\right]$	1

Thus after production of one dominant the probability that an F_2 dominant is homozygous rises from $\frac{1}{3}$ to $\frac{1}{2}$, after two dominants to $\frac{2}{3}$, after three to $\frac{4}{5}$, after four to $\frac{8}{9}$, etc.

Extraction of Statistics from Frequency Distributions

Mode and Limits

The simplest set of numbers for describing a frequency distribution of many classes consists of specifying the upper and lower limits of variability, and the positions of maximum frequency (the mode or modes) and of intervening minima, if multiple modes. In the usual unimodal case, the position of the mode gives some idea of location on the scale, the range between the limits indicates the amount of dispersion, and the position of the mode relative to the limits indicate the degree of asymmetry.

It is obvious, however, that these statistics suffice only for a rough preliminary description. The positions of the extreme variants of a tapering distribution may be expected to vary greatly among samples of the same size, but, in general, to spread more and more as sample size is increased. The position of the mode does not discriminate well among distributions with broad classes and is likely to vary greatly from sample to sample of the same distribution with narrow classes. If a smooth curve has been fitted to a distribution by a more reliable set of statistics, however, the location of such features as the maxima and minima and points of inflection may be very useful.

Description by Percentiles

A second system of numbers for describing frequency distributions is given ideally by the positions on the scale at which the frequencies in an indefinitely large sample are divided into certain specified proportions. These are termed percentiles or quartiles and will be symbolized here by Q with a subscript that indicates the percentage of the total frequency below.

The median, Med or Q_{50}, gives a measure of location. The average quartile deviation $QD = (1/2)(Q_{75} - Q_{25})$ gives a measure of dispersion. Asymmetry can be indicated by the ratio of the difference between the two quartile deviations from the median to the average quartile deviation,

$$2(Q_{75} - 2Q_{50} + Q_{25})/(Q_{75} - Q_{25}).$$

Other form statistics can be based on ratios involving other percentiles but are little used.

There are certain complications in deriving these from actual distributions. In ungrouped individual measurements, the median may be taken as the measurement of the middle individual if there is an odd number, and as somewhere in the interval between the two middle individuals if an even number. In the latter case, this usually defines it sufficiently if the numbers are reasonably large. Otherwise it may be taken arbitrarily as being halfway between these individuals. Similarly, with N individuals the quartiles may be chosen arbitrarily within the intervals in which the $(N/4)^{th}$ and $(3N/4)^{th}$ fall among the measurements in order of size.

In an array of class frequencies some form of interpolation within classes is necessary. Mere linear interpolation is usually all that is warranted in view of the random irregularities in the class frequencies, unless the total number is very large or the number of classes small. Otherwise, the running sum of the percentage frequencies up to each class limit may be plotted against the same limit, and the points fitted by a smooth curve. The abscissas at which the ordinates are 25%, 50%, and 75% of the total define Q_{25}, Q_{50}, and Q_{75}, respectively. Figure 7.1 shows such a graph for the statures of man

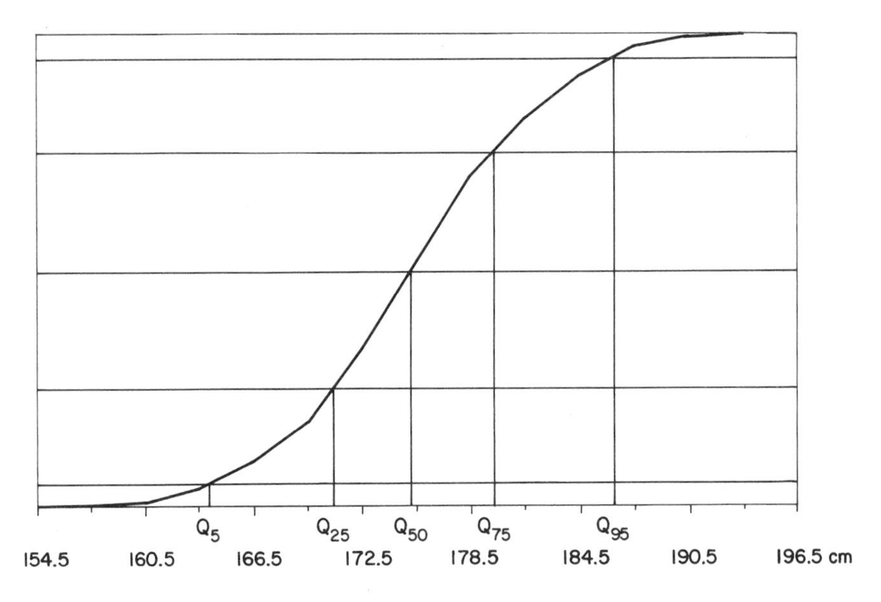

FIG. 7.1. Quartiles (Q_5, Q_{25}, Q_{50}, Q_{75}, and Q_{95}) in distribution of heights of 1,000 male students shown in Figure 6.2A.

of Table 6.1, Figure 6.2*A*, and shows the position of Q_5, Q_{25}, Q_{50}, Q_{75}, and Q_{95}. Figure 7.2 shows similar graphs for the amount of white in the coats of male guinea pigs of strains *DS*, 35, and 2 (Fig. 6.5*C*), and the positions of Q_{25}, Q_{50}, and Q_{75}. Percentiles are of little interest in meristic variability unless frequencies are treated as if they pertain to class ranges of an underlying continuous distribution of factor complexes.

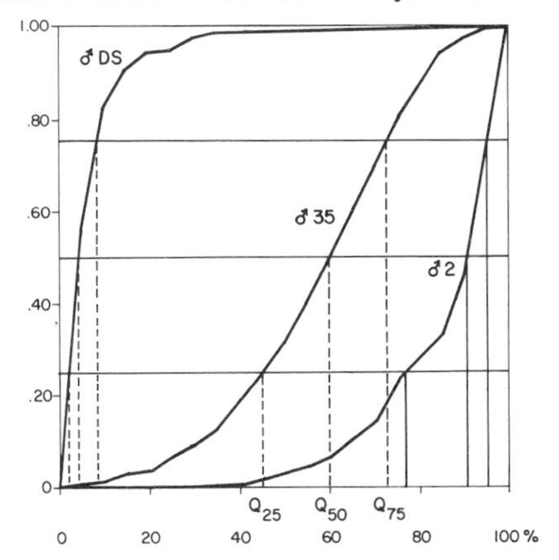

FIG. 7.2. Median Q_{50} and quartiles (Q_{25}, Q_{75}) in distributions of male guinea pigs of three strains, shown in Figure 6.5C.

Description by Moments

The Mean

The most frequently used statistic of location is the arithmetic average or mean. It is customary to represent this either by the symbol M with appropriate subscript or by putting a bar over the quantity averaged. Thus if V_i represents a single observed value on the scale, and N is the number of observation, then

$$M_V = \bar{V} = \frac{1}{N}\sum_{i}^{N} V_i. \tag{7.7}$$

If the observations are grouped into K classes about each midclass index, I_i, and the corresponding absolute frequencies are represented by f_i, the mean class index is

$$\bar{I} = \frac{1}{N}\sum^{K} I_i f_i, \qquad \sum f_i = \mathrm{N}. \tag{7.8}$$

I is identical with V if the distribution is meristic (and ungrouped), but if continuous, $\bar{I}$ in general differs from $\bar{V}$, and often considerably if the number of classes is small. The error in adopting $\bar{I}$ for $\bar{V}$ is unsystematic if the distribution is symmetrical but if not, there is a systematic error which should be avoided by calculation from ungrouped or at least finely grouped values.

If the class indexes have a common divisor (D), it may simplify the calculation to divide by this before averaging and then multiply by it afterward. A multiplier can always be shifted across a sum sign without changing the value of the expression.

$$(7.9) \qquad \bar{I} = \frac{1}{N}\sum^{K} If = \frac{D}{N}\sum^{K} I'f, \qquad I' = \frac{I}{D}.$$

It is also often convenient to subtract the same number, C, from all observations before averaging and add afterwards. Any expression after a sum symbol that involves plus or minus signs can be broken up into components at these signs.

$$(7.10) \qquad \bar{V} = \frac{1}{N}\sum^{N} V = \frac{1}{N}\sum^{N}(V' + C) = \bar{V}' + C, \qquad V' = V - C.$$

The Average Deviation

At first, the statistic of dispersion most logically associated with the mean would seem to be the average deviation from the mean, disregarding sign. If, however, an average absolute deviation is to be used, it is desirable that the deviations be taken from the value that makes it minimum. It can be easily shown that the sum of absolute deviations is least from the middle one of an odd number of observations or from anywhere between the two middle ones of an even number. Thus the average deviation is more properly associated with the median than with the mean.

The Standard Deviation

The statistic most frequently used for describing dispersion is the standard deviation (σ_V), defined as the square root of the mean squared deviation from the mean. The mean squared deviation itself has such important properties that it is convenient to give it a simple designation, the variance. The formula for the standard deviation is:

$$(7.11) \qquad \sigma_V = \sqrt{\frac{1}{N}\sum^{N}(V - \bar{V})^2}$$

In the case of grouped continuous variables the variance of midclass indexes is given by $\sigma_I^2 = (1/N)\sum^K (I - \bar{I})^2 f$. This may differ considerably from the variance of single observations, if the number of classes is small; and does so systematically even in symmetrical distributions of the usual bell-shaped form. Sheppard (1898) showed that the variance of a distribution that tapers off at both extremes can be calculated to a close approximation by subtracting one-twelfth of the squared class range (l) from σ_I^2.

$$(7.12) \qquad \sigma_V^2 \approx \frac{1}{N}\sum^{K}(I - \bar{I})^2 f - \frac{l^2}{12}.$$

Sheppard's correction must not be used with continuous distributions that do not taper off in both directions. Thus for a rectangular distribution, $y = 1/r$, continuous in the range $0 \leq x \leq r$, the variance is

$$\frac{1}{r}\int_0^r \left(x - \frac{r}{2}\right)^2 dx = \frac{r^2}{12}.$$

If grouped into K classes, each of class range l and frequency $1/K$, the variance of midclass indexes must be supplemented by that within the classes, $l^2/12$, since if $V = I + d$, d being the deviation from the midclass index, we have

$$\text{(7.13)}\quad \begin{aligned} \sigma_V^2 &= \frac{1}{N}\sum^N [I + d - \bar{I}]^2 = \frac{1}{N}\left[\sum (I - \bar{I})^2 + 2d\sum (I - \bar{I}) + \sum d^2\right] \\ &= \sigma_I^2 + \frac{l^2}{12} \quad \text{since } \sum (I - \bar{I}) = 0. \end{aligned}$$

Thus for a grouped rectangular distribution the same correction must be added to σ_I^2 that is subtracted for a grouped distribution that tapers off at both extremes.

The labor of calculation can be reduced by expanding the term $(V - \bar{V})^2$, or $(I - \bar{I})^2 f$, and making use of the properties of the sum symbol discussed earlier.

$$\text{(7.14)}\quad \begin{aligned} \sigma_V^2 &= \frac{1}{N}\sum^N (V - \bar{V})^2 = \frac{1}{N}\left[\sum^N V^2 - 2\sum^N \bar{V}V + \sum^N \bar{V}^2\right] \\ &= \frac{1}{N}\left[\sum^N V^2 - \bar{V}\sum^N V\right]. \end{aligned}$$

Again there may be some simplification by subtracting the same number, C, from all values and by dividing by a common divisor, D, before averaging. In this case there is no later correction for the subtraction since the *difference* $(V - \bar{V})$ is unaffected, but the division must be compensated for by multiplication of the average by D^2 in calculating σ_V^2 or by multiplication of $\sigma(V/D)$ by D.

The mean squared deviation is minimum from the mean since from any other point, $\bar{V} + d$, it is greater by d^2.

$$\text{(7.15)}\quad \begin{aligned} \frac{1}{N}\sum^N (V - \bar{V} - d)^2 &= \frac{1}{N}\left[\sum^N (V - \bar{V})^2 - 2d\sum^N (V - \bar{V}) + \sum^N d^2\right] \\ &= \sigma_V^2 + d^2. \end{aligned}$$

It must be supposed that the mean of a sample deviates, in general, in one direction or the other from the mean of the total population, and thus that the variance as estimated from the sample should be larger, on the average, by the variance of the means of samples, a quantity shown later to have the value $\sigma_{\bar{V}}^2 = \sigma_V^2/N$. An unbiased estimate of the variance $\sigma^2_{V(T)}$ of the total population from that of a sample $[\sigma^2_{V(S)}]$ is given by

$$\sigma^2_{V(T)} = \sigma^2_{V(S)} + \sigma^2_{V(T)}/N = \frac{N}{N-1}\sigma^2_{V(S)}. \tag{7.16}$$

The most convenient formula for estimating the standard deviation of the total population from the observations in a sample is thus:

$$\sigma_{V(T)} = \sqrt{\frac{1}{N-1}\left(\sum^{N} V^2 - \bar{V}\sum^{N} V\right)}. \tag{7.17}$$

There are certain other statistics of location and dispersion that are important. The geometric mean is the Nth root of the product of the N observed values. Its logarithm is merely an arithmetic mean on a logarithmic scale.

$$\log G_V = \frac{1}{N}\sum^{N} \log V. \tag{7.18}$$

The harmonic mean is the reciprocal of the arithmetic mean of the reciprocals of the observed values.

$$H_V = \frac{1}{\left[\frac{1}{N}\sum^{N}\frac{1}{V}\right]}. \tag{7.19}$$

The coefficient of variability measures relative instead of absolute variability. It is essentially the ratio of the standard deviation to the mean. Although ordinarily expressed as a percentage and used as a descriptive statistic, it is more convenient in theoretic use to treat it merely as a ratio.

$$C_V = \frac{\sigma}{\bar{V}}. \tag{7.20}$$

A system of descriptive statistics in which the mean and standard deviation have a natural place was based by Pearson (1894) on averages of higher powers of the deviations from the mean than the second power. The sth moment about an arbitrary value V_0 is defined by the expression $\mu'_s = (1/N)\sum(V_i - V_0)^s$. The mean is thus the first moment about the origin, and the standard deviation is the square root of the second moment about the mean. In a continuous distribution, represented by a smooth curve $y = f(x)$ and unit area $\int_{-\infty}^{+\infty} y dx = 1$, the sth moment about the mean is

$$\mu_s = \int_{-\infty}^{+\infty} (x - \bar{x})^s \, y dx. \tag{7.21}$$

In actual frequency distributions this is approximated by

$$\mu_s = \frac{1}{N} \sum (V - \bar{V})^s. \tag{7.22}$$

It is usually most convenient to use formulas in which this has been expanded and broken into components:

$$\mu_2 = \frac{1}{N} \left[\sum V^2 - \bar{V} \sum V \right]; \tag{7.23}$$

$$\mu_3 = \frac{1}{N} \left[\sum V^3 - 3\bar{V} \sum V^2 + 2\bar{V}^2 \sum V \right]; \tag{7.24}$$

$$\mu_4 = \frac{1}{N} \left[\sum V^4 - 4\bar{V} \sum V^3 + 6\bar{V}^2 \sum V^2 - 3\bar{V}^3 \sum V \right]. \tag{7.25}$$

To obtain abstract coefficients to describe form independently of the size of the scale unit, the moments are taken in standardized form,

$$\frac{1}{N} \sum \frac{(V - \bar{V})^s}{\sigma^s} = \frac{\mu_s}{\sigma^s}.$$

Since the values for samples become increasingly unreliable as s increases, only those in which $s = 3$ and $s = 4$ are ordinarily used. The former, μ_3/σ^3, is an index that is obviously zero if the distribution is symmetrical, positive with positive skewness, and negative with negative skewness. It is also obvious that among symmetrical distributions of a given standard deviation, ones in which there is a wide spread, compensated for by a narrow peak, have a larger value of μ_4/σ^4 than ones with relatively little spread but a broad peak. This statistic (β_2) may thus be used as an index of the aspect of form that Pearson called kurtosis. He used $\beta_1 = (\mu_3/\sigma^3)^2$ for his first form statistic. We shall use Fisher's (1925) symbols

$$\gamma_1 = \frac{\mu_3}{\sigma^3} = \pm\sqrt{\beta_1} \quad \text{and} \quad \gamma_2 = \frac{\mu_4 - 3\mu_2^2}{\mu_2^2} = \beta_2 - 3, \tag{7.26}$$

which seem preferable—the former because it permits the distinction between positive and negative skewness and is simpler than β_1, the latter for a reason that will be apparent later.

The calculation of the higher moments from samples are, as in the case of the variance, biased by being calculated from the deviations from the sample mean instead of from the mean of the total population. Thus the best estimate of $\mu_3(T)$ for total population is

$$\frac{N^2}{(N-1)(N-2)}\mu_3 \quad \text{(Thiele 1889)} \tag{7.27}$$

or

$$\frac{N}{N-3}\mu_3$$

to a first order. There are similar estimates for μ_4 and $\mu_4 - 3\mu_2{}^2$ for the total population, but these statistics are highly unreliable unless N of the sample is very large, in which case the corrections are of no practical importance. Moreover, the use of such estimates in the derivation of other formulas often leads to contradictions (Steffensen 1930).

As in the case of the variance, Sheppard has given a correction to be applied to estimates of μ_4 if the calculation has been based on midclass indexes of a continuous distribution that tapers off at the extremes. With class range l:

$$\mu_4(V) \approx \mu_4(I) - \frac{l^2}{2}\mu_2(I) + \frac{7}{240}l^4. \tag{7.28}$$

There is no correction for $\mu_3(V)$.

□ □

CHAPTER 8

Compound Distributions and Variables

Compound Distributions

The interpretation of patterns of variability depends on understanding the properties of compound distributions and compound variables. A *compound distribution* is merely the total of a number of elementary distributions in terms of absolute frequencies. It is the weighted average in terms of proportional frequencies. Using the means M_i and variances σ_i^2 of the component populations and taking the proportion of the total number in each component as its weight, W_i, the mean (M_T) and variance, σ_T^2, of the total may easily be seen to be as follows.

(8.1) $$M_T = \sum W_i M_i.$$

(8.2) $$\sigma_T^2 = \sum W_i(\sigma_i^2 + M_i^2) - M_T^2.$$

Compound Variability

Consider next the properties of a variable, V_T, that is a linear function of other variables, V_i. The C's are constant coefficients.

(8.3) $$V_T = C_T + C_{T_1} V_1 + C_{T_2} V_2 + \cdots + C_{T_K} V_K.$$

(8.4) $$\bar{V}_T = C_T + C_{T_1} \bar{V}_1 + C_{T_2} \bar{V}_2 + \cdots + C_{T_K} \bar{V}_K.$$

(8.5) $$(V_T - \bar{V}_T) = C_{T_1}(V_1 - \bar{V}_1) + C_{T_2}(V_2 - \bar{V}_2) + \cdots + C_{T_K}(V_K - \bar{V}_K).$$

It will be convenient to take the indicated multiples of the variables as the ones to be considered, $V_i' = C_{T_i} V_i$ and to represent the deviations from the means by d's, $d_T = \sum d_i$.

If $d_T = d_1 + d_2$, then dropping primes

(8.6) $$\mu_2(V_T) = \frac{1}{N}(d_1 + d_2)^2 = \frac{1}{N}\left[\sum d_1^2 + 2\sum d_1 d_2 + \sum d_2^2\right]$$
$$= \mu_2(V_1) + 2\mu_{11}(V_1, V_2) + \mu_2(V_2).$$

The average product of the deviation of two variables from their means is known as their product moment, symbolized as above by μ_{11} (Pearson 1895), or as their covariance, symbolized by cov (Fisher 1931). Similarly the variance may be symbolized by var or often merely by V, which will not be used here, however, because of the use of V for a variable.

(8.7) $$\text{var}\,(V_T) = \text{var}\,(V_1) + 2\,\text{cov}\,(V_1 V_2) + \text{var}\,(V_2).$$

The above product moment may be written $(1/N) \sum^{N_1} [d_1 \sum^{N_{2\cdot1}} d_{2\cdot1}]$, or $(1/N) \sum^{N_1} [d_1 N_{2\cdot1} d_{2\cdot1}]$, in which $d_{2\cdot1}$ represents a value of d_2 for a given d_1, $N_{2\cdot1}$ is the number of entries in such an array, and N_1 is the number of different classes of values of d_1 ($N = \sum^{N_1} N_{2\cdot1}$). If d_2 is independent of d_1, its distribution is the same, except for accidents of sampling, for all values of d_1 and hence the sum, $\sum d_{2\cdot1}$ differs from zero only by accidents of sampling. Thus the product moment does not differ significantly from zero if there is independence. In terms of squared standard deviations

(8.8) $$\sigma_T{}^2 = \sigma_1{}^2 + \sigma_2{}^2 \quad \text{if } V_1 \text{ and } V_2 \text{ vary independently}.$$

Two examples of this important formula have already been given: $\sigma_V{}^2 = \sigma_I{}^2 + (l^2/12)$ for a rectangular distribution in which $V = I + d$, and $\sigma^2_{V(T)} = \sigma^2_{V(S)} + \sigma^2_{\bar{V}(S)}$, the unbiased estimate of the variance of a total population from that of a sample, on the basis of analysis of the total deviation into two necessarily independent components, $(V - \bar{V}_T) = (V - \bar{V}_S) + (\bar{V}_S - \bar{V}_T)$.

The formula for the variance of a sum may obviously be extended to any number (K) of components:

(8.9) $$\mu_2(V_T) = \sum^K \mu_2(V_i) + 2 \sum \mu_{11}(V_i V_j), \qquad i < j;$$

(8.10) $$\mu_2(V_T) = \sum^K \mu_2(V_i) \quad \text{if all components are independent,}$$

or

$$\sigma_T{}^2 = \sum \sigma_i{}^2 \quad \text{in the last case.}$$

An analogous formula can be obtained for the third moments of a compound variable. If $d_T = d_1 + d_2$, then

(8.11) $$\frac{1}{N} \sum (d_T)^3 = \frac{1}{N} \left[\sum d_1{}^3 + 3 \sum d_1{}^2 d_2 + 3 \sum d_1 d_2{}^2 + \sum d_2{}^3\right];$$

(8.12) $$\mu_3(\bar{V}_T) = \mu_{30} + 3\mu_{21} + 3\mu_{12} + \mu_{03},$$

using Pearson's symbol μ_{st} for $(1/N) \sum d_1{}^s d_2{}^t$.

It is evident that $\sum d_1{}^2 d_2$ disappears if the mean values of d_2 for all d_1's are the same and similarly that $\sum d_1 d_2^2$ disappears if the mean values of d_1 for all d_2's are the same. Thus if d_1 and d_2 vary independently,

$$\mu_3(V_T) = \mu_3(V_1) + \mu_3(V_2). \tag{8.13}$$

This can be extended step by step to the sum of any number of independent variables: $V_{T_2} = V_{T_1} + V_3$, $V_{T_3} = V_{T_2} + V_4$, etc.

$$\mu_3(V_T) = \sum^K \mu_3(V_i) \quad \text{for } K \text{ independent components.} \tag{8.14}$$

In the case of the fourth moment of the sum of two variables,

$$\frac{1}{N}\sum (d_T)^4 = \frac{1}{N}\left[\sum d_1^4 + 4\sum d_1^3 d_2 + 6\sum d_1^2 d_2^2 + 4\sum d_1 d_2^3 + \sum d_2^4\right]. \tag{8.15}$$

$$\mu_4(V_T) = \mu_{40} + 4\mu_{31} + 6\mu_{22} + 4\mu_{13} + \mu_{04}. \tag{8.16}$$

The terms in which one deviation appears only to the first power (μ_{31} and μ_{13}) disappear if V_1 and V_2 vary independently for the same reason, as in the cases of μ_{11}, μ_{21} and μ_{12}.

$$\mu_{22} = \frac{1}{N}\sum d_1^2 d_2^2 = \frac{1}{N}\sum^{N_1}\left[d_1^2 \sum^{N_{2\cdot 1}} (d_{2\cdot 1}^2)\right]. \tag{8.17}$$

If V_1 and V_2 vary independently so that $\sum d_{2\cdot 1}^2$ is the same for all values of V_1 and thus equal to $N_{2\cdot 1}\,\mu_2(V_2)$, we have

$$\mu_{22} = \mu_2(V_1)\mu_2(V_2) = \mu_{20}\mu_{02}. \tag{8.18}$$

But

$$2\mu_{20}\mu_{02} = (\mu_{20} + \mu_{02})^2 - \mu_{20}^2 - \mu_{02}^2 = \mu_2^2(V_T) - \mu_{20}^2 - \mu_{02}^2 \tag{8.19}$$

Equation 8.16 can now be written

$$[\mu_4(V_T) - 3\mu_2^2(V_T)] = [\mu_4(V_1) - 3\mu_2^2(V_1)] + [\mu_4(V_2) - 3\mu_2^2(V_2)]. \tag{8.20}$$

Thus while the fourth moment of the total is not the sum of the fourth moments of two independent components, there is a function of fourth and second moments that is additive. Functions of the moments with this additive property were called semi-invariants by Thiele (1889) and given the symbol λ.

It is obvious that if this property applies to two independent components, it must apply to the sum of these and a third that is also independent, and by mathematical induction to the sum of any number.

(8.21) $$\lambda_4(V_T) = \sum^K \lambda_4(V_i) \quad \text{for } K \text{ independent components.}$$

In Thiele's terminology $\lambda_1 = M$, $\lambda_2 = \mu_2$, $\lambda_3 = \mu_3$ and $\lambda_4 = \mu_4 - 3\mu_2^2$. The formulas

(8.22) $$\lambda_5(V_T) = \sum \lambda_5(V_i) \quad \text{where } \lambda_5 = \mu_5 - 10\mu_3\mu_2,$$

and

(8.23) $$\lambda_6(V_T) = \sum \lambda_6(V_i) \quad \text{where } \lambda_6 = \mu_6 - 15\mu_4\mu_2 + 30\mu_2^3 - 10\mu_3^2$$

can readily be demonstrated in the same way, but his generalization that such functions are related to all moments (that exist) and his method of determining their values require a more powerful approach, which we will touch on later. In all of these cases, corrections based on the size of the sample must be applied to the crude determination from the latter to obtain unbiased estimates. Fisher (1931) has given the name cumulant to estimates obtained in this way. These corrections are negligible in individual cases in comparison with the sampling errors.

The form indexes $\gamma_1 = (\mu_3/\sigma^3)$ and $\gamma_2 = [(\mu_4/\sigma^4) - 3]$ can obviously be written as λ_3/σ^3 and λ_4/σ^4, respectively. In general, a series of form indexes can be defined by $\gamma_{n-2} = \lambda_n/\sigma^n$.

It is of interest to consider what happens to these in the distributions of a compound variable. Assume that there are K variables (simple or themselves compound) that have the same variance σ_i^2, but not necessarily the same form indexes.

(8.24) $$\begin{aligned}\gamma_{n-2}(V_T) &= \frac{\lambda_n(V_T)}{\sigma_T^n} = \frac{\sum^K \lambda_n(V_i)}{[K\sigma_i^2]^{n/2}} \\ &= \frac{\sigma^n \sum^K \gamma_{n-2}}{K^{n/2}\sigma^n} = \frac{\sum \gamma_{n-2}}{K^{n/2}} = \frac{\bar{\gamma}_{n-2}}{\sqrt{K^{n-2}}},\end{aligned}$$

or

(8.25) $$\gamma_n(V_T) = \frac{\bar{\gamma}_n}{\sqrt{K^n}}.$$

Thus, all form indexes of the compound variable tend to approach zero as K increases, with the important qualification that this holds only if all the components have finite moments. There are frequency distributions in which all moments (and hence all semi-invariants) are infinite. This, however, cannot be the case with the frequency distribution of measurable or meristic characters of organisms. Neither can it be the case with ratios of such characters unless the denominator can be zero. We thus arrive at the important conclusion that if a variable is the result of

a) many (K) factors or factor groups, all with finite moments,
b) that all make the same contributions (σ_i^2) to its variance,
c) and that are independent in occurrence,
d) and independent (additive) in amount of contribution,

the distribution is characterized by an approach to symmetry $[\gamma_1(\bar{V}_T) = \bar{\gamma}_1(V_i)/\sqrt{K}]$, and to zero kurtosis $[\gamma_2(V_T) = \bar{\gamma}_2(V_i)/K]$, and to zero for all higher indexes $[\gamma_3(V_T) = \gamma_3(V_i)/K\sqrt{K}$, etc.].

The Binomial Distribution

It should be possible to find the form of this limiting distribution from one in which the distributions of all the elementary variables are of the same type but in which an indefinitely large number are combined. The two-point distribution $[p(0) + q(1)]$, $p + q = 1$ is the simplest that can be chosen for an elementary variable. The mean and all higher moments about the origin obviously have the value q. The moments about the mean can be found from the formulas that have been given.

$$(8.26)\quad \begin{cases} \mu_1 = \lambda_1 = q, & M = q, \\ \mu_2 = \lambda_2 = pq, & \sigma = \sqrt{pq}, \\ \mu_3 = \lambda_3 = pq(p - q), & \gamma_1 = (p - q)/\sqrt{pq}, \\ \mu_4 = pq(1 - 3pq), & \gamma_2 = (1 - 6pq)/pq. \\ \lambda_4 = pq(1 - 6pq), & \end{cases}$$

The combination of n independent distributions of this sort may be written $[p(0) + q(1)]^n$ by the product principle of probabilities. As the semi-invariants of the total are merely the sums of those of the components and $\mu_4 = \lambda_4 + 3\lambda_2^2$, we have at once

$$(8.27)\quad \begin{cases} \lambda_1 = \mu_1 = nq, & M = nq, \\ \lambda_2 = \mu_2 = npq, & \sigma = \sqrt{npq}, \\ \lambda_3 = \mu_3 = npq(p - q), & \gamma_1 = (p - q)/\sqrt{npq}, \\ \lambda_4 = npq(1 - 6pq), & \gamma_2 = (1 - 6pq)/npq. \\ \mu_4 = npq(1 - 6pq) + 3n^2p^2q^2, & \end{cases}$$

As a classroom exercise, 10 coins were shaken up and thrown 100 times by each of 166 students in classes in biometry, giving 166 sets, none rejected. The observed frequencies (f_o) of the various numbers of heads in the 16,600 sets of ten are given in Table 8.1. The expected frequencies (f_c) are given by the expansion of $[\frac{1}{2}(0) + \frac{1}{2}(1)]^{10}$, rated up to a total of 16,600. Inspection indicates a fairly close agreement between observed and calculated frequencies and between the various statistics that are given and their theoretical values from the formulas given above. That the degree of agreement is about as close as is to be expected is shown by χ^2-tests indicated here but discussed later.

Table 8.2 gives a biological example. Detlefsen (1918) classified litters of mice from a backcross mating (*AaPpBb* × *aappbb*) according to the numbers of each of the dominants. The results in litters of 5, 6, and 7 (the three most numerous) are shown. Expectation (f_c) is according to the expansion of the binomial $N_L(\frac{1}{2}\,\text{Dom} + \frac{1}{2}\,\text{Rec})^L$ where L is size of litter. The differences between the observed and expected frequencies are small, and the appropriate statistical tests indicate no significance. This was true of all sizes of litter (table 8.3).

Table 8.4 gives the distribution of otocephalic guinea pigs in litters of the large branch of strain 13 (referred to earlier), descended from a single mating in the thirteenth generation of brother-sister mating, excluding the high-producing cluster from a mutation in the nineteenth generation. There were no significant differences among substrains or different sizes of litter in the incidence 5.5%. The expected distribution in litters of size L on the hypothesis that the probabilities of abnormal development are independent in

TABLE 8.1. Observed (f_o) and calculated (f_c) frequencies of heads, and derived statistics, from 16,600 sets of 10 coins.

No. of Heads	f_o	f_c	$f_o - f_c$	$\frac{(f_o - f_c)^2}{f_c}$	Sample	Theory
0	11	16.2	−5.2	1.67	$N = 16{,}600$	
1	160	162.1	−2.1	0.03	$\mu_1' = 5.0027$	5.0000 ± 0.0123
2	736	729.5	+6.5	0.06	$\mu_2 = 2.5040$	2.5000 ± 0.0260
3	1,923	1,945.3	−22.3	0.26	$\mu_3 = 0.1038$	0
4	3,419	3,404.3	+14.7	0.06	$\mu_4 = 17.6134$	17.5000
5	4,145	4,085.2	+59.8	0.88	$\lambda_4 = -1.196$	−1.2500
6	3,364	3,404.3	−40.3	0.48		
7	1,878	1,945.3	−67.3	2.33	$M = 5.0027$	5.0000 ± 0.0123
8	786	729.5	+56.5	4.38	$\sigma = 1.5824$	1.5811 ± 0.0082
9	156	162.1	−6.1	0.23	$\gamma_1 = 0.026$	0 ± 0.019
10	22	16.2	+5.8	2.08	$\gamma_2 = -0.191$	-0.200 ± 0.038
	16,600	16,600.0	0.0	12.46	$\chi^2 = 12.46$	Prob 0.26

TABLE 8.2. Number of mice showing the dominant character in litters of 5, 6, and 7 (the most numerous sizes) from a back cross (*AaPpBb* × *aappbb*). Litters are classified separately for each pair of alleles.

No. of Dominants	5			6			7		
	f_o	f_c	$f_o - f_c$	f_o	f_c	$f_o - f_c$	f_o	f_c	$f_o - f_c$
0	9	10.3	−1.3	3	5.9	−2.9	2	2.4	−0.4
1	47	51.6	−4.6	32	35.7	−3.7	23	16.9	+6.1
2	106	103.1	+2.9	99	89.3	+9.7	55	50.7	+4.3
3	103	103.1	−0.1	101	119.1	−18.1	79	84.5	−5.5
4	51	51.6	−0.6	98	89.3	+8.7	85	84.5	+0.5
5	14	10.3	+3.7	40	35.7	+4.3	48	50.7	−2.7
6				8	5.9	+2.1	16	16.9	−0.9
7							1	2.4	−1.4
Total	330	330.0	0	381	380.9	+0.1	309	309.0	0
χ^2			1.98			7.71			4.02
Probability			0.85			0.26			0.78

From data of Detlefsen (1918).

TABLE 8.3. Tests of frequencies in litters of all sizes in Detlefsen's data. *DF* refers to degrees of freedom.

Size of Litter	Litters	*DF*	χ^2	Probability
1	30	1	0.00	1.00
2	51	2	1.91	0.40
3	87	3	0.77	0.86
4	153	4	2.49	0.65
5	330	5	1.98	0.85
6	381	6	7.71	0.26
7	309	7	4.02	0.78
8	177	6	1.75	0.94
9	60	5	5.86	0.32
10	15	4	3.87	0.43
	1,593	43	30.36	0.923

From data of Detlefsen (1918).

TABLE 8.4. Frequency of otocephalic monsters in litters of sizes 2, 3, and 4, and 2 to 7 combined, in an inbred strain of guinea pigs descended from a single mating in the thirteenth generation of brother–sister mating but excluding a large cluster descended from a single mating in the 19th generation. This strain produced 203 monsters among 3,689 young (5.50%) including 17 in 245 singletons. The expected frequencies in litters of size L is taken as $N_L[0.055(\mathrm{O}) + 0.945(N)]^L$.

No. of Otos.	2(5.9%)			3(4.9%)			4(5.2%)			2–7(5.43%)		
	o	c	$o-c$	o	c	$o-c$	o	c	$o-c$	o	c	$o-c$
0	419	421.5	−2.5	389	381.4	+7.6	191	187.4	+3.6	1,023	1,017.1	+5.9
1	50	49.1	+0.9	60	66.6	−6.6	39	43.6	−4.6	159	167.5	−8.5
2	3	1.4	+1.6	2	3.9	−1.0	5	3.8	+1.0	12	10.1	+2.6
3				1	0.1		0	0.2		1	0.3	
	472	472.0	0	452	452.0	0	235	235.0	0	1,195	1,195.0	0
χ^2	1.86			1.06			0.80			1.12		

From data of Wright (1934).

litter mates and always 0.055 in this strain, are given by expansion $N_L(0.945 \text{ normal} + 0.055 \text{ otocephalic})^L$. Inspection of Table 8.4 shows that this is approximately realized, and this is borne out by the χ^2 test (Table 8.6). Thus there can be no important genetic nor tangible environmental differences (common to litter mates) with respect to otocephaly within this large strain.

Tables 8.5 and 8.6 deal similarly with the number of 4-toed guinea pigs in litters of different sizes in a large branch of strain 35, descended from a single mating in the twelfth generation of brother–sister mating. There was some genetic differentiation of substrains in this branch, but probably little or none in a later branch descended from a single mating in the twenty-second generation ("Chicago"). In this case, the large excess of litters with zero or high incidence and deficiency with intermediate incidence show that there are important nongenetic factors common to litter mates.

The Binomial Limit

The frequency distribution for which all of the form indexes are zero should occupy a central position in the theory of variability. It should be possible to derive its formula by considering the distribution of the sum of an indefinitely large number of independent, equally contributing elementary variables, irrespective of the nature of the latter. This formula was first derived by de Moivre (1733) as the limit of the binomial distribution

TABLE 8.5. Frequency of animals with 4 toes (instead of 3) on hindfeet in litters of 2, 3, and 4 in an inbred strain of guinea pigs, descended from a single mating in the twelfth generation of brother-sister mating which produced 31.1% with 4 toes among 1,976 young and the frequencies indicated in litters of 2 and 3 in a later group (Chicago), descended from a single mating in the twenty-second generation.

No. with 4 Toes	Beltsville									Chicago					
	2(35.7%)			3(34.2%)			4(25.5%)			2(34.9%)			3(49.7%)		
	o	*c*	*o − c*	*o*	*c*	*o − c*	*o*	*c*	*o − c*	*o*	*c*	*o − c*	*o*	*c*	*o − c*
0	146	118.4	+27.6	119	74.6	+44.4	44	28.6	+15.4	32	26.6	+5.4	17	7.5	+9.5
1	76	131.2	−55.2	63	116.4	−53.4	23	39.2	−16.2	18	28.7	−10.7	12	22.3	−12.3
2	64	36.4	+27.6	34	60.5	−26.5	11	20.2	−9.2	13	7.7	+5.3	14	22.0	−6.0
3				46	10.5	+35.5	10	4.6	+5.4				16	7.2	+8.8
4							5	0.4	+4.6						
5															
	286	286.0	0	262	262.0	0	93	93.0	0	63	63.0	0	59	59.0	0
χ^2	50.6	$DF = 1$		182.5	$DF = 2$		39.2	$DF = 2$		8.7	$DF = 1$		25.6	$DF = 2$	

From data of Wright (1934*c*).

TABLE 8.6. Totals from data in Tables 8.5 (4-toe) and 8.4 (otocephaly).

	4-TOE					OTOCEPHALY				
	Size of Litter	No. of Litters	DF	χ^2	Probability	Size of Litter	No. of Litters	DF	χ^2	Probability
Beltsville	2	286	1	50.6	$<10^{-6}$	1	262	1–	0.50	0.48
	3	262	2	182.5	$<10^{-6}$	2	472	2–	1.86	0.41
	4	93	2	39.2	$<10^{-6}$	3	452	2–	1.06	0.60
						4	235	2–	0.80	0.68
Chicago	2	63	1	8.7	0.003	5,6,7	36	2–	1.69	0.44
	3	59	2	25.6	$<10^{-6}$					
							1,457	8	5.91	0.66
			8	306.6	$<10^{-6}$	2–7	1,195	7	5.41	0.62

$[(1 - q)a + qA]^n$. It can be obtained from the difference between successive terms of the expansion.

$$(8.28) \quad y_{x+1} = \frac{n!}{(x+1)!(n-x-1)!}p^{n-x-1}q^{n+1} = \left(\frac{n-x}{x+1}\right)\frac{q}{p}\,y_x.$$

$$(8.29) \quad \Delta y = y_{x+1} - y_x = \left[\frac{nq - x - p}{(x+1)p}\right]y_x = -\left[\frac{x - nq + p}{npq + (x - \bar{x})p + p}\right]y_x.$$

On removing terms in the numerator and denominator that become of the second order as n is increased and putting $nq = \bar{x}$, $npq = \sigma_x^2$

$$(8.30) \quad \Delta y = -\left[\frac{x - \bar{x}}{\sigma_x^2}\right]y.$$

The slope of the limiting smooth curve is thus

$$(8.31) \quad \frac{dy}{dx} = -\left(\frac{x - \bar{x}}{\sigma_x^2}\right)y;$$

$$(8.32) \quad \log y = -\frac{1}{2}\left(\frac{x - \bar{x}}{\sigma_x}\right)^2;$$

$$(8.33) \quad y = Ce^{-(x-\bar{x})^2/2\sigma_x^2};$$

$$(8.34) \quad C = \frac{1}{\sigma\sqrt{2\pi}} \quad \text{since} \quad \int_{-\infty}^{+\infty} e^{-x^2/2\sigma_x^2}\,dx = \sigma\sqrt{2\pi}.$$

This is the normal probability curve.

Characteristic Functions

At this point, it is desirable to introduce a powerful method of dealing with frequency distributions that traces back to Laplace, the method of characteristic functions. For a full treatment the reader may be referred to Cramér (1946).

This method rests on the properties of complex operators of the type e^{itx} in which x refers to the scale of the frequency distribution and t is any real number. The characteristic function pertaining to a given frequency distribution is a function of the arbitrary real variable t.

(8.35) $$\varphi(t) = \sum_{-\infty}^{\infty} e^{itx_j} q_j \quad \text{for a discrete variable;}$$

(8.36) $$\varphi(t) = \int_{-\infty}^{+\infty} e^{itx} f(x)\, dx \quad \text{for a continuous variable.}$$

The meaning of these expressions can be grasped most easily by considering ones pertaining to simple discrete distributions. The simplest of all is the one-point distribution in which the entire frequency, $q = 1$, is located at one point on the scale, $x = x_1$.

(8.37) $$\varphi(t) = e^{itx_1} = \cos(tx_1) + i \sin(tx_1).$$

If the distribution is referred to the origin, the absolute value is 1, whatever the value of x_1, but the direction in the complex plane is at the angle tx_1 from the angle of real positive numbers.

Consider next the array $[q_1(C_1) + q_2(C_2) + \cdots + q_k(C_k)]$ in which the C's are qualitative designations of alternative categories, and the q's are their frequencies. Assume that scale values can be assigned these categories x_1 for C_1, etc.

(8.38) $$\varphi(t) = [e^{itx_1}q_1 + e^{itx_2}q_2 + \cdots + e^{itx_k}q_k].$$

The characteristic function, as a series of terms with a given value of t, is thus another way of specifying the frequency array. It has the advantage that the scale values are clearly indicated to be in a different dimension from the frequencies because of the two-dimensional character of complex numbers. Scale values are indicated by the direction of the unit vectors e^{itx_i}, etc., with t as the angular scale unit, and frequencies are indicated by the lengths of the radius vectors q_i, etc. Alternatively one may think of the frequencies as weights assigned to class values located along the circle of unit complex values.

The characteristic function of a linear function of x, $x' = a + bx$ can easily be found:

$$\varphi'(t) = \sum e^{it(a+bx_i)}q_i = e^{ita} \sum e^{i(bt)x_i}q_i = e^{ita}\varphi(bt). \tag{8.39}$$

Thus the scale unit t is multiplied by b, and the distribution is rotated through the angle indicated by e^{ita}. In particular, if a concrete variable is transformed to standard form, $x' = (x - \bar{x})/\sigma$, then

$$\varphi'(t) = e^{it\bar{x}/\sigma}\varphi\left(\frac{t}{\sigma}\right). \tag{8.40}$$

The characteristic function $\varphi_T(t)$ of a distribution that consists of the weighted sum of a number of components is merely the weighted sum of the characteristic functions φ_c, of the latter, as expected from the interpretation of the characteristic function as a frequency distribution along a unit circle in the complex plane. If

$$f_T(x) = \sum^k w_c q_c, \qquad \varphi_T(t) = \sum^k w_c\varphi_c(t). \tag{8.41}$$

In the case of compound variable in which each value x_T is the sum of contributions from independent components, x_c

$$\varphi_T(t) = \sum e^{it\sum x_c}(q_1 q_2 \cdots q_k) = \prod^k \varphi_c(t); \tag{8.42}$$

$$\log \varphi_T(t) = \sum^k \log \varphi_c(t). \tag{8.43}$$

As a simple example consider the gametic array $[\frac{1}{2}A + \frac{1}{2}a]$ with $\varphi_g(t) = \frac{1}{2}e^{it(A)} + \frac{1}{2}e^{it(a)}$ in which (A) and (a) measure the effect of the genes, and the zygotic array $[\frac{1}{2}A + \frac{1}{2}a]^2 = \frac{1}{4}A^2 + \frac{2}{4}Aa + \frac{1}{4}a^2$ with $\varphi_z(t) = \varphi_g{}^2(t) = \frac{1}{4}e^{2it(A)} + \frac{1}{2}e^{it(A+a)} + \frac{1}{4}e^{2it(a)}$. In the linear array, the phenotypes associated with the "products" of the qualitative term $(A^2, 2Aa, a^2)$ are not shown as functions of gametic effects. In the characteristic function the phenotypic effects are spelled out in the coefficient of it in the exponentials, provided that the effects of the genes are additive. This may or may not be an advantage, depending on whether or not the assumption of additivity is made.

If the summation indicated in the formula of the characteristic function for any given value of t is carried out, it ceases to represent an array and becomes a single complex number, the average of vectors e^{itx} if the q's are treated as weights. Since the absolute value of e^{itx} is always 1 and the q's are proportional frequencies $(\sum q = 1)$, the absolute value of $\varphi(t)$ can never exceed 1. The two-point distribution $[\frac{1}{2}a + \frac{1}{2}A]$ in which a and A

are assigned the values 0 and 1 respectively, yields a characteristic function $\varphi(t) = [\frac{1}{2} + \frac{1}{2}e^{it}]$ which describes a circle in the complex field through the values 0 and +1 as t increases. If centered at the origin by assigning the value $-\frac{1}{2}$ to a, $+\frac{1}{2}$ to A, $\varphi(t) = [\frac{1}{2}e^{-\frac{1}{2}it} + \frac{1}{2}e^{\frac{1}{2}it}] = \cos \frac{1}{2}t$ is always real and oscillates between the values +1 and −1 as t increases. It is obvious that it is always real in the case of a symmetrical distribution in which x is the deviation from the mean and that it cannot always be real with any asymmetrical distribution.

The properties of the characteristic function of a continuous distribution, $\varphi(t) = \int_{-\infty}^{+\infty} e^{itx} f(x)\, dx$, are analogous to those of a discrete distribution.

If $x' = a + bx$,

$$\varphi'(t) = e^{ita}\varphi(bt). \tag{8.44}$$

If $x' = (x - M)/\sigma$,

$$\varphi'(t) = e^{-itM/\sigma}\varphi\left(\frac{t}{\sigma}\right). \tag{8.45}$$

If $f_T(x) = \sum^k w_c f_c(x)\, dx$,

$$\varphi_T(t) = \sum^k [w_c \varphi_c(t)]. \tag{8.46}$$

If $x_T = \sum x_c$,

$$\log \varphi_T(t) = \sum^k \log \varphi_c(t) \quad \text{(if the } x_c\text{'s are independent)}. \tag{8.47}$$

One of the most important properties of the characteristic function, whether discrete or continuous, is that it is a moment-generating function. Repeated differentiation with respect to t and the assumption that t is so small that e^{itx} may be treated as 1, leads to the successive moments about the origin, multiplied merely by $(i)^s$ where s is the number of differentiations, provided that the moments exist. In the case of a continuous variable:

$$\varphi(t) = \int_{-\infty}^{+\infty} e^{itx} f(x)\, dx; \tag{8.48}$$

$$\frac{d\varphi(t)}{dt} = i \int_{-\infty}^{+\infty} x e^{itx} f(x)\, dx; \tag{8.49}$$

$$\frac{d^s\varphi(t)}{dt^s} = (i)^s \int_{-\infty}^{+\infty} x^s e^{itx} f(x)\, dx; \tag{8.50}$$

$$\frac{d^s\varphi(0)}{dt^s} = (i)^s \int_{-\infty}^{+\infty} x^s f(x)\, dx = (i)^s \mu'_s. \tag{8.51}$$

This property makes it possible to represent the characteristic function in the neighborhood of $t = 0$ by a power series with coefficients that depend in a simple way on the moments, provided again that the moments exist. By Maclaurin's theorem

$$\varphi(t) = 1 + \mu'_1(it) + \mu'_2 \frac{(it)^2}{2!} + \mu'_3 \frac{(it)^3}{3!} + \cdots + \mu'_s \frac{(it)^s}{s!} + \cdots. \tag{8.52}$$

As a simple example consider a rectangular distribution with range $2a$, referred to its mean as origin.

$$f(x) = \frac{1}{2a}, \qquad -a \le x \le +a. \tag{8.53}$$

$$\varphi(t) = \frac{1}{2a} \int_{-a}^{+a} e^{itx}\, dx = \frac{e^{ita} - e^{-ita}}{2ita} = \frac{\sin at}{at}. \tag{8.54}$$

The moments (central in this case) could be obtained by repeated differentiation. This rapidly leads to expressions of some complexity. It is, however, merely necessary to expand, $\varphi(t)$ in powers of t by any means, and take the coefficient of $(it)^s/s!$ to obtain each moment.

$$\begin{aligned} \varphi(t) &= \left[at - \frac{(at)^3}{3!} + \frac{(at)^5}{5!} - \frac{(at)^7}{7!} + \cdots\right]\Big/at \\ &= 1 + \frac{a^2}{3}\left(\frac{(it)^2}{2!}\right) + \frac{a^4}{5}\left(\frac{(it)^4}{4!}\right) + \frac{a^6}{7}\left(\frac{(it)^6}{6!}\right) + \cdots. \end{aligned} \tag{8.55}$$

The odd moments are all zero as expected.

$$\mu_2 = \frac{a^2}{3}, \quad \mu_4 = \frac{a^4}{5}, \quad \mu_6 = \frac{a^6}{7}, \quad \lambda_4 = -\frac{2\,a^4}{15}, \quad \gamma_2 = -\frac{6}{5}. \tag{8.56}$$

As another example consider the triangular distribution with range $2a$ and maximum ordinate $1/a$ at the origin.

$$f(x) = \frac{1}{a}\left(1 - \frac{x}{2a}\right), \qquad 0 \le x \le 2a: \tag{8.57}$$

$$\begin{aligned} \varphi(t) &= \frac{1}{2a^2t^2}\left[1 + 2iat - e^{2iat}\right] \\ &= 1 + \frac{2a}{3}(it) + \frac{(2a)^2}{6}\left(\frac{(it)^2}{2!}\right) + \frac{(2a)^3}{10}\left(\frac{(it)^3}{3!}\right) + \cdots \\ &\qquad + \frac{2(2a)^s}{(s+1)(s+2)}\left(\frac{(it)^s}{s!}\right) + \cdots. \end{aligned} \tag{8.58}$$

Thus the moments about the origin are

$$\mu'_s = \frac{2(2a)^s}{(s+1)(s+2)} \tag{8.59}$$

and

$$\begin{cases} \lambda_1 = M = \dfrac{2a}{3}, & \lambda_3 = \dfrac{(2a)^3}{135}, & \gamma_1 = \dfrac{2\sqrt{2}}{5}, \\ \lambda_2 = \sigma^2 = \dfrac{(2a)^2}{18}, & \lambda_4 = -\dfrac{(2a)^4}{540}, & \gamma_2 = -\dfrac{3}{5}. \end{cases} \tag{8.60}$$

A triangle ranging from the origin to $(-2b)$ with maximum ordinate at the origin, $1/b$, may be treated similarly:

$$f(x) = \frac{1}{b}\left(1 + \frac{x}{2b}\right), \qquad -2b \leq x \leq 0; \tag{8.61}$$

$$\begin{aligned} \varphi(t) &= \frac{1}{2b^2t^2}[1 - 2ibt - e^{-2ibt}] \\ &= 1 - \frac{2b}{3}(it) + \frac{(2b)^2}{6}\frac{(it)^2}{2!} - \frac{(2b)^3}{10}\frac{(it)^3}{3!} + \cdots \\ &\qquad + \frac{2(-2b)^s}{(s+1)(s+2)}\frac{(it)^s}{s!}. \end{aligned} \tag{8.62}$$

These may be combined to form the general triangular distribution by using weights of $a/(a + b)$ and $b/(a + b)$ to give a common maximum ordinate of $1/(a + b)$ at the origin.

$$f(x) = \begin{cases} \dfrac{1}{(a+b)}\left(1 - \dfrac{x}{2a}\right), & 0 \leq x \leq 2a, \\ \dfrac{1}{(a+b)}\left(1 + \dfrac{x}{2b}\right), & -2b \leq x \leq 0. \end{cases} \tag{8.63}$$

$$\begin{aligned} \varphi(t) = 1 &+ \frac{2}{3}\left(\frac{a^2 - b^2}{a+b}\right)(it) + \frac{4}{6}\left(\frac{a^3 + b^3}{a+b}\right)\frac{(it)^2}{2!} \\ &+ \frac{8}{10}\left(\frac{a^4 - b^4}{a+b}\right)\frac{(it)^3}{3!} + \frac{16}{15}\left(\frac{a^5 + b^5}{a+b}\right)\frac{(it)^4}{4!}. \end{aligned} \tag{8.64}$$

$$\begin{cases} \mu'_1 = \tfrac{2}{3}(a - b), \\ \lambda_2 = \tfrac{2}{9}(a^2 + ab + b^2), \\ \lambda_3 = \tfrac{4}{135}(a - b)(2a^2 + 5ab + 2b^2), \\ \mu_4 = \tfrac{16}{135}(a^2 + ab + b^2)^2, \\ \lambda_4 = -\tfrac{4}{135}(a^2 + ab + b^2)^2, \\ \gamma_2 = -\tfrac{3}{5}. \end{cases} \tag{8.65}$$

In the symmetrical case, $b = a$,

(8.66)
$$f(x) = \frac{1}{2a}\left(1 - \frac{1}{2a}|x|\right), \qquad -2a \leq x \leq 2a,$$

(8.67)
$$\begin{aligned}\varphi(t) &= \left[\frac{1}{4a^2t^2}\right][2 - e^{2iat} - e^{-2iat}] \\ &= \left[\frac{1}{2iat}(e^{iat} - e^{-iat})\right]^2 = \left[\frac{\sin(at)}{at}\right]^2.\end{aligned}$$

The central moments can readily be found from the expansion of the first of the expressions for $\varphi(t)$. From the last it may be seen that $\varphi(t)$ for the symmetrical triangular distribution with range $4a$ is the square of that for the rectangular distribution with range $2a$, both centered at their means, indicating that it is the distribution of a variable that consists of the sum of two independent variables, each with the above rectangular distribution. The errors in estimating an interval of time for which errors at both ends are uniformly distributed over a certain period show this distribution.

Exponential distributions can be treated similarly.

(8.68)
$$f(x) = ae^{-ax}, \qquad 0 \leq x.$$

(8.69)
$$\varphi(t) = \frac{a}{a - it} = 1 + \left(\frac{1}{a}\right)(it) + \frac{2!}{a^2}\frac{(it)^2}{2!} + \frac{3!}{a^3}\frac{(it)^3}{3!} + \cdots,$$

and

(8.70)
$$f(x) = be^{bx}, \qquad x \leq 0.$$

(8.71)
$$\varphi(t) = \frac{b}{b + it} = 1 - \left(\frac{1}{b}\right)(it) + \frac{2!}{b^2}\frac{(it)^2}{2!} - \frac{3!}{a^3}\frac{(it)^3}{3!} + \cdots.$$

Again $\varphi(t)$ can be found for the compound distribution with a common maximum ordinate at the origin by using weights of $b/(a + b)$ and $a/(a + b)$ for two components. Of most interest is the symmetrical case, known as Laplace's distribution.

(8.72)
$$f(x) = \frac{a}{2}e^{-a|x|}.$$

(8.73)
$$\begin{aligned}\varphi(t) &= \frac{a}{2(a - it)} + \frac{a}{2(a + it)} = \frac{a^2}{a^2 + t^2} \\ &= 1 + \frac{2!}{a^2}\frac{(it)^2}{2!} + \frac{4!}{a^4}\frac{(it)^4}{4!} + \cdots.\end{aligned}$$

The odd moments are all zero, as expected from symmetry.

$$\mu_2 = \frac{2!}{a^2}, \quad \mu_4 = \frac{4!}{a^4}, \quad \mu_6 = \frac{6!}{a^6}, \ldots. \tag{8.74}$$

One may often expand log $\varphi(t)$ in powers of (it), by Maclaurin's theorem, assuming t to be in the neighborhood of zero, and using λ's for the coefficients.

$$\log \varphi(t) = \lambda_1(it) + \lambda_2 \frac{(it)^2}{2!} + \lambda_3 \frac{(it)^3}{3!} + \cdots + \lambda_s \frac{(it)^s}{s!}. \tag{8.75}$$

The coefficients are of great importance in the theory of compound variables because of the relation $\log \varphi_T(t) = \sum \log \varphi(t)$. By equating the coefficients of powers of (it) on the two sides of this last equation, it follows that each of the λ's of the compound variable must be the sum of the corresponding λ's of the components. The λ's are, in short, Thiele's semi-invariants. Their values can be found by repeated differentiation, $(d^s/dt^s) \log \varphi(t) = i^s\lambda_s$ but this usually leads to great complexity. As in the case of the moments, they can be read off directly from the expansion of log $\varphi(t)$ in a power series.

In the case of the rectangular distribution

$$f(x) = \frac{1}{2a}, \qquad -a \leq x \leq a, \tag{8.76}$$

$$\begin{aligned} \log \varphi(t) &= \log \sin at - \log at \\ &= \left[\log at - \frac{1}{6}(at)^2 - \frac{1}{180}(at)^4 - \frac{1}{2835}(at)^6 - \cdots\right] - \log at \\ &= \frac{a^2}{3}\frac{(it)^2}{2!} - \frac{2a^4}{15}\frac{(it)^4}{4!} + \frac{16a^6}{63}\frac{(it)^6}{6!}. \end{aligned} \tag{8.77}$$

Thus

$$\lambda_2 = \frac{a^2}{3}, \quad \lambda_4 = \frac{-2a^4}{15}, \quad \lambda_6 = \frac{16a^6}{63} \quad \text{as before.} \tag{8.78}$$

This is a highly platykurtic distribution ($\gamma_2 = -6/5$).

Since the symmetrical triangular distribution discussed above is given by the sum of two variables, each with this rectangular distribution, its semi-invariants are all just twice those of the rectangular distribution. Its form indexes γ_s are $(1/2)^{s/2}$ times those of the latter.

In the case of Laplace's distribution,

$$f(x) = (a/2)e^{-|ax|}, \tag{8.79}$$

$$\varphi(t) = \frac{1}{1 + (t/a)^2}, \tag{8.80}$$

$$\log \varphi(t) = -\log\left[1 + \left(\frac{t}{a}\right)^2\right] \tag{8.81}$$
$$= \frac{2}{a^2}\frac{(it)^2}{2!} + \frac{12}{a^4}\frac{(it)^4}{4!} + \frac{240}{a^6}\frac{(it)^6}{6!} + \cdots.$$

$$\lambda_2 = \frac{2}{a^2}, \quad \lambda_4 = \frac{12}{a^4}, \quad \lambda_6 = \frac{240}{a^6}, \tag{8.82}$$

which agree with the values derived from $\varphi(t)$. This is a highly leptokurtic distribution ($\gamma_2 = 3$).

The values of the semi-invariants can be found in terms of the moments by equating two expressions for $\log \varphi(t)$:

$$\log\left[1 + \sum \frac{\mu'_s}{s!}(it)^s\right] = \sum\left[\frac{\lambda_s}{s!}(it)^s\right] \tag{8.83}$$

On differentiating both sides with respect to t,

$$\frac{\mu'_1(i) + \mu'_2(i^2t) + \frac{\mu'_3}{2!}(i^3t^2) + \frac{\mu'_4}{3!}(i^4t^3) + \cdots}{1 + \mu'_1(it) + \frac{\mu'_2}{2!}(it)^2 + \frac{\mu'_3}{3!}(it)^3 + \frac{\mu'_4}{4!}(it)^4 + \cdots} \tag{8.84}$$
$$= \lambda_1(i) + \lambda_2(i^2t) + \frac{\lambda_3}{2!}(i^3t^2) + \frac{\lambda_4}{3!}(i^4t^3).$$

By clearing the fraction and equating coefficients of powers of t, a series of expressions can be written by which each λ may be expressed in terms of μ's. It may be noted at once that $\lambda_1 = \mu'_1$. By putting $\mu'_1 = \lambda_1 = 0$, the semi-invariants can be expressed in terms of the central moments.

$$\left\{\begin{array}{ll} \lambda_2 = \mu_2, & \lambda_5 = \mu_5 - 10\mu_2\mu_3, \\ \lambda_3 = \mu_3, & \lambda_6 = \mu_6 - 15\mu_4\mu_2 - 10\mu_3^2 + 30\mu_2^3. \\ \lambda_4 = \mu_4 - 3\mu_2^2, & \end{array}\right. \tag{8.85}$$

These agree with those calculated directly (8.21 to 8.23) and can be more easily extended. More important, however, is the evidence that semi-invariants exist to as high a degree as the moments.

It has been shown earlier that the form coefficients of the distribution of the sum of k independent variables with equal variances and finite sth moments tend to zero as the number of components increases. The limiting form, if all moments of the components exist, should be that of the distribution of which all semi-invariants above the second are zero.

$$\log \varphi(t) = \lambda_1(it) + \lambda_2\frac{(it)^2}{2!}. \tag{8.86}$$

It has been shown that the compound binomial $[p(a) + q(t)]^k$ approaches the normal probability curve as k increases. Since the distribution of the sum of independent variables should approach this same distribution, irrespective of the distribution of the components (if these have finite moments), the normal curve should be the limit in all cases. This can be tested by finding its characteristic function.

$$f(x) = \frac{1}{\sigma\sqrt{2\pi}} \exp\left(-\frac{1}{2}\frac{(x - M)^2}{\sigma^2}\right); \tag{8.87}$$

$$\begin{aligned} \varphi(t) &= \frac{1}{\sigma\sqrt{2\pi}} \int_{-\infty}^{+\infty} \exp\left(itx - \frac{1}{2}\frac{(x - M)^2}{\sigma^2}\right) dx \\ &= \left[\frac{1}{\sigma\sqrt{2\pi}} \int_{-\infty}^{+\infty} \exp\left(-\frac{1}{2\sigma^2}[x - (M + it\sigma^2)]^2\right) dx\right] \\ &\qquad \times \exp\left(\frac{1}{2\sigma^2}(2Mit\sigma^2 - t^2\sigma^4)\right) \\ &= \exp\left(Mit - \frac{\sigma^2 t^2}{2}\right). \end{aligned} \tag{8.88}$$

$$\log \varphi(t) = M(it) - \frac{\sigma^2 t^2}{2} = \lambda_1(it) + \lambda_2 \frac{(it)^2}{2!}. \tag{8.89}$$

Thus the normal probability distribution is the one for which all semi-invariants above the second are zero.

If any component of a compound variable has infinite moments and thus infinite form indexes above a certain degree, the corresponding form indexes of the compound variable must also be infinite. Nevertheless, the departure from normality in the main portion of the distribution may be small. The conditions are discussed by Cramér (1946).

It was shown above that the normal probability distribution, arrived at as the limiting distribution for the sum of an indefinitely large number of independent two-point variables, is in fact the distribution for which all semi-variants above the second are zero. It would be of interest to be able to derive the formula of this limiting distribution directly from this condition. This requires an inversion formula for deducing the distribution that has a given characteristic function. There is such a formula:

$$f(x_a) = \frac{1}{2\pi} \int_{-\infty}^{+\infty} e^{-itx_a} \varphi(t)\, dt \quad \text{where } \varphi(t) = \int_{-\infty}^{+\infty} e^{itx} f(x)\, dx. \tag{8.90}$$

The formula for the normal probability curve can readily be derived from the condition,

$$\log \varphi(t) = \lambda_1(it) + \frac{\lambda_2(it)^2}{2!}.$$

Properties of the Normal Probability Distribution

From the foregoing considerations it is obvious that the normal probability distribution must occupy a central position in the theory of variability. Its importance was described by Galton (1889, p. 66) as follows:

I know of scarcely anything so apt to impress the imagination as the wonderful form of cosmic order expressed by the law of frequency of error. The law would have been personified by the Greeks if they had known of it. It reigns with serenity and complete self-effacement amidst the wildest confusion. The larger the mob, the greater the apparent anarchy, the more perfect is its sway. It is the supreme law of unreason.

It is a symmetrical bell-shaped curve of infinite range, with points of inflection above and below the mean at one standard deviation. Although it cannot be integrated in terms of ordinary functions, the portion of the area below any given point on the scale can be calculated from expansion of the formula into an infinite series. Tables are available in various forms. It is convenient to take the deviation from the mean in standard form, $x_i = (V_i - M)/\sigma$ and represent the probability integral by the symbol pri x_i. This indicates the probability (p_i) of values below this deviation.

$$(8.91) \qquad p_i = \text{pri } x_i = \int_{-\infty}^{x_i} \frac{1}{\sqrt{2\pi}} e^{(-\frac{1}{2})x^2}\, dx.$$

The inverse probability integral is thus $\text{pri}^{-1}\, p = x_i$.

Following are certain important percentiles.

$$\begin{aligned}
&Q_{75} \text{ (upper quartile)} &&= \text{pri}^{-1}\,(0.75) &&= +0.6745\\
&Q_{50} \text{ (median)} &&= \text{pri}^{-1}\,(0.50) &&= \quad 0\\
&Q_{25} \text{ (lower quartile)} &&= \text{pri}^{-1}\,(0.25) &&= -0.6745\\
&Q_{2\cdot5} &&= \text{pri}^{-1}\,(0.025) &&= -1.9600\\
&Q_{0\cdot5} &&= \text{pri}^{-1}\,(0.005) &&= -2.5758\\
&Q_{0\cdot05} &&= \text{pri}^{-1}\,(0.0005) &&= -3.2905
\end{aligned}$$

The term probit p (from Bliss 1935) is often used for $(5 + \text{pri}^{-1}\, p)$ to avoid negative values in practical work.

A formula for expressing any central moment in terms of lower ones can be derived at once by using the differential equation $dy/dx = -xy/\sigma^2$ in an integration by parts.

(8.92)
$$\begin{aligned}\mu_n &= \int_{-\infty}^{+\infty} x^n y\, dx = -\sigma^2 \int_{x=-\infty}^{+\infty} x^{n-1}\, dy \\ &= -\sigma^2[x^{n-1}y]_{x=-\infty}^{+\infty} + (n-1)\sigma^2 \int_{-\infty}^{+\infty} x^{n-2} y\, dx = (n-1)\mu_2\mu_{n-2}\end{aligned}$$

since

$$[x^{n-1}y]_{x=-\infty}^{+\infty} = 0.$$

The odd central moments are all 0.

(8.93)
$$\begin{cases} \mu_0 = 1, & \mu_6 = 15\mu_2{}^3, \\ \mu_2 = \sigma^2, & \mu_8 = 105\mu_2{}^4, \\ \mu_4 = 3\mu_2{}^2, & \mu_{10} = 945\mu_2{}^5, \text{ etc.} \end{cases}$$

All of the semi-invariants above the second are, of course, 0.

The mean of the lower tail of the unit normal distribution,

$$z = \frac{1}{\sqrt{2\pi}} e^{-\frac{1}{2}x^2},$$

with tail frequency p_i takes the simple form

(8.94)
$$\frac{\int_{-\infty}^{x_i} xz\, dx}{\int_{-\infty}^{x_i} z\, dx} = -\frac{z_i}{p_i}.$$

The mean of the upper tail with tail frequency $q_i = 1 - p_i$ is similarly z_i/q_i. That of the class between abscissas x_1 and x_2 and thus with area $q_2 - q_1$ is $(z_2 - z_1)/(q_2 - q_1)$. Many available tables give z, p (or q), and sometimes z/p or z/q for values of x or inversely.

In normal distributions with standard deviation σ, the above means must be multiplied by σ. The average deviation, ignoring sign (mean of half normal distribution), is $AD = (z_0/0.5)\sigma = \sqrt{2/\pi}\,\sigma = 0.7979\sigma$.

A normal curve can be fitted to data in several ways.

1. *Frequency method.* Let f_i represent an observed class frequency. $N(= \sum f_i)$ the total number of cases.

Find $x_i = (V_i - M)/\sigma$ for each class limit.

Look up $p_i = \text{pri}\ x$ for each x_i.

Find Δp_i the successive differences of the p_i's.

Find $f_{i(c)} = N\,\Delta p_i$ the calculated class frequency.

2. *Ordinate Method.* Less accurately, the observed class frequencies can be identified with the elements of area in the histogram (midclass ordinates each multiplied by the class range, l).

Find $x_i = (V_i - M)/\sigma$ for each midclass abscissa.

Look up z_i for each x_i.

Find $f_{i(c)} = Nlz_i/\sigma$.

3. *Least-squares method.* Another possibility in this case is to fit the logarithms of the frequencies (again treated as midclass ordinates multiplied in each case by the class range) by the method of least squares.

$$\text{(8.95)}\qquad \begin{aligned}\log_\epsilon f_i &= \log_\epsilon\left(\frac{N}{\sigma\sqrt{2\pi}}\right) - \frac{1}{2}\left(\frac{V_i - M}{\sigma}\right)^2 + \delta_i \\ &= \left[\log_\epsilon\left(\frac{N}{\sigma\sqrt{2\pi}}\right) - \frac{M^2}{2\sigma^2}\right] + \left(\frac{M}{\sigma^2}\right)V_i - \left(\frac{1}{2\sigma^2}\right)V_i^2 + \delta_i.\end{aligned}$$

These observation equations must be weighted by f_i to obtain the best fit to the observed frequencies. The estimates for the constant term and for the coefficients of V and V^2 permit estimates of N, M, and σ.

4. *Transformation method.* Finally the normal curve can be rectified by using the transformation

$$\text{(8.96)}\qquad \text{pri}^{-1}\, p_i = \frac{V_i - M}{\sigma}.$$

On plotting the inverse probability integral of the proportion (p_i) of the total number to each class limit (V_i) against the latter, the points should fall approximately in a straight line if the normal curve is appropriate. The mean (M) and standard deviation (σ) are given by the intercepts with the axes, $\text{pri}^{-1}\, p = 0$ at which $V = M$, and $V = 0$ at which $\text{pri}^{-1}\, p = -M/\sigma$. The weight of each observation by itself is proportional to z^2/pq, but unfortunately the observations are correlated. This method is more suitable for exploration of the nature of the variability than for precise fitting. It applies only to complete distributions, except that it would be possible by repeated trials of values of N to approximate that which most nearly rectifies $\text{pri}^{-1}\, p$ as a function of V. It is of course a less satisfactory method of fitting complete distributions than (1) but will be useful in another connection.

The Poisson Distribution

There is another limiting form of the binomial distribution that is important. If the frequency of one alternative, A, is very small, but the number of events, n, is so large that the mean, $M = nq$ is considerable, the formula $[(1 - q) + q(A)]^n$ can be written in the form $[1 - (M/n)(1 - A)]^n$, which approaches $e^{-M(1-A)}$ as n becomes indefinitely large; recalling that the qualitative symbol, A, may be treated as if an algebraic quantity. The expansion

$$(8.97) \qquad e^{-M}\left[1 + M(A) + \frac{M^2(A)^2}{2!} + \cdots + \frac{M^k(A)^k}{k!}\right]$$

gives the frequencies of successive numbers of occurrences of A. This is the Poisson distribution. If non(A), frequency $(1 - q)$, is represented by (a) the general term has the qualitative factor, $(a)^{n-k}$.

The characteristic function of the above binomial in which non(A) is assigned the value zero and (A) one is

$$[(1 - q) + qe^{it}] = \left[1 - \frac{M}{n}(1 - e^{it})\right].$$

The compound of n such variables is $[1 - (M/n)(1 - e^{it})]^n$. As n is indefinitely increased, this approaches $e^{M(e^{it}-1)}$.

$$(8.98) \qquad \log\varphi(t) = M(e^{it} - 1) = M(it) + M\frac{(it)^2}{2!} + M\frac{(it)^3}{3!} + \cdots + M\frac{(it)^k}{k!}.$$

Thus all semi-invariants have the value M. The best estimate of this parameter (by the method of maximum likelihood) is the mean, $M = \sum Vf/\sum f$. This distribution is markedly asymmetrical ($\gamma_1 = 1/\sqrt{M}$) if the mean is small, but all the form indexes approach zero and the distribution approaches normality (except for its discrete character) as the mean increases.

The Poisson distribution is of great importance as an approximation to the distribution of random independent events where the number of possible events is very much larger than the mean. Thus it applies to red blood cells in equal squares on a haemocytometer slide, provided that the blood is spread uniformly and there is no clumping. Similarly it applies to the numbers of bacterial colonies on plates where the requisite of independence is met. It applies also to the variations in density of large organisms that are moving about in a uniform environment wholly independently of each other.

There is, however, a strong likelihood in such cases that the elementary events are not wholly independent. There may be a tendency to agglutination of red blood cells or of bacteria and toward aggregation of large organisms. The environment may not be uniform and may favor aggregation in some regions. On the other hand, there may be a tendency toward repulsion, as is exhibited in the defense of territories by many birds.

Table 8.7 shows counts of blood cells of a fowl in 160 squares (10 blocks of 16 squares each) on a haemocytometer slide made by Dr. Mary Juhn. There appears to be good agreement with the expected values. One of the most

TABLE 8.7. Observed distribution (f_o) of numbers of various entities in space in comparison with those calculated (f_c) on the hypothesis of randomness.

CELLS					BEETLES					SPIDERS				
	Squares					Cubes					Quadrats			
V	f_o	f_c	f_o-f_c	$\frac{(f_o-f_c)^2}{f_c}$	V	f_o	f_c	f_o-f_c	$\frac{(f_o-f_c)^2}{f_c}$	V	f_o	f_c	f_o-f_c	$\frac{(f_o-f_c)^2}{f_c}$
0	3	3.9	−0.9	0.21	0	237	246.3	−9.3	0.35	0	53	49.2	+3.8	0.29
1	18	14.6	+3.4	0.85	1	161	147.3	+13.7	1.27	1	19	23.9	−4.9	1.00
2	27	27.0	0	0	2	45	44.1	+0.9	0.02	2	5	5.8		
3	26	33.4	−7.4	1.64	3	3	8.8			3	2	1.0	+1.1	0.18
4	33	30.9	+2.1	0.14	4	2	1.3	−5.3	2.73	4	1	0.1		
5	23	22.9	+0.1	0	5	0	0.2			5	0			
6	20	14.2	+5.8	2.37										
7	6	7.5	−1.5	0.30										
8	3	3.5	−0.5	0.07										
9	0	1.4												
10	1	0.5	−1.1	0.58										
11	0	0.2												
N	160	160	0.0	6.16	N	448	448.0	0.0	4.37	N	80	80.0	0.0	1.47
M	3.706	3.706		$DF=8$	M	0.598	0.598		$DF=2$	M	0.487	0.487		$DF=1$
σ^2	3.592	3.706		Probability 0.63	σ^2	0.536	0.598		Probability 0.12	σ^2	0.683	0.487		Probability 0.23
Red blood cells of fowl on squares on haemocytometer slide (Juhn).					Beetles (*Tribolium confusum*) in cubes in jar of flour (Park 1933).					Spiders (*Argenna subnigra*) in quadrats (10 × 10 × 3 cm.) in uniform field (Baklemischev).				

important properties of the Poisson distribution is its cumulative character. If the classes are grouped into larger classes of equal size, these have a Poisson distribution with the parameter $\sum M$ instead of M. Thus in the 10 blocks in Table 8.7 the counts were 70, 56, 76, 60, 49, 53, 55, 51, 61, 62, with mean 59.3 and variance 72.0, which does not differ significantly.

Park (1933) dissected into equal cubes the flour in a container in which flour beetles (*Tribolium confusum*) were wandering. The distribution of the numbers in cubes seems to agree well with the Poisson distribution with the same mean.

Baklemischev (cf. Carpenter 1939) determined the number of animals of various sorts in quadrats, 10 × 10 cm, 3 cm deep, in a uniform field. Some, such as the spiders in Table 8.7, were distributed approximately at random. Others, however, showed more or less marked tendencies to aggregate and thus an excess of quadrats with none or many, and deficiency of intermediate numbers, for example an Oribatid, *Notaspis coleoptratus*, with mean 3.95 and variance 17.95 among quadrats, and earthworms, *Eisenia nordenskiöldi*, with mean 1.30 and variance 3.14.

Cole (1946) determined the numbers of animals under similar boards in a uniform field. A species of spider again showed a random distribution, but others such as the diplopod species (Table 8.8) showed a pronounced tendency to aggregation.

Haldane (1931) analyzed the frequencies of chiasmata in the longest and in five short chromosomes of *Vicia faba*, as reported by Maeda, to test whether they were occurring at random. The data for the short chromosomes in Table 8.8 show that there is a marked excess of ones with three or four chiasmata and a deficiency of the others. Chiasmata clearly exhibit interference. The indicated tests of significance are discussed later.

Vector Compounds of Normal Variables

Vector sums of variations of independent normal distribution are important in several different ways. These frequency functions were introduced by Pearson (1900*a*) for dealing with goodness of fit and are also important in estimating the reliability of standard deviations. By the product principle of probabilities, the element of joint frequency for two independent normal variables is as follows:

$$df(V_1, V_2) = \frac{N}{2\pi\sigma_1\sigma_2} \times \left[\exp - \frac{1}{2}\left[\left(\frac{V_1 - M_1}{\sigma^1}\right)^2 + \left(\frac{V_2 - M_2}{\sigma_2}\right)^2\right] dV_1\, dV_2\right]. \tag{8.99}$$

TABLE 8.8. Further observed (f_o) distributions of numbers of various entities in space in comparison with those calculated (f_c) on the hypothesis of randomness.

SPIDERS

V	Boards f_o	Boards f_c	$f_o - f_c$	$\frac{(f_o - f_c)^2}{f_c}$
0	159	156.9	+2.1	0.03
1	64	66.7	−2.7	0.11
2	13	14.2	−1.2	0.10
3	4	2.0}	+1.8	1.47
4	0	0.2}		
N	240	240.0	0.0	1.71
M	0.425	0.425		$DF=2$
σ^2	0.455	0.425		Probability 0.44

Spiders counted under boards (Cole 1946).

DIPLOPODS

V	Boards f_o	Boards f_c	$f_o - f_c$	$\frac{(f_o - f_c)^2}{f_c}$
0	128	100.6	+27.4	7.46
1	71	95.5	−24.5	6.29
2	34	45.4	−11.4	2.86
3	11	14.4	−3.4	0.80
4	8	3.4}		
5	5	0.6}	+11.9	34.54
6	3	0.1}		
N	260	260	0.0	51.95
M	0.950	0.950		$DF=3$
σ^2	1.669	0.950		Probability 0.000,000

Diplopods counted under boards (Cole 1946).

CHIASMATA

V	Chromosomes f_o	Chromosomes f_c	$f_o - f_c$	$\frac{(f_o - f_c)}{f_o}$
0	0	7.3	−7.3	7.30
1	2	26.6	−24.6	22.75
2	21	48.6	−27.6	15.67
3	104	59.2	+44.8	33.90
4	106	54.0	+52.0	50.07
5	42	39.5	+2.5	0.16
6	6	24.0	−18.0	13.50
7		12.5	−12.5	12.50
8		5.7	−5.7	5.70
9		2.3}		
10		0.9}	−3.6	3.60
11		0.3}		
12		0.1}		
N	281	281.0	0.0	165.15
M	3.651	3.651		$DF=8$
σ^2	0.849	3.651		Probability 0.000,000

Numbers of chiasmata in short chromosome of *Vicia faba* (Haldane 1931).

It is convenient to express this in terms of standardized deviations,

$$x_i = \frac{V_i - M_i}{\sigma_i},$$

and proportional frequencies

$$df_{12} = \frac{1}{2\pi} e^{-(\frac{1}{2})(x_1^2 + x_2^2)} \, dx_1 \, dx_2 .$$

This is a radially symmetrical surface in which the contours of equal frequency are concentric circles with radii $\chi = \sqrt{x_1^2 + x_2^2}$. It is convenient to represent it in polar coordinates

$$df = \frac{1}{2\pi} e^{-\frac{1}{2}\chi^2} \, d\chi \, d(\text{arc}) = \frac{1}{2\pi} \chi e^{-\frac{1}{2}\chi^2} \, d\chi \, d\omega, \tag{8.100}$$

in which $d\omega$ is the element of angle through which χ rotates and $d(\text{arc}) = \chi \, d\omega$ is the element of arc on the circle of equal frequencies to which χ relates. For the total element of frequency for a given value of χ

$$df = \frac{1}{2\pi} \int_0^{2\pi} \chi e^{-\frac{1}{2}\chi^2} \, d\chi \, d\omega = \chi e^{-\frac{1}{2}\chi^2} \, d\chi . \tag{8.101}$$

This is a very asymmetrical distribution, ranging from 0 to ∞. An example is the frequency array of distances from the bullseye on a target, assuming similar normal distributions vertically and horizontally about the bullseye.

The corresponding frequency distribution of deviations from the mean in one dimension, ignoring sign, is obviously a half-normal distribution with mean at $\sqrt{(2/\pi)}$ in terms of a unit standard deviation of the total distribution. In the three-dimensional case, $\chi = \sqrt{x_1^2 + x_2^2 + x_3^2}$.

$$df = \frac{1}{2\pi\sqrt{2\pi}} \chi^2 e^{-\frac{1}{2}\chi^2} \, d\chi \, d\omega , \tag{8.102}$$

in which $d\omega$ is an element of solid angle and $\chi^2 \, d\omega$ is the element of surface on the sphere of equal frequencies to which χ relates. With n dimensions, $\chi = \sqrt{\sum_1^n x^2}$.

$$df = \frac{1}{(\sqrt{2\pi})^n} \chi^{n-1} e^{-\frac{1}{2}\chi^2} \, d\chi \, d\omega . \tag{8.103}$$

By integrating over the shell of equal frequency to obtain the frequency distribution of χ:

$$df = C_n \chi^{n-1} e^{-\frac{1}{2}\chi^2} \, d\chi . \tag{8.104}$$

The value of C_n required to make the total frequency unity can readily be found by means of the reduction formula used in determining the moments of the normal frequency distribution.

$$(8.105)\quad \begin{cases} C_n = \dfrac{n}{n(n-2)\cdots 3\cdot 1}\sqrt{\dfrac{2}{\pi}} = \dfrac{2^{(n-1)/2}[(n-1)/2]!\sqrt{2/\pi}}{(n-1)!} & \text{if } n \text{ is odd,} \\ C_n = \dfrac{n}{n(n-2)\cdots 4\cdot 2} = \dfrac{1}{2^{(n-2)/2}[(n-2)/2]!} & \text{if } n \text{ is even.} \end{cases}$$

The moments about 0 can readily be found from the reduction formula.

$$(8.106)\quad \mu'_1 = M = \begin{cases} \dfrac{2^{(n-1)/2}[(n-1)/2]!C_n}{(n-1)!} & \text{if } n \text{ is odd,} \\ \dfrac{(n-1)!}{2^{(n-2)/2}[(n-2)/2]!}\sqrt{\dfrac{\pi}{2}}\,C_n & \text{if } n \text{ is even.} \end{cases}$$

$$(8.107)\quad \begin{cases} \mu'_2 = n, & \lambda_2 = n - M^2 = \sigma^2, \\ \mu'_3 = (n+1)M, & \lambda_3 = M(1 - 2\sigma^2) = \gamma_1\sigma^3, \\ \mu'_4 = n(n+2), & \lambda_4 = 2n(1-n) + M^2(8n - 4 - 6M^2) \\ & \quad = \gamma_2\sigma^4. \end{cases}$$

The way in which this frequency distribution approaches normality with increasing n may be seen from Table 8.9. Kurtosis is rapidly lost and the mean and standard deviation are close to their limiting values with $n = 30$, but considerable asymmetry persists up to this point.

TABLE 8.9. Mean (M), standard deviation (σ), and γ_1 and γ_2 for the distribution of χ with various values of n.

n	M	σ	γ_1	γ_2
1	0.7979	0.6028	0.9953	0.8692
2	1.2533	0.6551	0.6311	0.2527
3	1.5958	0.6734	0.4857	0.1082
4	1.8800	0.6824	0.4057	0.0593
10	3.0843	0.6978	0.2375	0.0083
20	4.4166	0.7026	0.1630	0.0020
30	5.4318	0.7041	0.1318	0.0013
$n \to \infty$	$\sqrt{n}$	0.7071	0	0

Compounds of Squared Normal Variables

The distribution of squared deviations in a unit normal curve can be written at once. The frequency distribution for u $(=x^2)$ can be found at once from that for x, noting that the frequencies for u correspond to those for both x and $-x$. $[f(u)\,du = 2f(x)\,dx]$ and that $dx/du = 1/(2\sqrt{u})$.

If

$$f(x)\,dx = \frac{1}{\sqrt{2\pi}}e^{-\frac{1}{2}x^2}\,dx, \qquad f(u)\,du = \frac{1}{\sqrt{2\pi u}}e^{-\frac{1}{2}u}\,du,$$

then

$$\varphi(t) = \frac{1}{\sqrt{2\pi}}\int_0^\infty u^{-\frac{1}{2}}e^{u[it-\frac{1}{2}]}\,du. \tag{8.108}$$

The integral is of the form $\int_0^\infty x^n e^{-ax}\,dx$ which equals $\Gamma(n+1)/a^{n+1}$.

$$\varphi(t) = \frac{\Gamma(\frac{1}{2})}{\sqrt{2\pi}\,[\frac{1}{2} - it]^{1/2}} = (1-2it)^{-1/2} \quad \text{since } \Gamma(\tfrac{1}{2}) = \sqrt{\pi}. \tag{8.109}$$

The moments about the origin can be read off as the coefficients of $(it)^s/s!$ in the expansion of this.

$$\begin{aligned}\varphi(t) = 1 + (it) + \frac{3\,(it)^2}{2!} + 15\,\frac{(it)^3}{3!} + 105\,\frac{(it)^4}{4!} + \cdots \\ + [1\cdot 3\cdots(2s-1)]\,\frac{(it)^s}{s!}.\end{aligned} \tag{8.110}$$

The semi-invariants can be derived from these or read off from the coefficient of $(it)^s/s!$ in the expansion of $\log\varphi(t)$.

$$\begin{aligned}\log\varphi(t) &= -\tfrac{1}{2}\log(1-2it) \\ &= (it) + 2\frac{(it)^2}{2!} + 8\frac{(it)^3}{3!} + 48\frac{(it)^4}{4!} + \cdots + 2^{s-1}(s-1)!\frac{(it)^s}{s!}.\end{aligned} \tag{8.111}$$

Thus

$$\begin{cases} \lambda_1 = 1, & M = 1, \\ \lambda_2 = 2, & \sigma = \sqrt{2}, \\ \lambda_3 = 8, & \gamma_1 = 2\sqrt{2}, \\ \lambda_4 = 48, & \gamma_2 = 12. \end{cases} \tag{8.112}$$

This is an extremely asymmetrical distribution, with asymptote at 0.

Letting $\chi_n^2 = \sum^n u = \sum^n \chi^2$ be the sum of contributions from independent variables with the same properties, the semi-invariants of the compound distribution

$$f\left(\sum^n u\right) = \frac{1}{2^{n/2}\Gamma(n/2)}\left(\sum u\right)^{(n-2)/2}e^{-\frac{1}{2}\sum u} \tag{8.113}$$

are simply n times these for one of the variables

$$(8.114)\quad \begin{cases} \lambda_1 = n, & \lambda_4 = 48n, & \gamma_1 = (2\sqrt{2})/\sqrt{n}, \\ \lambda_2 = 2n, & M = n, & \gamma_2 = 12/n, \\ \lambda_3 = 8n, & \sigma = \sqrt{2n}, & \text{Mode} = n - 2. \end{cases}$$

γ_1 and γ_2 are so enormous for the elementary variable that n must be very large before there is much approach to normality.

n	M	σ	γ_1	γ_2	Mode
1	1	1.414	2.828	12	none
2	2	2.000	2.000	6	none
3	3	2.449	1.633	4	1
10	10	4.472	0.894	1.2	8
30	30	7.745	0.516	0.4	28
100	100	14.142	0.283	0.12	98
300	300	24.49	0.163	0.04	298
1000	1000	44.72	0.089	0.012	998
$n \to \infty$	n	$\sqrt{2n}$	0	0	n–2

There is less approach to normality in the distribution of χ^2 with 300 components (or in fact 460) than in that of χ with 30 components. The smallest number at which there is a mode is 3.

The Gamma Distribution

The χ^2 distribution is a special case of an important class known as gamma distributions.

$$(8.115)\qquad f(x) = \frac{a^n}{\Gamma(n)} x^{n-1} e^{-ax}, \qquad x > 0$$

$$(8.116)\qquad \varphi(t) = \left(1 - \frac{it}{a}\right)^{-n}$$

The semi-invariants can be read off from the coefficient of $(it)^s/s!$ in the expansion of $\log \varphi(t)$.

$$(8.117)\quad \begin{cases} \lambda_1 = n/a, & M = n/a, \\ \lambda_2 = n/a^2, & \sigma = \sqrt{n}/a, \\ \lambda_3 = 2n/a^3, & \gamma_1 = 2/\sqrt{n}, \\ \lambda_4 = 6n/a^4, & \gamma_2 = 6/n. \\ \lambda_s = (s-1)!n/a^4, & \end{cases}$$

CHAPTER 9

Tests of Significance

The extraction of descriptive statistics from data is of little value unless these can be supplemented by numbers that indicate their degrees of reliability. Such an index can be obtained by considering the statistic itself as a variable and determining what its standard deviation would be in an array of samples of the same size from an indefinitely large population from which the given sample may be considered to be drawn at random.

Alternative Categories

Consider first an experiment in which the proportional frequencies of two alternatives constitute the array $[p_s(A) + q_s(a)]$ in a sample of size N. Let p_t be the proportion of A's in the theoretical infinite population. It has already been shown (Eq. 8.27) that the expectation for the standard deviation of number of A's in samples of size N is $\sqrt{Np_tq_t}$ and that the distribution rapidly approaches normality as N increases. In terms of proportions, the expected standard deviation is $\sqrt{p_tq_t/N}$. Since deviations of 2.0σ are expected to occur with a frequency of less than 2.5% in each direction, the value of p_s in a single reasonably large sample will usually not differ so much from true p_t as to have much effect on the estimate of the standard deviation. Thus it is customary to give the estimate from a sample in the form

$$p_s \pm \sqrt{\frac{p_sq_s}{N}} \tag{9.1}$$

with the implication that true p_t is within $p_s \pm 2\sqrt{p_sq_s/N}$ with a chance of error of only about 5% or within $p_s \pm 2.6\sqrt{p_sq_s/N}$ with a chance of error of only about 1%.

If there is an a priori probability, p_t, it is more accurate, however, to state that the expectation from a sample of size N is

$$p_t \pm \sqrt{\frac{p_tq_t}{N}} \tag{9.2}$$

and find the chance that accidents of sampling might give a result as different from p_t in one direction or the other as observed p_s. Practically, it makes no appreciable difference in the conclusion that p_s does or does not differ significantly from the a priori value, whether one uses $\sqrt{p_s q_s / N}$ or $\sqrt{p_t q_t / N}$, if N is reasonably large and p_s is not too close to 0 or 1. If p_t is close to 0.5, $p_s q_s$ and $p_t q_t$ differ very little since if $p_s = p_t + \delta_p$, $p_s q_s = p_t q_t + \delta_p(1 - 2p_t) + (\delta_p)^2$ or $p_t q_t + (\delta_p)^2$, if p_t is exactly 0.5.

An illustration may be drawn from the coin-tossing experiment referred to in chapter 8. Each student shook up and tossed 10 coins 100 times. In each set the expectation for the numbers of heads (N_H) was $500 \pm \frac{1}{2}\sqrt{1{,}000}$ or 500 ± 15.81. The actual standard deviation of the 166 observed sets $[\sigma_H = \sqrt{1/165 \sum^{166} (N_H - 500)^2}]$ came out 15.23. That this is in good agreement with the theoretical standard deviation may be seen from its own standard error, which was brought out later to be 0.87.

This formula has frequent application in genetics in the comparison of observed ratios with theoretical Mendelian ones.

Table 9.1 gives data on F_2 and backcross tests of progeny of matings between dark-eyed intense guinea pigs of genotype $CCPP$ and pink-eyed

TABLE 9.1. Observed frequencies (f_o) in the backcross and F_2 generations from the cross between guinea pigs, $CCPP \times c^x c^x pp$ in comparison with those calculated (f_c).

	f_o	f_c	SE	$f_o - f_c$	$\frac{f_o - f_c}{\text{SE}}$	$\frac{(f_o - f_c)^2}{f_c}$
$Cc^xPp \times c^xc^xpp$						
Cc^xPp	154	141.25	± 10.29	+12.75	+1.24	1.15
Cc^xpp	144	,,	,,	+2.75	+0.27	0.05
c^xc^xPp	144	,,	,,	+2.75	+0.27	0.05
c^xc^xpp	123	,,	,,	−18.25	−1.77	2.36
Total	565	565.0		0.0		$\chi^2 = 3.61$
$Cc^xPp \times Cc^xPp$						
$C\text{–}P\text{–}$	696	686.8	+ 17.33	+9.2	+0.53	1.23
$C\text{–}pp$	229	228.9	± 13.64	+0.1	+0.01	0.00
$c^xc^xP\text{–}$	217	,,	,,	−11.9	−0.87	0.62
c^xc^xpp	79	76.3	± 8.46	+2.7	+0.32	0.10
Total	1,221	1,220.9		+0.1		$\chi^2 = 1.95$

TABLE 9.2. Tests of the segregation and recombination ratios from the data in Table 9.1.

	f_o	f_c	SE	$f_o - f_c$	$\frac{f_o - f_c}{\text{SE}}$	$\frac{(f_o - f_c)^2}{f_c}$	
$Cc^xPp \times c^xc^xpp$							
Cc	298	282.5	± 11.89	+15.5	+1.30	0.85	
cc	267	,,	,,	−15.5	−1.30	0.85	1.70
Pp	298	,,	,,	+15.5	+1.30	0.85	
pp	267	,,	,,	−15.5	−1.30	0.85	1.70
$CP/cp + cp/cp$	277	,,	,,	+5.5	−0.46	0.11	
$Cp/cp + cP/cp$	288	,,	,,	− 5.5	+0.46	0.11	0.22
						$\chi^2 =$	3.62
$Cc^xPp \times Cc^xPp$							
C–	925	915.75	± 15.13	+9.25	+0.61	0.09	
cc	296	305.25	,,	−9.25	−0.61	0.28	0.37
P–	913	915.75	± 15.13	−2.75	−0.18	0.01	
pp	308	305.25	,,	+2.75	+0.18	0.02	0.03
C–P– + $ccpp$	775	763.12	± 16.92	+11.88	+0.70	0.18	
C–pp + ccP–	446	457.88	,,	−11.88	−0.70	0.31	0.49
						$\chi^2 =$	0.89

dilutes c^xc^xpp (in which c^x represents alleles of the c series that dilute coat color but do not reduce the eyes to pink, as does the lowest allele c^a). The phenotypes presented no difficulty in classification. Tests of F_1 females (with 227 offspring) and of F_1 males (with 338) gave similar results and are here combined for simplicity. Each of the classes is here tested separately with respect to its deviation from Mendelian expectation. There is no significant suggestion of a real difference from expectation in the frequency of any of the genotypes in either the backcross or F_2. The test in the last column is considered later.

A more analytic procedure is to test the segregation ratios and the recombination ratio separately (Table 9.2). There is no indication of any real deviations from expectation in either segregation or randomness of assortment in either the backcross or F_2. It should be noted, however, that although both segregation and recombination are revealed without complication in the backcross, this is not the case with F_2.

If there is no a priori expectation, the most accurate form of statement on reliability of a statistic is in terms of Neyman's (1941) confidence interval. One specifies the risks of error (e_1, e_2) one is willing to accept in each direction.

This risk is often taken as 2.5% in each direction (5% level) or as 0.5% in each direction (1% level), but need not be the same in the two directions. The value of the statistic that would yield sample values exceeding that observed with the frequency e_1 is the lower limit of the confidence interval, whereas the value that would yield sample values less than that observed with frequency e_2 is the upper limit of this interval.

In the case of alternatives, let p_s be the observed frequency in the sample under consideration, p_1 and p_2 the values relative to which p_s is k standard errors above and below, respectively.

$$p_1 + k\sqrt{\frac{p_1(1 - p_1)}{N}} = p_s = p_2 - k\sqrt{\frac{p_2(1 - p_2)}{N}} \tag{9.3}$$

from which p_1 and p_2 are given by

$$\frac{N}{N + k^2}\left[\left(p_s + \frac{k^2}{2N}\right) \mp k\sqrt{\frac{p_s(1 - p_s)}{N} + \frac{k^2}{4N^2}}\right]. \tag{9.4}$$

If

$$p_s = 1/2,$$

then

$$p_1 = \frac{1}{2} - \frac{k}{2}\sqrt{\frac{1}{N + k^2}} \qquad p_2 = \frac{1}{2} + \frac{k}{2}\sqrt{\frac{1}{N + k^2}}.$$

The confidence limits in this case are symmetrically related to the observed value but are slightly closer together than those indicated by the crude formula $p_s \pm k\sqrt{p_s(1 - p_s/N}$, the ratio being about $(1 - k^2/2N)$.

There is moderate asymmetry if $p_s = 0.25$. Thus with $p_s = 0.25$, $N = 100$, $k = 2$, the limits $p_1 = 0.171$ and $p_2 = 0.348$ differ considerably from the values 0.163 and 0.337 indicated by $0.25 \pm 2\sqrt{3/1{,}600}$.

If $p_s = 0$, the ordinary formula $\sigma_p = \sqrt{p_s q_s/n}$ breaks down, becoming zero although obviously it is not at all certain the true value of p is zero. The upper confidence limit $k^2/(N + k)^2$ represents the situation correctly.

The ordinary formula for the standard error of difference between the values of a statistic in two populations, $\sqrt{SE_1^2 + SE_2^2}$ is represented by the formula $\sqrt{p_1q_1/N_1 + p_2q_2/N_2}$ in the case of proportions. This also is clearly inadequate if the proportion in one population is zero. Thus if p_1 is not zero, but the standard error is very small because of large N_1, while p_2 is zero and N_2 is small, the formula might seem to indicate a highly significant difference that obviously has no justification. An essentially correct conclusion is indicated, however, if p is calculated from the combined population as if it were homogeneous, and the question that is asked is the probability that such a population may give samples differing as much as observed, with a standard deviation of the difference of

$$\sqrt{pq(1/N_1 + 1/N_2)}. \tag{9.5}$$

In the case cited this would not be dominated by the value of p_1 in the large population but by the small number N_2 of the sample. The most accurate evaluation would come from applying the elementary theory of probability to the actual frequencies as samples from a single population.

Maximum Likelihood

It is assumed above that the best estimate of p_t provided by a sample is the frequency p_s in the latter. There are many cases, however, in which the best estimate of a parameter that is of interest is not given directly by any statistic of the sample. We may note here a method due to Gauss and developed further by Edgeworth (1908) and by Fisher (1921) who called it the method of maximum likelihood. Assume that the probability of occurrence of some event is theoretically a function of a parameter ϑ, the value of which it is desired to estimate. Given a set of n independent observations, V, for each of which the probability is of the form $\text{Prob}(V_i) = f_i(\vartheta)$, the probability of the observed set in terms of ϑ_1, a particular value of ϑ, is given by

$$\text{Prob}(\text{Set}|\vartheta_1) = \prod_{i=1}^{n} f_i(\vartheta_1).$$

There is a certain value of ϑ at which this probability is maximum. Since the continuous curve of probabilities relative to varying ϑ is not itself a probability distribution in the absence of knowledge of the distribution of values of ϑ, Fisher proposed to substitute "likelihood" (L) for probability.

$$\begin{aligned} L &= \prod_{i=1}^{n} f_i(\vartheta) \\ \log L &= \sum_{i=1}^{n} \log f_i(\vartheta) \\ \frac{\partial \log L}{\partial \vartheta} &= \sum_{i=1}^{n} \frac{\partial \log f_i(\vartheta)}{\partial \vartheta} = 0 \quad \text{at maximum } L. \end{aligned} \tag{9.6}$$

The solution $\vartheta = \hat{\vartheta}$ is the maximum likelihood estimate of ϑ. If the array of likelihoods for a given sample but varying ϑ is normal, relative to the scale of ϑ's, and L is the maximum likelihood

$$\begin{aligned} L &= \hat{L} \exp\left[-\frac{1}{2}\left(\frac{\vartheta - \hat{\vartheta}}{\sigma_\partial}\right)^2\right] \\ \log L &= \log \hat{L} - \frac{1}{2}\left(\frac{\vartheta - \hat{\vartheta}}{\sigma_\partial}\right)^2 \\ \frac{\partial \log L}{\partial \vartheta} &= -\frac{\vartheta - \hat{\vartheta}}{\sigma_\vartheta^2} \\ \frac{\partial^2 \log L}{\partial \vartheta^2} &= -\frac{1}{\sigma_\vartheta^2} \end{aligned} \tag{9.7}$$

This gives a basis for estimating the variance of the likelihood array in the observed sample. It is not the variance of estimates of ϑ in the probability distribution of such estimates from multiple sets of n observations, but it can be shown that it is a close approximation.

The simplest example is a sample from the array of alternatives

$$[N_1(A) + N_2(a)], \qquad N_1 + N_2 = N.$$

Let p be the unknown probability of (A).

$$\begin{aligned} L &= p^{N_1}(1-p)^{N_2}, \\ \log L &= N_1 \log p + N_2 \log (1-p), \\ \frac{d \log L}{dp} &= \frac{N_1}{p} - \frac{N_2}{1-p} = 0 \end{aligned} \tag{9.8}$$

giving $p = N_1/N$ as expected.

$$\begin{aligned} \frac{d^2 \log L}{dp^2} &= -\frac{N_1}{p^2} - \frac{N_2}{(1-p)^2} = -\left(\frac{N}{p} + \frac{N}{1-p}\right) \\ &= -\frac{N}{p(1-p)} = -\frac{1}{\sigma_p^2}. \\ \sigma_p^2 &= \frac{p(1-p)}{N} \end{aligned} \tag{9.9}$$

in agreement with the previous results.

Estimates of recombination percentages from backcross data are merely a special case of this, but estimates from F_2 data are less obvious. In the following (Fisher 1938), the values for oogenesis c_e and spermatogenesis (c_s) are not estimated separately but merely the products below. The subscripts C and R are used for coupling and repulsion, respectively.

$$\vartheta_{\mathrm{C}} = (1 - c_e)(1 - c_s) \qquad\qquad \vartheta_{\mathrm{R}} = c_e c_s$$

Genotype	$\frac{AB}{ab} \times \frac{AB}{ab}$	$\frac{Ab}{aB} \times \frac{Ab}{aB}$
A–B–	$N_{AB} = (N/4)(2 + \vartheta_{\mathrm{C}})$	$N_{AB} = (N/4)(2 + \vartheta_{\mathrm{R}})$
A–bb	$N_{Ab} = (N/4)(1 - \vartheta_{\mathrm{C}})$	$N_{Ab} = (N/4)(1 - \vartheta_{\mathrm{R}})$
aaB–	$N_{aB} = (N/4)(1 - \vartheta_{\mathrm{C}})$	$N_{aB} = (N/4)(1 - \vartheta_{\mathrm{R}})$
$aabb$	$N_{ab} = (N/4)\vartheta_{\mathrm{C}}$	$N_{ab} = (N/4)\vartheta_{\mathrm{R}}$

In either case

$$L = \left(\frac{2+\vartheta}{4}\right)^{N_{AB}} \left(\frac{1-\vartheta}{4}\right)^{(N_{Ab}+N_{aB})} \left(\frac{\vartheta}{4}\right)^{N_{ab}}.$$

$$\log L = N_{AB} \log\left(\frac{2+\vartheta}{4}\right) + (N_{Ab} + N_{aB}) \log\left(\frac{1-\vartheta}{4}\right) + N_{ab} \log\frac{\vartheta}{4} \tag{9.10}$$

$$\frac{d \log L}{d\vartheta} = \frac{N_{AB}}{2+\vartheta} - \frac{(N_{Ab}+N_{aB})}{1-\vartheta} + \frac{N_{ab}}{\vartheta} = 0 \quad \text{to be solved for } \vartheta.$$

$$\frac{d^2 \log L}{d\vartheta^2} = -\left[\frac{N_{AB}}{(2+\vartheta)^2} + \frac{(N_{Ab}+N_{aB})}{(1-\vartheta)^2} + \frac{N_{ab}}{\vartheta^2}\right] = -\frac{1}{\sigma_\vartheta^2}. \tag{9.11}$$

The best estimate of the variance would be obtained by using the theoretical numbers instead of the observed ones. On making this substitution,

$$\frac{1}{\sigma_\vartheta^2} = \frac{N}{4}\left[\frac{1}{2+\vartheta} + \frac{2}{1-\vartheta} + \frac{1}{\vartheta}\right]. \tag{9.12}$$

If it is assumed that the recombination percentages are the same in oogenesis and spermatogenesis, $c_e = c_s = c = \sqrt{\vartheta_R} = 1 - \sqrt{\vartheta_C}$ and σ_c can be estimated from the approximate relation $\sigma_{f(\vartheta)} = \sigma_\vartheta \, df(\vartheta)/d\vartheta$. Thus $\sigma_c = (d\sqrt{\vartheta}/d\vartheta)\, \sigma_\vartheta = [1/(2\sqrt{\vartheta})]\, \sigma_\vartheta$.

TABLE 9.3. F_2 coupling and F_2 repulsion data from crosses between varieties of sweet pea.

	COUPLING				REPULSION			
Phenotype	f_o	f_c	$f_o - f_c$	$\frac{(f_o - f_c)^2}{f_c}$	f_o	f_c	$f_o - f_c$	$\frac{(f_o - f_c)^2}{f_c}$
Purple, long pollen	4,831	4,821.7	+9.3	0.018	226	210.8	+15.2	1.10
round pollen	390	392.3	−2.3	0.013	95	103.5	−8.5	0.70
Red, long pollen	393	392.3	+0.7	0.001	97	103.5	−6.5	0.41
round pollen	1,338	1,345.7	−7.7	0.044	1	1.3	−0.3	0.07
Totals	6,952	6,952.0	0	0.076	419	419.1	−0.1	2.28
	$\vartheta_C = 0.7743 \pm 0.0074$				$\vartheta_R = 0.0122 \pm 0.0106$			
	$c = 0.1201 \pm 0.0042$				$c = 0.1104 \pm 0.0481$			

There are other ways of estimating ϑ from F_2 data, but unless essentially equivalent to the maximum likelihood estimate, they are less efficient and the standard errors of estimate are larger. Linkage was discovered by Bateson and Punnett in 1906. Their cross involved a gene for flower color and one for pollen shape. In the first data the dominant genes entered together (coupling); some repulsion data were obtained later. The data in Table 9.3 were given by Punnett (1923).

Chi Square

It is important to be able to test not only the deviation of a single frequency from an expected value but also to make a single overall test of a system of frequencies. Pearson applied the χ^2 distribution, discussed in chapter 8, to this end. The correct number of degrees of freedom is due to Fisher (1922).

One of two alternative frequencies is distributed according to the binomial $[p(A) + q(a)]^N$ with mean $Np(A)$, variance $Np(1 - p)(A)^2$, and range from 0 to N(A). A case in which there is no appreciable constraint by this upper limit can be represented by making N indefinitely large and $p(=M/N)$ correspondingly small. A Poisson distribution is approached with mean M and variance and all higher semi-invariants also M. Consider two or more such frequencies, varying wholly independently, each about a mean that is sufficiently large that asymmetry measured by $\gamma_1(=1/\sqrt{M})$ and kurtosis measured by $\gamma_2(=1/M)$ may be ignored. The distribution of each may be represented by an approximately normal curve but one of which the variance is the same as the mean. Use f_{oi} and f_{ci} for the observed and calculated frequencies, respectively. It is convenient to represent the deviations in standard forms $\chi_i = (f_{oi} - f_{ci})/\sigma = (f_{oi} - f_{ci})/\sqrt{f_{ci}}$. For a set of deviations:

$$\chi^2 = \sum\left[\frac{(f_o - f_c)^2}{f_c}\right]. \tag{9.13}$$

The joint distribution for n variables has already been discussed (chap. 8).

The probability of exceeding a given value of χ is given by integrating from this value to infinity if the total frequency is taken as 1.

$$prob_\chi = C\int_\chi^\infty \chi^{n-1}e^{-\frac{1}{2}\chi^2}\,d\chi \Big/ C\int_0^\infty \chi^{n-1}e^{-\frac{1}{2}\chi^2}\,d\chi. \tag{9.14}$$

Tables are available that give the probability of exceeding given values for each number of independent variables (degrees of freedom from 1 to 30) at particular probability levels (0.001, 0.01, 0.02, 0.05, etc. [Fisher and Yates 1938]). As noted earlier the mean value approaches $\sqrt{n}$, its variance approaches $\sqrt{1/2}$, and its form approaches normality with values of n greater than 30. Thus values of $\sqrt{2}\,\chi$ may be taken in such cases as deviating normally from $\sqrt{2n}$ with unit standard deviation. There is a slight bias, however, in calculating $\bar{\chi}$ from $\sqrt{\overline{\chi^2}}$. The variance of χ is given by $\overline{\chi^2} - (\bar{\chi})^2 = n - (\bar{\chi})^2 = \frac{1}{2}$ from which $(\bar{\chi})^2 = n - \frac{1}{2}$. Thus Fisher uses $\sqrt{2\chi^2} - \sqrt{2n - 1}$ as the quantity that has a unit standard deviation in calculating the probabilities of values of χ^2 for n greater than 30.

In practical applications, systems of wholly independent variables are rarely encountered. If nothing else, the sum of all frequencies is nearly always accepted as a datum. For two varying frequencies the joint values are restricted to a line passing through the origin, that on which the sum of the deviations is zero. The distribution of χ is thus one dimensional, half a normal distribution, instead of two dimensional. Similarly with three varying frequencies having values restricted to those for which the sum of the deviations is zero, the joint values are restricted to a plane passing through the origin. The shells of equal probability are concentric circles instead of concentric spheres. Any additional condition that is linear with respect to the frequencies and satisfied by $\chi = 0$ reduces the distribution of joint values by an additional dimension and correspondingly reduces the numbers of degrees of freedom with which one enters the table. The mean central moments, $\mu_s = \sum (V - \bar{V})^s f$, are examples of such conditions. Thus two degrees of freedom are lost in fitting a Poisson distribution because of acceptance of N and M from the data, and three in the case of the normal distribution because of acceptance of N, M, and σ^2.

The assumption that the frequencies vary normally becomes an increasingly poor approximation as the theoretical mean value of the frequency in question decreases. It is accordingly desirable to group classes with small theoretical values so that the total is not too small. Pearson set a lower limit of 3.

Data that may be used to illustrate applications of the χ^2 method have been given earlier. In the coin-tossing experiment (Table 8.1), the value of χ^2 for the 11 classes came out 12.46. Since the only restriction in the distribution $(\frac{1}{2}H + \frac{1}{2}T)^N$ that is drawn from the data is the total number $N = 16{,}600$, there are ten degrees of freedom with probability 0.26 of as great a value by accidents of sampling.

Similarly, only one degree of freedom is lost in fitting the number of dominants in segregating litters of mice by the binomial $(\frac{1}{2}D + \frac{1}{2}R)^L$. The values of χ^2 for all sizes of litter are given in Table 8.3. There is good fit in each case.

The value of χ^2 from all of the litters may be added to give a total χ^2 to be interpreted with a value of n equal to the sum of the degrees of freedom of the experiments. For the number of dominants, this total is so small that the overall probability is about 0.92. This is higher than one would usually expect, but not so high that it casts any doubt on the validity of the data.

For the number of otocephalic monsters, the value of χ^2, the numbers of degrees of freedom, and the approximate probability are given for each size of litter in Table 8.6. The probabilities are a little high because the number of degrees of freedom is given as only one less than the number of classes (after grouping), although each contributes something to the overall estimate of frequency of otocephaly, 5.50% in the strain. An additional degree of freedom is subtracted in the grand total because of the acceptance of the above frequency. Thus the probability 0.66 for the entire body of data is based on the correct number of degrees of freedom. The figures for litters of 2 to 7 are given separately (Prob 0.62) to point out that there is no appreciable tendency toward clustering of otocephalics in litters.

The data on the distribution of 4-toed guinea pigs in litters of various sizes in another inbred strain (Table 8.6) reveals a very different situation. The conspicuous clustering within litters has already been noted. The percentage was calculated separately in each case so that two degrees of freedom are lost. The χ^2 values are so large in all but one case that the probabilities of being due to accidents of sampling are utterly negligible. Even in the litters of 2 from the Chicago branch the probability is only 0.003.

Among the fits to Poisson distribution (Tables 8.7 and 8.8), those for red blood cells, beetles in cubes of flour, and spiders, as observed by Baklemischev and by Cole, agree reasonably well. The distributions of diplopods under boards and of chiasmata do not agree at all, however, for opposite reasons. The data from the two-factor backcross and F_2 in Table 9.1 all show good fit by the χ^2 test, with one less degree of freedom than the number of classes because of acceptance of the total number.

An important property of χ^2 is illustrated by comparison of the values from the four classes of the backcross experiment ($\chi^2 = 3.61$, $n = 3$) and the values from the separate analyses of the two segregations and recombination (each with $n = 1$). The sum of the three values of χ^2 in these is exactly the same, algebraically, as the total χ^2 for the four classes, and the total degrees of freedom in both cases is three. These relations do not hold for F_2. The fact that the three cleavages of the backcross data are completely independent (orthogonal) permits separation of the three degrees of freedom and determination of their separate contributions to χ^2 exactly. The F_2 data, on the other hand, do not permit three orthogonal cleavages.

In the coupling and repulsion data of Table 9.3 from which recombination was estimated by the method of maximum likelihood, there are two degrees of freedom each. The fit was extraordinarily good in the coupling data

($\chi^2 = 0.076$, Prob 0.97), but not especially good in the repulsion data ($\chi^2 = 2.28$, Prob 0.24).

An important application of χ^2 is in testing whether two empirical distributions can be interpreted as samples of a single population. In this case, the marginal totals are accepted and give the basis for calculating theoretical values that maintain the same proportionality among the columns and among the rows. With K columns and L rows the number of degrees of freedom is $(K - 1)(L - 1)$.

The simplest case is that of a 2×2 table with one degree of freedom. Table 9.4 compares guinea pigs of a certain phenotype "silver white" (genotype *sisi dmdm*) with littermates that were not of this phenotype, with respect to whether they were born dead or alive. Note that the contributions

TABLE 9.4. Comparison of numbers of still and live births among silver-white guinea pigs (*sisi, dmdm*) with those among littermates, not of this phenotype.

	AT BIRTH			DEAD			
	Dead (f_o)	Alive	Total	%Dead	f_c	$f_o - f_c$	$\frac{(f_o - f_c)^2}{f_c}$
Silver white	43	109	152	28.29	40.13	+2.87	0.205
Others	51	153	204	25.00	53.87	−2.87	0.153
Total	94	262	356	26.40	94.00	0	0.358

$\chi^2 = \frac{356}{262} \times 0.358 = 0.486 \qquad \chi = 0.697 \qquad$ Prob ca. 0.49

to χ^2 from those that died must be supplemented by the contribution for those born alive by multiplying by 356/262. There is no significant difference between silver-whites and nonsilver-white littermates in mortality at birth. In Table 9.5 the live-born young in the preceding table are compared with respect to mortality up to weaning at 21 days. The mortality of the silver whites between birth and weaning is very significantly greater than that of their nonsilver-white littermates.

These data can also be tested by comparing the mortality percentages of the two types by the formula presented earlier. In the case of mortality at birth

$$t = \frac{0.2829 - 0.2500}{\sqrt{0.2640(1 - 0.2640)(1/152 + 1/204)}} = 0.697.$$

TABLE 9.5. Comparison of numbers of silver-white guinea pigs that died between birth and weaning, or survived, with those of littermates not of this phenotype.

	LIVE-BORN			DIED			
	Died (f_o)	Weaned	Total	% Died	f_c	$f_o - f_c$	$\frac{(f_o - f_c)^2}{f_c}$
Silver white	40	69	109	36.70	28.29	+11.71	4.847
Others	28	125	153	18.30	39.71	−11.71	3.453
Total	68	194	262	25.95	68.00	0	8.300

$$\chi^2 = \frac{262}{194} \times 8.300 = 11.209 \qquad \chi = 3.35 \qquad \text{Prob} = 0.0008$$

This is identical with $\chi = \sqrt{0.486} = 0.697$.

In the case of postnatal mortality before weaning

$$t = \frac{0.3670 - 0.1830}{\sqrt{0.2595(1 - 0.2595)(1/_{109} + 1/_{153})}} = 3.35.$$

This is identical with $\chi = \sqrt{11.209} = 3.35$.

There is in fact algebraic identity of t and χ. It is to be noted that the area beyond the deviation χ from the origin on half a normal curve (the whole distribution of χ) corresponds to the areas beyond deviations of $+t$ and $-t$ in the complete normal distribution.

It is possible to calculate the exact probabilities of any sets of frequencies in the four cells of the 2×2 table from elementary probability theory. This gives a step distribution of which the normal distributions of the preceding methods are only approximations. The probability of a deviation greater than t (or χ) corresponds to the probability of a deviation greater than that observed plus about half of the observed class. The full probability of as great a deviation as that observed is thus greater by about half the probability of the observed class. A correction suggested by Yates (1934) is often made by subtracting 0.5 from the positive differences ($f_o - f_c$) of all frequencies in the 2×2 table. If this is done χ^2 is reduced to 0.332 ($\chi = 0.58$, Prob 0.56 instead of 0.49) in the case of mortality at birth, and to 10.27 ($\chi = 3.21$, prob 0.0013 instead of 0.0008).

Table 9.6 shows the numbers of guinea pigs with three toes or four toes on the hind feet in inbred strain no. 35, according to month of birth. In

TABLE 9.6. Analysis of numbers of 3-toed and 4-toed guinea pigs of strain 35 born in each month in the Beltsville colony.

Month	3-toed	4-toed f_o	Total	% 4-toed	4-toed f_c	$f_o - f_c$	$\frac{(f_o - f_c)^2}{f_c}$
1	89	71	160	44.4	49.7	+21.3	9.13
2	90	59	149	39.6	46.3	+12.7	3.48
3	83	65	148	43.9	46.0	+19.0	7.85
4	92	49	141	34.8	43.8	+5.2	0.62
5	97	39	136	28.7	42.3	−3.3	0.26
6	121	39	160	24.4	49.7	−10.7	2.30
7	175	56	231	24.2	71.8	−15.8	3.48
8	125	51	176	29.0	54.7	−3.7	0.25
9	106	34	140	24.3	43.5	−9.5	2.08
10	164	51	215	23.7	66.8	−15.8	3.74
11	109	52	161	32.3	50.0	+2.0	0.08
12	111	48	159	30.2	49.4	−1.4	0.04
	1,362	614	1,976	31.1	614.0	0.0	33.31

$$\chi^2 = \frac{1,976}{1,362} \times 33.31 = 48.33 \qquad DF = 11 \qquad \text{Prob} = 0.000,002$$

this case there are eleven degrees of freedom $(2 - 1) \times (12 - 1)$. It is again only necessary to calculate contributions to χ^2 for one column. To allow for the contributions from the other column, the sum obtained for one column may be multiplied by the ratio of the total number to the number in the column not used.

There is no question about the significance of the heterogeneity revealed by $\chi^2 = 48.33$ with eleven degrees of freedom. Moreover, this significance takes no account of the grouping of signs of differences $(f_o - f_c)$ indicating excess of polydactyly in the six months, November to April inclusively with one minor exception, and deficiency in all months from May to October inclusively. This grouping largely justifies a condensation into a 2×2 table according to these two six-month periods.

With Yates correction, χ^2 is reduced to 32.2, χ to 5.67. The grouping reduces the probability by about one-hundred-fold, but as it was negligibly small without grouping, this does not affect the interpretation. In other cases, however, it might.

Table 9.8 shows the colors of the sires and dams of 6,000 Shorthorn calves (1,000 males and 1,000 females chosen at random from each of the British, American, and Canadian herd books for 1920).

TABLE 9.7. Condensation of Table 9.6.

Months	3-toed	4-toed (f_o)	Total	% 4-toed	4-toed (f_c)	($f_c - f_o$)	$\frac{(f_o - f_c)^2}{f_c}$
November–April	574	344	918	37.5	285.3	+58.7	12.08
May–October	788	270	1,058	25.5	328.7	−58.7	10.48
	1,362	614	1,976		614.0	0	22.56

$$\chi^2 = \frac{1{,}976}{1{,}362} \times 22.56 = 32.7 \qquad \chi = 5.72 \qquad DF = 1 \qquad \text{Prob} = 10^{-8}$$

With $\chi^2 = 12.15$ and four degrees of freedom, there is significant departure from random mating (Prob = 0.017). It is to be noted, however, that nearly 58% of χ^2 is due to the smallest class, white × white, with 14 matings where 28 are expected. If there had been 24 matings of this sort, the probability would have risen to more than 0.05 with no clear evidence of nonrandom

TABLE 9.8. Comparison of frequencies of matings among Shorthorn cattle of different colors with those expected under random mating.

Dam \ Sire	Red	Roan	White	Total	Red	Roan	White	Total
	Observed (f_o)				Under random mating (f_c)			
Red	1,252	1,717	315	3,284	1,243	1,751	290	3,284
Roan	892	1,310	201	2,403	909	1,282	212	2,403
White	126	173	14	313	118	167	28	313
Total	2,270	3,200	530	6,000	2,270	3,200	530	6,000
	$f_o - f_c$				$(f_o - f_c)^2/f_c$			
Red	+9	−34	+25	0	0.06	0.66	2.16	2.88
Roan	−17	+28	−11	0	0.32	0.63	0.57	1.52
White	+8	+6	−14	0	0.54	0.21	7.00	7.75
Total	0	0	0	0	0.92	1.50	9.73	12.15

$\chi^2 = 12.15$; $DF = 4$; Prob ca. 0.017

mating. It may be concluded that there is a slight tendency to avoid white × white matings, but as less than 0.5% are expected even under random mating, the departure from random mating is very slight.

The fact that values of χ^2 from any number of wholly separate experiments can always be added to give a joint χ^2 for the whole body of data, to be interpreted with a total *DF* equal to the sum of the degrees of freedom of the experiments, applies a very severe test to data that seemed to fit the results of the separate experiments satisfactorily. The probability of the total χ^2 may well be so low that at least some qualification of the hypothesis may be necessary. On the other hand, one may have the problem of accounting for a probability of 0.99 or more. An excessively good fit usually indicates a bias somewhere. There may have been an unconscious tendency to interpret doubtful observations in accordance with the theory. There may have been a biased exclusion of unsatisfactory results for what seemed good reasons in each case. Circular reasoning may have been involved in allocating cases, such as those in which multifactorial genotypes are assigned to parents in part on the basis of their progeny and the goodness of fit of the ratios among the latter is under test (cf. Wright 1935*b*, Wright and Chase 1936, Fisher 1936). In a few cases, analysis has indicated deliberate manipulation of the data (Philip and Haldane 1939). Only the most rigorous objectivity in classifying results can stand up under extensive testing by χ^2. Conversely, the testing of a body of work by this means requires equal objectivity on the part of the tester.

Some Pitfalls in Interpretation

The interpretation of probabilities is a subject that has many pitfalls. A very small probability does not necessarily require rejection of the hypothesis on which it was based. Thus if one records the sequence of heads and tails in 1,000 throws of a coin, one gets a result that a priori would have had a probability of $\frac{1}{2}^{1,000}$, or less than 10^{-300}, of occurring. Having occurred, it does not arouse any suspicion of the hypothesis on which the probability is based since whatever sequence is actually obtained, its a priori probability was the same. The probability of any event that involves a long chain of independent events is always very small. A low probability is ordinarily used as a reason for rejection of a hypothesis only if it refers to a simple alternative.

There are usually considerations other than the particular data that should be taken into account. One of the students in the coin-tossing experiment discussed in chapter 8 reported only 453 heads in the 1,000 throws, and another 544 heads. These are 2.97 and 2.78 times the standard deviation with probabilities of being exceeded of 0.0015 and 0.0027 in the same directions and thus decidedly improbable in themselves. There were, however, 166 sets. The expected number of cases that will deviate as much as 47 one way or the other is 0.49, and the expected number that will deviate as much as 44 one way or the other is 0.90. Thus one actual case of the former and one of the latter are not especially surprising. Again a report of 500 heads is the most probable single result. It is, however, a result with deviation from expectation that has a probability of 1.00 of being equalled or exceeded. Does this make it improbably good? The probability of getting just this result is $1{,}000!/[500!]^2 2^{-1{,}000} = 0.0252$. Thus the expectation is 4.2 cases in 166 sets. There actually were 5 cases.

Another difficulty concerns the use that should be made of a priori probabilities and the application of Bayes' theorem. A test for linkage in the guinea pig was made from backcrosses of the type $RrPxpx \times rrpxpx$. There were 178 recombinants in 422 offspring or 42.2% (Wright 1941). Since the deviation from 50% is 3.25SE, indicating a probability of about 0.00057 of as great a deviation in this direction by accidents of sampling, there is strong evidence for linkage. There is, however, an unusually heavy burden of proof on linkage in the guinea pig because of the presence of 32 autosomes (neither locus was sex linked). The chance that two different autosomes were involved and produced as few as 42.2% recombinants is about $^{31}/_{32}$ (0.00057) = 0.00053. Since experience with other organisms indicates that recombination percentages are distributed rather uniformly between 0% and 50% the probability that only one autosome would be involved and would show recombination as great as 42.2% is $^{1}/_{32} \times [(0.500 - 0.422)/0.500] = 0.0049$. From this viewpoint the probability that 2 genes would show such weak linkage as observed is only some 9 times as great as the probability that 2 genes would show as few recombinants as observed, although in different chromosomes, but both probabilities are slight. So far no allowance has been made for the fact that 10 loci were being tested for linkage in every possible combination at this time. The test of RPx was the only one of the 45 tests that gave any suggestion of linkage in fairly adequate numbers. The chance of observing either of the above phenomena was thus much greater than indicated above and that for observing weak linkage is no longer small (about 0.22). It may be added that some repulsion data of a rather unsatisfactory sort and considerable additional later data confirmed the conclusion that there is weak linkage of R and Px (final deviation 3.9 SE; prob = 0.00005 in the observed direction).

The systematic testing of a large number of loci for linkage is a project in which an individual geneticist may find it useful to use a method for arriving at an equable decision on when to stop in each case. One might decide to produce a certain constant number of backcross offspring in every test. This, however, would involve raising more than enough for the purpose of a survey in cases in which strong linkage is immediately apparent but not enough, perhaps, for reliable identification of cases in which there may be recombination of 40% or more. The sequential method developed by Wald (1947) seems to be the most generally suitable. We will consider only the simplest application, that aimed at irrevocable acceptance or rejection of an interpretation in each of a number of cases.

As a more typical example of its use consider the case of a seed inspector who has the task of deciding whether carload lots of clover seed have more or less than a legally permissible admixture of thistle seed. No amount of testing of samples, short of a count of the entire carload, would be enough for a certain decision if the actual admixture is very close to the legal limit. The procedure under Wald's method is to set limits, p_1 and p_2, slightly below and above the legal limit covering a range within which it is considered that an erroneous decision is unimportant. Samples are examined until the accumulated total percentage is so much below the upper limit p_2 that there is an acceptably small probability α of the error of accepting a lot in which p is really greater than p_2 or until the accumulated percentage falls so much above the lower limit p_1 that there is an acceptably small probability β of the opposite error of rejecting a lot in which p is really smaller than p_1. The decision thus rests on the choice of four parameters, p_1, p_2, α, β, on a priori grounds. If the true probability array is $[p_1(A) + (1 - p_1)a]$ the probability of a sample with the array $[m(A) + (n - m)(a)]$ is proportional to $p_1{}^m(1 - p)^{n-m}$.

Tests of Moments

The mean of a random sample of size N, $M_s = (1/N) \sum^N V_i$ may be looked upon as a compound variable in which all components are drawn from the same total population and are weighted by $1/N$. The semi-invariants of an indefinitely large array of such sample means may thus be written at once from the additive property of these statistics.

$$\lambda_k(M_s) = \frac{1}{N^k} \lambda_k\left(\sum^N V_i\right) = \frac{1}{N^{k-1}} \lambda_k(V_i). \tag{9.15}$$

The mean of an indefinitely large array of sample means approaches the mean of the total population. The variance of such an array of sample means approaches σ^2/N. If variance of the sample, $\sigma_s{}^2$, is the best available estimate of σ_t^2, the true variance of the total population, the best estimate of the standard error of means is given by the very important formula

$$\text{SE}_M = \frac{\sigma_s}{\sqrt{N}}. \tag{9.16}$$

It is evident from the higher semi-invariants of sample means that the distribution of these approaches normality with an increase in the size of the sample, irrespective of the form of distribution of the total population, provided that all moments of the latter are finite.

The semi-invariants of sample means of functions of the variables can be found similarly. Of special importance are the sample moments about a fixed origin $\mu'_{g(s)} = (1/N) \sum^N (V_i - a)^g$ using g here to indicate which moment.

$$\lambda_k(\mu'_{g(s)}) = \frac{1}{N^{k-1}} \lambda_k(V_i - a)^g. \tag{9.17}$$

The mean in an indefinitely large array of samples approaches the value of the moment in question in the total population. The variance of such an array approaches $(1/N)[\mu'_{2g} - (\mu'_g)^2]$.

If the population mean is assumed to be the origin, these formulas give the semi-invariants of the central moments of samples in terms of the central moments of the total population. The most important case is that of the sample variances

$$\sigma^2_{\mu_{2(s)}} = \frac{1}{N}\left[\mu_4 - \mu_2{}^2\right]. \tag{9.18}$$

Use of the moments of the sample as estimates of those of the total population introduces a bias which, however, is very small if the sample is large (cf. Cramér 1946).

The variance of sample standard deviations can be found approximately from the relation $\delta\sigma_s = \delta\sqrt{\mu_{2(s)}} = [1/(2\sqrt{\mu_2})]\,\delta\mu_2$.

$$\sigma^2_{\sigma_s} = \frac{1}{4\sigma^2}\left(\frac{\mu_4 - \mu_2{}^2}{N}\right). \tag{9.19}$$

The formulas for the variances of the form statistics of samples ($\gamma_{1(s)}$ and $\gamma_{2(s)}$) involve such high moments (up to μ_6 in the case of $\sigma^2\gamma_{1(s)}$ and up to μ_8 in the case of $\sigma^2\gamma_{2(s)}$) that reliable calculation from actual data is ordinarily out of the question. In the special case in which the distribution of the total population is normal these reduce to $\sigma^2(\gamma_{1(s)}) = 6/N$, $\sigma^2(\gamma_{2(s)}) = 24/N$ respectively, disregarding the corrections for uncertainty of the mean that are negligible in cases in which N is large enough to warrant estimation of γ_1 and γ_2 at all. The principal use of the derived standard errors is as a basis for judging whether γ_1 and γ_2 of the sample deviate significantly from zero.

$$\text{(9.20)} \qquad \text{SE}\ \gamma_1 = \sqrt{6/N} \quad \text{on hypothesis that } \gamma_1 = 0, \gamma_2 = 0.$$

$$\text{(9.21)} \qquad \text{SE}\ \gamma_2 = 2\sqrt{6/N} \quad \text{on hypothesis that } \gamma_1 = 0, \gamma_2 = 0.$$

If the distribution of the total population is normal ($\mu_4 = 3\mu_2{}^2$) the standard error of the variance reduces to approximately

$$\text{(9.22)} \qquad \text{SE}\ \sigma_s{}^2 = \sqrt{2/N}\,\sigma^2.$$

Differences cannot ordinarily be interpreted by use of the normal probability integral, however, because the distribution of sample variances is very far from normal even with rather large numbers. It is, indeed, obvious that the distribution of sample variances is of the χ^2-type if the distribution of individual observations is normal.

On the other hand, the standard error of the standard deviation is of the χ-type in this case and thus approaches normality sufficiently closely in samples of 30 or more for use of the normal probability integral.

$$\text{(9.23)} \qquad \text{SE}\ \sigma = \sigma/\sqrt{2N}$$

on the hypothesis of normality of the observation.

The mean, variance, and standard deviation were determined for each of the sets of 100 throws of 10 coins referred to earlier. The actual statistics of the 166 sets may be compared with the theoretical values. Those for the 16,600 throws of 10 coins are repeated here for convenience. The standard errors of the observed values are based on these.

The agreement between the observed statistics of the 166 sets and the expected is very good, and it makes no substantial difference whether the Gaussian corrections are made or not. As noted, the effect of a correction is always of a lower order than the standard error. There is an inevitable bias in calculating the standard deviation as the square root of the unbiased estimate of the variance since the difference between the mean variance $\overline{\sigma^2}$, and the square of the mean standard deviation $(\bar{\sigma})^2$ where

$$\sigma^2 = \frac{\sum V^2 - \bar{V} \sum V}{N - 1}$$

and $\sigma = \sqrt{\sigma^2}$ is essentially the variance of standard deviations,

$$\sigma_\sigma{}^2 = \frac{\sum \sigma^2 - \bar{\sigma} \sum \sigma}{N - 1},$$

TABLE 9.9. Analysis of coin-tossing data of Table 8.1.

Formula	SE	No Correction	Gaussian Correction	SE_0	Theoretical Value	SE_T	$\frac{Dev}{SE_0}$	$\frac{Dev}{SE_T}$
Total Array $N = 16{,}600$	$\mu_3 = 0.1038$ (theory 0.00)	$\mu_4 = 17.61$ (theory 17.50)						
$M = \frac{\sum V}{N}$	$\frac{\sigma}{\sqrt{N}}$	5.0027		0.0123	5.0000	0.0123	+ 0.22	+ 0.22
$\sigma^2 = \frac{\sum V^2 - M \sum V}{N - 1}$	$\sqrt{\frac{\mu_4 - \sigma^4}{N}}$	2.5040	2.5041	0.0261	2.5000	0.0260	+ 0.16	+ 0.16
$\sigma = \sqrt{\sigma^2}$	$\frac{1}{2\sigma} SE_{\sigma^2}$	1.5824	1.5824	0.0083	1.5811	0.0082	+ 0.16	+ 0.16
$\gamma_1 = \frac{\mu_3}{\sigma^3}$	$\sqrt{6/N}$	0.026	0.026	0.019	0	0.019	+ 1.37	+ 1.37
$\gamma_2 = \frac{\mu_4 - 3\sigma^4}{\sigma^4}$	$2\sqrt{6/N}$	− 0.191	− 0.191	0.038	− 0.200	0.038	− 0.24	− 0.24
Array of Means $M = \frac{\sum V}{100}$ $N_M = 166$								
$\overline{M} = \frac{\sum M}{N_M}$	$\frac{\sigma_M}{\sqrt{N_M}}$	5.0027		0.0119	5.0000	0.0123	+ 0.22	+ 0.22
$\sigma_M^2 = \frac{\sum M^2 - \overline{M} \sum M}{N_M - 1}$	$\sqrt{\frac{2}{N_M}} \sigma_M^2$	0.02319	0.02333	0.00256	0.02500	0.00274	− 0.66	− 0.61
$\sigma_M = \sqrt{\sigma_M^2}$	$\frac{\sigma_M}{\sqrt{2N_M}}$	0.1523	0.1528	0.0087	0.1581	0.0087	− 0.67	− 0.61
Array of Standard Deviations $\sigma = \sqrt{\frac{\sum V^2 - M \sum V}{100}}$ or $\sqrt{\frac{\sum V^2 - M \sum V}{99}}$, $N_\sigma = 166$								
$\bar{\sigma} = \frac{\sum \sigma}{N_\sigma}$	$\frac{\sigma_\sigma}{\sqrt{N_\sigma}}$	1.5711	1.5790	0.0088	1.5772	0.0087	− 0.70	+ 0.21
${\sigma_\sigma}^2 = \frac{\sum \sigma^2 - \bar{\sigma} \sum \sigma}{N_\sigma - 1}$	$\sqrt{\frac{2}{N_\sigma}} \sigma_\sigma^2$	0.01255	0.01275	0.00140	0.01250	0.00137	+ 0.04	+ 0.18
$\sigma_\sigma = \sqrt{\sigma_\sigma^2}$	$\frac{\sigma_\sigma}{\sqrt{2N_\sigma}}$	0.1120	0.1129	0.0062	0.1118	0.0061	+ 0.03	+ 0.18

the latter is a real quantity, the empirical value of which agrees excellently with its theoretic value. Pearson usually ignored the Gaussian corrections. According to Steffensen (1930), "It appears thus that neither of the two systems of presumptive values of frequency constants which have so far been proposed, respectively by Thiele and Tschuprow, is free from contradictions and that a strong case can be made even against the time-honored Gaussian formula $\bar{m}_2 = [n/(n-1)]m^2$." Correct determination of the number of degrees of freedom is, however, of prime importance in interpreting χ^2 and certain other statistics discussed later.

If the value of a statistic μ has been determined from a number of samples, the best estimate of the grand average is given by $\sum w\mu/\sum w$ where the weight of each determination is given by $w = 1/\sigma_\mu{}^2$. Thus in the case of a mean of means, $w = N/\sigma^2$. The standard error is given by $\sqrt{1/\sum w}$. It should be emphasized that such estimates are not legitimate unless suitable tests indicate that the samples can properly be treated as drawn from a single population.

Poisson Distributions

The theoretical identity of the mean and variance of a Poisson distribution provides a delicate means of testing for deviations from random dispersion. If the numbers are reasonably large the test can be made by comparing $(\sigma - \sqrt{M})$ with its standard error. The variance of means is $\sigma^2/N = M/N$ and thus that of $\sqrt{M}$ is $(1/4M)\sigma_M{}^2 = 1/4N$. The variance of variances is $(\mu_4 - \mu_2{}^2)/N = (M + 2M^2)/N$ and thus that of σ is $(1/4M)\sigma^2_{\sigma^2} = (1 + 2M)/4N$. The deviations of the standard deviation, σ, from its expected value $\sqrt{M}$, are independent of the latter, $\sigma^2_{(\sigma - \sqrt{M})} + \sigma^2{}_{\sqrt{M}} = \sigma_\sigma{}^2$. Thus

$$\sigma^2_{(\sigma-\sqrt{M})} = \sigma_\sigma{}^2 - \sigma^2{}_{\sqrt{M}} = \frac{M}{2N}, \qquad t = \frac{\sigma - \sqrt{M}}{\sqrt{M/2N}}. \tag{9.24}$$

An alternative procedure is to compare the frequency in each cell with the mean as its expected frequency, by the χ^2 method. As the value of the latter (M) is here constant, the formula for χ^2 reduces to

$$\frac{1}{M}\sum(V - M)^2 f_o = \frac{N-1}{M}\sigma^2,$$

with $(N - 1)$ degrees of freedom. It has been noted that with 30 or more degrees of freedom $(DF = N - 1)$, the quantity $\sqrt{2\chi^2} - \sqrt{2n-1}$ may be treated as a normal deviant with unit standard deviation. In the present case this differs from $t = \sqrt{2N/M}(\sigma - \sqrt{M})$ by quantities that are of lower order than the standard error and so are of no importance in interpreting individual cases. The most accurate test is, however, to calculate χ^2 as above and interpret accordingly.

The probabilities of differences between standard deviation and square root of mean (or the essentially equivalent χ^2 test of the variance) for the three examples in Table 8.7 agree fairly well with those for the χ^2 test of the class frequencies in two of the cases. In that of Baklemischev's spiders, the class frequencies agree well enough with a Poisson distribution (Prob 0.23) but the standard deviation seems to be significantly greater than $\sqrt{M}$ (2.36 SE, Prob 0.02). These are not tests of quite the same thing. The χ^2 test for the class frequencies does not take into account the distribution of signs of differences. Only one of the three cases in Table 8.7 agrees.

Small Numbers

The standard deviation of a sample becomes a highly unreliable substitute for that of the total population if the number of observations is small. M. S. Gosset (who wrote under the nom de plume of "Student") brought out that even if there is good reason to believe that the total distribution is normal, the distribution of the ratio of deviations of the sample means to the standard error

$$t = \frac{\bar{V} - m}{\sigma_{\bar{V}}} \quad \text{where } \bar{V} = \frac{1}{N}\sum^{N} V \quad \text{and} \quad \sigma_{\bar{V}}^2 = \frac{\sum^{N}(V - \bar{V})^2)}{N(N-1)}$$

is leptokurtic instead of normal.

$$\text{(9.25)} \qquad y = y_o\left(1 + \frac{t^2}{n}\right)^{-\frac{(n+1)}{2}}$$

where n is the number of degrees of freedom, here $(N - 1)$.

"Student" calculated the portion of the area of the curve of unit area below values of t for 0 to 6.0 at intervals of 0.1 and values of n from 1 to 20 (cf. Yule and Kendall (1937)). Fisher and Yates (1938) give the value of t that has a specified probability of being exceeded in one or the other direction (0.001, 0.01, 0.02, 0.05, etc.) for degrees of freedom from 1 to 30 and certain larger values. $P_F = 2(1 - P_S)$ where P_S and P_F are the probabilities given by "Student" and Fisher, respectively.

The conventional borderline of significance where the number of observations is large enough to justify use of the normal probability integral is at $t = 1.96\sigma$, which has a probability of 0.05 of being exceeded in one or the other direction. With 20 degrees of freedom this value implies a corresponding probability of 0.064. Similarly, the values of t of 2.58, 3.29, and 3.89 that imply probabilities of 0.010, 0.001, and 0.0001, respectively on the basis of the normal curve, imply probabilities of 0.018, 0.0036, and 0.0008 on the basis of "Student's" curve with 20 degrees of freedom. There is evidently some danger of misinterpreting significance if n is as small as 20. Serious errors of interpretation are not likely with n greater than 30.

The most important application of "Student's" distribution is to a series of paired observations. It is easier to insure that the conditions underlying two observations shall be virtually identical in all but one respect under investigation than that this shall be true throughout two large sets. Paired observations may, of course, be expected to be correlated if other conditions vary from pair to pair, but the correlation term in the formula for the variance of a difference ($\sigma^2_{M_1-M_2} = \sigma^2_{M_1} + \sigma^2_{M_2} - 2\sigma_{M_1}\sigma_{M_2}r_{M_1M_2}$) is automatically taken care of by using the set of differences as the individual observations.

As an example consider the following mean weights at 33 days of the 2 heaviest of 5 inbred strains of guinea pigs that were compared over a period of 9 years (Table 9.10).

The weights of both improved in moderately close parallelism. On the average, strain 13 was 11.0 grams heavier with a standard error of 3.365

TABLE 9.10. Mean weights at 33 days of age of guinea pigs in two inbred strains over a period of nine years.

Year	Strain 13	Strain 35	Difference
1916	190.7	195.4	−4.7
1917	206.3	206.7	−0.4
1918	239.3	223.0	+16.3
1919	240.4	227.2	+13.2
1920	259.0	232.4	+26.6
1921	228.6	207.1	+21.5
1922	234.2	230.3	+3.9
1923	256.3	245.9	+10.4
1924	257.0	244.8	+12.2
M	234.64	223.64	+11.0
σ_M	7.815	5.827	3.365
t			3.27
n			8
Prob			0.011

grams. The ratio $t = 3.27$ with 8 degrees of freedom indicates the probability of obtaining as great and as variable a series of difference is only about 0.011.

This result may be compared with the results of a test of this same difference on the hypothesis that these are merely two independent series of determinations from two populations. The conventional formula for the standard error of the difference between the means of two such populations is as noted $\sigma_{M_1 - M_2} = \sqrt{\sigma_{M_1}^2 + \sigma_{M_2}^2}$. As the samples are small it is desirable, if it seems legitimate, to obtain a joint estimate of the variance from Fisher's (1925) formula

$$\sigma^2 = \frac{\sum (V_1 - M_1)^2 + \sum (V_2 - M_2)^2}{N_1 + N_2 - 2}. \tag{9.27}$$

The formula for the standard error is then

$$\text{SE}_{\bar{\Delta}} = \frac{M_1 - M_2}{\sigma\sqrt{1/N_1 + 1/N_2}}, \tag{9.28}$$

yielding $\sigma_{\bar{\Delta}} = 9.75$, $t = 11.0/9.75 = 1.13$, which with 16 degrees of freedom indicates a probability of about 0.28 that accidents of sampling could yield as great a difference between two samples from the same populations. The advantage of the pairing by years is clearly brought out.

The legitimate use of "Student's" formula depends on a well-founded a priori belief that the distribution of the array of differences does not differ seriously from normality. The conclusions may be wholly erroneous if the distribution is strongly leptokurtic.

The calculation of a joint variance as above may lead to an incorrect conclusion if the numbers are very different, and if the variance of the array with small numbers is really much greater than that with large numbers. The latter dominates in the determination of the joint variance and, by leading to a gross underestimate of the variability of the small array, may indicate a significant difference where there is no appreciable real difference. Finally, considerable care should be exercised with variables that are discrete or, if continuous, grouped in broad classes, if the number of degrees of freedom is extremely small. As an extreme case, assume that two matings of one inbred strain of guinea pig have each produced a litter of two, and two matings of another strain have each produced a litter of three. As litters of two or three are much the most frequent sizes in inbred strains, it is obvious that very little significance can be attributed to this result. Yet the joint variance within matings is zero, t is infinite, and there are two degrees of freedom.

On the other hand, there is an important class of paired comparisons in which use of the method without correcting for known departures from normality may indicate less significance than is warranted. This is in cases in which the effect of factor differences is a function of position on the scale. The appropriate transformation of scale increases the consistency of the differences and thus increases t.

TABLE 9.11. Comparison of median percentages of white in males and females of seven large inbred arrays of guinea pigs and of the inverse probability integrals of these figures.

	Median Percentage of White				pri^{-1}(Med)		
	No.	♂	♀	Δ	♂	♀	Δ
39	659	14.4	25.5	11.1	−1.063	−0.659	+0.404
35	1460	62.2	68.9	6.7	+0.311	+0.493	+0.182
2	1430	73.8	83.9	10.1	+0.637	+0.990	+0.353
32	872	84.2	89.2	5.0	+1.003	+1.237	+0.234
2_N	1650	93.2	94.6	1.4	+1.491	+1.607	+0.116
13	1278	95.1	96.3	1.2	+1.655	+1.787	+0.132
13_N	1680	97.0	98.6	1.6	+1.881	+2.197	+0.316
$\bar{\Delta}$				+5.300			+0.2481
$\sigma_{\bar{\Delta}}$				1.577			0.0424
t				3.36			5.85
DF				6			6
Prob				0.016			*ca.* 0.001

$$\bar{\Delta} = (1/7)\sum \Delta \qquad \sigma^2_{\bar{\Delta}} = \tfrac{1}{42}\left(\sum \Delta^2 - \bar{\Delta}\sum \Delta\right)$$

Table 9.11 shows the median percentages of white in males and females of seven large inbred arrays of guinea pigs, including two branches of two of the strains, the differences, and the corresponding figures on using the inverse probability transformation, one that has been shown to normalize approximately the extremely asymmetrical distributions of this character. The females were whiter than the males in all cases (average, 5.30%) but the differences varied enormously. The value of t is 3.36, which with six degrees of freedom gives a probability of 0.016 of being exceeded by accidents of sampling. On using the inverse probability transformation, the value of t

is raised to 5.85 and the probability becomes 0.001. There is little doubt of the reality of the difference on using percentages, but whatever doubt exists is reduced many fold by transforming the scale.

Standard Errors of Percentiles

The standard error of a percentile is closely related to that of the corresponding proportion of the total frequency below it. Let $y = f(x)$ be the ordinate of the frequency distribution of unit area that separates the proportion p below and $(1 - p)$ above. For small deviations $\delta p = y\delta x$. Since the variance of p for given x is $p(1 - p)/N$, that of x for given p is

$$(9.29) \qquad \sigma_x{}^2 = \frac{\sigma_p{}^2}{y^2} = \frac{p(1 - p)}{Ny^2}.$$

Thus the standard error of the median is $1/(2y\sqrt{N})$ and that of either of the quartiles is $\sqrt{3}/(4y\sqrt{N})$. In the normal frequency distribution $y_{\text{med}} = 1/(\sigma\sqrt{2\pi})$ and thus $\sigma_{\text{med}}^2 = \pi\sigma^2/2N = 1.5708\sigma^2/N$

$$(9.30) \qquad \sigma_{\text{med}} = \frac{1.2533\sigma}{\sqrt{N}}.$$

Similarly $y_Q = 0.31777/\sigma$, $\sigma_Q{}^2 = 1.8568\sigma^2/N$,

$$(9.31) \qquad \sigma_Q = \frac{1.3626\sigma}{\sqrt{N}}.$$

The standard error of the difference between two values of x is affected by the correlation that exists between those within samples.

$$(9.32) \qquad \sigma_{x_2-x_1}^2 = \sigma_{x_2}^2 - 2\mu_{11}(x_1\,x_2) + \sigma_{x_1}^2 = \sigma_{x_2}^2 - 2b_{x_2x_1}\sigma_{x_1}^2 + \sigma_{x_1}^2.$$

As indicated here, the product moment $\mu_{11}(x_1x_2)$ is equal to the $b_{x_2x_1}\,\sigma_{x_1}^2$ where $b_{x_2x_1}$ is the regression x_2 on x_1 (average change in x_2 for a unit change in x_1).

If two abscissas x_1 and x_2 are taken as fixed, an increment, δp_1, of the lower-tail frequency due to sampling implies a decrement of the same amount in the frequency above this point, which on the average is apportioned to the area between x_1 and x_2 and that above x_2, proportionately. Thus the average change in $(1 - p_2)$ for a given change in p_1 is $\delta(1 - p_2)/\delta p_1 = -(1 - p_2)/(1 - p_1)$. Thus

$$(9.33) \qquad b_{p_2p_1} = \frac{\delta p_2}{\delta p_1} = \frac{(1 - p_2)}{(1 - p_1)}$$

since $\delta(1 - p_2) = -\delta p_2$. For given p_1 and p_2 but variable x_1 and x_2,

$$b_{x_2x_1} = \frac{\delta x_2}{\delta x_1} = \frac{\delta p_2/y_2}{\delta p_1/y_1} = \frac{y_1(1 - p_2)}{y_2(1 - p_1)}. \tag{9.34}$$

Thus

$$\begin{aligned} \sigma^2_{x_2 - x_1} &= \sigma^2_{x_2} - \frac{2y_1(1 - p_2)}{y_2(1 - p_1)}\sigma^2_{x_1} + \sigma^2_{x_1} \\ &= \frac{1}{N}\left[\frac{p_2(1 - p_2)}{y_2^2} - \frac{2p_1(1 - p_2)}{y_1y_2} + \frac{p_1(1 - p_1)}{y_1^2}\right]. \end{aligned} \tag{9.35}$$

In the quartile deviation $QD = \frac{1}{2}(Q_{75} - Q_{25})$,

$$\sigma^2_{QD} = \frac{1}{64N}\left[\frac{3}{y_2^2} - \frac{2}{y_1y_2} + \frac{3}{y_1^2}\right]. \tag{9.36}$$

For a normal frequency distribution

$$\sigma^2_{QD} = \frac{0.6189\sigma^2}{N}, \qquad \sigma_{QD} = \frac{0.7867\sigma}{\sqrt{N}}. \tag{9.37}$$

If the distribution is not normal the estimation of the ordinate at a given percentile from the crude data is often unsatisfactory, apart from the uncertainty of the abscissa of the percentile. Better estimates can be made if a smooth curve can be fitted to the data. This was done for the earlier data of four inbred strains of guinea pigs with respect to percentage of white in the coat (Table 9.12). It may be seen that the higher median percentage of white in the females is highly significant in all of these cases.

TABLE 9.12. Comparison of the median amounts of white in the coat patterns of male and female guinea pigs of four inbred strains.

Strain		N	x_{med}	y_{med}	Δ $x_♂ - x_♀$	σ_Δ	Δ/σ_Δ
39	♂	364	0.144 ± 0.0111	2.359	0.111	0.0205	5.42
	♀	295	0.255 ± 0.0172	1.693			
35	♂	751	0.622 ± 0.0100	1.825	0.067	0.0138	4.85
	♀	709	0.689 ± 0.0095	1.971			
2	♂	745	0.738 ± 0.0144	1.272	0.101	0.0172	5.89
	♀	685	0.839 ± 0.0093	2.046			
32	♂	405	0.842 ± 0.0118	2.105	0.050	0.0135	3.69
	♀	467	0.892 ± 0.0066	3.480			

These results make it possible to compare the methods of describing frequency distributions by the methods of moments and percentiles. In the normal frequency distribution the variance of the median can be expressed in terms of the standard error of the mean (σ^2/N).

$$\sigma^2_{\text{med}} = \frac{\pi\sigma^2}{2N} = \frac{\pi\sigma^2_M}{2} = 1.5708\sigma^2_M. \tag{9.38}$$

Thus it requires 57% more data to locate the median as narrowly as the mean.

Since the quartile deviation is 0.6745σ in the normal distribution, its variance must be expressed in terms of $(0.6745\sigma)^2/2N$ to compare its precision with that of the standard deviation with variance ($\sigma^2/2N$)

$$\sigma^2_{QD} = \frac{2.7207(QD)^2}{2N}. \tag{9.39}$$

It requires 172% more data to determine the quartile deviation with the same relative precision as the standard deviation.

Application of the method of maximum likelihood indicates that the normal curve can be fitted most efficiently by means of its moments. This method is not, however, in general the most efficient in fitting distributions that deviate from normality (Fisher 1921). The method of moments in fact breaks down completely in distributions that are so leptokurtic that the second moment is infinite.

The frequencies of intersections of a line, which takes all directions from a fixed point with equal probabilities, along equal intervals of another line at distance a from the fixed point is given by $f(x) = Na/\pi(a^2 + x^2)$, which is known as Cauchy's distribution.

This distribution has a finite area (N) but an indeterminate mean and infinite second moment. Yet the median is obviously at zero and the percentiles are obviously given by $x = a \tan[(0.5 - p)\pi]$ by the mode of construction. The quartiles are thus at $+a$ and $-a$. The value of x which is exceeded by 2.5% of the area in the same direction as $12.706a$, by 0.5% is $63.66a$, and by 0.05% is $636.6a$. The last compares with 3.29σ or $4.88QD$ in normal distributions.

The length of life after an experimental treatment that kills nearly all individuals in a relatively short time, but from which a few recover to live a normal span, is an example of a highly leptokurtic distribution that can be described better by median and other percentiles than by mean and standard deviation. Apart from this the percentile method has an advantage in distributions that deviate widely from normality but that may be normalized by a suitable transformation of scale, in that any individual observation is located at the same percentile irrespective of the scale. On the other hand, the great majority of frequency distributions that arise in genetics are ones of limited range, finite moments for which the moment system is much superior to the percentile system as a means of description.

Threshold Di- and Trichotomies

On the assumption of an underlying normal distribution of factor complexes, the type of observed distribution that deviates most from normality is the threshold dichotomy. In this case, the deviation of the threshold (t) from the mean on the hypothetical scale of additive effects, in terms of the standard deviation σ on the latter, is given by $(t - M)/\sigma = \text{pri}^{-1}\, p$. Conversely taking $t = 0$ as origin,

(9.40) $$\frac{M}{\sigma} = -\,\text{pri}^{-1}\, p.$$

This permits comparison of the means of different dichotomies on the assumption that the standard deviations of their underlying distributions are the same.

More can be done in cases in which there is both a threshold and a ceiling and thus a trichotomy, since both means and standard deviations on the underlying scale may be compared. Let t_1 and t_2 be the location of threshold and ceiling on the hypothetical scale and p_1 and p_2 the proportions below t_1 and t_2, respectively. Then

(9.41) $$\frac{t_1 - M}{\sigma} = \text{pri}^{-1}\, p_1 = x_1 \quad \text{and} \quad \frac{t_2 - M}{\sigma} = \text{pri}^{-1}\, p_2 = x_2;$$

(9.42) $$\frac{t_2 - t_1}{\sigma} = \text{pri}^{-1}\, p_2 - \text{pri}^{-1}\, p_1 = x_2 - x_1.$$

It is convenient to take the interval between t_1 and t_2 as the unit of measurement on the transformed scale and to take t_1 as the origin ($t_1 = 0$, $t_2 = 1$).

(9.43) $$\sigma = \frac{1}{\text{pri}^{-1}\, p_2 - \text{pri}^{-1}\, p_1} = \frac{1}{x_2 - x_1}.$$

(9.44) $$M = -\,\sigma\, \text{pri}^{-1}\, p_1 = -\,\frac{x_1}{x_2 - x_1}.$$

The standard errors of σ and M and their covariances are easily deduced from the standard errors and correlations of the percentiles x_1 and x_2, letting

y_1 and y_2 be the ordinates of the unit normal curve at abscissas x_1 and x_2, respectively.

$$(9.45)\quad \sigma_{x_1}^2 = \frac{p_1(1 - p_1)}{Ny_1^2}, \quad \sigma_{x_2}^2 = \frac{p_2(1 - p_2)}{Ny_2^2}, \quad \mu_{x_1x_2} = \frac{p_1(1 - p_2)}{Ny_1y_2}.$$

$$(9.46)\quad \sigma_\sigma^2 = \frac{\sigma^4}{N}\left[\frac{p_1(1 - p_1)}{y_1^2} + \frac{p_2(1 - p_2)}{y_2^2} - 2\,\frac{p_1(1 - p_2)}{y_1y_2}\right].$$

$$(9.47)\quad \sigma_M^2 = \frac{\sigma^4}{N}\left[x_2^2\,\frac{p_1(1 - p_1)}{y_1^2} + x_2^2\,\frac{p_2(1 - p_2)}{y_2^2} - 2x_1x_2\,\frac{p_1(1 - p_2)}{y_1y_2}\right].$$

$$(9.48)\quad \mu_{M\sigma} = \frac{\sigma_4}{N}\left[-\,x_2\,\frac{p_1(1 - p_1)}{p_1^2} - x_2\,\frac{p_2(1 - p_2)}{y_2^2} + (x_1 + x_2)\,\frac{p_1(1 - p_2)}{y_1y_2}\right].$$

As an example, consider a strain of guinea pigs (35*C*) derived from a single mating after 22 generations of brother-sister mating in which 57.9% were 3-toed, 35.9% intermediate, and 6.2% perfectly 4-toed, in a total of 356 individuals.

$$\sigma = \frac{1}{\text{pri}^{-1}\,0.938 - \text{pri}^{-1}\,0.579} = \frac{1}{1.538 - 0.199} = 0.747 \pm 0.06,$$

$$M = -\,0.747 \times 0.199 = -\,0.149 \pm 0.04.$$

CHAPTER 10

Deviations from Normality

Pearson's System of Frequency Distributions

Frequency distributions may be treated either from an empirical or an analytic standpoint, or from a combination. We are concerned here with interpretation and with forms that reduce the differences within families of distributions to differences in as few parameters as possible, rather than with the best possible fitting of isolated distributions.

Attention will, therefore, merely be called to the system of frequency curves, covering all values of β_1 and β_2, that was developed by Karl Pearson. This system was based on a generalization of the differential equation of the normal probability curve, $dy/dx = -(x - M)y/\sigma^2$.

$$\frac{dy}{dx} = \frac{(x + a)y}{b_0 + b_1 x + b_2 x^2}. \tag{10.1}$$

Integration leads to three major types that cover the field but that reduce to special types at the boundaries and in symmetrical cases.

In dealing with actual distributions, assignment to type is based on a function of β_1 and β_2 that serves as a criterion. The observed frequencies may be fitted in a wide range of cases by means of parameters based on β_1 and β_2 and, hence, on the first four moments. Unfortunately, as shown by Fisher (1921), fitting by this method may be highly inefficient, except for distributions that deviate only slightly from normality, and it becomes impossible in strongly leptokurtic distributions that can be fitted otherwise. A full discussion of the method has been given by Elderton (1938).

Compound Distributions

Compound distributions may obviously be of the most diverse sorts. We will restrict discussion to certain cases in which the components are all normal distributions. Assume n normal distributions to which weights, w_i, are assigned.

$$(10.2)\qquad y = \sum_{i=1}^{n} \frac{w_i}{\sigma_i\sqrt{2\pi}} \exp\left[-\frac{1}{2}\left(\frac{x_i - M_i}{\sigma_i}\right)^2\right], \qquad \sum w_i = 1.$$

The characteristic function is as follows:

$$(10.3)\qquad \varphi(t) = \sum^{n} \{w_i \exp [itM_i + \tfrac{1}{2}(it)^2\sigma_i^{2}]\}.$$

On expanding and taking the coefficients of $(it)^k/k!$ we find

$$(10.4)\qquad \mu_1' = \sum w_i M_i,$$

$$(10.5)\qquad \mu_2' = \sum w_i[M_i^2 + \sigma_i^2],$$

$$(10.6)\qquad \mu_3' = \sum w_i[M_i^3 + 3M_i\sigma_i^2].$$

$$(10.7)\qquad \mu_4' = \sum w_i[M_i^4 + 6M_i^2\sigma_i^2 + 3\sigma_i^4].$$

The central moments and form indexes may be calculated from these. We will consider only certain special cases that are especially instructive.

Case 1.—Two normal distributions with the same mean but with different standard deviations. In this case $w_2 = 1 - w_1$, $\gamma_1 = 0$. Taking the mean as origin:

$$(10.8)\qquad \mu_2 = w_1\sigma_1^2 + w_2\sigma_2^2,$$

$$(10.9)\qquad \mu_4 = 3w_1\sigma_1^4 + 3w_2\sigma_2^4,$$

$$(10.10)\qquad \lambda_4 = 3w_1w_2[\sigma_1^2 - \sigma_2^2]^2,$$

$$(10.11)\qquad \gamma_2 = 3w_1w_2\left[\frac{(\sigma_1^2 - \sigma_2^2)}{w_1\sigma_1^2 + w_2\sigma_2^2}\right]^2.$$

Let $w_2 = Kw_1$, $\sigma_2^2 = L\sigma_1^2$:

$$(10.12)\qquad \sigma_T^2 = w_1(KL + 1)\sigma_1^2,$$

$$(10.13)\qquad \gamma_2 = 3K\left(\frac{L - 1}{KL + 1}\right)^2.$$

There is only one component if $K = 0$ or ∞ and the components are identical if $L = 1$, all giving $\gamma_2 = 0$. Assume $\sigma_2^2 > \sigma_1^2$.

For given L, γ_2 is maximum if $K = 1/L$. For given K, γ_2 increases as L increases above 1. If most of the distribution consists of a component with small standard deviation, but there is a small component with large standard

TABLE 10.1. γ_2 for diverse values of $L(=\sigma_2^2/\sigma_1^2)$ and $K(=w_2/w_1)$.

L \ K	0	⅛	¼	½	1	2	4	8	∞
∞	–	24	12	6	3	1.5	0.75	0.38	0
8	0	4.59	4.08	2.94	1.81	1.02	0.54	0.28	0
4	0	1.50	1.69	1.50	1.08	0.67	0.37	0.20	0
2	0	0.24	0.33	0.38	0.33	0.24	0.15	0.08	0
1	0	0	0	0	0	0	0	0	0

deviation (K small, L large), there may be marked leptokurtosis, whereas in the converse case kurtosis is usually negligible. Figure 10.1 shows the distribution for components, one of which has twice the standard deviation of the other ($L = 4$), but the one with large standard deviation contributes only one fourth as much ($K = 1/4$), the same amount ($K = 1$), or four times as much ($K = 4$). The first (top) is highly leptokurtic ($\gamma_2 = 1.69$), the second (middle) moderately leptokurtic ($\gamma_2 = 1.08$), and the last (bottom) so slightly ($\gamma_2 = 0.37$) that it would require 687 individuals to demonstrate a deviation from normality by the criterion 2 SE in contrast with 34 and 82 in the other two cases (SE $\gamma_2 = \sqrt{24/N}$, N for 2 SE $= 96/\gamma_2^2$).

Something like this situation could account for the highly leptokurtic distribution in any direction in space found after release of *Drosophila pseudoobscura* at a certain place in a homogeneous terrain (Table 10.2; Dobzhansky and Wright 1947). This is from a later experiment, with a wider range of traps, than that illustrated in Figure 6.1.

TABLE 10.2. Standard deviation in meters, variance, and γ_2 along line of traps through point of release from one to six days after release.

Day	Range of Traps	σ	σ^2	γ_2
1	1,000m	64m	4,050	10.0
2		85m	7,250	4.7
3	1,200m	119m	14,200	4.9
4		124m	15,400	4.8
5		153m	23,400	3.1
6		169m	28,400	2.4

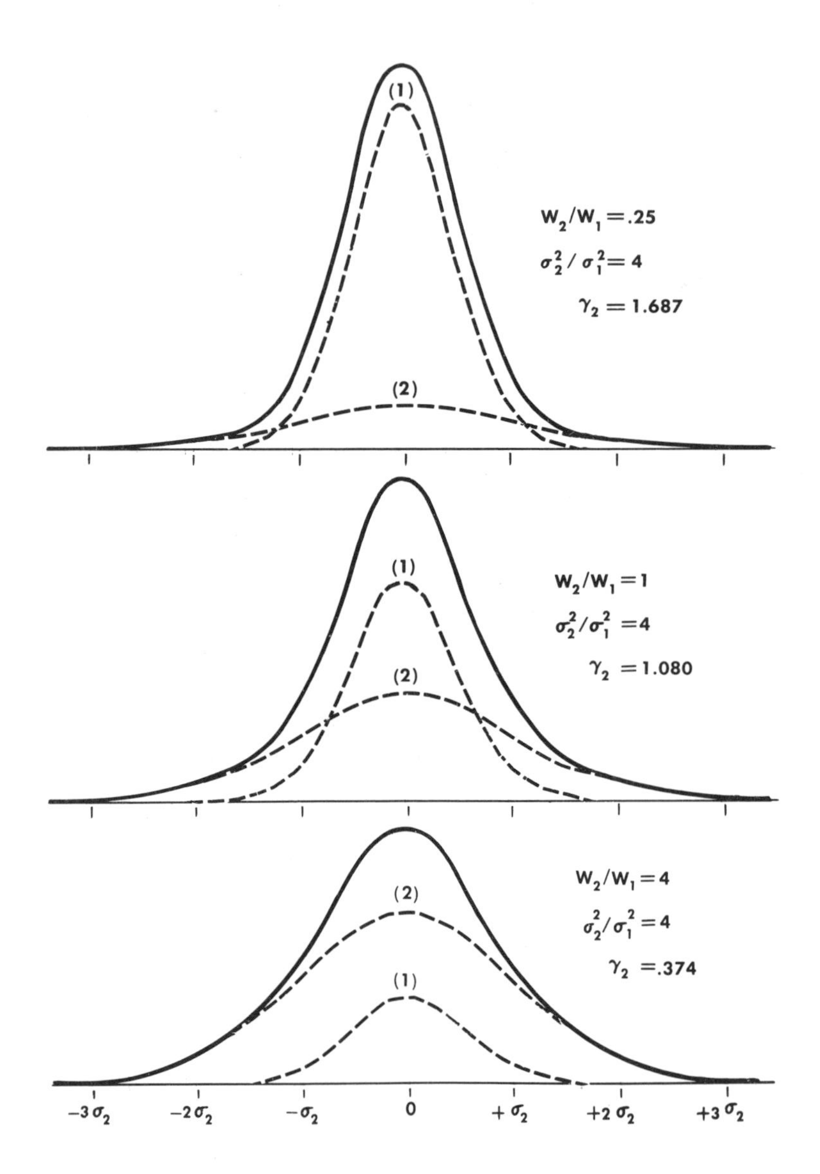

FIG. 10.1. Compound distributions in which one component (2) has four times the variance of the other (1). The ratios of the frequencies, w_2/w_1 are 0.25, 1 and 4 from top to bottom, giving decreasing kurtosis.

Note that the variance increased in roughly linear fashion and that the kurtosis decreased from the enormous value of 10.0 after one day, both in rough agreement with expectation from the daily compounding of variability. It is evident that most of the flies merely fluttered about locally each day, since 87% of the recaptured flies were within 70 meters of the point of release after one day, and 86% were within 210 meters even after six days. A few, however, were carried long distances on each day. The range of recaptured flies reached so nearly to the extremes of the line of traps (500 meters in each direction) by the second day that it was deemed necessary to extend the lines to 600 meters in each direction. It is probable that some had gone beyond this by the sixth day and thus that the true kurtosis was somewhat greater.

Case 2.—Two normal distributions with the same standard deviations (σ_r) but with different means.

If the frequencies are equal ($w_2 = w_1$) and the origin is taken at the mean of the compound distribution ($\delta M_1 = -\delta M_2$), the difference between the component means is $d = 2\delta M_2$, the variance σ^2 is $(d^2/4) + \sigma_r^2$, $\gamma_1 = 0$, and $\gamma_2 = -d^4/8\sigma^4$. Figure 10.2 (left) shows cases in which $d = \sigma_r$, $2\sigma_r$, and $3\sigma_r$

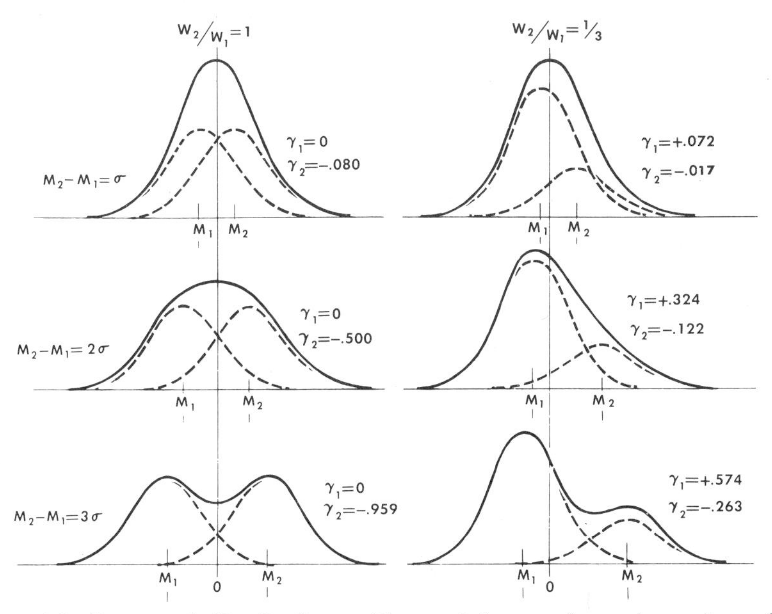

FIG. 10.2. Compound distributions with equal frequencies $w_2/w_1 = 1$, on left side in unequal frequencies, $(w_2/w_1) = 1/3$ on right side. The means are separated by 1σ, 2σ, and 3σ from top to bottom. The effects on skewness, γ_1, and kurtosis, γ_2, are indicated.

respectively. In the first (top), γ_2 is so small (-0.08) that it would require some 15,000 individuals to detect a deviation from normality at 2 SE. The second case (middle) ($d = 2\sigma_r$, $\gamma_2 = -0.5$) is of interest as that in which the components intersect at their points of inflection and thus is the greatest separation compatible with a single mode. It requires 384 individuals to detect deviation from normality at 2 SE. The distribution in the case of $d = 3\sigma_r$ (bottom) is strongly bimodal ($\gamma_2 = -0.96$). Even so, it would require about 104 individuals to demonstrate that the deviation from normality is significant at 2 SE. It is evident that rather large numbers are needed for a critical study of the forms of frequency distributions.

The distributions of F_2 populations in which there is segregation of one major factor ($w_2/w_1 = 1/3$) and overlapping due to residual normal variability is of interest. If the means are separated by $d = 4/3\delta M_2$, $\sigma^2 = 3d^2/16 + \sigma_r^2$, $\gamma_1 = +3d^3/32\sigma^3$, and $\gamma_2 = -3d^4/128\sigma^4$. Figure 10.2 (right) again shows cases of $d = \sigma_r$, $2\sigma_r$, and $3\sigma_r$, respectively. In the first (top), both asymmetry ($\gamma_1 = +0.072$) and kurtosis ($\gamma_2 = -0.0166$) are negligible, unless the numbers are enormous (N for 2 SE = 4,572 and 350,000, respectively). If $d = 2\sigma$ (middle) there is rather obvious asymmetry ($\gamma_1 = +0.324$) although N for 2 SE is 229. Kurtosis is still usually negligible ($\gamma_2 = -0.122$, N for 2 SE = 6,400). With $d = 3\sigma$ (bottom), there is marked asymmetry and slight bimodality ($\gamma_1 = +0.574$), N for 2 SE = 73. Kurtosis ($\gamma_2 = -0.263$) is much less than in the symmetrical case and requires 1,390 individuals for detection at 2 SE.

The distributions of length and diameter of fruits of squashes (Fig. 6.6*A* and *B*) illustrate F_2's with and without sufficient separation of the means of the segregants to insure bimodality. There is the complication that the standard deviations are markedly different.

The standard deviations of the two components in Forficula (Fig. 6.6*D*) also differ. In this case the frequencies of the two alternative conditions vary in relation to the size of the insects. The case of Xylotrupes (Fig. 6.6*E*) is more complicated since the means of the alternative patterns of growth of horns are correlated with body size.

Compound Variability

Compounds of two distributions with the same standard deviation could equally well be treated as compound variables in which each value is the sum of a contribution of a major factor of the type $(pa + qA)$ and a normal deviation. The semi-invariants of the total can be obtained as the sum for these two components. Since λ_3 and λ_4 are both zero for the normal component, these semi-invariants for the total are the same as for the binomial component ($\lambda_3 = pq(p - q)d^3$, $\lambda_4 = pq(1 - 6pq)d^4$, and $\gamma_1 = pq(p - q)d^3/\sigma^3$, $\gamma_2 = pq(1 - 6pq)d^4/\sigma^4$ where $\sigma^2 = pqd^2 + \sigma_r^2$). These agree with the results obtained by treating these as compound distributions.

It has already been shown that a distribution that depends on multiple independent additive factors with equal effects and equal frequencies of alternatives, $(pa + qA)^n$, rapidly approaches normality as n is increased. With γ_2 equal to $-2/n$ in the symmetrical case $(0.5a + 0.5A)^n$ (indicating platykurtosis), it requires $24n^2$ observation for significant deviation from normality at 2 SE (Fig. 10.3, bottom row). Kurtosis disappears altogether if $p = 0.211$ or 0.789 and is thus very slight in the case of F_2 segregation. It may become very great, however, and positive, with extreme asymmetry of the frequencies of the elementary alternatives and approaches $1/\sigma^2$. It is the same whatever the directions of effect of the rare alternatives.

Thus the symmetrical distribution $(pa + qA)^{n/2}(qb + pB)^{n/2}$ in which a and b contribute in the same direction, has the same kurtosis as $(pa + qA)^n$. Figure 10.3 (top row) shows the distribution $(0.1a + 0.9A)^2(0.9b + 0.1B)^2$ with $\gamma_2 = 1.278$, and $(0.1a + 0.9A)^4(0.9b + 0.1B)^4$ with γ_2 half as great

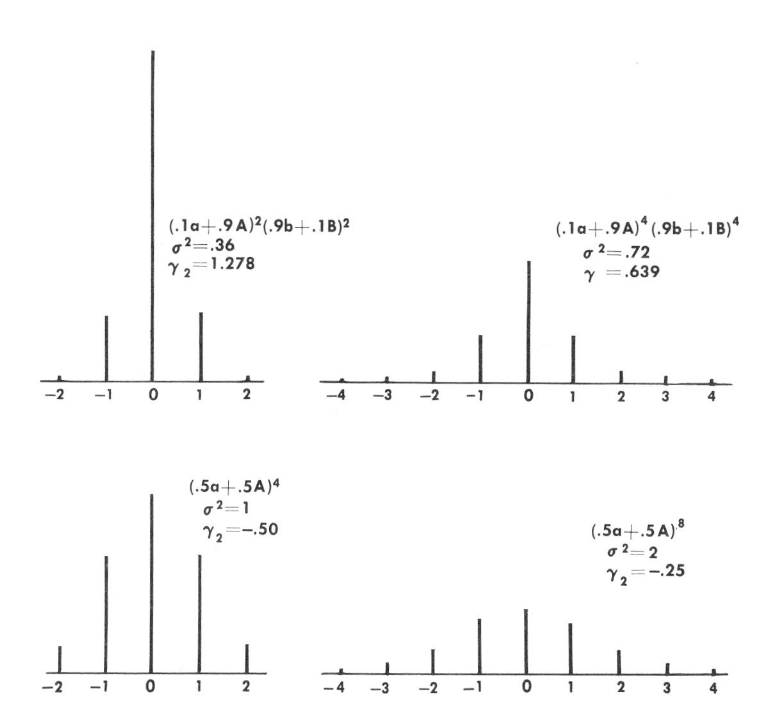

FIG. 10.3. Variables compounded of multiple equal factors as indicated.

(0.639). Kurtosis could be demonstrated at 2 SE with only about 59 observations in the former but would require 235 in the latter. With p (or q) as small as 0.01, kurtosis becomes enormous as already noted.

It might be supposed from this that if variability at most of the loci depends on rare segregants, the resulting distribution would be highly leptokurtic. This, however, does not allow for the disproportionately great contribution to variability and to kurtosis of a small number of factors, the alternatives of which are more or less equal in frequency. It will be shown later that in small populations with low mutation rates and no selection, the distribution of gene frequencies is expected to be of the type $\varphi(q) = C/q(1 - q)$ in which frequencies become indefinitely great as q approaches 0 or 1. The variance contributed by one factor is $2q(1 - q)A^2$.

$$(10.14) \qquad \gamma_2 = \frac{\sum \sigma^4 \gamma_2}{[\sum \sigma^2]^2} \approx \frac{\int_0^1 2q(1 - q)[1 - 6q(1 - q)]\varphi(q)\, dq}{[2 \int_0^1 q(1 - q)\varphi(q)\, dq]^2}.$$

On substituting $\varphi(q) = C/q(1 - q)$,

$$(10.15) \qquad \gamma_2 = \frac{2C \int_0^1 (1 - 6q + 6q^2)\, dq}{[2C \int_0^1 dq]^2} = 0.$$

Thus far from being highly leptokurtic, the total distribution is exactly normal.

Another interesting case is that in which there are an indefinitely large number of factors with effects falling off from that of the leading one in geometric progression. We will consider only the symmetrical case in detail $[0.5a + 0.5A][0.5b + 0.5B][0.5c + 0.5C]$, with effects (A), $r(A)$, $r^2(A)$, etc.

$$(10.16) \qquad \begin{aligned} \lambda_1 &= \frac{(A)}{2(1 - r)}, & \lambda_3 &= 0, \\ \lambda_2 = \sigma^2 &= \frac{(A)^2}{4(1 - r^2)}, & \lambda_4 &= -\frac{(A)^4}{8(1 - r^4)}. \end{aligned}$$

$$(10.17) \qquad \gamma_1 = 0, \qquad \gamma_2 = -2\left(\frac{1 - r^2}{1 + r^2}\right).$$

Table 10.3 shows the standard deviation in terms of the effect of the leading factor, γ_2, and the number of observations $(96/\gamma_2^2)$ required to detect significant kurtosis at 2 SE.

The distributions become increasingly platykurtic with decrease in r. If $r = 0.5$ and there is a finite number of factors, the combinations of factors do not overlap but are equally spaced along the scale. The distribution approaches a rectangular form (Fig. 10.4 [top left]). With any smaller value

TABLE 10.3. Gene effects in geometric progression, (ratio r), in backcross and F_2. Standard deviation as multiple of effect of leading factor, γ_2, and number of individuals for significance of latter at 2 SE. All gene frequencies 0.5.

	F_1			F_2		
r	$\sigma/(A)$	γ_2	N for 2 SE	$\sigma/(A)$	γ_2	N for 2 SE
0.9	1.147	−0.210	2,179	1.622	−0.105	8,712
0.8	0.833	−0.439	498	1.179	−0.220	1,992
0.7	0.671	−0.685	205	1.010	−0.342	819
0.6	0.625	−0.941	108	0.884	−0.471	433
0.5	0.577	−1.200	67	0.817	−0.600	267
0.4	0.546	−1.448	46	0.786	−0.724	183
0.3	0.524	−1.670	34	0.741	−0.835	138
0.2	0.510	−1.846	28	0.722	−0.923	113
0.1	0.502	−1.960	25	0.711	−0.980	100
0	0.500	−2.000	24	0.707	−1.000	96

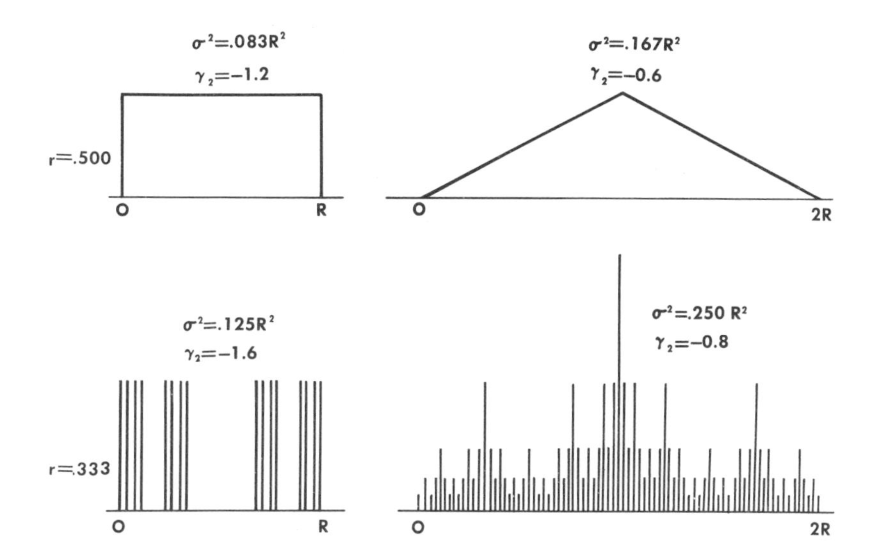

FIG. 10.4. Variables compounded of factors that fall off in geometric progression: $r = 0.5$ (*top row*), $r = 0.333$ (*bottom row*), two equally frequent alternatives in each factor (*at left*), three alternatives with frequencies in ratio 1:2:1 (*at right*).

of r than 0.5 the distribution becomes discontinuous. The case of $r = \frac{1}{3}$ with four factors is shown in Figure 10.4 (bottom left). The classes split into two with 1:1 ratio of frequencies with each additional factor but without substantially closing the gaps in the distribution. The distribution is so platykurtic even with r as large as 0.7 that the deviation from normality is significant at 2 SE with N as small as 205.

The preceding situation is found in a backcross population for a character in which there are independent additive effects of the factors with effects in geometric progression. In F_2 of the same cross, and dominance, the distribution is of the type $(0.5a + 0.5A)^2(0.5b + 0.5B)^2(0.5c + 0.5C)^2 \cdots$. The standard deviation in terms of the effect of the leading factor, γ_2, and the number of observations for significant kurtosis at 2 SE are shown for this case also in Table 10.3 (right). There is much more approach to normality than in the preceding case. If $r = 0.5$, the distribution is triangular (Fig. 10.4 [top right]) and it requires 267 observations to demonstrate at 2 SE that it is not normal. At $r = \frac{1}{3}$, the combinations do not overlap but are equally spaced, whatever the number of factors, and thus tend to close all gaps with an indefinitely large number. The distribution does not approach a smooth curve, however, since the frequency of any class is either one half or twice that of either adjacent class, however close these classes may be (Fig. 10.4 [bottom right]).

This closes our discussion of distributions dependent on independent additive effects. With multiple equal factors, deviation from normality is not likely to be easily detected. With effects in geometric series, discontinuity occurs if the ratio is less than 0.5. There may be marked platykurtosis if the frequencies of alternatives, especially of the leading factors, are equal and r is less than 0.7 or even 0.8. With unequal frequencies of the leading alternatives and irregularity among the others, kurtosis tends to disappear, but there may be marked asymmetry. If there are two (or more) leading factors with equal effect, there is much more approach to normality. But even if there is only one major factor and residual normal variability that is responsible for as much as half the total variance, deviations from normality are not easy to detect.

Correlations in Values of Components

The general treatment of the effects of correlation in the values of the components of a compound variable is very complicated since all possible product moments of all orders must be taken into account. We will merely present certain general considerations and go briefly into modification of the binomial and Poisson distributions that arise when there is correlation in the values of the elementary events.

If the components are correlated because of common factors included in a multiplicity of independent ultimate factors, among which none are of outstanding importance, the latter may be considered to be the components of the variable in question. There may well be a close approach to normality.

If, however, the components are correlated because of a single factor that acts on them in such a way that the effects on the variable are all in the same direction, this factor will have an outstanding effect on the latter. If it, itself, varies normally the variables will tend to vary normally. One may, for example, consider the length of an insect as the sum of the lengths of the head, thorax, and abdomen. These are practically certain to be highly correlated because of the common effect of general size, but as size is usually affected by a multiplicity of genetic and environmental factors, total length may be expected to vary normally. If, on the other hand, the common factor back of the correlated components does not vary normally, this will be reflected to a large extent in the form of distribution of the total variable.

Correlated Binomial Factors

Clusters of n events, each event with the probability array $[(1 - q)a + qA]$ for the alternatives, is distributed according to the expansion of the expression $[(1 - q)a + qA]^n$ if there is no correlation in the occurrence of plus or minus factors. An excess of clusters consisting largely or wholly of one of the alternatives indicates a correlation in occurrence. Such distributions may be fitted by assuming variation of the value of q according to some rule, $F(q)$. The most convenient distribution, limited to the range 0 to 1, is the Beta distribution.

$$(10.18) \qquad F(q) = y = \frac{\Gamma(c)}{\Gamma(c\bar{q})\Gamma c(1 - \bar{q})}\, q^{c\bar{q}-1}(1 - q)^{c(1-\bar{q})-1}.$$

This has the following moments about zero, derivable by use of the formula

$$\int_0^1 x^{n-1}(1 - x)^{n-1}\, dx = B(m, n) = \frac{\Gamma(m)\Gamma(n)}{\Gamma(m + n)}$$

and $\Gamma(x + 1) = x\Gamma(x)$. The corresponding semi-invariants are given below using $s = 1 - q$

$$
(10.19)\quad \left\{
\begin{aligned}
&\int_0^1 y\,dq = 1. \\
\mu_1' &= \int_0^1 qy\,dq = \bar{q}, && \lambda_1(q) = \bar{q}. \\
\mu_2' &= \int_0^1 q^2 y\,dq && \lambda_2(q) = \frac{\bar{p}\bar{q}}{c+1}. \\
&= \frac{\bar{q}(c\bar{q}+1)}{(c+1)}, \\
\mu_3' &= \int_0^1 q^3 y\,dq && \lambda_3(q) = \frac{2\bar{p}\bar{q}(\bar{p}-\bar{q})}{(c+1)(c+2)}. \\
&= \frac{\bar{q}(c\bar{q}+1)(c\bar{q}+2)}{(c+1)(c+2)}, \\
\mu_4' &= \int_0^1 q^4 y\,dq && \lambda_4(q) = \frac{6\bar{p}\bar{q}[c+1-(5c+6)\bar{p}\bar{q}]}{(c+1)^2(c+2)(c+3)}. \\
&= \frac{\bar{q}(c\bar{q}+1)(c\bar{q}+2)(c\bar{q}+3)}{(c+1)(c+2)(c+3)},
\end{aligned}
\right.
$$

If $c = 0$, these are the moments of the binomial $(\bar{p}a + \bar{q}A)$. If c is large, the distribution approaches normality but with little variability about $\bar{q}$. Let $f(x)$ be the frequency of the class with xA's among the clusters of n.

$$
(10.20)\quad f(x) = \binom{n}{x} \int_0^1 q^x p^{n-x} F(q)\,dq
$$

$$
= \binom{n}{x} \frac{\Gamma(c)}{\Gamma(c\bar{p})\Gamma(c\bar{q})} \int_0^1 q^{c\bar{q}+x-1} p^{c\bar{p}+n-x-1}\,dq
$$

$$
(10.21)\quad = \binom{n}{x} \frac{\Gamma(c)\Gamma(c\bar{q}+x)\Gamma(c\bar{p}+n-x)}{\Gamma(c\bar{q})\Gamma(c\bar{p})\Gamma(c+n)}.
$$

With a single component ($n = 1$), $f_0 = \bar{p}$, $f_1 = \bar{q}$. The mean is $\bar{q}$ and the variance $= \bar{p}\bar{q}$ as expected. For the sum of n components with correlation r between any two, the mean is $n\bar{q}$, the variance is $\sigma_x^2 = n[1 + (n-1)r]\bar{p}\bar{q}$ giving

$$
(10.22)\quad r = \frac{1}{n-1}\left[\frac{\sigma_x^2}{n\bar{p}\bar{q}} - 1\right]
$$

from which r can be calculated from observed distributions.

The frequencies for random pairs of components within any cluster may be derived and equated to the frequencies in terms of the covariance, $r\bar{p}\bar{q} = f_0 - \bar{p}^2 = \bar{p}\bar{q} - 0.5f_1 = f_2 - \bar{q}^2$,

$$(10.23)\quad \begin{cases} f_0 = \dfrac{\bar{p}(c\bar{p}+1)}{(c+1)} = \bar{p}^2 + \bar{p}\bar{q}r \\ f_1 = \dfrac{2c\bar{p}\bar{q}}{(c+1)} = 2\bar{p}\bar{q} - 2\bar{p}\bar{q}r \\ f_2 = \dfrac{\bar{q}(c\bar{q}+1)}{(c+1)} = \bar{q}^2 + \bar{p}\bar{q}r. \end{cases}$$

From any of these,

$$(10.24)\quad c = \frac{1-r}{r}.$$

The frequencies of successive classes among clusters of n can be calculated from the formulas

$$(10.25)\quad \begin{cases} f_0 = \dfrac{\Gamma(c)\Gamma(c\bar{p}+n)}{\Gamma(c+n)\Gamma(c\bar{p})} = \dfrac{c\bar{p}(c\bar{p}+1)(c\bar{p}+2)\cdots(c\bar{p}+n-1)}{c(c+1)(c+2)\cdots(c+n-1)}, \\ f_1 = f_0\left[\dfrac{nc\bar{q}}{c\bar{p}+n-1}\right], \\ f_2 = f_1\left[\dfrac{(n-1)(c\bar{q}+1)}{2(c\bar{p}+n-2)}\right], \\ f_x = f_{x-1}\left[\dfrac{(n-x+1)(c\bar{q}+x-1)}{x(c\bar{p}+n-x)}\right]. \end{cases}$$

Following are the distributions of 4-toed guinea pigs in three cases already presented (Table 8.5) as examples of wide departure from the binomial expectation. The values of r and c are derived from 10.22 and 10.24, respectively. Agreement is reasonably good in all cases on allowing for correlation within litters. There was some substrain heterogeneity in the Beltsville data that would contribute to the correlation within litters, but little or none in the Chicago data. Even in the former, however, the evidence indicated that the most important factor common to littermates was environmental—primarily the age of the mother, but seasonal effects as well, as brought out in previous chapters.

Hypergeometric Series

The effect of a uniform negative correlation between components may be illustrated by the case of drawing without replacement from a limited total (n) in which two alternatives are present in the array $[p(a) + q(A)]$. The correlation is $-1/(n-1)$. The distribution in samples of given size, $K \leq n$, is known as the hypergeometric series. The frequency $f(x)$ of a given

TABLE 10.4. Comparison of observed and calculated frequencies of 4-toed guinea pigs (strain 35) in litters of 3 and 4 at the Beltsville colony and of litters of 3 at the Chicago colony.

4-TOED PER LITTER	BELTSVILLE						CHICAGO		
	Litters of 3			Litters of 4			Litters of 3		
	Obs.	Calc.	Obs. − Calc.	Obs.	Calc.	Obs. − Calc.	Obs.	Calc.	Obs. − Calc.
0	119	121.9	−2.9	44	44.7	−0.7	17	16.7	+0.3
1	63	54.3	+8.7	23	20.8	+2.2	12	13.0	−1.0
2	34	42.7	−8.7	11	13.5	−2.5	14	13.0	+1.0
3	46	43.1	+2.9	10	8.9	+1.1	16	16.3	−0.3
4				5	5.1	−0.1			
Litters	262	262.0	0.0	93	93.0	0.0	59	59.0	0.0
$\bar{q}$	0.3422			0.2554			0.4972		
σ^2	1.2855			1.5049			1.3686		
r	0.4518			0.3262			0.4125		
c	1.214			2.0660			1.424		
χ^2	3.43			0.85			0.165		
DF	1.0			2.0			1.0		
Prob	0.05–0.10			0.50–0.70			0.50–0.70		

From data of Wright (1934).

number, x, of alternative A can be expressed as follows by the theory of probability.

$$f_x = \binom{nq}{x}\binom{np}{K-x}\Big/\binom{n}{K}. \tag{10.26}$$

As the mean expectation for each element A is q, the mean for a sample of size K is Kq. The variance of each element is pq.

$$\bar{x} = Kq. \tag{10.27}$$

$$\sigma_x^2 = Kpq - \frac{K(K-1)}{n-1}pq = Kpq\frac{n-K}{n-1}. \tag{10.28}$$

The variance is thus smaller, relative to the mean, Kq, than is that of a binomial for samples of the same size.

A case that arises in genetics is that of segregation in polyploids (ignoring the effects of crossing over between the locus and centromere). In K-ploid gametes from a $2K$-ploid zygote, $n = 2K$. Thus in a hexaploid of constitution A^4a^2, the gametic array is as follows:

$$(10.29)\quad \begin{cases} Aaa \ \binom{4}{1}\binom{2}{2}\Big/\binom{6}{3} = 0.20 \\ AAa \ \binom{4}{2}\binom{2}{1}\Big/\binom{6}{3} = 0.60 \\ AAA \binom{4}{3}\binom{2}{0}\Big/\binom{6}{3} = 0.20. \end{cases}$$

The hypergeometric series can be generalized to cover cases in which there are more than two alternatives.

$$(10.30)\quad f_{x_1 x_2 x_n} = \binom{nq_1}{x_1}\binom{nq_2}{x_2}\cdots\binom{n_{q_n}}{x_n}\Big/\binom{n}{K}.$$

The segregation of multiple alleles in polyploids comes under this head (with the same qualification as above).

The Negative Binomial

It has been noted that in the case of rare events, the frequencies in large random samples are expected to be distributed according to the Poisson distribution. A tendency toward clustering of events implies that the mean expectation varies from sample to sample. It is convenient to assume that these means are distributed from 0 to $+\infty$ in a gamma distribution.

$$(10.31)\quad f_M = \frac{a^k}{\Gamma(K)} M^{K-1} e^{-aM}$$

with mean K/a and variance K/a^2.

The distribution of events is the sum of Poisson distributions weighted by $f(m)$.

$$(10.32)\quad \begin{aligned} F_{(x)} &= \frac{1}{x!}\int_0^\infty M^x e^{-M} f_{(M)}\, dM \\ &= \frac{a^k}{x!\Gamma(K)}\int_0^\infty M^{x+k-1} e^{-M(1+a)}\, dM, \end{aligned}$$

but

$$\Gamma(K) = \int_0^\infty M^{K-1} e^{-M}\, dM \text{ by definition;}$$

$$\begin{aligned} \Gamma(x+K) &= \int_0^\infty [(1+a)M]^{x+K-1} e^{-M(1+a)}\, d(1+a)M \\ &= (1+a)^{x+K}\int_0^\infty M^{x+K-1} e^{-M(1+a)}\, dM. \end{aligned}$$

Thus

$$F_{(x)} = \frac{\Gamma(x + K)}{x!\Gamma(K)} \left(\frac{a}{1 + a}\right)^K \left(\frac{1}{1 + a}\right)^x. \tag{10.33}$$

The array for successive numbers (x) of the events under consideration (A) is as follows:

$$\left(\frac{a}{1+a}\right)^K \left[(A)^0 + K\left(\frac{1}{1+a}\right)(A)^1 + K\frac{(K+1)}{2!}\left(\frac{1}{1+a}\right)^2 (A)^2 + \frac{K(K+1)(K+2)}{3!}\left(\frac{1}{1+a}\right)^3 (A)^3\right] + \cdots. \tag{10.34}$$

This is the expansion of

$$\left(\frac{a}{1+a}\right)^K \left[1 - \frac{1}{1+a}(A)\right]^{-K}.$$

It can be put in a form that brings about an interesting analogy with the binomial distribution $[(1 - q') + q'(A)]^K$ by putting $1/a = q$, giving $[(1 + q) - q(A)]^{-K}$ the so-called negative binomial distribution (or Pascal distribution). The semi-invariants of the negative and positive binomial distribution are compared below.

$$(10.35)\quad \begin{array}{lll} & [(1+q) - q(A)]^{-K} & [(1-q') + q'(A)]^K \\ \lambda_1 = M = & Kq & Kq' \\ \lambda_2 = \sigma^2 = & Kq(1+q) & Kq'(1-q') \\ \lambda_3 = & Kq(1+q)(1+2q) & Kq'(1-q')(1-2q') \\ \lambda_4 = & Kq(1+q)[1+6q(1+q)] & Kq'(1-q')[1-6q'(1-q)']. \end{array}$$

Since the variance of the means (λ_2 of the gamma distribution above) is $K/a^2 = Kq^2$, the portion of the total variance due to heterogeneity is $q/(1 + q)$ and that due to random variability is $1/(1 + q)$. The parameters K and q can be estimated from the mean and variance of an observed distribution $q = (\sigma^2 - M)/M$, $K = M/q = M^2/(\sigma^2 - M)$.

Table 10.5 gives the distribution of a diplopod. *Scytonotus quadrata* as observed by Cole (1946), under boards in a homogeneous region (Table 8.8). With mean 0.9500 and variance 1.6693, there was wide deviation from the Poisson expectation in the direction that indicates a tendency toward aggregation. In this case $q = 0.7572$, $K = 1.2547$, $a = 1.3207$.

The distribution can be fitted very well under these assumptions. Many, but not all, distributions that depart from the Poisson expectation because of a tendency to aggregation can be fitted in this way. Other formulas with a rational basis have been suggested (cf. Neyman 1939).

TABLE 10.5. Comparison of observed and calculated frequencies of the diplopod, *Scytonotus quadrata*, under boards.

	o	c	$o - c$	$\frac{(o - c)^2}{c}$
0	128	128.2	−0.2	0.00
1	71	69.3	+1.7	0.04
2	34	33.7	+0.3	0.03
3	11	15.7	−4.7	1.41
4	8	7.0	+1.0	0.14
5	5	3.3	+1.7	0.88
6	3	2.8	+0.2	0.01
	260	260.0	0.0	2.51

From data of Cole (1946).

$\chi^2 = 2.51$ DF $= 4$ Prob, 0.50–0.70

Distributions in which there is less clustering than expected from random distributions may again in some cases be fitted by a hypergeometric distribution, which as noted implies a negative correlation in occurrence.

Independent Nonadditive Effects

As brought out in chapter 5, the analysis of actual characters indicates that the effects of any gene replacement depend on the complex of factors, genetic and environmental, with which it is combined. Effects are thus, in general, not additive. It is important to consider the consequences of interaction on frequency distributions.

First, however, it should be noted that interaction within a system that, as a whole, plays an independent and minor role in determining the variability of a character does not cause appreciable deviation from normality. Each interaction system may be treated as a unit. If there are enough such systems that are additive with each other, the total distribution approaches normality. It is of primary importance to consider interaction patterns that apply to the entire set of factors that affect the character.

Normalization

The first procedure that suggests itself is to assume that there is a scale on which all factors, genetic and environmental, are approximately additive and to make use of Laplace's principle that a variable compounded additively of many small independent contributions shows an approximately normal distribution, irrespective of the frequency distributions of the separate components. A transformation function is then to be sought that normalizes all distributions of the set under consideration as far as possible.

Galton, as long ago as 1879, noted that the logarithms of measurements of organisms may be more appropriate than the measurements themselves on the hypothesis that growth factors tend to contribute constant percentage increments rather than constant absolute ones. It is to be expected that the distribution of a variable to which each factor contributes a continually greater amount as the base to which it is applied increases would show positive skewness on the ordinary scale of measurements, but may become symmetrical and approximately normal on a scale of logarithms of the measurements.

A general theory can be based on the invariance of frequencies of corresponding classes under any transformation. If δx and $\delta x'$ are the corresponding class ranges on the actual and transformed scales, respectively, and y and y' the ordinates at the corresponding scale values x and x' at the mid ranges, then $y\ \delta x = y'\ \delta x'$ (for small δx). The relation between ordinates is given by $y = y'\ dx'/dx$. The observed effects of factors are given by the expression $F(x) = dx/dx'$.

If it be assumed that the conditions are such that a normal distribution is to be expected on the scale x', on which elementary factors have additive effects,

$$(10.36) \qquad y = \frac{1}{F(x)\sqrt{2\pi}\,\sigma'} \exp\left[-\frac{1}{2}\left(\frac{x' - \bar{x}'}{\sigma'}\right)^2\right].$$

McAlister (1879) at Galton's suggestions, derived the formula for the case of normalization by means of the logarithmic transformation.

$$(10.37) \qquad \begin{aligned} F(x) &= \frac{dx}{dx'} = x \\ x' &= \log x \\ y &= \frac{1}{x\sqrt{2\pi}\sigma'} \exp\left[-\frac{1}{2}\left(\frac{\log x - M'}{\sigma'}\right)^2\right], \\ M' &= \overline{\log x}, \quad \sigma' = \sigma_{\log x}. \end{aligned}$$

The semi-invariants of the "logarithmico-normal" distribution can be derived from the characteristic function.

$$\varphi(t) = \int_{-\infty}^{+\infty} e^{itx} y\,dx = \int_{-\infty}^{+\infty} \frac{1}{\sqrt{2\pi}\,\sigma'} \exp\left[ite^{x'} - \frac{1}{2}\left(\frac{x' - M'}{\sigma'}\right)^2\right] dx'$$

(10.38)

$$= \int_{-\infty}^{+\infty} \left(1 + ite^{x'} + \frac{(it)^2}{2!} e^{2x'} + \frac{(it)^3}{3!} e^{3x'} + \cdots\right) \times \frac{1}{\sqrt{2\pi}\,\sigma'} \exp\left[-\frac{1}{2}\left(\frac{x' - M'}{\sigma'}\right)^2\right] dx'.$$

But

$$\exp\left[Kx' - \frac{1}{2}\left(\frac{x' - M'}{\sigma'}\right)^2\right] = \exp\left[-\frac{1}{2}\left(\frac{x' - M' - K\sigma'^2}{\sigma'}\right)^2\right] \exp\left[KM' + \frac{K^2\sigma'^2}{2}\right].$$

Thus

(10.39)

$$\varphi(t) = 1 + it \exp[M' + \tfrac{1}{2}\sigma'^2] + \frac{(it)^2}{2!} \exp[2M' + \tfrac{4}{2}\sigma'^2] + \frac{(it)^3}{3!} [3M' + \tfrac{9}{2}\sigma'^2] + \cdots.$$

The moments about zero are given by the coefficients of $(it)^K/K!$. The central moments can be derived as usual. It is convenient to put C^2 for $\exp(\sigma'^2) - 1$.

(10.40)

$$\begin{cases} M = \exp[M' + \tfrac{1}{2}\sigma'^2], & M' = \log M - \tfrac{1}{2}\sigma'^2, \\ \sigma^2 = C^2 M^2, & \sigma'^2 = \log(1 + C^2). \\ \gamma_1 = C(3 + C^2), & \\ \gamma_2 = C^2(16 + 15C^2 + 6C^4 + C^6), & \end{cases}$$

The mean and variance on the transformed scale can thus be estimated from observed mean (M) and observed coefficient of variability, $C^2 = \sigma^2/M^2$, by the following formulas in terms of $\log_{10}$:

(10.41) $$M' = \overline{\log_{10} x} = \log_{10} M - \tfrac{1}{2}\log_{10}(1 + C^2);$$

(10.42) $$\sigma'^2 = \sigma^2_{\log_{10} x} = 0.4343 \log_{10}(1 + C^2).$$

These formulas depend on the assumption that the distribution actually is normal on the transformed scale. If it is not, they are more or less in error.

The logarithmic transformation can be given greater flexibility at the expense of an additional parameter by taking deviations from some other origin than zero. On replacing x by $x - L$, we may find $C^2 = \sigma^2/(M - L)^2$ from the equation $C^3 + 3C - \gamma_1 = 0$ and calculate M' and C^2 by the above formulas except that x is replaced by $(x - L)$ and C is the coefficient of variability relative to deviations from L instead of zero. It may be noted that L may be negative as well as positive.

It is evident from the formulas for γ_1 and γ_2 that a simple reasonable relation between factor effects and scale may lead to extreme deviations from normality. There are other rational transformation functions that may produce great deviations of other sorts. This suggests the desirability of systematic methods for finding normalizing functions.

Relation of Variability to Mean

A second criterion for a transformation is absence under it of correlation between variability and mean among populations that are believed to be comparable in causes of variation. Pearson's coefficient of variability $(CV = 100\sigma/M)$ was devised in recognition of the fact that relative variabilities are usually more nearly the same than absolute ones (1896).

The chief difficulty with precise use of this criterion lies in the choice of comparable populations. The most obvious method is to restrict comparisons to closely inbred strains and their first crosses, reared under the same environmental conditions as far as possible. There is the complication that close inbreeding may break down homeostatic controls and actually result in increased susceptibility to environmental disturbances. Comparison of variabilities in inbred strains and their F_1's often shows, however, that the above difficulty is of minor importance. Table 10.6 shows two cases given by East and Hayes (1911) in which coefficients of variability are much more uniform than standard deviations in widely different parental inbred strains and their F_1's.

TABLE 10.6. Comparisons of means (M), standard deviation (σ) and coefficients of variability (100 C) of two characters in two strains of maize and their F_1's.

	M	σ	100 C
Ear length in maize (cm.)			
Inbred strain (no. 60)	6.6 ± 0.07	0.81 ± 0.051	12.3 ± 0.78
(no. 54)	16.8 ± 0.12	1.87 ± 0.088	11.1 ± 0.53
$F_1(60 \times 54)$	12.1 ± 0.12	1.51 ± 0.088	12.5 ± 0.72
Weight of 25 kernels (gm.) from same cross			
Inbred strain (no. 60)	2.7 ± 0.03	0.39 ± 0.024	14.4 ± 0.90
(no. 54)	8.3 ± 0.11	1.21 ± 0.074	14.5 ± 0.96
$F_1(60 \times 54)$	4.6 ± 0.06	0.64 ± 0.041	13.9 ± 0.91

From data of East and Hayes (1911).

In Table 10.7 two tomato varieties differed more than 55-fold in weight of fruit (Powers 1942). The coefficient of variability and the standard deviation of logarithms (which Powers also calculated) are at least much more alike than are the standard deviations of the actual weights.

TABLE 10.7. Comparison of means (M), standard deviations (σ), and coefficients of variability (100 C), and of means ($M' = \overline{\log_{10} V}$) and standard deviations ($\sigma' = \sigma_{\log_{10} V}$) of logarithms of weights of tomatoes in two strains and their F_1's.

	M	σ	100 C	M'	σ'
Weight of tomatoes (gm.)					
Red currant	0.92 ± 0.032	0.23	25	-0.053	0.114
Danmark	51.2 ± 3.5	15.4	30	1.671	0.145
F_1	5.5 ± 0.32	1.5	27	0.707	0.134

From data of Powers (1942).

There are, on the other hand, cases in which coefficients of variability of different strains are not closely similar. In Table 10.8 (height of maize plants from East and Hayes [1911]), the standard deviation of the taller strain is actually somewhat less than that of the shorter, and that of F_1 is decidedly greater than either, although its height is somewhat less than that of the taller strain.

TABLE 10.8. Comparison of means (M), standard deviations (σ), and coefficients of variability (100 C) of height in two strains of maize and their F_1's.

	M	σ	100 C
Height of maize plants (in.)			
Inbred strain (no. 5)	68.2 ± 0.41	6.49 ± 0.32	9.5 ± 0.42
(no. 6)	101.2 ± 0.40	5.07 ± 0.28	5.0 ± 0.28
F_1 (5 × 6)	94.5 ± 0.74	8.21 ± 0.52	8.7 ± 0.55

From data of East and Hayes (1911).

For thorough study of the relation of variability to scale, data should be obtained from many comparable populations with means distributed over as wide a range as possible. Such material provides the basis for finding an empirical expression, $F(M)$, for the standard deviation in terms of the mean. The standard deviation may be taken as an indicator of the relation, $F(x)$ between a given elementary factor on the observed scale, x, and that on the desired scale, x', on which the effect is to be the same at all grades as far as possible.

$$F(x) = \frac{dx}{dx'}. \tag{10.43}$$

$$x' = \int \frac{dx}{F(x)}. \tag{10.44}$$

The most convenient functional relation for empirical fitting is usually a power series, $\sigma = a + bM + cM^2 + \cdots$. This brings us back to the simple logarithmic transformation as the case in which $\sigma = bM$, indicating $F(x) = x$, $x' = \log x$. Next in simplicity is the case in which a satisfactory fit is given by a straight line, $\sigma = a + bM$, that does not necessarily pass through the origin. This case has already been considered.

This method may be illustrated by an analysis of data presented by Emerson and Smith (1950) on a very extensive study of number of rows in maize. The unit of variation is really the number of double rows. Table 10.9 shows means, standard deviations, and coefficients of variability of the inbred lines with which they worked (six 8-rowed, three 10-rowed, thirteen 12-rowed) and of the four first crosses between 8-rowed and 12-rowed for which the distributions were reported. The coefficients of variability are more uniform than the standard deviations.

There is, however, much spread at any given value of the mean, and analysis of the 12-rowed strains shows that the actual variance of standard deviations (unweighted) is 14 times as great as the average squared standard error. Thus there were highly significant differences among these strains in spite of 8 to 21 generations of self-fertilization and actual absence of genetic variability as demonstrated by failure of selection. This illustrates the point that a comparison of standard deviations in two large inbred strains may not give an adequate basis for transformation of scale because of inherent differences in variability that would be found even between ones with the same mean grade of the character.

The regression formula for standard deviation in terms of means among all of the inbreds and F_1's of Table 10.9, giving all equal weight because of the relative unimportance of sampling errors where all numbers are large, comes out $-0.271 + 0.130M$. This suggests the transformation function $x' = \log(x - 2.1)$, in which $2.1 = 0.271/0.130$.

TABLE 10.9. Means (M), standard deviations (σ), and coefficients of variability 100 C of number of rows on ears of maize in 22 inbred lines and 4 first crosses. The second column refers to number of generations of self-fertilization.

Line	Gen.	No.	M	σ	100 C	Line	Gen.	No.	M	σ	100 C
X	6	186	7.55	1.18	15.6	*b*	18	520	11.69	1.37	11.7
1	5	408	7.80	0.76	9.7	IV	15	495	11.70	1.20	10.3
XII	3	129	7.92	0.68	8.6	III	17	467	11.88	1.13	9.5
VIII	4	183	7.95	0.36	4.5	*A*	13	452	11.93	1.19	10.0
51	8	439	7.99	0.54	6.8	2	8	286	11.93	1.59	13.3
XIII	4	136	8.17	0.57	7.0	*B*	18	601	11.95	1.24	10.4
E	15	396	10.07	0.92	9.1	II	18	226	11.95	1.35	11.3
V	19	612	10.39	1.22	11.7	*VII*	19	905	11.98	1.14	9.5
D	19	553	10.51	1.20	11.4	*VI*	13	359	12.10	1.25	10.3
						39	11	382	12.17	1.05	8.6
F_1 (1 × 2)		351	9.03	1.09	12.1	*G*	18	585	12.26	1.21	9.9
F_1 (1 × 39)		676	9.69	1.05	10.8	4	10	325	12.51	1.38	11.0
F_1 (51 × 2)		412	10.41	1.43	13.7	*c*	21	665	12.53	1.42	11.3
F_1 (51 × 39)		682	11.34	1.09	9.6						

From data of Emerson and Smith (1950).

All possible crosses were made within each of the groups of inbreds, 8-rowed, 10-rowed, and 12-rowed (except for strain *c*), and between all 8- and 10-rowed strains. Plus and minus selection was carried on for at least one generation (F_3) and usually for several more generations of inbreeding from the F_2 generations of the 66 possible crosses between 12-rowed strains, excluding strain *c*. The results in all cases in which more than 200 individuals were scored in both selection experiments are shown here in Table 10.10. One case, *AG*, is included in which there were less than 200 because it was part of the foundation for later plus selection experiments. Four of those based on crosses between F_5 and F_6 of two-strain crosses produced over 100 offspring in arrays with similar means; they are shown as 4*L*. Seven inbred strains were brought together in three lines (7*L*) by crossing F_1's of 4*L* (Table 10.11). The populations in these selection experiments should be less homogeneous genetically than the inbred strains or their F_1's but should be fairly comparable and cover a wider range of means.

The regression of standard deviations on means in these data comes out $+0.279 + 0.096M$, indicating a transformation function $x' = \log (x + 2.9)$. The estimated standard deviations are greater than those from the inbreds

TABLE 10.10. Means (M), standard deviations (σ), and coefficients of variability (100 C) of number of rows on ears of maize from lines selected in minus and plus directions from crosses between inbred 12-rowed strains.

	Minus				Plus			
	No.	M	σ	100 C	No.	M	σ	100 C
III × 4	289	13.06	1.53	11.7	257	13.19	1.56	11.8
2 × 39	327	11.63	1.44	12.4	359	12.54	1.51	12.0
A × B	277	10.08	1.12	11.1	282	12.90	1.40	10.9
III × VII	408	10.27	0.91	8.9	413	13.40	1.39	10.4
G × VII	203	10.10	1.14	11.3	286	13.51	1.37	10.1
A × G	177	10.34	1.37	13.2	222	14.22	1.23	8.6
IV × VI	302	9.42	1.53	16.2	342	13.34	1.67	12.5
IV × VII	515	10.45	1.27	12.2	498	14.56	1.55	10.6
III × VI	218	9.47	1.16	12.2	203	13.59	1.95	14.3
VI × VII	356	9.62	1.06	11.0	411	13.92	1.55	11.1
B × b	394	9.84	1.49	15.1	394	14.40	1.89	13.1
III × IV	357	9.56	1.30	13.6	269	14.33	1.71	11.9
II × 39	236	9.34	1.21	13.0	204	15.26	2.09	13.7

From data of Emerson and Smith (1950).

TABLE 10.11. Means (M), standard deviations (σ), and coefficients of variability (100 C) of number of rows on ears of maize from lines selected in the plus direction for crosses involving four (4L) or seven (7L) of the inbred lines 1.

	No.	M	σ	100 C	Origin
4L–1	120	15.57	1.76	11.3	(A × B) × (III × IV)
4L–2	115	15.83	1.99	12.6	(A × B) × (VI × VII)
4L–4	122	16.59	1.95	11.8	(A × G) × (VI × VII)
4L–5	119	17.26	1.62	9.4	(III × IV) × (VI × VII)
7L–1	127	20.66	2.28	11.0	4L(1 × 2)(3 × 5)
7L–2a	103	22.49	2.20	9.8	4L(1 × 2)(4 × 5)
7L–2b	154	22.10	2.74	12.4	4L(1 × 2)(4 × 5)

From data of Emerson and Smith (1950).

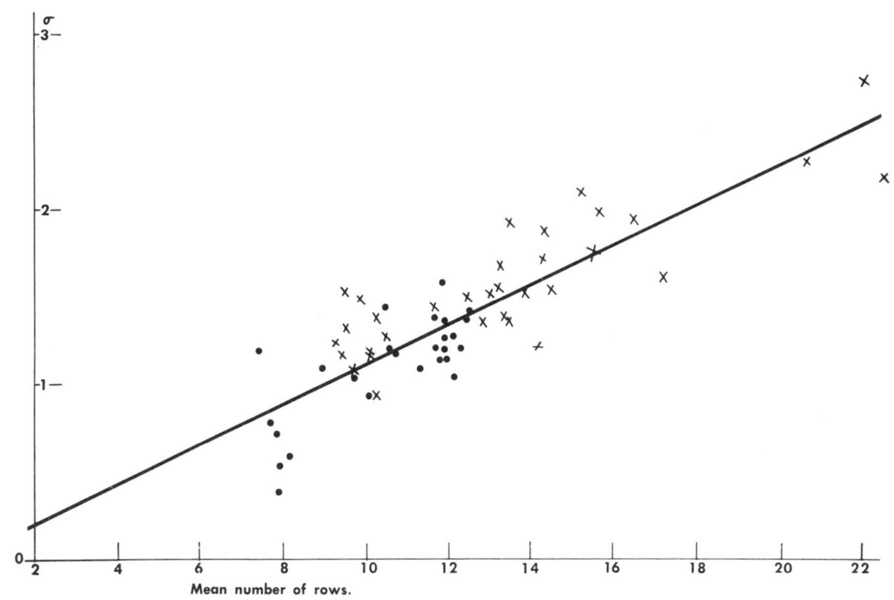

FIG. 10.5. Regression of standard deviation on mean number of rows from genetically uniform lots of maize (22 inbred strains, 4 first crosses represented by dots, 33 less uniform strains, selected and inbred a few generations represented by crosses).

up to about 16 rows, but the difference is not great. Figure 10.5 shows the relations of standard deviation to mean in all of the 59 population in Tables 10.9 to 10.11. The regression here is $-0.031 + 0.115M$, indicating that a transformation function $x' = \log X$ would be reasonably satisfactory.

The relation between the variabilities and means of different measurements of the same individual may be of interest. Edgar Anderson measured length and width of both sepals and petals on a large number of individuals of the blue flags, *Iris versicolor* and *Iris virginica*, throughout their ranges. The number, means, standard deviations, and coefficients of variability are shown in Table 10.12.

On plotting the eight standard deviations against the corresponding means, there is a fairly close approach to a straight line over the range covered. The regression $\sigma = 0.193 + 0.0908M$ indicate the transformation

$$x' = \log (x + 2.12).$$

Rasmussen (1933) found evidence of damping of variability on approach to an upper limit in the case of certain crop yields. He suggested that this could be represented by a relation of the type $x = L_2 - e^{-x'}$ in which L_2

TABLE 10.12. Comparisons of means (M), standard deviations (σ), and coefficients of variability ($100\,C$) of four floral characters of two species of *Iris*.

	Iris virginica (1,580)			*Iris versicolor* (560)		
	M(cm.)	σ	$100\,C$	M(cm.)	σ	$100\,C$
Petal length	4.829	0.697	14.4	3.702	0.525	14.2
Petal width	1.672	0.344	20.6	1.269	0.334	26.3
Sepal length	5.904	0.717	12.1	5.490	0.599	10.9
Sepal width	2.625	0.414	16.0	2.733	0.407	14.9

From data of Anderson (1928).

is the upper limit and x' is the transformed measurement. This is equivalent to a reversed logarithmic transformation.

$$F(x) = L_2 - x. \tag{10.45}$$

$$x' = \log (L_2 - x). \tag{10.46}$$

There is presumably also damping on approach to a lower limit in such cases. Damping jointly by upper and lower limits can be represented by $F(x)$ in the form:

$$F(x) = (x - L_1)(L_2 - x) = -x^2 + (L_1 + L_2)x - L_1 L_2, \tag{10.47}$$

$$x' = \log \left(\frac{x - L_1}{L_2 - x}\right). \tag{10.48}$$

There are other functions that are sometimes useful. In a study of spotting in inbred strains of guinea pigs (Wright 1920), it was found that the distribution of frequencies on a scale of percentage of white showed strong positive skewness if the median percentage was well below 50%, although no animals were wholly self-colored, but strong negative skewness if the median was well above 50%, although very few were self-white (cf. Fig. 6.5).

On plotting upper and lower quartiles, as measures of variability, against the scale value half-way between median and quartile in 34 distributions (males and females of 17 inbred strains), there was rough fit to the ordinates, z, of a normal probability curve, corresponding to the percentage of white (x) treated as the tail percentage.

$$z = \frac{dx}{dx'} \quad \text{where} \quad x = \text{pri } x' = \int_{-\infty}^{x'} z\, dx' = \int_{-\infty}^{x'} \frac{1}{\sqrt{2\pi}} e^{-\frac{1}{2}x'^2}\, dx'. \tag{10.49}$$

Thus

$$x' = \text{pri}^{-1} x. \tag{10.50}$$

This inverse probability transformation was given a rational interpretation by the hypothesis that there is a normal distribution of potentialities for color among elementary areas of the coat. This transformation gave a fairly satisfactory normalization of the distributions.

The formula of the curve on the original scale of percentages, on the assumption of normality on the transformed scale is as follows, noting that

$$\frac{d}{dx}\,\text{pri}^{-1}\,x = \sqrt{2\pi}\,\exp\left[\frac{1}{2}\,(\text{pri}^{-1}\,x)^2\right] \tag{10.51}$$

$$y = \frac{1}{\sigma'}\exp\left\{\frac{1}{2}\left[(\text{pri}^{-1}\,x)^2 - \left(\frac{\text{pri}^{-1}\,x - M'}{\sigma'}\right)^2\right]\right\}. \tag{10.52}$$

Figure 10.6 is a conspectus of forms taken by this curve with different values of M' and σ' (Wright 1926).

There is of course, no necessity for a normal distribution of potentialities for color in the case above, so that other forms of this auxiliary distribution should be considered. In particular, if there are variates with 0% or 100% occurrence, it is indicated that the auxiliary distribution, z, has a finite lower or upper limit. We will consider first the case of an auxiliary distribution that is triangular with its lower limit at $x' = -1$, apex at $x' = a$, and its upper limit at $x' = +1$.

$$\left\{\begin{array}{l}
z = 0 \quad \text{if} \quad x' < -1, \quad x = 0 \\
z = \dfrac{1 + x'}{1 + a} \quad \text{if} \quad -1 < x' < a, \\
\qquad x = \dfrac{1}{2(1 + a)}\,(1 + 2x' + x'^2), \quad x' = \sqrt{2(1 + a)\,x} - 1 \\
z = 1 \quad \text{if} \quad x' = a, \quad x = \dfrac{1 + a}{2} \\
z = \dfrac{1 - x'}{1 - a} \quad \text{if} \quad a < x' < 1, \\
\qquad x = \dfrac{1}{2(1 - a)}\,(1 - 2a + 2x' - x'^2), \\
\qquad x' = 1 - \sqrt{2(1 - a)(1 - x)} \\
z = 0 \quad \text{if} \quad 1 < x', \quad x = 1.
\end{array}\right. \tag{10.53}$$

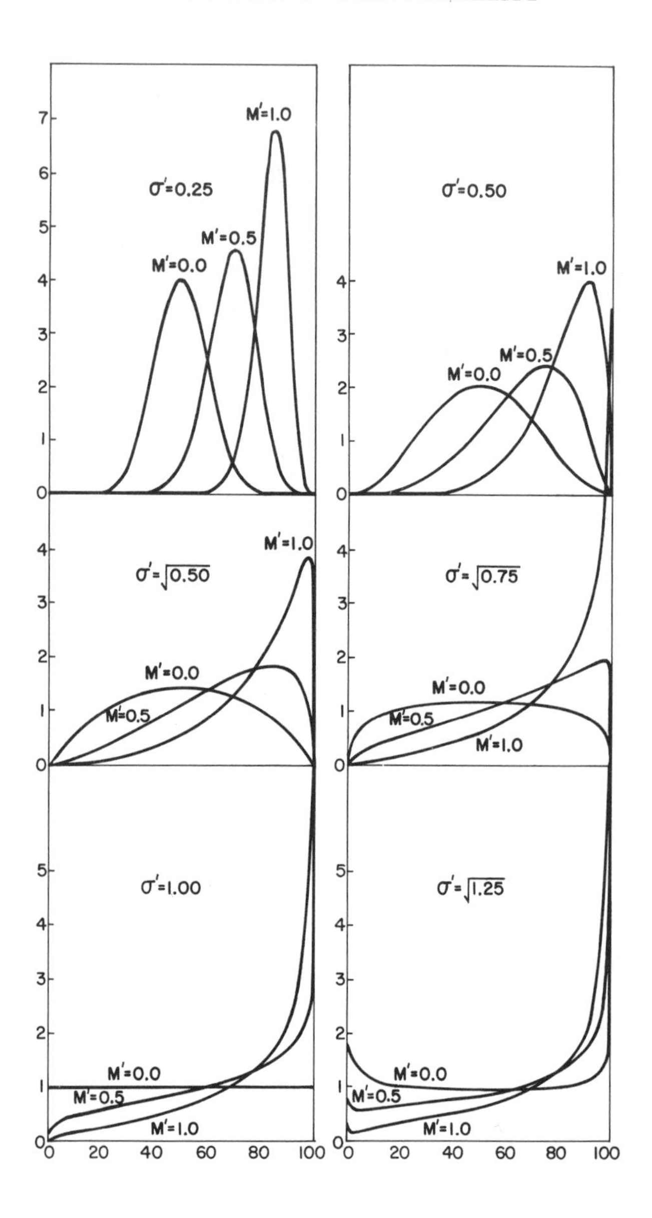

FIG. 10.6. Conspectus of forms taken by the curve $y = (1/\sigma') \exp \{\tfrac{1}{2}(\text{pri}^{-1}x)^2 - [(\text{pri}^{-1}x - M')/\sigma']^2\}$ where M' is the mean and σ' is the standard deviation on the transformed scale. (Redrawn from Fig. 2, Wright 1926.)

Thus

$$(10.54)\quad \begin{cases} z = F(x) = \sqrt{\dfrac{2x}{1+a}}, & 0 \le x \le \dfrac{1+a}{2} \\ z = F(x) = \sqrt{\dfrac{2(1-x)}{1-a}}, & \dfrac{1+a}{2} \le x \le 1. \end{cases}$$

As noted above, spotted guinea pigs (*ss*) are occasionally self-white, and one inbred strain produced many in a late branch. Thus a better agreement with the data can be obtained by assuming a finite upper limit of the auxiliary distribution. Moreover, although *ss* has never been observed to be completely self-colored, self-colored animals are common among heterozygotes, *Ss*, and are usual among *SS*. Thus a finite lower limit is also indicated. We will return to this later.

In the extreme case in which the auxiliary distribution is rectangular ($z = 1$ between $x' = 0$ and $x' = 1$, but $z = 0$ if $x' < 0$ or if $x' > 1$; $x = x'$ between these limits, but $x = 0$ if $x' \le 0$, and $x' = 1$ if $x' \ge 1$). In this case the observed distribution is the same as the transformed one between these limits

$$\left(y = y' = \frac{1}{\sigma'\sqrt{2\pi}} \exp\left[-\frac{1}{2}\left(\frac{x' - M'}{\sigma'}\right)^2\right]\right),$$

where M' and σ' are mean and standard deviation on the unlimited transformed distribution, but all frequencies for $x' < 0$ are grouped in one class $x = 0$, and all for $x' > 1$ are grouped in one class $x = 1$ (or 100%). This is the case of a normal distribution truncated, without damping at 0% and 100%.

Interaction of Major Factors

A third mode of approach to the problem of transformation of scale is to find that function of the scale that causes the effects of two or more major differentials (unitary or compound, genetic or environmental) to become additive. Zeleny (1920) applied a criterion of this sort as well as others, in justifying his scale of factorial values, which was essentially a logarithmic scale, for facet counts in *Drosophila*. He found that the differences in facet numbers at 15° and 25° in three strains with widely different means (ultra bar (eye), low-selected bar, and high-selected bar) were relatively consistent on the factorial scale, although varying nearly 6-fold in absolute terms.

TABLE 10.13. Average numbers of facets in eyes of three genetically different strains of *Drosophila melanogaster* at 15° and 25° with comparison of differences and ratios at these temperatures.

	15°	25°	Difference	Ratio 25°/15°
Ultra bar	51.5	25.2	26.3	0.489
Low-selected bar	189.0	74.2	114.8	0.393
High-selected bar	269.8	120.5	149.3	0.447

From data of Zeleny (1920).

On the other hand, this change in temperature causes a change of about 10% in the opposite direction in normal eyes (with more than 800 facets).

As another example, consider the mean weights at 33 days in litters of from one to four in five inbred strains of guinea pigs (Wright and Eaton 1929; Tables 10.14–10.17).

The difference between the heaviest and lightest strains for given size of litter, or between litters of one and four in given strains, are far from constant (litters of five were rather few in number). Logarithms of the weights

TABLE 10.14. Mean weight of guinea pigs at 33 days in grams (1916–24) according to strain and size of litter.

Strain	Size of litter						
	1	2	3	4	(5)	(1) − (4)	log (1) − log (4)
13	299.5	264.0	226.8	203.6	185.5	95.9	0.167
35	276.8	249.5	221.1	188.7	172.7	88.1	0.166
32	254.3	237.9	198.8	186.1	184.5	68.2	0.135
39	231.3	217.4	188.3	176.3	174.8	55.0	0.118
2	213.4	195.0	171.8	155.4	153.5	58.0	0.138
(13) − (2)	86.1	69.0	55.0	48.2	32.0	37.9	
log (13) − log (2)	0.147	0.132	0.121	0.118	0.082		0.029

From data of Wright and Eaton (1929).

TABLE 10.15. Values of $x' = \log_{10}(x - 80)$ where x is mean weight of guinea pig at 33 days in grams according to strain and size of litter.

Strain	Size of litter					
	1	2	3	4	(5)	(1) − (4)
13	2.341	2.265	2.167	2.092	2.023	0.249
35	2.294	2.229	2.150	2.036	1.967	0.258
32	2.241	2.198	2.075	2.026	2.019	0.215
39	2.180	2.138	2.035	1.984	1.977	0.196
2	2.125	2.061	1.963	1.877	1.866	0.248
(13) − (2)	0.216	0.204	0.204	0.215	0.157	0.001

From data of Wright and Eaton (1929).

were used in dealing with this character, but the differences still show significant trends. Trend is largely eliminated by using the transformation $x' = \log(x - 80)$. The subtraction of 80 grams is based here merely on the extreme cases. The determination of the best scale would require consideration of much other data.

There were marked differences in the records of the strains in different years. There was a rising trend from 1916 to 1924, in spite of increasing inbreeding, due undoubtedly to improving conditions. The differences between

TABLE 10.16. Mean weight of guinea pigs at 33 days (in grams) corrected for the size of litter.

Year	13	35	32	39	2	(13) − (2)
1916	190.7	195.4	181.8	170.6	148.2	42.5
1917	206.3	206.7	195.6	185.1	160.9	45.4
1918	239.3	223.0	198.9	197.0	174.9	64.4
1919	240.4	227.2	212.5	204.5	182.1	58.3
1920	259.0	232.4	215.0	215.5	177.6	84.4
1921	228.6	207.1	204.3	200.9	162.9	65.7
1922	234.2	230.3	217.7	197.3	185.9	48.3
1923	256.3	245.7	227.0	223.4	191.6	64.7
1924	257.0	244.8	223.3	229.2	189.9	67.1
(24) − (16)	66.3	49.4	41.5	58.6	41.7	24.6

From data of Wright and Eaton (1929).

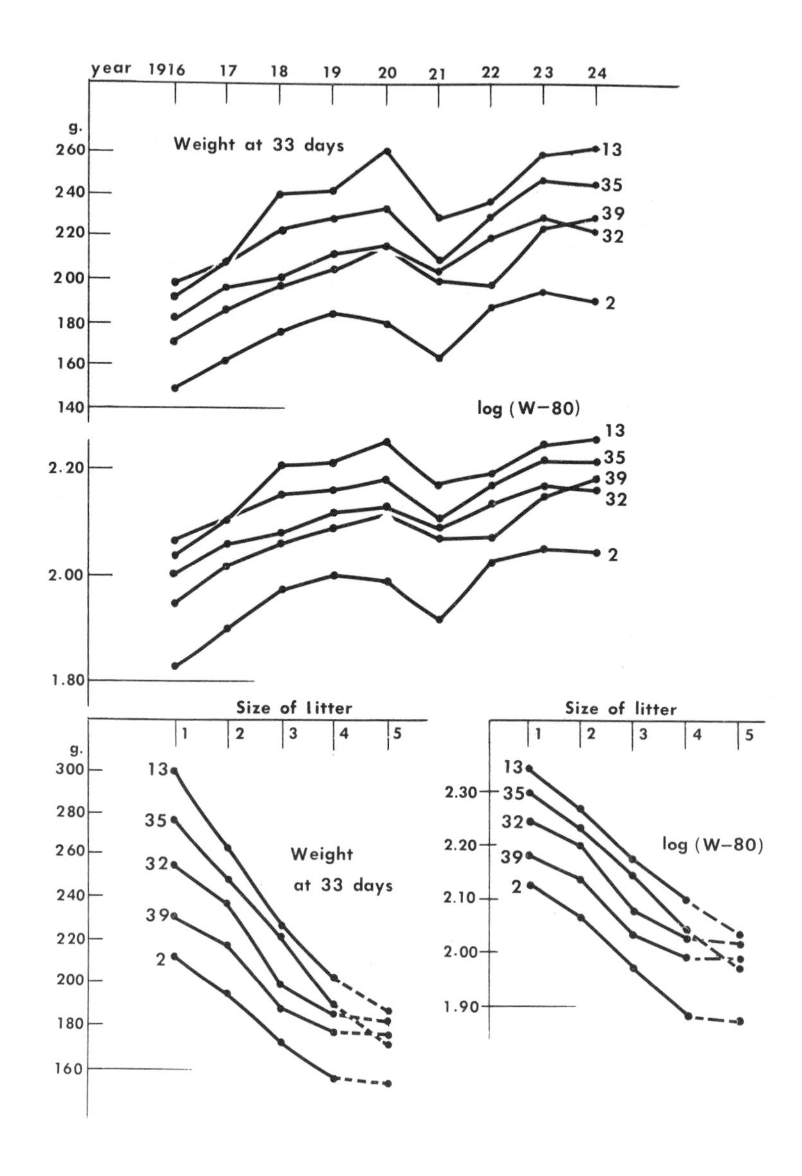

FIG. 10.7. Weight at 33 days (W) of guinea pigs of 5 inbred strains by year at top and below this the more nearly parallel values of $\log_{10}(W - 80)$. W plotted against size of litter in lower left and $\log_{10}(W - 80)$ plotted similarly at lower right. (From data of Wright and Eaton 1929.)

TABLE 10.17. Value of $x^1 = \log_{10}(x - 80)$ where x is mean weight of guinea pig at 33 days (in grams), corrected for the effect of size of litter, according to strain and year.

Year	13	35	32	39	2	(13) − (2)
1916	2.044	2.066	2.008	1.957	1.834	0.210
1917	2.101	2.103	2.063	2.022	1.908	0.193
1918	2.202	2.155	2.074	2.068	1.977	0.255
1919	2.205	2.168	2.122	2.095	2.009	0.196
1920	2.253	2.183	2.130	2.132	1.989	0.264
1921	2.172	2.104	2.095	2.082	1.919	0.253
1922	2.188	2.177	2.139	2.069	2.025	0.163
1923	2.246	2.220	2.167	2.157	2.048	0.198
1924	2.248	2.217	2.168	2.174	2.041	0.207
(24) − (16)	0.204	0.151	0.160	0.217	0.207	−0.003

From data of Wright and Eaton (1929).

large and small strains was on the whole less in the earlier years than in the later ones. Taking the extreme cases, we again see that the transformation $x' = \log_{10}(x - 80)$ is satisfactory (cf. Fig. 10.7).

When, however, we compare the variability of the weights within litters in one of the inbred strains (no. 35) (Table 10.18), we find that the standard deviations are much more uniform than the coefficients of variability and still more so than such coefficients relative to $(M - 80)$.

TABLE 10.18. Comparison of means (M), standard deviations (σ), and coefficients of variability relative to 0 and to 80g of weight of guinea pigs at 33 days according to size of litter.

Litter size	No.	M	σ	$100\frac{\sigma}{M}$	$100\frac{\sigma}{(M-80)}$
1	80	276.8	36.8	13.3	18.7
2	382	249.5	37.2	14.9	21.9
3	530	221.1	37.2	16.8	26.4
4	205	188.7	40.4	21.4	37.2

From data of Wright and Eaton (1929).

This illustrates that no transformation of scale can be expected to make all factor effects strictly additive. This case will be considered further later.

Mendelian Expectations

A fourth way to approach the problem is to find the transformation that agrees best with the theoretical relations of the means of F_2 and backcross populations to those of parental strains (preferably isogenic) and F_1, on the assumption of additive effects. This approach contrasts especially with the second in that it is based on genetic instead of predominantly on non-genetic factors and is thus of especial importance in population genetics. We will, however, defer consideration to a later chapter on the genetics of quantitative variability.

This chapter has considered deviations from normality that may result from heterogeneity among the variances and among the means of components of a compound distribution. The former may lead to strong leptokurtosis and the latter to bimodality and extreme platykurtosis. Gross heterogeneity, however, can usually be avoided.

In the case of a compound variable, strong leptokurtosis is theoretically possible where one alternative of each component is rare, but this may be overcome by a very small number of components with more or less equally frequent alternatives. Extreme platykurtosis may arise where the effects of binomial components are in geometric progression, but it requires that the leading one determine something like one-half or more of the total variance.

The effects of correlations in the values of binomial components may bring about wide deviations from a binomial distribution, and there may be wide deviation for an otherwise expected Poisson distribution by clustering of rare events. This complication is not likely to be important for the distribution of characters of organisms.

The general conclusion with respect to frequency distributions of quantitatively varying characters from which gross heterogeneity has been removed is that there should be a strong bias toward normality, provided that factors have additive effects.

The last portion of the chapter was devoted to criteria for transformation functions that would eliminate nonadditivity as far as possible. The first criterion for a transformation function is that it actually normalize the forms of distributions. The second is based on determination of the function of comparable standard deviations of distributions that makes them independent of the mean as far as possible. The third is based on the discovery of a function that removes interaction effects of specific factors or factor groups, genetic or environmental. Finally, the theoretical relations among the means and variances of segregating generations (F_2, backcross) following a cross, to those of the parental strain and their F_1's was referred to as fourth criterion for transformation.

CHAPTER 11

Analysis of Actual Frequency Distributions

It is often difficult to recognize from inspection of a distribution whether it differs appreciably from normality. The calculation of the form statistics, γ_1 and γ_2 is a tedious process. A much simpler method, based on percentiles rather than moments, is available for a survey of distributions and for testing possible transformation functions if transformation of scale seems indicated.

The first step is to plot the running sum of the proportional frequencies up to each class limit, against the latter. If the distribution is normal, we have, by definition, $\text{pri}^{-1}\, p = (V - M)/\sigma$, a linear relation. Any systematic deviation from a straight line is immediately apparent, provided that the size of the population is large enough (several hundred at least) to give a reasonably smooth curve.

If there is a close approach to a straight line, the mean and standard deviation can be estimated from the intercepts on the axes since $V = M$ for $\text{pri}^{-1}\, p = 0$ and $\text{pri}^{-1}\, p = -M/\sigma$ for $V = 0$. Precise determination of the parameters, however, should be based on the moments because of the complications of weighting in fitting a straight line in this case. It may be noted that little weight is to be put on deviations from linearity where $\text{pri}^{-1}\, p$ exceeds 2.6 (or even 2) in absolute value in judging the approach to normality unless the numbers are very large.

In making a survey by this method, it is desirable to consider first the sorts of deviation from linearity due to gross heterogeneity. In Figure 11.1*A*, $\text{pri}^{-1}\, p$ is plotted against V for the three distributions of Figure 10.1, which are leptokurtic because compounded of two normal distributions with the same mean but different variances. In all three, one component has four times the variance of the other ($L = \sigma_2^2/\sigma_1^2 = 4$). If the more variable component is also four times as numerous as the other ($K = 4$), there is little divergence from the line derived from its distribution alone ($K = \infty$). If, however, the more variable component is only one-fourth as numerous as the other ($K = 0.25$), there is wide divergence from linearity in the graph. This

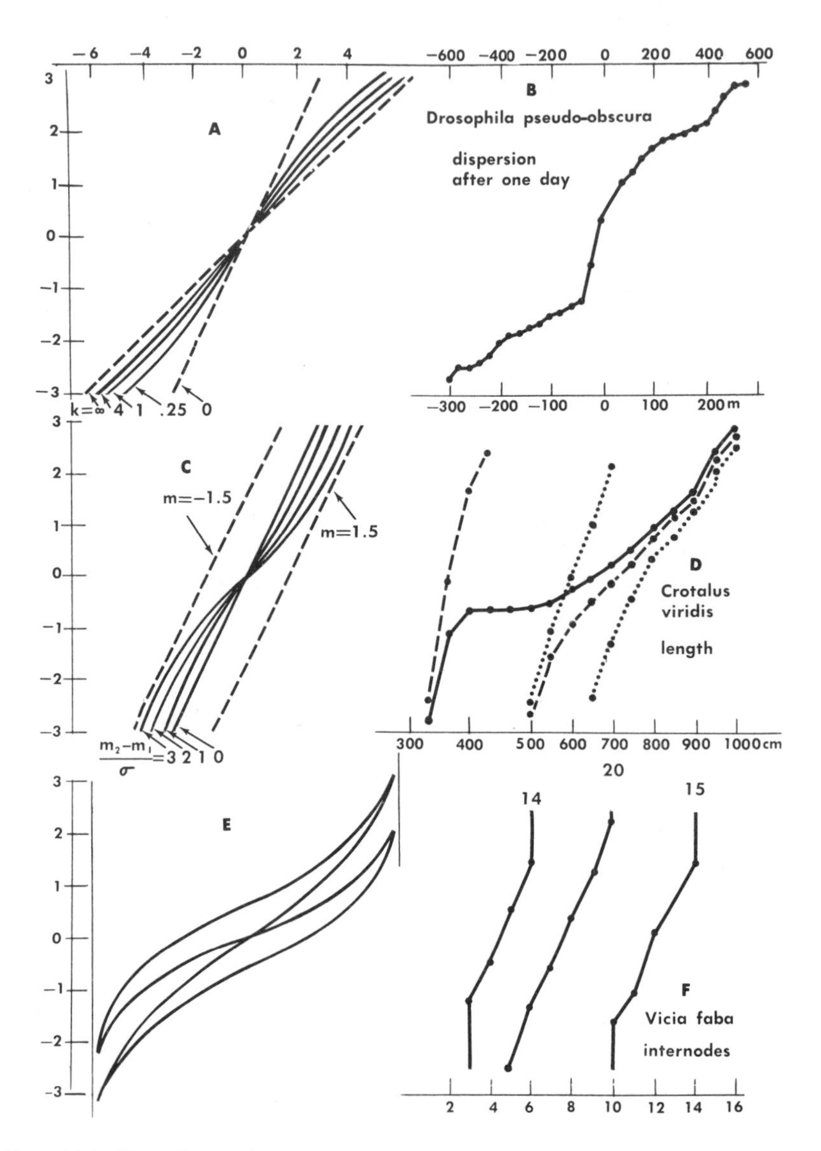

FIG. 11.1. Transformations ($\text{pri}^{-1}p$) of various theoretical distributions (*left*) and observed ones (*right*). *A* from Fig. 10.1; *B* from Fig. 6.1; *C* from Fig. 10.2; *D* from Fig. 6.6*C*; *E* in part from Fig. 10.4*A*, *B*; *F* from Fig. 6.11*B*. Further description in text.

shows the relatively steep slope based on the abundant but less variable component in the neighborhood of the mean, but it approaches parallelism with the line based on the more variable component in the outlying regions.

In the dispersion of *Drosophila pseudoobscura* after release from a station in a homogeneous terrain (Fig. 6.1), the values of $\text{pri}^{-1}\, p$ (Fig. 11.1*B*) yield a curve of the type discussed above, but the ratio of standard deviations of the two hypothetical components is about six ($L = 36$). As most of the flies in the middle three stations (74% of the total) must be assigned to the group with small standard deviation, K is about one-third.

The pattern of deviations is naturally opposite in platykurtic distributions—with low slope of $\text{pri}^{-1}\, p$ in the middle, steep near the ends. Figure 11.1*C* shows the graphs for the three distributions compounded of two equally numerous and equally variable components with means separated by one, two, or three standard deviations (of Fig. 10.2*A*, *B*, *C*). There is little deviation from the straight line yielded by a single component if the separation of means is equal to σ, whereas deviation is considerable with a separation of 3σ. The lines representing the two components in this last case are shown by broken lines. Cases in which the components are not equally frequent (such as those of Fig. 10.2*D*, *E*, *F*) give similar curves except that the point of intersection is not at $\text{pri}^{-1}\, p = 0$.

The distribution of lengths of rattlesnakes (*Crotalus viridis*) of Figure 6.6*C* (data of Klauber 1937) is analyzed by this method in Figure 11.1*D*. The solid line shows $\text{pri}^{-1}\, p$ plotted against V. In this case the distribution of snakes less than a year old hardly overlaps that of the older ones, and separate graphs of $\text{pri}^{-1}\, p$ can readily be made (broken lines). Distribution for the younger snakes is nearly straight in the range $\text{pri}^{-1}\, p = \pm 2$, but that for the older snakes again indicates a compound distribution with separated means. As the distribution itself indicates bimodality, a rough cleavage can be made by dividing the low intermediate class equally and exchanging two individuals in the adjacent classes. The graphs for the yearlings come out almost straight, but the residual class is again of the type that indicates a compound distribution with separate means.

Figure 11.1*E* shows the graphs for four distributions of finite range. All necessarily turn down sharply at the lower limit and turn up sharply at the upper limit. The flatter one that passes through $\text{pri}^{-1}\, p$ halfway between the limits is for a rectangular distribution (Fig. 10.4*A*) The steeper one through this same point is for an isosceles triangle (Fig. 10.4*B*). The uppermost curve is for a right triangle with the apex at the lower limit, while the lowest one is for a right triangle with the apex at the upper limit. These illustrate platykurtic distributions of a very different type from these compounded of normal distributions with different means.

The distributions of internode number in inbred lines of *Vicia faba* (data of Sirks 1932) are the most triangular of those illustrated in chapter 6 (Fig. 6.11*B*). The inverse probability graphs are shown in Figure 11.1*F*. The middle one (20) is fairly straight and extends so far in both directions that there is good agreement with a normal distribution. The other two, however, both stop well above $\text{pri}^{-1}\, p = -2$ and below $\text{pri}^{-1}\, p = +2$, and thus seem to be of sharply limited range.

If there is no indication of gross heterogeneity, it becomes appropriate to search for a transformation of scale to remove systematic nonadditive effects as far as possible. Obviously if M' and σ' are the mean and standard deviation on a scale on which the distribution becomes normal, we have $\text{pri}^{-1}\, p = (V' - M')/\sigma'$. Various transformations can readily be tested not only for isolated distributions, but for families of related distributions. The degree of success can perhaps be judged best by plotting the successive slopes, $\Delta\, \text{pri}^{-1}\, p/\Delta V'$. After the trend has been eliminated as far as possible, $1/\sigma'$ may be estimated from the average. Since 95% of the frequencies of the distribution curve fall within the range $\text{pri}^{-1}\, p = \pm 2$, and 99% within the range $\text{pri}^{-1}\, p = \pm 2.6$, little attention need be paid (as noted earlier) to entries outside of ± 2.6 or even ± 2, unless the numbers are very large. The mean on the transformed scale may be estimated from the intercept of the best line of slope σ with the axis, $\text{pri}^{-1}\, p = 0$. As for primarily normal distributions, it is best where practicable to calculate the parameters from moments. In the important class of logarithmic transformations $V' = \log (V' - L)$, it has been noted (chap. 10) that the parameters L, M', and σ' may all be derived from the first three moments of the untransformed variates. Having obtained estimates of M' and σ' in any way, estimated values of $\text{pri}^{-1}\, p\ (= V' - M')/\sigma')$ can be derived for each transformed class limit V' from which estimated values of p and, in two more steps, estimates of the class frequencies ($f = N\Delta p$) can readily be made and tested for goodness of fit.

Figures 11.2*A*, *B*, *C* illustrate three common types of graph on plotting $\text{pri}^{-1}\, p$ against observed measurements. The graph (11.2*B*) for height of 1,000 male Harvard students, from Figure 6.2*A* (Castle 1916*b*), is so nearly straight that no transformation is indicated. The normal probability curve gives an excellent fit. The plot for weights of the same men, Figure 11.2*C* from Figure 6.2*B*, gives a fairly smooth curve that is concave to the right. As shown below (Fig. 11.2*E*), there is a much closer approach to a straight line on plotting $\text{pri}^{-1}\, p$ against the logarithms of the weights. There is still, however, a systematic concavity to the right. This is eliminated on plotting against $\log (V - 18)$, weight (V) being in kilograms.

If the size factors in this case were producing certain percentage

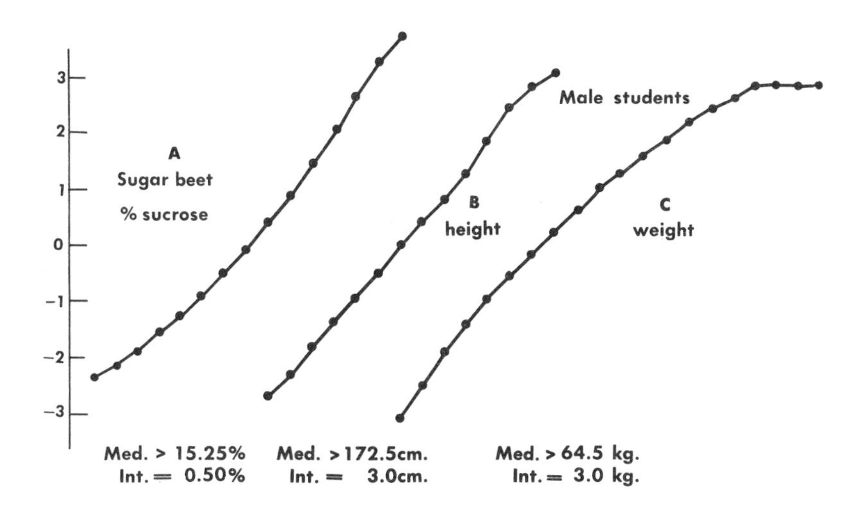

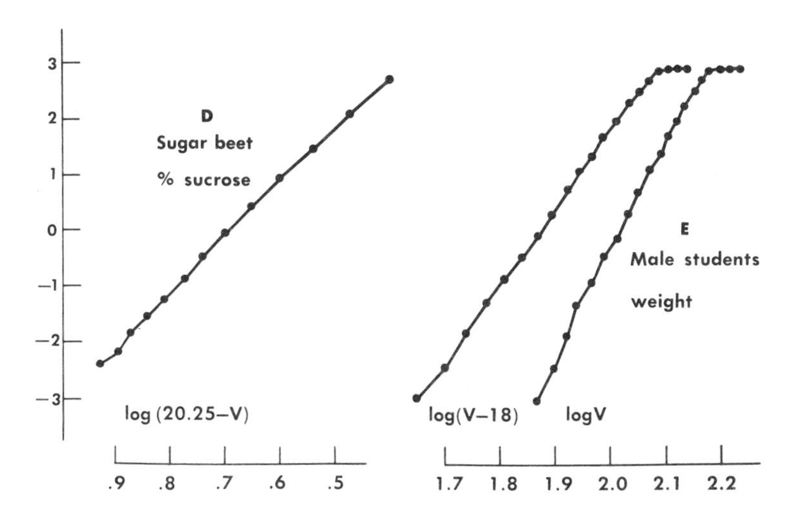

FIG. 11.2. Transformations ($\text{pri}^{-1} p$) of three continuous distributions with negative skewness (sugar beet, *A* from Fig. 6.3*C*), almost no skewness (human height, *B* from Fig. 6.2*A*), and positive skewness (human weight, *C* from Fig. 6.2*B*). Attempts at rectifications of the skewed distributions, by use of logarithmic transformations of scale, are shown below (*D*, *E*).

increments above a lower limit, it would be expected that similar logarithmic transformations would be needed to normalize the distributions of both height and weight. Yet any transformation gives a slightly poorer fit in the case of height than none at all.

In the sucrose content of sugar beets there is concavity to the left (Fig. 11.2*A* from Fig. 6.3*C*, data of De Vries 1909). An approximately straight line is arrived at by plotting against log $(20.25 - V)$ where V is the percentage of sucrose (Fig. 11.2*D*). This seems to be a case of the kind suggested by Rasmussen (1933) in which there is damping of variability on approach to an upper limit.

Transformations are most useful in connection with families of distribution. Figure 11.3*A* shows plots of $\text{pri}^{-1}\, p$ against the actual petal and sepal measurements obtained by Edgar Anderson (1928) in two species of *Iris* (the distributions for one of which were shown in Fig. 6.2*E*). The lines diverge from bottom to top of the figure, indicating that the greater measurements vary more than the smaller ones. Moreover, all but one or two show systematic concavity to the right. The transformation $V' = \log V$ brings about overcorrection in both respects, causing convergence at the top and concavity to the left. It has indeed already been noted that there is a regression of standard deviation on mean but not in full proportion to the latter. This regression indicated the transformation $V' = \log (V' + 2.12)$. Its use brings about parallelism as far as possible in the lines of Figure 11.3*C*. It also straightens them out as well as any single formula can. The question remains as to why the factors have effects that are intermediate between constancy on the scale of centimeters and constancy with respect to percentage increments.

In Figure 11.3*B* $\text{pri}^{-1}\, p$ is plotted against weight at 33 days in litters of one to four in a closely inbred strain of guinea pigs (no. 35) (Fig. 6.2*D*, Wright and Eaton 1929). The straightness and the close approach to parallelism of the lines (implying uniform standard deviations, table 10.18) suggest that the factors (nongenetic) that cause variations within litters act additively and that the arrays of factors are the same in litters of all sizes. Yet, as brought out earlier, the differences in strain and size of litter (including the means of the present data) interact nonadditively in such a way as to require a super-logarithmic transformation (log $[W - 80]$), Table 10.15, and this is essentially true also of differences in strain and year. Moreover, the standard deviations of strains rise markedly with strain average. There seems to be no doubt that for the most part the factors that affect early growth in guinea pigs tend to produce disproportionately great percentage increments with increase in size and that the constancy suggested by the results in Figure 6.2*D* is in some way spurious. It is indeed obvious that the arrays of factors

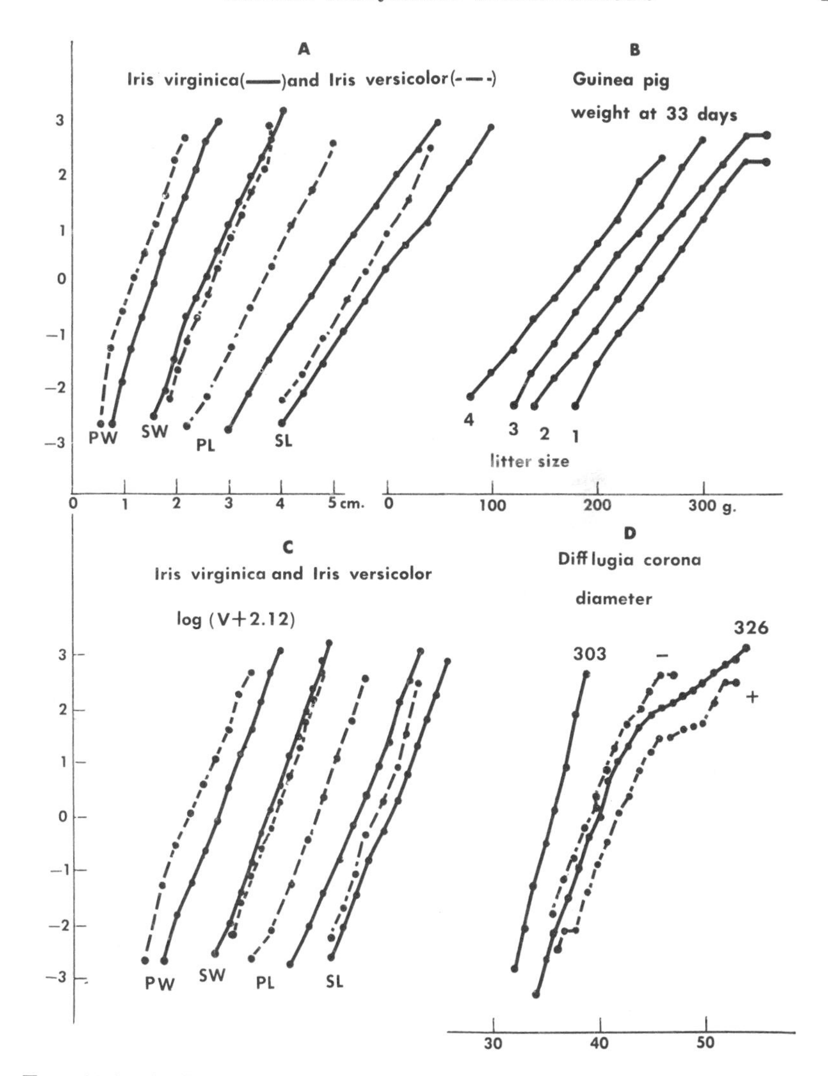

FIG. 11.3. *A*, Transformations ($\text{pri}^{-1}p$) of distributions of lengths (*L*) and widths (*W*) of petals (*P*) and sepals (*S*) of two species of *Iris* (one in Fig. 6.2*E*). *B*, Transformations ($\text{pri}^{-1}p$) of weights of guinea pigs at 33 days (Fig. 6.2*D*). *C*, Attempt at rectification of the *Iris* distributions by mean of a logarithmic transformation of scale. *D*, Transformations ($\text{pri}^{-1}p$) of distributions of diameters of shells of *Difflugia corona* of two clones, 303 and 326, and subclones of the latter selected in opposite directions.

operating within litters of different sizes are not alike. Competition among littermates is not a factor in litters of one, but becomes increasingly important in larger litters. Apparently the greater stress in large litters results in greater variability in weight than expected on the simple transformation theory, and the greater stress in large animals in litters of a given size just offsets the greater absolute gains that they would otherwise be expected to make. Newborn guinea pigs are extraordinarily large in relation to their mothers. These considerations do not apply on comparing different strains for variability or in comparing strain averages with respect to litters of different sizes or different years. Nevertheless, it is remarkable that a balancing of influences on growth rates of the sorts indicated should result in lines that are so nearly straight and so nearly parallel as in Figure 6.2*D*.

The diameters of shells of *Difflugia* in clone 303 (Fig. 11.3*D*) published by Jennings (1916) yield about as straight a plot of $\text{pri}^{-1}\, p$ as can be expected from the number, but those for clone 326, although fairly straight up to a certain size, diverge widely beyond this. This is not the sort of divergence that can be rectified by a logarithmic transformation. Moreover, the numbers are so large for this clone (2,375) that the divergence is undoubtedly real. Small portions of the clone were subject to plus and minus selection. These show the same phenomenon. It appears that beyond diameters of about 42 scale units, factors tend to have a runaway effect on size.

Sucrose content of sugar beets, discussed above, is an example of a quantitatively varying physiological character. Several others are dealt with in Figure 11.4. Burks's data on the intelligence quotients of foster and own children in a somewhat restricted specified population (Fig. 11.4*B* from Fig. 6.3*B*) are fitted by parallel straight lines, indicating normal distributions as well as can be judged from the rather small numbers, except for some indication of heterogeneity at the extremes.

The data of Harris and Benedict (1919) on basal metabolism (calories per square meter of surface in men and women) are also hardly numerous enough to be adequate for our purpose (Fig. 11.4*A* from Fig. 6.3*A*). Those for men are fitted as well as can be expected by a straight line. Those for women are somewhat improved by a logarithmic transformation. All that can be said is that both are of the usual near normal type.

Data on the speed of American trotting horses was tabulated from the records in the Trotting Register by Galton (1895) (Fig. 11.4*C* from Fig. 6.3*E*). As noted earlier, a mile record of 2 min 30 sec was required for entry, and it was obvious that this severely truncated the distribution. Moreover, those entered in each year as 2:29 or more were obviously unreliable. The mode in 1892–93 was apparently only slightly below the entry requirement. The logarithms of the ordinates of a normal curve give a parabola. Pearson

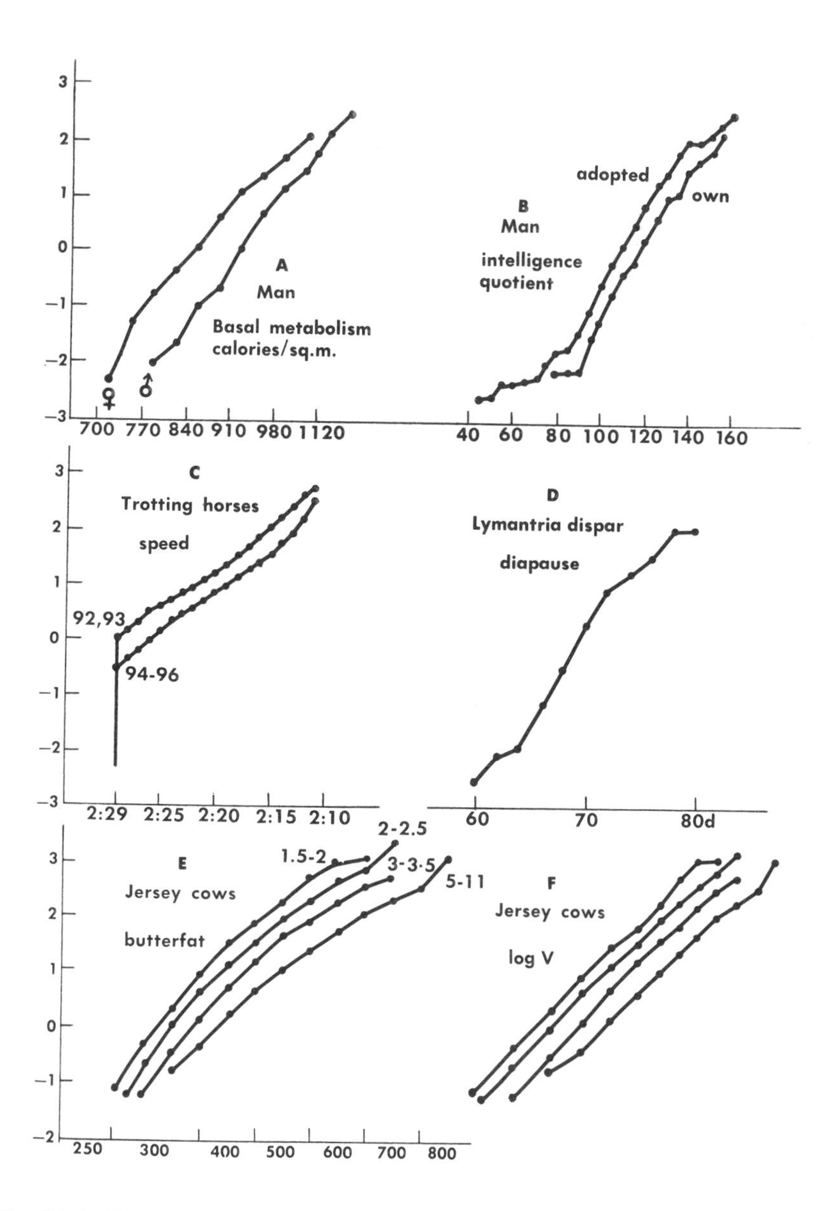

FIG. 11.4. Transformations ($\text{pri}^{-1}\,p$) of distributions of a number of physiological variates illustrated in Fig. 6.3. Those for butter-fat production (E) are rectified by means of a logarithmic transformation of scale (F).

(1902) fitted such parabolas by the method of moments and obtained 2:28.86 and 2:28.81 as the means in 1892 and 1893, respectively. It appears that the amount of truncation was only slightly less than 50%. In Figure 11.4*C*, $\text{pri}^{-1}\,p$ has been plotted against speed in feet per second under the hypothesis of 50% truncation, which gives a fairly close approach to a straight line. It may be added that the approach is somewhat closer in terms of feet per second than in terms of seconds for a mile. The records for 1894–95 showed a distinctly higher mode, probably due to improvement in the sulkies, and truncation of only about 30%. The plot of $\text{pri}^{-1}\,p$ against speed on this basis departs from straightness in the way expected from the triangular appearance of the distribution.

The data on butterfat production by original-entry Jersey cows in various age classes (American Jersey Cattle Club) (Fig. 11.4*E*, *F* from Fig. 6.3*F*) are much more adequate from the standpoint of numbers, but again they are incomplete. Cows less than three years old must produce 250 lbs in 365 days to be entered in the records. This minimum rises to 360 lbs at 5 years. The truncated portion must be estimated in some way before $\text{pri}^{-1}\,p$ can be calculated. It was obvious from the forms of the distribution (Fig. 6.3*F*) that the amount of truncation was not very great in this case and that there was marked positive skewness. Davidson (1928) fitted the logarithms of the observed frequencies to the parabola derived from the logarithmic normal curve and modified the weights to be given the entries in order to minimize χ^2. In fitting frequency distributions by least squares, the squared residuals should be weighted by the reciprocals of the frequencies ($1/f_o$ in practice) to make $\chi^2 = (f_o - f_c)^2/f_c$ minimum. In fitting logarithms of frequencies, Δf in the primary curve becomes $\Delta f/f$ and the squared residuals are $(\Delta f/f)^2$. Thus these squared residuals should be weighted by f_c (f_o in practice). The total frequency N' as well as M' ($= \overline{\log V}$) and σ' ($= \sigma_{\log V}$) could be estimated. The closeness of fit to the logarithmico-normal curve was such that the probability for χ^2 in the 19 age classes was in all cases greater than 0.10. To avoid confusion, all age classes from 5 years to 11 years (in which M' and σ' were substantially constant) are combined in Figure 11.4*E* and *F* and only three of the younger half-year age class are shown. It may be seen that $\text{pri}^{-1}\,p$ is in all cases an almost linear function of log V and that the lines are substantially parallel. This held for all of the age classes. Clearly the factors that affect butterfat production tend to have constant percentage effects.

The graph for length of diapause in gypsy moths collected by Goldschmidt (1933) in Portici, Italy, like those for human IQ and basal metabolism, is only roughly linear, but, like these, the numbers are rather inadequate (Fig. 11.4*D* from Fig. 6.3*D*).

Figure 11.5 shows graphs for a number of morphological indexes. The case with the most adequate numbers is that for relative head length in rattlesnakes, *Crotalus viridis viridis* (Fig. 11.5A from Fig. 6.4A). Klauber (1938) found systematic change in the relation of head (H) to total length (L) with increasing size (heterauxetic growth). To avoid this he derived an index from the regression coefficient that was independent of length $H = 6.968 + 0.03553\ L$. The coefficient of variability is rather small so that it makes little difference whether $\text{pri}^{-1}\,p$ is plotted against V or $\log V$, but there is a slight concavity with the former (solid line) which disappears with the latter.

The graphs for ratio of length to greatest diameter in a population of snails, *Limnaea palustris* (Fig. 11.5B from Fig. 6.4B; data of Mozley 1935), and of frons to head width in a population of flies, *Musca vicina* (Fig. 11.5C from Fig. 6.4C; data of Peffley 1953), are close to linearity in the range between -2 to $+2$ of $\text{pri}^{-1}\,p$. The situation was essentially the same in other smaller populations reported by these authors.

Figures 11.5D and 11.5E (from Figs. 6.4E and 6.4D, respectively) present the graphs for relative tail length (tail/body) and percentage width of the dark tail stripe in three subspecifically different local populations of *Peromyscus maniculatus* reported by Sumner (1920). For relative tail length no transformation would improve the straightness and parallelism of the plots of $\text{pri}^{-1}\,p$ against V, which are surprisingly good considering the far from adequate numbers. There is more curvature and less parallelism for tail stripe. The curvature in the Victorville population suggests damping with approach to the lower limit, 0%, but the numbers are inadequate for placing confidence in this.

Figure 11.6 deals with distributions of percentages that are more adequate in numbers. In the percentage of black (as opposed to red, yellow, and white) required to match skin color in the highly variable Jamaican population (Fig. 11.6A from Fig. 6.5A), studied by Davenport (1913), the extreme positive skewness (median at about 20%) clearly indicates damping of variability as the limit, 0%, is approached but never reached. The plot of $\text{pri}^{-1}\,p$ against percentage shows a different sort of curvature from that for human weight. The use of $V' = \text{pri}^{-1}\,V$ as a transformation function gives a fairly good approach to a straight line, and on estimating M' and σ' from the intercepts, it was found that an acceptable fit was obtained (Fig. 11.6C) (Prob $= 0.10$, from Wright 1926). These data were used by Pearson (1914) to illustrate graduation by means of his type I frequency curve (a beta curve). The fit is slightly closer than with the above transformation but involves two more parameters.

The function, $V' = \text{pri}^{-1}\,V$ is not the only one that can be used empirically where there is complete damping at the limits. This is also true of $\log [V/(1 - V)]$ and of $[\cos^{-1}(1 - 2V) - \pi/2]$, the latter suggested by

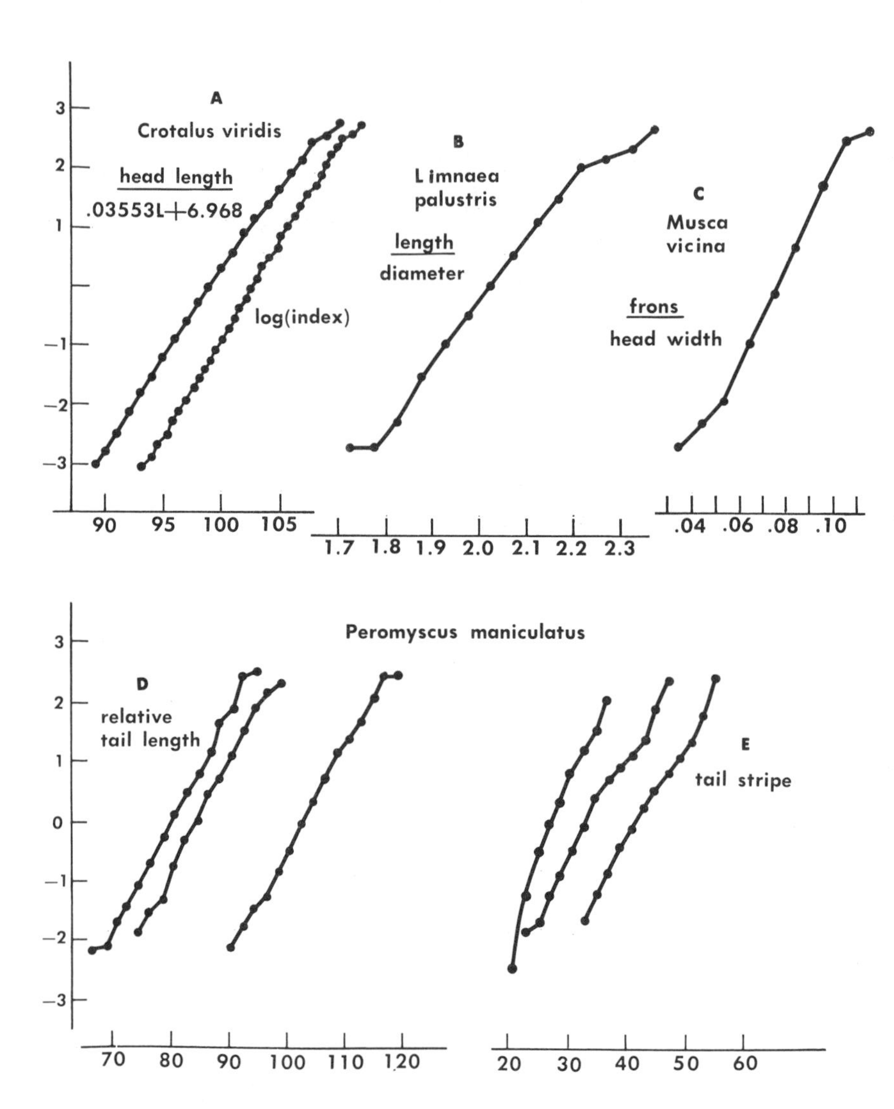

FIG. 11.5. Transformations ($\text{pri}^{-1}p$) of a number of distributions of indexes from Fig. 6.4.

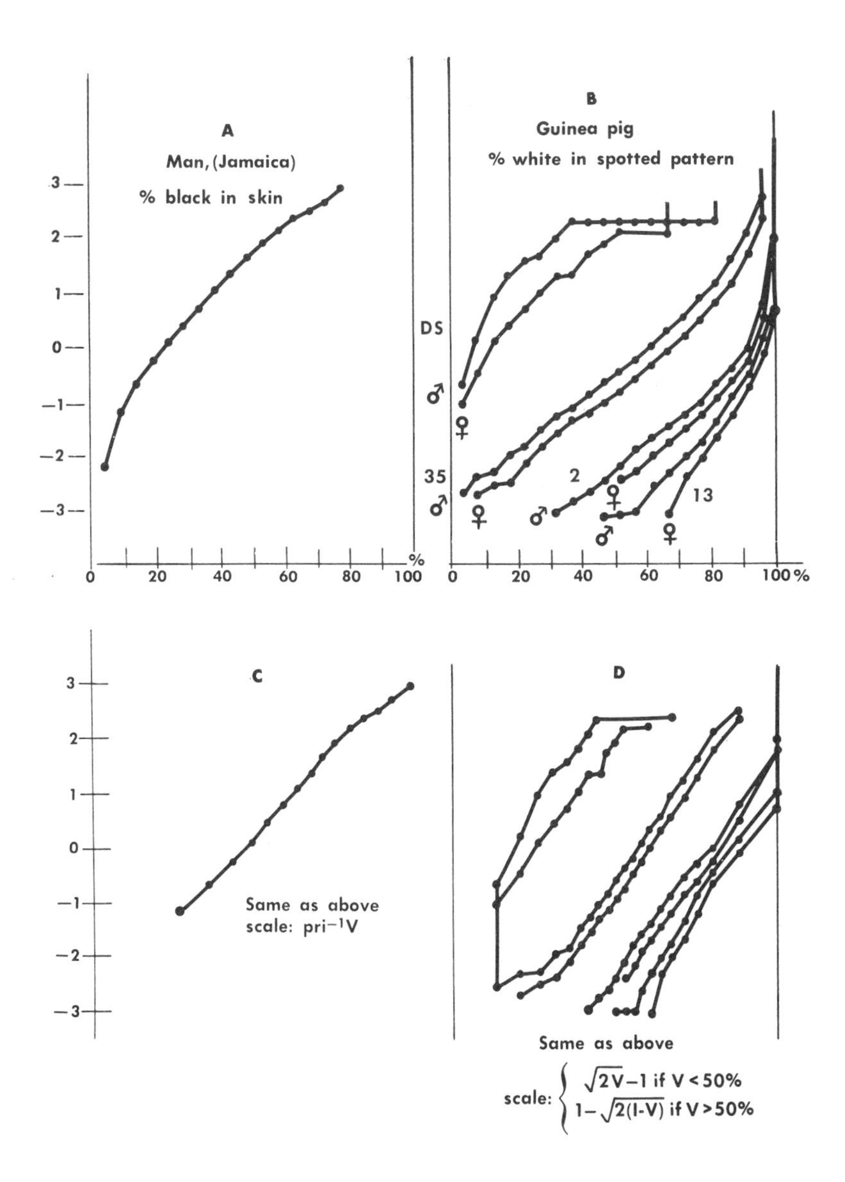

FIG. 11.6. Transformations ($\mathrm{pri}^{-1}p$) of distributions of human skin color (Fig. 6.5), percentage of white in the coats of guinea pigs (Fig. 6.5), and rectifications by transformations of scale.

Fisher (1922*a*) as that on which the sampling variance of a binomial distribution, proportional to $p(1 - p)$, is made uniform. On comparison of these transformations, all made tangent to the linear relation $V' = V - 0.50$ at $V = 0.50$, it is found that none deviates much from linearity between the limits $V = 25\%$ and 75%. The inverse cosine transformation deviates least and is only about 57% greater than the linear values at the limits 0% and 100%. The inverse probability transformation deviates much more from linearity, and the logarithmic one still more. Both of the latter approach $-\infty$ at 0% and $+\infty$ at 100%. These relations are exhibited in Table 11.1.

TABLE 11.1. Comparison of four transformations of scale, applicable to distributions of frequencies of percentages.

	V as Percentage of Range						
V'	0%	5%	25%	50%	75%	95%	100%
$V - 0.50$	−0.5000	−0.4500	−0.2500	0	0.2500	0.4500	+0.5000
$\frac{1}{2}[\cos^{-1}(1-2V) - \pi/2]$	−0.7854	−0.5598	−0.2618	0	0.2618	0.5598	+0.7854
$1/\sqrt{2\pi}\,\text{pri}^{-1}\,V$	$-\infty$	−0.6562	−0.2691	0	0.2691	0.6562	$+\infty$
$\frac{1}{4}\log V/(1 - V)$	$-\infty$	−0.7361	−0.2746	0	0.2746	0.7361	$+\infty$

In cases in which V is a percentage of actual events, an underlying probability distribution is implied for these events. The percentage that occurs represents the tail frequency of this distribution above the threshold between nonoccurrence and occurrence for individual events. In this case there is a rational basis for testing the transformation $V' = \text{pri}^{-1}\,V$. It was noted earlier that a study of the relation between quartile deviation and its midpoint in the highly asymmetrical distributions of percentage of white in the coats of inbred strains of guinea pigs suggested both this transformation function and the interpretation that there is a normal distribution of color tendencies among elementary areas of the coat (Wright 1920). The scale of percentages of white was transformed directly on this basis in this paper. In a later paper (Wright 1926) $\text{pri}^{-1}\,p$ was plotted against $\text{pri}^{-1}\,V$, and M' and σ' were estimated from the intercepts. Among 10 distributions (males and females separately of 5 inbred strains), 6 were fitted satisfactorily (Prob > 0.05), two less satisfactorily (Prob > 0.01) and two very poorly. The poorness of fit may, however, have been due to gross heterogeneity since only one of the strains (35) was at this time descended from an adequate number of generations of common brother-sister matings, and there was indeed considerable differentiation among substrains in some of the other cases.

We have, however, noted another difficulty that was not appreciated at the time, since there were no animals completely devoid of white even in a strain with median percentage of 14% in the males and relatively few self-whites (about 9%) in a strain with median at 96%. In its later history, however, the latter (strain 13) produced 24% pure white. All of these strains were genotype *ss*. Populations of genotype *Ss* range from ones with 0% to 100% self-colored, and *SS* are usually 100% colored, but not always. Thus the condition of complete damping of variability at the limits of the range need not be realized. An underlying probability distribution of limited range is clearly required for the interpretation of some of these populations. A rectangular distribution does not do at all since the distributions, excluding the classes at 0% and 100%, were not at all normal. On plotting $\text{pri}^{-1}\, p$ against V (Fig. 11.6*C*), those with medians below 50% show strong concavity to the right and those with medians above 50% strong concavity to the left (Fig. 11.6*B*, some from Fig. 6.5*C*). An underlying parabolic distribution gives a rather good fit for the whitest strain (13) but concavity to the left for the others. The hypothesis of an underlying symmetrical triangular distribution gives the straightest lines for the group of distributions as a whole (Fig. 11.6*D*). The transformation function is $V' = \sqrt{2V} - 1$ if V is less than 0.50, and $1 - \sqrt{2(1 - V)}$ if greater (both giving $V' = 0$ if $V = 0.50$). Graphs are shown for males and females of three strains (35, 2, 13), each descended in the data presented here from many generations of common brother-sister mating and shown by analysis to be almost isogenic with respect to spotting in spite of the enormous range of variability. Strain *DS* derived its spotting factor from strain 2 but its modifiers largely from repeated backcrossing to a self-colored closely inbred strain *D*. Although the numbers are hardly adequate in this case, it is brought out that there is approximately the same slope as in the other cases. Thus the same array of varying factors (almost wholly nongenetic in all these cases) is indicated in all.

It may be seen that strain 13 tends to show concavity to the right and strain 2 slight concavity to the left in Figure 11.6*D*. No transformation could fit both. There is, however, no necessity that the underlying distribution of color tendencies in areas of the skin be exactly the same. The high incidence of spots in strain 2 in certain regions (nose, rump) in which they do not occur in strain 13 shows that the patterns actually do differ somewhat.

There is a serious inconsistency in the greater difference between males and females of *DS* (borne out by a similarly great difference in other less homogeneous but more numerous populations with small, median amounts of white) and the small differences between the medians in the other cases. The application of our third criterion for transformation functions, the interaction effects of strain and a particular factor—here sex—indicate a different function than that indicated by form and variability, the first and second criteria. It can only be concluded that particular interaction effects may be more or less out of line with the trend.

In Figure 11.7 we return to the analysis of bimodal distributions, but in this case of ones that do not involve the combining of separable populations. We will consider first a type of distribution of percentages that is of a very different sort from those just discussed.

The characteristic type of distribution of percentage frequency of occurrence of lethality of homozygous chromosomes extracted from wild populations of *Drosophila* was illustrated by the data of Pavan *et al.* (1958) for *D. willistoni* (Fig. 6.7*B*). All such distributions have been markedly bimodal. The plot of $\text{pri}^{-1}\,p$ against percentage of lethality of homozygous second chromosomes relative to that of heterozygotes in such chromosomes shows the characteristic pattern for bimodal distributions (Fig. 11.7*D*). On omitting all below 55% of control, the graph is approximately straight up to $\text{pri}^{-1}\,p = 2.5$ and indicates approximate normality in this main portion of the distribution. In this case, far from a damping of variability as the limit 0% is approached, there seems to be a runaway process below 15% which can be interpreted as almost a complete breakdown of homeostatic developmental mechanisms. It was noted (Fig. 6.7*C*) that length of life as a character usually shows this type of distribution: high larval or infant mortality, followed first by a minimum and later by a high peak.

Figure 11.7*C* (from Fig. 6.7*A*) gives an analysis of the distribution of a morphological character that is of somewhat this type. This is Grüneberg's (1951) case of the distribution of length of the third lower molar in a particular inbred strain of mice (*CBA*). If the 12% in which this tooth was wholly absent are omitted, $\text{pri}^{-1}\,p$ plotted against length does not deviate widely from linearity and thus indicates approximate normality. The mean is, however, considerably lower and the variability (indicated by the reciprocal of the slope) much greater than in the normal strain *C57Bl*. Apparently something in the genotype of the *CBA* strain brings about general instability and a runaway tendency toward complete failure at the lower end of the distribution of factor combinations.

The distributions of length of forceps in earwigs (Diakanov 1925; Fig. 11.7*A* from Fig. 6.6*D*) and of horns in the beetle Xylotrupes (Bateson and Brindley 1892; Fig. 11.7*B* from Fig. 6.6*E*) were discussed earlier as examples of bimodality with respect to morphological characters, not interpretable as due to heterogeneity in strain, age, or environment but rather to the presence of two or more discontinuous developmental norms (probably

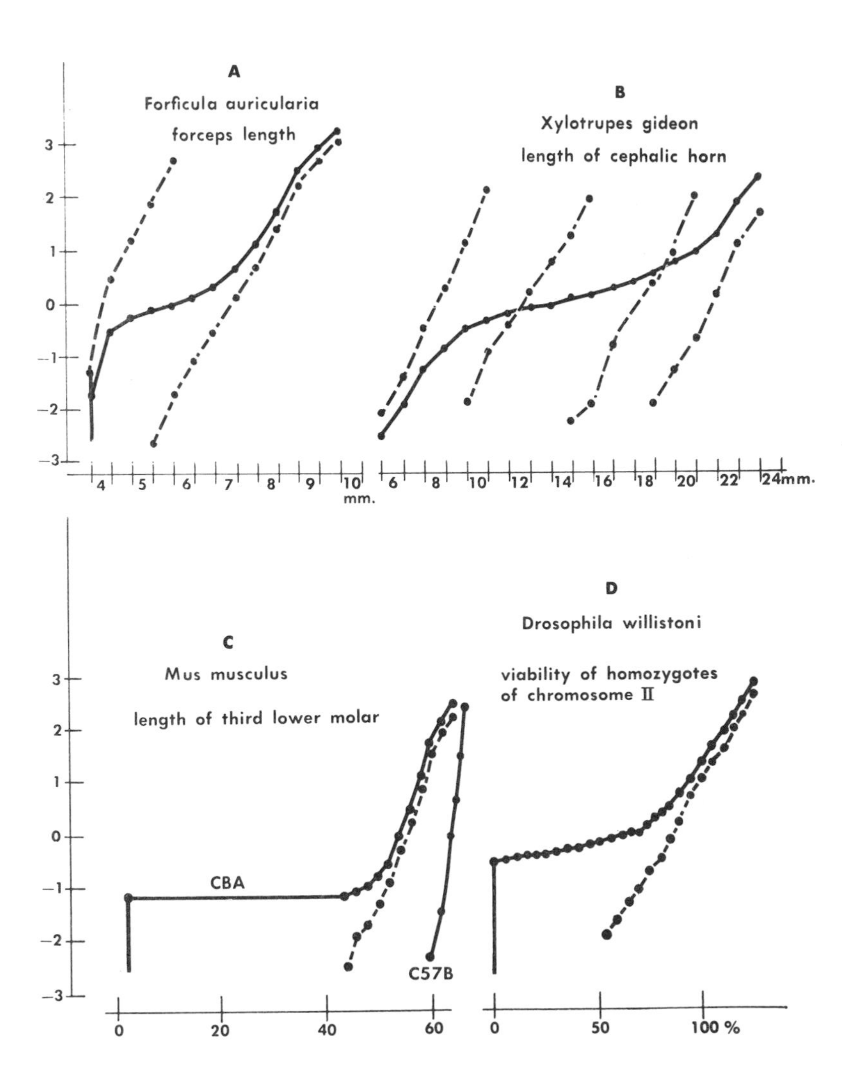

Fig. 11.7. Transformations ($\text{pri}^{-1}p$) of bimodal and multimodal distributions for Figs. 6.6 and 6.7, and of components of these (broken lines).

number of molts before heterauxetic growth ceases). The distributions can readily be split into roughly normal components, indicated by dotted lines in Fig. 11.7*A*, *D*). Only two of these are indicated in Forficula (Fig. 11.7*A*), but at least four in Xylotrupes (Fig. 11.7*B*).

Meristic Variability

Meristic variability differs from continuous quantitative variability in the presence of a natural scale. The focus of interest is the relation of this to the underlying scale of factor complexes. That it need not be close is shown by the distribution of tentaculocysts in the jelly fish Ephyra (Browne 1895) shown in Figure 6.9*A* and analyzed in Figure 11.8*B*. This is representative of the large class of distributions in which most individuals show a certain "normal" number of elements, but a few in both directions show more or less divergence.

In this figure each number, V, is represented as centered in a unit range $V \pm 0.5$, and $\text{pri}^{-1}\, p$ is plotted against these limits. This gives a series of steps (dotted line) since there is no change within a range. This mode of representation becomes confusing, however, if it is desired to plot a family of distributions together. It is more convenient to connect the midpoints of the ranges. This method is followed in all other cases.

If the distribution of factor complexes is normal, the intervals on the observed scale must be assigned values proportional to the steps, $\Delta\, \text{pri}^{-1}\, p$, centered in the corresponding numbers. In the case of Ephyra, the values are roughly equal for all numbers except 8 which is 5.7 times the average for the others. It must be supposed that there are powerful regulatory processes in development that insure this number over a wide range of factor complexes, but that if this range is exceeded, the degree of divergence from 8 reflects fairly accurately the further divergence of the factor complex. The scale can be transformed only by making the interval over which 8 tentaculocysts are determined 5.7 times as much as the other class ranges.

The number of ray florets in a population of *Chrysanthemum segatum* (Fig. 6.9*C*) is analyzed in Figure 11.8*D*. The situation is very similar to that in Ephyra. In other strains, however, de Vries (1910) found that the normal number was 21 instead of 13.

The common category in which there is variability in nature on only one side of a highly standardized normal number is illustrated by Payne's (1920) selection for increase beyond 4 in the number of dorsocentral bristles of *Drosophila melanogaster* (Fig. 6.9*B*), and by de Vries' (1910) selection for increase beyond 5 in the number of petals of *Ranunculus bulbosus* (Fig.

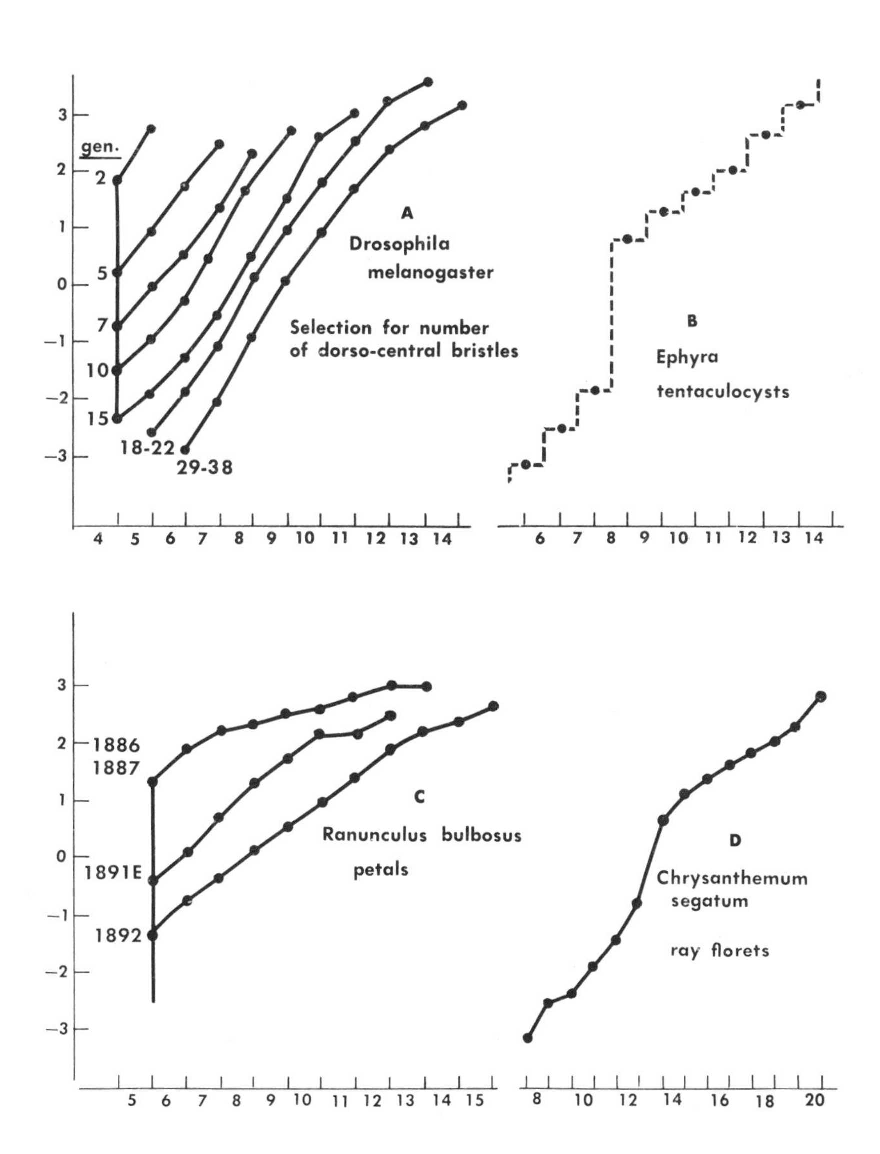

FIG. 11.8. Transformations ($\text{pri}^{-1}p$) of meristic distributions from Fig. 6.9, in some of which one class has become strongly standardized.

6.9*D*). The graphs of $\text{pri}^{-1} p$ against number in Payne's data (Fig. 11.8*A*) tend to be parallel rather than divergent, which is contrary to expectation if the factors tended to produce constant percentage increments. In the earlier generations the lines are either straight or convex to the left, but these give way to reversed curves and ultimately convexity to the right. These changes reflect a secondary standardization of the number 8 at which there is the greatest step in $\text{pri}^{-1} p$.

The situation in *Ranunculus bulbosus* (de Vries 1910) is somewhat similar (Fig. 11.8*C*). The upper graph refers to a wild population. Its low slope probably reflects genetic heterogeneity. The two selected populations yield lines that are approximately parallel and straight with no indication of secondary standardization.

A number of the distributions that were presented earlier (Fig. 6.8, 6.10) are analyzed in Figure 11.9. For the number of spines of the protozoan *Difflugia corona* (Jennings 1916), there is of course a threshold at zero. The plots for different clones (Fig. 11.9*A*) are roughly parallel and straight up to 6 spines, but the steps become shorter beyond this. As with shell diameter there seems to be something of a runaway process beyond a certain point, although the correlation between the two characters was small (+0.21 in clone 326).

The number of mammae in fetal swine (Parker and Bullard 1913) almost has a threshold at 10, but in contrast with the dorsocentral bristles of unselected *Drosophilas* and the petals of the unselected buttercups, the mode is higher (12). The somewhat zigzag form of the plots (Fig. 11.9*C*) is due to a moderately strong correlation (+0.55) between the numbers on the right and left sides which causes an excess of even numbers.

Among the other graphs, two (number of infralabial scales in *Crotalus viridis viridis* (Klauber 1941), and number of rays in the lower valve of the scallop *Pecten* (Davenport 1904; Fig. 6.8*G*) are close to straight lines (Fig. 11.9*E*, *F*), whereas the number of supralabials (Fig. 11.9*D* from Fig. 6.10*C*) in the same rattlesnakes, number of scales on the fourth toe of the lizard, *Cnemidophorus tesselata* (Fig. 11.9*G* from Fig. 6.10*F*), number of rays in the anal fins of *Menidia menidia* (Hubbs and Ramey 1946; Fig. 11.9*H* from Fig. 6.8*D*), and number of teeth in a clone of *Difflugia corona* (Fig. 11.9*B* from Fig. 6.8*C*), all show an unduly long step near the mode. Klauber demonstrated the significant leptokurtosis of the distribution of supralabials in contrast with the insignificant departures from normality in the infralabials of the same large rattlesnake population by calculation of γ_2 and its standard error. He similarly found close approach to normality of the number of subcaudals, but significant leptokurtosis in the case of *Cnemidophorus*.

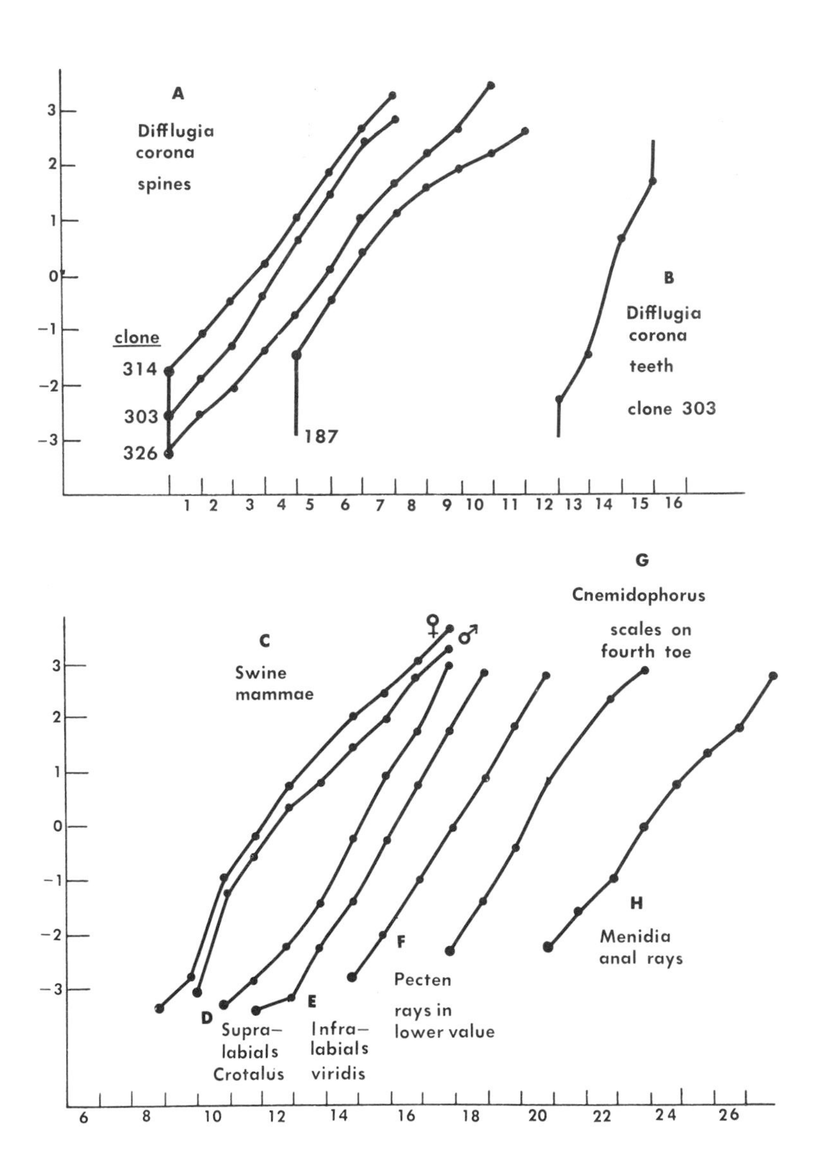

Fig. 11.9. Transformations ($\text{pri}^{-1}p$) of several meristic distributions from Figs. 6.8 and 6.10.

Figure 11.10 analyzes several meristic distributions in which the number of elements is relatively great. The plot of pri^{-1} for number of vertebrae in the fish *Nototropis* (Hubbs 1922; Fig. 11.10*B* from Fig. 6.8*B*) is virtually straight, indicating a normal distribution, but the coefficient of variability is so small that transformation $V' = \log V$ makes little difference.

The graph for the number of dorsal scale rows on the snake *Pituophis catenops* (Klauber 1941) is remarkable for its zigzag character, reflecting the multimodal nature of the distribution due to strong correlation between the numbers to right and left of the median dorsal row (Fig. 11.10*A* from Fig. 6.10*E*). The curvature of lines connecting alternate numbers suggests the appropriateness of some sort of logarithmic transformation. These lines become virtually straight on the scale $V' = \log (V - 21)$ (dotted, odd and even numbers separate). The total scale count of the fish *Boleosoma nigrum* (Hubbs and Ramey [1946], data on subspecies *B. olmstedi*, Fig. 6.10*A*) yields a plot, 11.10*C*, against pri^{-1} p that is strongly concave to the right, but that can be largely rectified by the transformation $V' = \log (V - 37)$, as shown by the broken line.

In general, distributions that concern a large number of undifferentiated elements can be treated practically as if they were distributions of a continuous variable. Thus, in his studies of number of ommatidia in a bar-eyed strain of *Drosophila melanogaster*, Zeleny (1922) found, as already noted, that all criteria indicated additivity on a logarithmic scale. The plot of pri^{-1} p against log V in an unselected strain is roughly linear as shown in Figure 11.10*D* from Fig. 6.5*B* (mode at about 54 facets) but gives some indication of heterogeneity. A similar graph after 32 or more generations of plus selection (mode at about 175 facets) on Zeleny's factorial (logarithmic) scale showed much closer linearity and increased slope. There was, however, rather serious departure from linearity on this scale in a strain produced by 32 or more generations of minus selection (mode at about 44 facets), although again with increased slope. The transformation implied by the factorial scale was probably less satisfactory in this region.

Figure 11.11*C* (cf. Fig. 6.11*A* for four distributions) shows plots of number of rows of kernels of maize against pri^{-1} p (data of Emerson and Smith 1950). The decreasing slope as the mean number increases reflects the increasing standard deviation which, according to the analysis presented earlier seems to justify a logarithmic transformation. There is, however, an apparent discrepancy in the approximate straightness of the lines. There is certainly no indication of any concavity to the right such as would be expected from multiplicative action of the nongenetic factors responsible for variability among plants of the same genotype. Yet those same factors produce proportionately greater variability in conjunction with a genotype that gives a

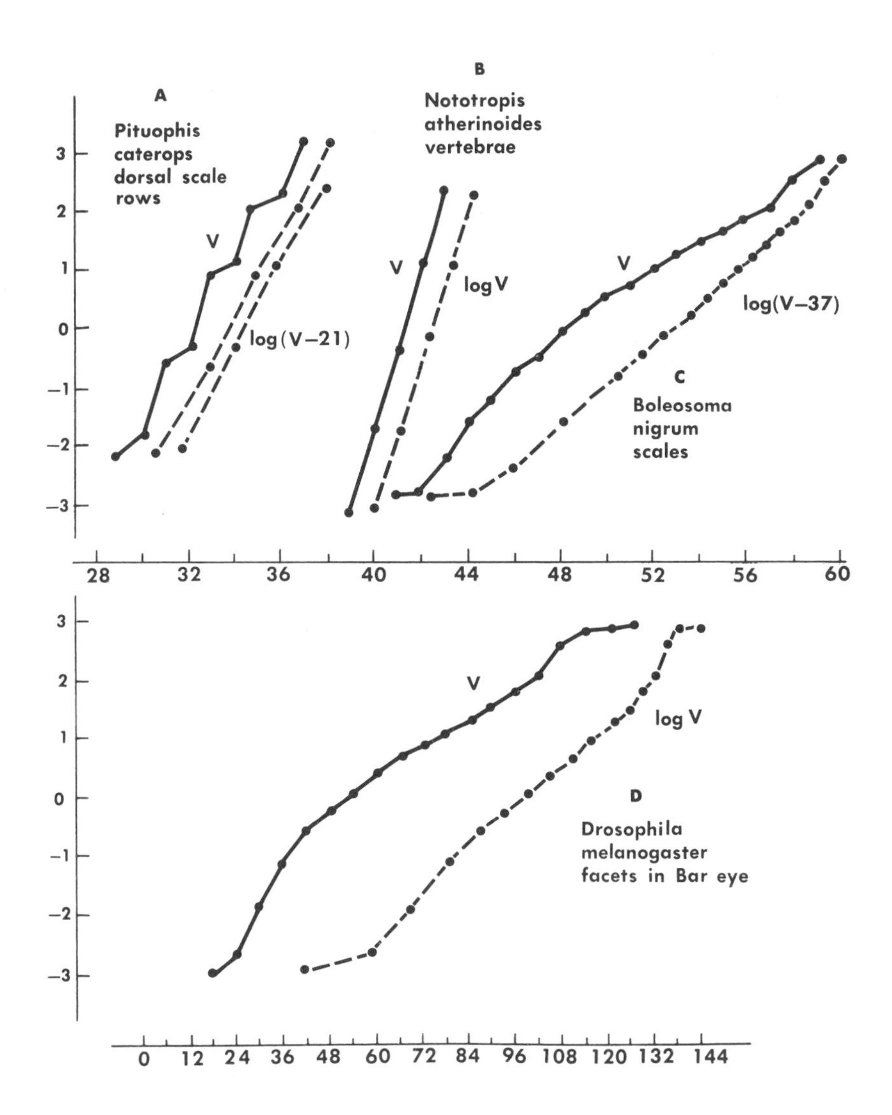

FIG. 11.10. Transformations (pri^{-1}p) of several meristic distributions from Figs. 6.5, 6.8, and 6.10, without and with logarithmic rectification of scale.

higher mean. It is as if a tendency to positive skewness from the multiplicative effect were balanced by damping with an approach to an upper limit characteristic of each genotype. This illustrates in a different way from the case of weight of guinea pigs at 33 days that an interpretation of variability by one criterion may disagree with that by another criterion. Any use of transformation functions requires a careful consideration of which aspects of variability it is desirable to normalize as far as possible for the purpose at hand.

The distributions of number of seeds per fruit and of seed weight in *Crinum longifolium*, described by Harris (1912) (Fig. 11.11*A*, *B*, from Fig. 6.11*C*, *D*) were referred to in chapter 6 as among the most aberrant encountered in the absence of obvious gross heterogeneity. The plots of pri^{-1} against number and weight are rather similar in showing strong concavity to the right in the lower portion, followed by approximate linearity or slight reversal of curvature. Use of a simple logarithmic transformation $V' = \log V$ straightens out the lower portion but gives strong concavity to the left in the upper portion. The assumption of damping at an upper limit as well as at zero ($V' = \log (V/(L - V))$) permits approximate straightening throughout, as shown by the lines to the right.

Finally, graphs are given in Figure 11.11*D*, *E*, *F*, *G* for a number of cases of brood size. The graph for size of litter of swine (data of Parker and Bullard (1913), Fig. 11.11*F* from Fig. 6.12*B*) shows a nearly linear relation to $\text{pri}^{-1}\, p$ and thus indicates a practically normal distribution. Clutch size of starlings 11.11*E* (data of Lack 1948; Fig. 6.12*C*) also does not deviate much from linearity, but the large increment between 4 and 5 eggs suggests some standardization of 5. The graph for the rattlesnakes (Fig. 11.11*G*) (Klauber 1936, Fig. 6.12*D*) on the other hand, suggests platykurtosis in its middle portion, probably reflecting the already discussed multimodality in size. In guinea-pig litters, graphs are shown in Figure 11.11*D* for two inbred strains (2, 13) with widely different means and an especially vigorous crossbred group. The distributions for the two extremes were illustrated in Figure 6.12*A*. Litters of one are so common in guinea pigs that it is probable that the array of factors responsible for variation in number may lead to failure to produce any, although factors of a different sort are probably much more important when mated females have failed to produce a litter. The initial portions of the curves (through litters of 3) could be made nearly straight by assuming about 2% in class 0. The curve for strain 2 continues nearly straight, that for 13 curves slightly to the right, and that for the vigorous crossbreds curves much more, but not in a way rectified by a logarithmic transformation. The physiological interval corresponding to each additional fetus seems to become progressively less beyond litters of 3 or 4.

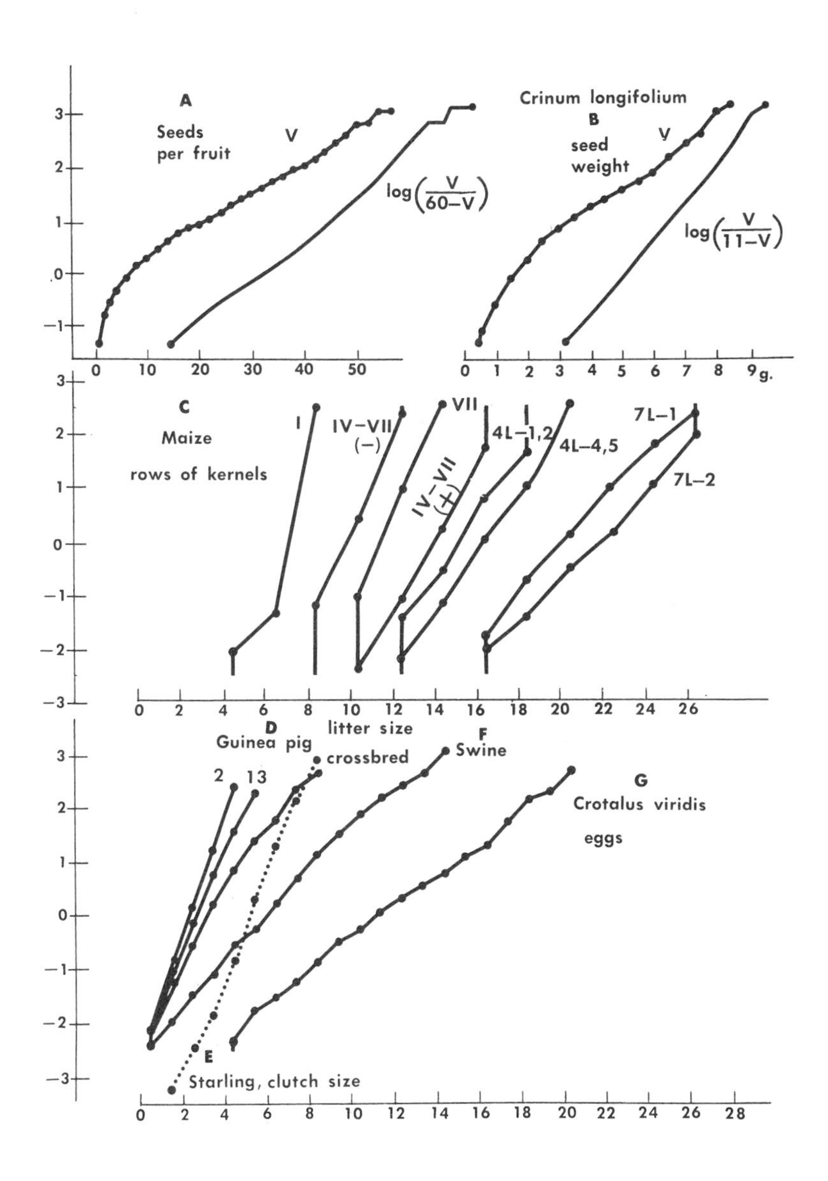

FIG. 11.11. Transformations ($\text{pri}^{-1}p$) of largely meristic distributions for Figs. 6.11 and 6.12. Rectification by logarithmic transformation of scale are attempted in two characters of *Crinum longifolium* at the top.

Summary

In this survey an attempt has been made to present diverse types of distributions. A more random assembly would undoubtedly bring out more forcibly the central position occupied by the normal distribution and the prevalence of near normal distributions that can be satisfactorily normalized by use of some sort of a logarithmic transformation, most frequently one which corrects for damping by a lower limit ($\log V$, $\log (V \pm L)$, but occasionally by an upper limit ($\log (L - V)$ or by both $\log [(V \pm L_1)/(L_2 - V)]$). This applies especially to continuous variability.

The most important complication in continuous variability within a range far from lower or upper limits is due to gross heterogeneity for the presence of major alternatives. The special case of distributions with respect to percentage of occurrence involves complications near the limits 0% and 100% that vary accordingly to the amount of damping at these limits. There may be every gradation, from an absence of damping associated with transcendence of the limits by classes at 0% and 100% that represent the tails of a completely normal distribution, to complete damping by these limits and thus to a complete absence of classes at 0% and 100%, but with a piling up of classes near one or both limits.

There is a class of cases that is relatively unusual for ordinary measurements in which the distribution consists of two almost or quite discrete parts, a near normal distribution, and a class of complete or nearly complete failures of the character. The array of factors underlying a character not only may have a threshold below which it wholly fails, but also a ceiling at which it is so fully developed that further increase in the factors produces no additional effect. A meristic character has a succession of thresholds and ceilings for each element. The variability of such a character often exhibits normal or near normal distributions similar to those of a continuous variable, indicating that the intervals between thresholds correspond to physiologically equivalent steps. This is not always the case, however. It is, indeed, rather common for the modal class to be unduly high for a normal distribution, indicating that its number of elements covers a wider interval physiologically, or from another standpoint is under some degree of homeostatic control.

The most important frequency distributions for evolutionary theory are those with respect to selective value. Unfortunately selective value is such a complex character involving not only viability and fecundity, but such characters as success in securing mates, absence of tendency to move out of the territory, and so on, that it has not been practicable to determine it for individuals.

The distribution of such components as length of life and fecundity have, however, been determined in many cases. The distribution for any index of viability typically consists of two almost discrete parts, a portion of complete or nearly complete failures, and a near normal distribution. This applies especially to percentages of genotypes that are born alive and to length of life under controlled conditions. Age at death under natural conditions is likely to depend so much on accidents that the two components of the distribution tend to merge. The distribution of fecundity of individuals under favorable conditions is also likely to consist of two discrete portions: a group of sterile individuals and near normal distribution among those that are fertile. The distribution under natural conditions leads to consideration of the important question of the effective size of the population.

More important than the distribution of selective value or of any component by itself is the distribution of selective value in relation to each character. In most laboratory experiments on selection, this corresponds more or less closely to the distribution of the character itself, since desirability of saving for breeding is correlated positively with grade in plus selection and negatively with minus selection. In a population that has been living for many generations under similar conditions, however, it is safe to assume that natural selection operates against types that deviate in either direction from an optimum very near the mean. The increase in the size of the horse since the eocene has been described as an explosively rapid process. But an increase of 50 inches in 50,000,000 years means an average change of only a few millionths of an inch per generation. This must always have been very small in comparison with the standard deviation of segregating genotypes. It is unlikely that the optimum height ever differed appreciably from the mean.

If selective value falls off directly with deviation from the mean in the case of a variable with a normal distribution, the distribution of selective value is a half-normal distribution and thus very asymmetrical. If, as is more likely, it falls off approximately as the square of the deviation from the mean, it is enormously more asymmetrical, being that for χ^2 with one degree of freedom. If the character is compounded of elementary characters that are separately selected the distribution becomes that of χ^2 with a corresponding number of degrees of freedom and is still extremely asymmetrical unless the number of components is very great.

It might be expected that some functional characters would depend on harmonious effects of underlying characters in such a way that they would themselves exhibit distributions of the above sort. There is no such tendency in the speed of trotting horses, in milk production, or in IQ in man. Distributions of number of dorsocentral bristles in *Drosophila* and of petals in *Ranunculus* are of an extremely asymmetrical type in nature, but the results of selection experiments indicate that these are to be explained by homeostatic standardization.

The distributions of seeds per fruit and seed weight in *Crinum* are also extremely asymmetrical and can be fitted more or less satisfactorily by Gamma distributions. On the other hand, the modes, one or two fruits in one case with range up to 70, weight of 0 to 0.50 gm. with a range up to 9 gm., can hardly be interpreted as optimal. It is most probable that both of these characters are determined by factors with essentially multiplicative effects, but over such extraordinary ranges of variability that there is an unusual amount of damping at both ends.

The conclusion that after elimination of gross heterogeneity, most variability depends on additive compounding on an appropriate scale by no means implies that factor interactions are absent. It does not require the compounding of very many independent groups of factors to give an approach to normality, and there may be a great deal of interaction within such groups. Moreover, the character on which selection operates directly, overall selective value, almost necessarily depends on deviation from an intermediate optimum in nature and thus must involve interaction of the most extreme sort.

□ □

CHAPTER 12

Description of Multivariate Distributions

The Method of Least Squares

The varying characteristics of individuals in a population can only be isolated from each other artificially. The traditional method of relating one variable to others is to find the best estimate for it on the basis of known values of the others by means of Gauss's method of least squares.

Let $y_e = f(x_1, x_2, \cdots)$ be the chosen function with m parameters to be estimated from a set of n observations of y and the x's ($n > m$). The subscripts a to n in equations 12.1 refer to these observations. The differences between observed values of y and the fitted values, y_e, are represented by δ's.

n Observation Equations

$$\left\{\begin{array}{l} y_a = f(x_{1a}, x_{2a}, \cdots) + \delta y_a \\ y_b = f(x_{1b}, x_{2b}, \cdots) + \delta y_b \\ y_n = f(x_{1n}, x_{2n}, \cdots) + \delta y_n. \end{array}\right. \tag{12.1}$$

One of the ways Gauss used in arriving at a solution is now known as the method of maximum likelihood, applied on the assumption that the δ's are normally distributed. On this assumption, the probability of any given difference, δy_i, is proportional to $\exp\left[-\tfrac{1}{2}(\delta y_i/\sigma_{\delta y_i})^2\right]$. The joint probability for the n observations (if independent) is proportional to the product of their separate probabilities. If this is maximum, its logarithm

$$\left[C - \left(\frac{1}{2}\right)^n \sum \left(\frac{\delta y_i}{\sigma_{y_i}}\right)^2\right]$$

is also maximum. Thus the set of parameters that makes $\sum (\delta y/\sigma_y)^2$ minimum is the one with the maximum joint probability for the observed set.

If there are multiple observed y's for a given set of x's, the weight (w) to be assigned this set is K/σ_y^2. There is often, however, no known basis for weighing other than the number of cases, K. In fitting frequency distributions, however, it may be assumed that the sampling variability is of the Poisson type with variance equal to the theoretical mean of the given class. The quantity minimized in this case is thus

$$\sum \left[\frac{(f_o - f_c)^2}{f_c}\right] = \chi^2$$

where f_o and f_c are the observed and theoretical frequencies of a class.

The values of the parameters, $c_1, c_2, \cdots, c_m$, which minimize $\sum w(\delta y)^2$, may be found by partial differentiation with respect to each. Note that the "constants" of the function are treated as variables while the "variables", y, x_1, x_2, etc., are here represented by observed values and are thus constants in the m "normal equations" of the type:

$$\frac{\partial \sum w(\delta y)^2}{\partial c_i} = 0. \tag{12.2}$$

The type of function fitted most easily is that in which the parameters enter linearly. We assume one term (c_o), which is free of the variables, and s terms in which one parameter in each enters as a coefficient of a variable. The number of parameters is thus $m = s + 1$. Summations are according to the number of observation equations.

$$y = c_0 + c_1x_1 + c_2x_2 + \cdots + c_sx_s + \delta y. \tag{12.3}$$

$$\sum w(\delta y)^2 = \sum w[y - c_0 - c_1x_1 - c_2x_2 - \cdots - c_sx_s]^2. \tag{12.4}$$

The x's need not be independent of each other. Some or all may, indeed, be merely different functions of the same observed variable.

$$\frac{\partial \sum w(\delta y)^2}{\partial c_i} = -2 \sum wx_i(y - c_0 - c_1x_1 - c_2x_2 - \cdots - c_sx_s) = 0. \tag{12.5}$$

The coefficients may be arranged conveniently as below for solution of the set of $(s + 1)$ simultaneous linear equations.

$$\begin{array}{cccccc} \underline{c_0} & \underline{c_1} & \underline{c_2} & \cdots & \underline{c_s} & \\ \sum w & \sum wx_1 & \sum wx_2 & \cdots & \sum wx_s & = \sum wy \\ \sum wx_1 & \sum wx_1^2 & \sum wx_1x_2 & \cdots & \sum wx_1x_s & = \sum wx_1y \\ \sum wx_2 & \sum wx_1x_2 & \sum x_2^2 & \cdots & \sum wx_2x_s & = \sum wx_2y \\ \vdots & \vdots & \vdots & & \vdots & \vdots \\ \sum wx_s & \sum wx_1x_s & \sum wx_2x_s & \cdots & \sum wx_s^2 & = \sum wx_sy. \end{array} \tag{12.6}$$

The symmetry of the set at the left about the diagonal from upper left to lower right makes for easy solution. Gauss's method consisted in eliminating at each step the variable in the first column by multiplying each of the terms in the other columns by whatever comes to be in the upper-left corner, and in each case subtracting the product of the term at the top of the column and at the beginning of the row. The entry at the top of the first row and first column of the new set, after eliminating the original first row and first column is thus $(\sum w \sum x_1^2 - \sum wx_1 \sum wx_1)$, and the next term in the top row becomes $(\sum w \sum wx_1x_2 - \sum wx_1 \sum wx_2)$. The resulting reduced set has the same sort of symmetry as the first, permitting continuation of the process until only one equation is left, with one parameter to be determined. When this has been done the others may be determined by substitution in reverse order.

Where the number of equations is small, it is often more convenient to solve by means of determinants, using Cramer's method. In discussing this, the number of parameters will be reduced by one by expressing each variable in terms of its deviation from its mean. The y's and x's below are to be understood in this sense. This removes the first row and first column.

Let

$$\begin{vmatrix} \sum wx_1^2 & \sum wx_1x_2 & \cdots & \sum wx_1x_s \\ \sum wx_1x_2 & \sum wx_2^2 & \cdots & \sum wx_2x_s \\ \vdots & \vdots & & \vdots \\ \sum wx_1x_s & \sum wx_2x_s & \cdots & \sum wx_s^2 \end{vmatrix} = D. \tag{12.7}$$

Let D_{ij} be the minor made by removing row i, column j, and let C_{ij} be the cofactor $(-1)^{i+j}D_{ij}$. Because of symmetry $C_{ji} = C_{ij}$.

$$\begin{cases} c_1 = [C_{11} \sum wx_1y + C_{12} \sum wx_2y + \cdots + C_{1s} \sum wx_sy]/D \\ c_2 = [C_{12} \sum wx_1y + C_{22} \sum wx_2y + \cdots + C_{2s} \sum wx_sy]/D \\ \vdots \qquad \vdots \qquad\qquad \vdots \qquad\qquad \vdots \\ c_s = [C_{1s} \sum wx_1y + C_{2s} \sum wx_2y + \cdots + C_{ss} \sum wx_sy]/D. \end{cases} \tag{12.8}$$

The standard deviation of estimates of y with $\sum k$ individual entries is given by

$$\mathrm{SE}_y = \sqrt{\frac{\sum w(y - y_e)^2}{\sum k - s - 1}}. \tag{12.9}$$

The simplest systematic method of fitting a curvilinear relation of one variable to another is usually by means of a power series.

$$y = c_0 + c_1x + c_2x^2 + \cdots + c_sx^s. \tag{12.10}$$

The normal equations are as follows:

$$\begin{array}{cccccc} \underline{c_0} & \underline{c_1} & \underline{c_2} & & \underline{c_s} & \\ \sum w & \sum wx & \sum wx^2 & \cdots & \sum wx^s & = \sum wy \\ \sum wx & \sum wx^2 & \sum wx^3 & \cdots & \sum wx^{s+1} & = \sum wxy \\ \vdots & \vdots & \vdots & & \vdots & \vdots \\ \sum wx^s & \sum wx^{s+1} & \sum wx^{s+2} & \cdots & \sum wx^{2s} & = \sum wx^sy. \end{array} \tag{12.11}$$

With only one parameter, c_0, the method yields the arithmetic mean as the value from which the sums of squares (and hence variance) is minimum.

$$c_0 = \frac{\sum wy}{\sum w} = \bar{y}. \tag{12.12}$$

The most important case in biological work is undoubtedly the fitting of a straight line, $y = c_0 + c_1 x$.

$$c_1 = \frac{\sum w \sum wxy - \sum wx \sum wy}{\sum w \sum wx^2 - (\sum wx)^2} = \frac{\sum wxy - \bar{x} \sum wy}{\sum wx^2 - \bar{x} \sum wx}. \tag{12.13}$$

$$c_0 = \frac{\sum wy}{\sum w} - c_1 \frac{\sum wx}{\sum w} = \bar{y} - c_1 \bar{x}. \tag{12.14}$$

In fitting power series of higher degree it may be noted that if the weights are all the same, if the x's are at equal intervals, and if the midpoint is taken as origin, the sums of odd powers of x all become zero, with considerable simplification of the calculation. The set of equations falls in two. Following is the solution for a second degree parabola with such data.

$$(12.15)\quad \left\{\begin{array}{cccclcl}
\underline{c_0} & \underline{c_1} & \underline{c_2} & & & & \\
n & 0 & \sum x^2 & = \sum y & & c_2 = & \dfrac{\sum x^2 y - \bar{y} \sum x^2}{\sum x^4 - (\sum x^2)^2/n} \\
0 & \sum x^2 & 0 & = \sum xy & & c_0 = & \bar{y} - \dfrac{(\sum x^2) c_2}{n} \\
\sum x^2 & 0 & \sum x^4 & = \sum x^2 y & & c_1 = & \sum xy / \sum x^2 .
\end{array}\right.$$

It is often useful to bring about linearity of a transformation by a transformation of one or both scales. If the weights to be assigned values of y are decided before transformation of scale, they need transformation. Primes are used below to indicate transformed values.

Let

$$\frac{\delta y'}{\delta y} = \frac{dy'}{dy}, \qquad \delta y = \delta y' \frac{dy}{dy'}. \tag{12.16}$$

The quantity $\sum w(\delta y)^2 = \sum w[\delta y'/(dy'/dy)]^2$ is to be minimized. This is equivalent to minimizing $\sum w'(\delta y')^2$ if $w' = w/(dy'/dy)^2$. It may, however, be more desirable to minimize the squared deviations of the transformed values than those of the original ones.

As an example of use of transformation, consider the relations between the amounts of melanin extracted from given weights of guinea-pig hair (determined colorimetrically) and grades of intensity, assigned to the animals at birth on the basis of a series of barely distinguishable skins (white = 0, black = 21) (Wright and Braddock 1949). Table 12.1 shows the relations in two series, dark-eyed sepias (genotype *EBP*) and pink-eyed pale sepias (genotype *EB pp FF*), which have pigment granules of very different types. The variations within each series were due primarily to albino alleles (C, c^k, c^d, c^r, c^a) but secondarily to various modifiers (*Dm*, *dm*; *Si*, *si* in both series, *Mp*, *mp* in the second).

TABLE 12.1. Transformation of colorimetric determinations of melanin in hair samples of guinea pigs with empirical grades of intensity from 0 (white) to 21 (black).

Dark-eyed Sepia				Pink-eyed Sepia			
Grade (x)	No.	$y = \overline{\log (M + 6)}$	σ_y	Grade (x)	No.	$y = \overline{\log (M + 6)}$	σ_y
21	28	2.075	0.062	12	1	1.522	—
20	13	2.029	0.055	11	5	1.480	0.030
19	22	1.923	0.077	10	6	1.473	0.063
18	10	1.877	0.066	9	7	1.417	0.047
17	9	1.806	0.084	8	2	1.306	0.016
16	9	1.776	0.055	7	4	1.228	0.052
15	7	1.729	0.037	6	2	1.194	0.033
14	9	1.669	0.062	5	2	1.250	0.040
13	5	1.642	0.074	4	5	1.091	0.016
12	2	1.597	0.078	3	5	1.033	0.040
11	1	1.496	—	2	5	0.975	0.028
				1	2	0.954	1.000
				(0)	(12)	(0.915)	(0.014)

Dark sepia $y = 0.8765 + 0.0564x$
Pale sepia $y = 0.8688 + 0.0578x$
Total $y = 0.8768 + 0.0564x$

From data of Wright and Braddock (1949).

The difference between each of the visual grades (x) and that preceding was found to correspond roughly to a certain percentage increment in colorimetric value (M). This suggested the use of $\log_{10} M$ in standardizing the visual grades, but this was found to give a systematic overcorrection. The use of $y = \log_{10} (M + 6)$ was found to give a closer approach to linearity than $\log_{10} (M + 5)$ or $\log_{10} (M + 7)$. The data from the two series were fitted separately as well as combined, using the numbers as weights. The white samples (which yielded melanoids on extraction) were excluded.

It may be seen that the two series yielded very nearly the same line (Fig. 12.1). As will be discussed later, the fit is as close as can be reasonably expected in the light of the variability within grades in the dark sepias and in the total. It is not as satisfactory in the case of the pale sepias (Prob

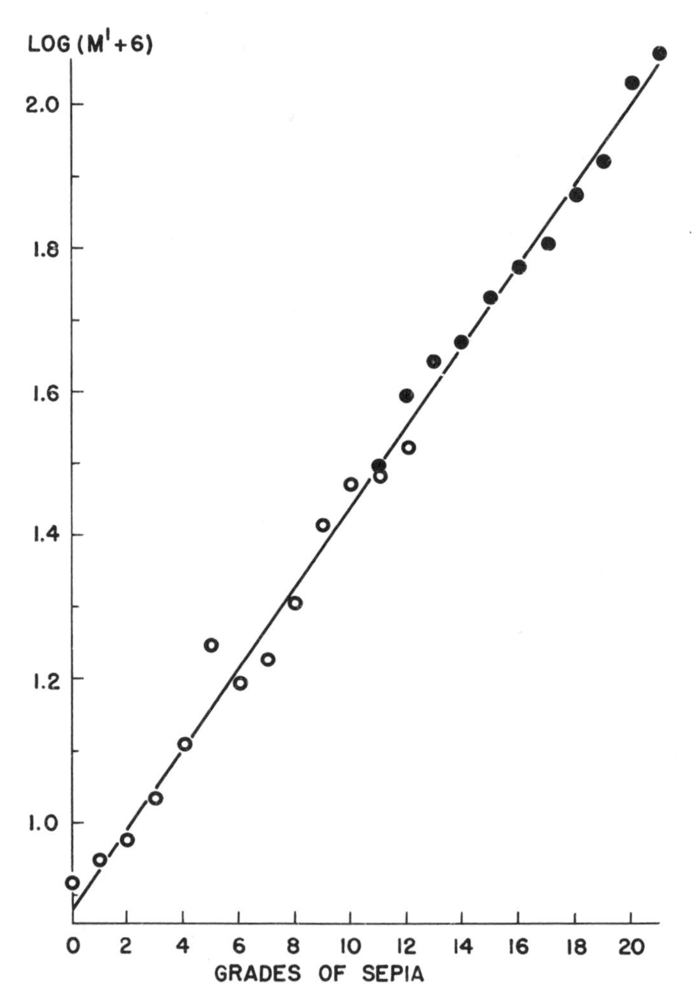

FIG. 12.1. Relation of $\log_{10}(M' + 6)$ to visual grade of sepia (from white [0] to black [21]) as fitted by the method of least squares, where M' is the average colorometric estimate of the concentration of melanin in hair samples of the guinea pig. (From data of Wright and Braddock 1949.)

between 0.01 and 0.05). It is apparent from the graph, however, that no smooth curve could fit the data better than a straight line, even in this case.

The use of this method in testing transformations of scale, $f(x)$, for their capacity to normalize distributions has been referred to earlier,

$$(\text{pri}^{-1}\, p = [(f(x) - M')]/\sigma').$$

The inverse probability transformation has also been found useful in relating mortality (p) to dosage (x) in treatment of populations with noxious agents. An approximately linear relation is found on using the equation, $\text{pri}^{-1}\, p = c_0 + c_1 \log x$. Bliss introduced the expression "probit" for $(\text{pri}^{-1}\, p + 5)$ to avoid the negative values taken by the inverse probability integral of percentages less than 50. The procedure to meet certain difficulties is discussed by Bliss (1935).

Trigonometric series have been used very extensively in the physical sciences in dealing with complex cycles. They have been found less useful in the biological sciences, but the fitting of a simple sine curve to seasonal cycles is sometimes useful and very simple if there are equally spaced intervals (x) over a cycle, and the observations (y) at these are assigned equal weights. Assume n intervals.

$$(12.17)\qquad \begin{aligned} y &= c_0 + a \sin\left(b + \frac{2\pi}{n} x\right) + \delta y \\ &= c_0 + a\left[\sin b \cos\left(\frac{2\pi}{n} x\right) + \cos b \sin\left(\frac{2\pi}{n} x\right)\right] + \delta y \\ &= c_0 + c_1 x_1 + c_2 x_2 \quad \text{in which } x_1 = \cos\left(\frac{2\pi}{n} x\right), \quad c_1 = a \sin \text{b} \\ &\qquad\qquad\qquad\qquad\qquad\quad x_2 = \sin\left(\frac{2\pi}{n} x\right), \quad c_2 = a \cos b. \end{aligned}$$

Normal Equations

$$(12.18)\qquad \left\{ \begin{array}{cccl} \underline{c_0} & \underline{c_1} & \underline{c_2} & \\ n & \sum x_1 & \sum x_2 & = \sum y \\ \sum x_1 & \sum x_1^2 & \sum x_1 x_2 & = \sum x_1 y \\ \sum x_2 & \sum x_1 x_2 & \sum x_2^2 & = \sum x_2 y. \end{array} \right.$$

It can readily be seen that $\sum x_1 = \sum x_2 = \sum x_1 x_2 = 0$ from the symmetry of $\cos [(2\pi/n)x]$ and $\sin [(2\pi/n)x]$ and their orthogonality. Thus the set falls into three separate parts which give the solutions directly.

$$(12.19)\qquad c_0 = \bar{y}, \quad c_1 = \frac{\sum x_1 y}{\sum x_1^2}, \quad c_2 = \frac{\sum x_2 y}{\sum x_2^2}.$$

The original parameters can easily be derived: $a = \sqrt{c_1^2 + c_2^2}$ is the amplitude, and $b = \sin^{-1}(c_1/a)$ is the epoch, the distance below the origin at which $(y - \bar{y}) = 0$.

As an example, Table 12.2 shows the relation between percentage with little toes in the inbred strain of guinea pigs, no. 35, and month of birth. The fit is shown in Figure 12.2.

TABLE 12.2. Numbers of guinea pigs of an inbred strain (no. 35) with 3 or with 4 toes on the hind feet, fitted by a sine curve.

Month (x)	3-toed	Y_0 4-toed	Total	y_0 %4-T	x_1 $\cos\left(\frac{2\pi}{12}x\right)$	x_2 $\sin\left(\frac{2\pi}{12}x\right)$	y_C %4-T	Y_C No. 4-T	$Y_0 - Y_C$
1	89	71	160	44.4	+0.866	+0.500	39.3	62.9	+8.1
2	90	59	149	39.6	+0.500	+0.866	40.5	60.4	−1.4
3	83	65	148	43.9	0	+1.000	39.4	58.3	+6.7
4	92	49	141	34.8	−0.500	+0.866	36.2	51.0	−2.0
5	97	39	136	28.7	−0.866	+0.500	31.7	43.2	−4.2
6	121	39	160	24.4	−1.000	0	27.3	43.6	−4.6
7	175	56	231	24.2	−0.866	−0.500	24.0	55.3	+0.7
8	125	51	176	29.0	−0.500	−0.866	22.7	40.0	+11.0
9	106	34	140	24.3	0	−1.000	23.8	33.4	0.6
10	164	51	215	23.7	+0.500	−0.866	27.1	58.2	−7.2
11	109	52	161	32.3	+0.866	−0.500	31.5	50.7	+1.3
12	111	48	159	30.2	+1.000	0	36.0	57.2	−9.2
	1,362	614	1,976	379.5			379.5	614.2	−0.2

From data of Wright (1934c).

$$c_0 = \sum y_0/12 = 379.5/12 = 31.62$$
$$c_1 = \sum x_1 y/\sum x_1^2 = 26.16/6 = 4.360 \qquad a = 8.920$$
$$c_2 = \sum x_2 y/\sum x_2^2 = 46.69/6 = 7.782 \qquad b = \sin^{-1} 0.489 = 29.25^\circ$$
$$y = 31.62 + 4.360 \cos[(2\pi/12)x] + 7.782 \sin[(2\pi/12)x]$$
$$y = 31.62 + 8.92 \sin(29.25 + 30x)^\circ$$

As this is a frequency distribution, the ultimate test is the goodness of fit χ^2. With $\chi^2 = 12.1$ and 9 degrees of freedom, there is a probability of 0.20 that accidents of sampling might give a worse fit. Although the use of weighting should yield a somewhat smaller χ^2, it could not improve the fit significantly. Compare this result with that of Table 9.6.

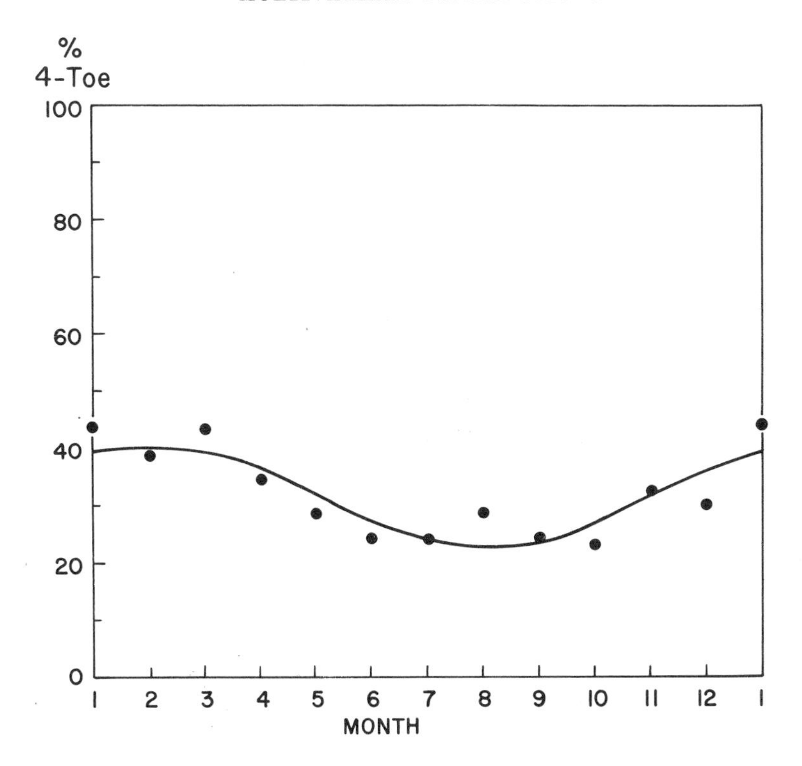

FIG. 12.2. Average monthly percentages (over many years) of the occurrence of the little toe in an inbred strain of guinea pigs, as fitted by the method of least squares. (From data of Wright 1934*c*.)

Correlation

The method of least squares was devised primarily for making the best determination of one quantity for given values of one or more others under conditions of such nearly complete determination that deviations might be interpreted as mere errors of observation. Such control of variation is more characteristic of astronomy and the experimental physical sciences than of biology. Galton (1888) suggested the need for a coefficient to describe the degree of correlation between variables on a universal scale in studies of data in which much real variation occurs. The theory was developed especially by Pearson (1904). The coefficient that is most widely useful is the average product of the deviations of the variables from their means in standard form.

$$r_{xy} = \frac{1}{N} \sum^{n} \left(\frac{x - \bar{x}}{\sigma_x}\right)\left(\frac{y - \bar{y}}{\sigma_y}\right). \tag{12.20}$$

This may be expressed in the following form, which is usually more convenient for calculation.

$$(12.21) \qquad r_{xy} = \frac{\sum xy - \bar{x} \sum y}{\sqrt{(\sum x^2 - \bar{x} \sum x)(\sum y^2 - \bar{y} \sum y)}}.$$

No Gaussian correction is applied to the variances since the uncertainty of the means affects the numerator as well as the denominator. If there is grouping of a continuous variate, the variates may be midclass values and the terms in the summations must be weighted by the appropriate frequencies. Sheppard's correction, where appropriate, is applied to the variances to obtain the best estimate of the true value.

The product-moment correlation is useful in several somewhat distinct ways.

(1) It measures correlation on a scale ranging from 1 for perfect positive linear correlations to 0 in the absence of linear correlation and to -1 for perfect negative linear correlation. It may be seen that if $(y - \bar{y}) = c(x - \bar{x})$, where c is a positive constant, $\sigma_y = c\sigma_x$ and $r_{xy} = +1$. Similarly, if $(y - \bar{y}) = -c(x - \bar{x})$, $\sigma_y = c\sigma_x$, but $r_{xy} = -1$. If $\sum (x - \bar{x})(y - \bar{y}) = 0$, $r_{xy} = 0$.

This is not the only statistic that provides such a scale. There are statistics that agree at the three points -1, 0, and $+1$ without even rough agreement elsewhere (cf. r_{xy}^3), but if one has become familiar with the degree of relationship implied by such values of r_{xy} as 0.10, 0.50, 0.90, etc., the specification of the correlation coefficient conveys valuable information about the population under consideration.

It should further be noted that the coefficient may be zero without independence. There may be complete dependence of one variable on the other, e.g., $y = \sin x$ between $x = 0$ and $x = 2\pi$, but an average product of the deviations from the means that is zero. The correlation coefficient is of most value where relations are linear or nearly so.

(2) The correlation coefficient is useful because of its close relation to the slope of the linear estimates of one in terms of the other by the method of least squares. The formula for the correlation need merely be multiplied by the ratio of the standard deviation of the dependent to that of the independent variable to give the formula (12.13) arrived at for the parameter c_1 in fitting $y = c_0 + c_1 x$ to the data. It is customary to use the symbol b_{yx} for the slope of this "regression" line.

$$(12.22) \qquad b_{yx} = \frac{(\sum xy - \bar{y} \sum x)}{(\sum x^2 - \bar{x} \sum x)} = r_{xy} \frac{\sigma_y}{\sigma_x}.$$

(12.23) $$y = \bar{y} + b_{yx}(x - \bar{x}) + \delta_{yx}$$

where δ_{yx} is the deviation from the line.

There are two regression lines, that of the *means* of y relative to the *individual* values of x given above, and that of the *means* of x relative to the *individual* values of y given by $b_{xy} = r_{xy}\,\sigma_x/\sigma_y$. Pearson, for example, found a correlation of $+0.50$ between parent (p) and offspring (o) in stature in human data. Since the standard deviations are essentially the same in both generations, both b_{po} and b_{op} have the value $+0.50$. It is at first sight somewhat confusing that sons deviate from average height only half as much as did their fathers and that fathers deviate only half as much as their sons. The paradox is resolved by noting that the dependent variable in each is a *class* average that is related to a *specified* value of the independent variable.

The regression coefficient is concrete, whereas the correlation is abstract. A correlation of -0.673 between size of litter (L) and mean birth weight (W) in litters in a stock of guinea pigs ($\sigma_L = 1.26$, $\sigma_W = 18.7$) implies a regression $b_{WL} = -10.0$ gm./pig.

(3) The correlation coefficient is also closely related to another concrete coefficient that is useful on its own account. This is the product moment $\mu(xy) = (1/N)\sum(x - \bar{x})(y - \bar{y})$ from which the correlation is derived by dividing by the product of the two standard deviations. The product moment in a heterogeneous population may be analyzed into the sum of the product moment of the weighted means of the subpopulations and the average product moment within these (Wright 1917). Because of this additive property, Fisher later renamed this statistic the covariance in analogy with his term variance (Fisher 1918) for the squared standard deviation.

(12.24) $$\text{cov}_{xy} = r_{xy}\sigma_x\sigma_y.$$

(4) Assuming linearity, the squared correlation coefficient measures the portion of the variance of either of the two variables that is controlled directly or indirectly by the other in the sense that it gives the ratio of the variance of the means of one for given values of the other to the total variance of the former. In the equation $(y - \bar{y}) = b_{yx}(x - \bar{x}) + \delta_{yx}$ the deviations of y from its mean are analyzed into two necessarily uncorrelated components.

If $\sigma^2_{y\cdot x}$ is used for the average variance of deviations, δ_{yx}, from the regression we may write

(12.25) $$\sigma^2_y = b^2_{yx}\sigma^2_x + \sigma^2_{y\cdot x} = r^2_{xy}\sigma^2_y + \sigma^2_{y\cdot x},$$

(12.26) $$\sigma^2_{y\cdot x} = \sigma^2_y(1 - r^2_{xy}).$$

It is sometimes convenient to use the symbol $\sigma^2_{y(x)}$ for the portion of the variance of y due to x in the above sense.

$$\sigma^2_{y(x)} = r^2_{xy}\sigma^2_y. \tag{12.27}$$

(5) The distributions of variables V_1 and V_2, if normal, are specified by the parameters $\overline{V}_1$ and σ_1, $\overline{V}_2$ and σ_2, respectively. The correlation coefficient, r_{12}, completes the set necessary for specifying the bivariate normal distribution, assuming a linear relation and that the residual variability of each about the appropriate regression line is independent of the other variate.

Let x and y be measured as deviations from their means. Then

$$f_x = \frac{1}{\sigma_x\sqrt{2\pi}} \exp\left[-\frac{1}{2}\left(\frac{x}{\sigma_x}\right)^2\right]$$

is the frequency distribution of x and

$$f_{y\cdot x} = \frac{1}{\sigma_{y\cdot x}\sqrt{2\pi}} \exp\left[-\frac{1}{2}\left(\frac{\delta_{y\cdot x}}{\sigma_{y\cdot x}}\right)^2\right]$$

is that of y for given x.

$$f_{y\cdot x} = \frac{1}{\sigma_y\sqrt{(1-r^2_{xy})2\pi}} \exp\left[-\frac{\tfrac{1}{2}(y - b_{yx}x)^2}{\sigma^2_y(1-r^2_{xy})}\right],$$

$$f_{xy} = f_x f_{y\cdot x} = \frac{1}{2\pi\sigma_x\sigma_y\sqrt{1-r^2_{xy}}} \times \exp\left\{-\frac{1}{2(1-r^2_{xy})}\left[\left(\frac{x}{\sigma_x}\right)^2 + \left(\frac{y}{\sigma_y}\right)^2 - 2\frac{xy}{\sigma_x\sigma_y}r_{xy}\right]\right\}. \tag{12.28}$$

The use of the correlation coefficient as a parameter in the formula for the bivariate normal distribution does not imply that its usefulness is restricted to such distributions. It will be shown in volume II to be useful in comparing the correlations between relatives with respect to single loci and thus with respect to discrete variates which may be highly asymmetrical. It measures degree of determination in the sense indicated, irrespective of normality. It measures the slopes of the regression lines on scales for which the standard deviations are the units, again irrespective of normality. There is no requirement of normality in such formulas as those for the variances of sums or differences in which it is a convenient parameter.

$$\sigma^2_{(x_1+x_2)} = \sigma^2_1 + \sigma^2_2 + 2\sigma_1\sigma_2 r_{12}. \tag{12.29}$$

$$\sigma^2_{(x_1-x_2)} = \sigma^2_1 + \sigma^2_2 - 2\sigma_1\sigma_2 r_{12}. \tag{12.30}$$

It may be noted that cov_{12} may be substituted for $\sigma_1\sigma_2 r_{12}$ in these formulas.

The most frequent cause of confusion in the interpretation of correlation coefficients arises from treating them as if they were intended to describe some absolute property of variables. A correlation coefficient is always a property of the population as well as of the variables and derives its usefulness from this. The regression coefficients of a causally dependent variable on a variable factor may in certain cases reveal essentially the same contribution of the latter per unit of change in widely different populations in which the correlation coefficients differ correspondingly. This has led to a rather frequently expressed view that the regression coefficient should completely supplant the correlation coefficient in the statistical treatment of related variables. In general, however, regression coefficients as well as correlations vary from population to population. In a random-bred strain of guinea pigs, for example, the correlation between parent and offspring with respect to spotting was 0.214 ± 0.018. This was also the regression of offspring on parent because of essential identity of the standard deviations in the two generations. In another population (descended from a single mating after seven generations of brother-sister mating), the correlation and regression were both $+0.014 \pm 0.022$. The difference in both is a reflection of the difference in the nature of the populations. The abstract correlation coefficient and the concrete regression are both useful when properly understood. They correspond to different modes of interpretation which taken together give a more penetrating grasp of a situation than either gives by itself.

The standard error of the regression coefficient, b_{yx}, based on N observations may easily be shown to be

$$(12.31) \qquad \frac{\sigma_{y \cdot x}}{\sigma_x\sqrt{n-2}} \quad \text{in which} \quad \sigma_{y \cdot x} = \sigma_y\sqrt{1 - r_{xy}^2} \quad \text{or} \quad \sqrt{\sigma_y^2 - b_{yx}^2\sigma_x^2}.$$

The standard error of the correlation coefficient is approximately

$$\frac{1 - r^2}{\sqrt{n-1}}.$$

The distribution of values of r in small samples, however, is far from normal owing to the limitation of the range at -1 and $+1$. There is no such limitation in the case of b. It is best, therefore, to test the reality of a relation by the significance of the regression coefficient. Where there is no doubt of significance, the difference between two correlations can ordinarily be judged by using the standard error of the difference. If a more thorough test is required, use can be made of a transformation of r, proposed by Fisher (1921), $z = \frac{1}{2}[\log_e (1 + r)/(1 - r)]$, which has an approximately normal distribution with standard error $1/\sqrt{n-3}$. This transformation is also valuable in averaging determinations from a number of small samples.

Although the distributions of the variables need not be normal for certain uses of the correlation coefficient, it is often highly desirable for purposes of comparison that the scales of quantitatively varying characters be normalized and thus made additive as far as possible with respect to the effects of elementary factors. Transformation of scale may be used in such cases. Karl Pearson (1913) devised methods for estimation of the correlation on a hypothetical underlying bivariate normal distribution in cases in which one or both variables are given only in broad categories. Unfortunately the standard errors of these coefficients are usually very complicated and dependent on the assumption of underlying normality. The reality of correlations is best tested from the untransformed data. The value of the statistics is on the descriptive and especially the comparative side where the numbers and other conditions are such that there is no question of statistical significance for those that are not close to zero. One case is that in which y is distributed normally while x is given only in nonmetric categories, but is of such a nature that the assumption of an underlying bivariate normal distribution seems warranted. The mean value of x is taken as 0 and its standard deviation unity. Let p_a and p_b represent the proportions of x below the upper and lower limits of a given class. The abscissas of the limits are $x_a = \text{pri}^{-1}\,p_a$ and $x_b = \text{pri}^{-1}\,p_b$. The ordinates are z_a and z_b of the unit normal distribution. Then $\bar{x}_c = (z_b - z_a)/(p_b - p_a)$ is the mean of this class, as brought out earlier (8.94). Let $\bar{y}_i$ be the theoretical mean of y at x_i, and $\bar{y}_c$ that at $\bar{x}_c$. Then $b_{yx} = (\bar{y}_i - \bar{y})/x_i$ for any x_i.

(12.32)
$$b_{yx} = \frac{\bar{y}_c - \bar{y}}{\bar{x}_c} = r_{\bar{x}(c)y}\frac{\sigma_y}{\sigma_{\bar{x}(c)}}.$$

Thus

(12.33)
$$r_{xy}\frac{\sigma_y}{\sigma_x} = r_{\bar{x}(c)y}\frac{\sigma_y}{\sigma_{\bar{x}(c)}}.$$

and

(12.34)
$$r_{xy} = \frac{r_{\bar{x}(c)y}}{\sigma_{\bar{x}(c)}}.$$

The underlying correlation coefficient, r_{xy}, can thus be estimated from the observed correlation between y and the values of $\bar{x}_c$, which is somewhat less than 1 (the value assumed for σ_x).

The most extreme case is that in which there are only two observed classes of x. The estimated coefficient is known as the biserial coefficient of correlation (Pearson 1909).

Let $x_1 = -z/p$ and $x_2 = z/q$ be the theoretical mean values of x, and $\bar{y}_1$ and $\bar{y}_2$ the observed mean values of y in the lower and upper categories respectively.

$$b_{yx} = \frac{\bar{y}_2 - \bar{y}_1}{x_2 - x_1} = (\bar{y}_2 - \bar{y}_1)\frac{pq}{z}. \tag{12.35}$$

$$r_{xy} = \frac{b_{yx}}{\sigma_y} = \left(\frac{\bar{y}_2 - \bar{y}_1}{\sigma_y}\right)\frac{pq}{z}. \tag{12.36}$$

In the case in which both variables are given by broad categories there are certain theoretical difficulties. If there is a considerable number of categories and all have small frequencies for at least one of the variable for example y, a normalized scale can be obtained from category means $y_c = (z_a - z_b)/(p_b - p_a)$ or the mid points $y'_c = (x_a + x_b)/2$ (except for the end classes). In the former, the correlation coefficients are theoretically too small and can be corrected approximately by applying broad category corrections to both variables, $r_{xy} = r_{x(c)y(c)}/\sigma_{x(c)}\sigma_{y(c)}$. In the latter, the correction for y would be in the opposite direction. Sheppard's correction would be in order if the class intervals of y were equal. In either case the correction for y is small, if there are many small categories all with small frequencies.

A special case is that in which both variables are given merely as dichotomies but an underlying bivariate normal distribution is indicated. Let a, b, c, and d be the proportions in the four categories with $p_1 = (a + b) > 0.50$, $p_2 = (a + c) > 0.50$, $q_1 = 1 - p_1 = c + d$, and $q_2 = 1 - p_2 = b + d$. The abscissas and ordinates of the unit normal curves at the points of dichotomy are x_1, x_2, and y_1, y_2, respectively. Pearson (1901) showed that the "tetrachoric" coefficient of correlation could be obtained from:

$$(d - q_1q_2)/z_1z_2 = r + (1/2)x_1x_2r^2 + (1/6)(x_1^2 - 1)(x_2^2 - 1)r^3 + (1/24)x_1x_2(x_1^2 - 3)(x_2^2 - 3)r^4 + (1/120)(x_1^4 - 6x_1^2 + 3)(x_2^4 - 6x_2^2 + 3)\ldots.$$

Components of Variability

It was brought out in chapter 8 that the basic principle of the theory of variability is that the squared standard deviation of the sum of independent variables is equal to the sum of the squared deviations of the latter. This principle makes it of interest to analyze variability in actual cases into the contributions of tangible components.

One of the uses of the correlation coefficient noted earlier is that its square measures the degree of determination of a variable by another that is linearly related. If the relations between graded variables are not linear, any given deviation of one $(y - \bar{y})$ may still be analyzed into the deviations from the grand average of the mean of its class, as specified by a given value of the other variable, and the necessarily uncorrelated deviations from this class mean. Pearson's (1905) squared correlation ratio (η^2) is the ratio of the variance of means of y, for the various classes of x, to its total variance and thus can be interpreted as a measure of degree of determination applicable to nonlinear relations.

$$\eta_{yx}^2 = \frac{\sigma_{y(x)}^2}{\sigma_y^2}. \tag{12.37}$$

$$1 - \eta_{yx}^2 = \frac{\sigma_{y \cdot x}^2}{\sigma_y^2} \tag{12.38}$$

The correlation ratio is equal to the coefficient of correlation in principle if the corresponding regression lines are linear, but if they are not, it is larger.

There is, indeed, no necessity that the independent variable x be a graded variable. If the data can be subdivided into x-classes on any basis, the degree of determination of y, by whatever is used as the basis for subdivision, is given by the ratio of the variance of class means of y to its total variance, since the deviations of the individual values of y's from their class means are necessarily independent of the deviations of the latter means from the grand average.

TABLE 12.3. Estimation of the average correlation, r_{BY}, between birth weight (B) and year weight (Y) within 24 inbred strains of guinea pigs from the variances and covariances for the total population and for the array of strain means.

Population	σ_B^2	σ_Y^2	σ_B	σ_Y	$\sigma_B\sigma_Y$	$r_{BY}\sigma_B\sigma_Y$	r_{BY}
Total (560)	130.53	14,852	11.43	121.9	1392	522.2	$+0.375 \pm 0.024$
Strain means (24)	20.50	4,837	4.53	69.6	315	198.4	$+0.630 \pm 0.083$
Average family (deduced)	110.03	10,015	10.49	100.1	1050	323.8	$+0.308 \pm 0.026$

From data of Wright (1917).

Table 12.3 (Wright 1917) illustrates the analysis of the variances of birth weight (B) and year weight (Y) in an array of inbred strains of guinea pigs into the contributions of 24 weighted family means and the deviations about these means. It also illustrates the additive property of the product moment (or covariance) by analyzing that for birth weights and year weights in the total population into the contribution of the weighted family means and that of the individual deviations within the families. This analysis made possible comparison of the average correlation between birth weight and year weight within a family (+0.308), with the correlation between the means of these variables (+0.630).

A portion of the observed variance of means is obviously to be expected from mere accidents of sampling. The corrections to be made on this account are referred to as Gaussian. If the standard errors of the family means were all the same, this spurious portion would be the squared standard error of the mean, $\mathrm{SE}_M^2 = \sigma_{2\cdot 1}^2/K$, where $\sigma_{2\cdot 1}^2$ is the variance within families and K is the number of individuals in the family. Thus the squared correlation ratio needs a correction which was, indeed, incorporated into Pearson's full formula, although not in a way that was always exact.

Assume that there are L classes and that the number of individuals is the same (K) in all classes. Let $M_s = (1/K) \sum^K x$ be the sample mean and M_t the true mean of a class. Let $\bar{M}_s = (1/L) \sum^L M_s$ be the mean of the means of the L classes. The corresponding mean of the L true means is $\bar{M}_t$.

The true variance within classes is

$$\sigma_{x\cdot c(t)}^2 = \frac{1}{NLK} \sum^N \sum^L \sum^K (X - M_t)^2 \tag{12.39}$$

with indefinitely large N.

An unbiased estimate is given by

$$E(\sigma_{x\cdot c}^2) = \frac{1}{L} \frac{\sum^L \sum^K (X - M_s)^2}{(K - 1)} \tag{12.40}$$

The theoretical variance of the means can be obtained on either of two hypothesis: (1) that the L classes are all that there are (subdivision exhaustive), or (2) that they are a random sample of an indefinitely large number of such classes. The former is more appropriate, for example, where the classes consist of the 2 sexes or the 12 months of the year. The latter is more appropriate for samples of organisms deliberately chosen at random locations in the range of the species. In many cases, either point of view may be taken. A number of laboratory strains may be treated either as the whole array that is to be considered or as a sample of a hypothetical total.

Consider first the case of classes that are exhaustive. The quantity $(M_s - \bar{M}_t)$ can be analyzed into independent components in two ways.

(1) $(M_s - \bar{M}_t) = (M_s - \bar{M}_s) + (\bar{M}_s - \bar{M}_t)$,

$$\frac{1}{NL} \sum^N \sum^L (M_s - \bar{M}_t)^2 = \frac{1}{NL} \sum^N \sum^L (M_s - \bar{M}_s)^2 + \frac{1}{NL} \sum^N \sum^L (\bar{M}_s - \bar{M}_t)^2.$$

But

$$\frac{1}{NL}\sum^N\sum^L(\bar{M}_s - \bar{M}_t)^2 = \frac{1}{NL^2}\sum^N\sum^L(M_s - M_t)^2$$

$$= \frac{1}{NL^2K}\sum^N\sum^L(X - M_t)^2 = \frac{1}{LK}\sigma^2_{x\cdot c(t)}.$$

Thus

$$\frac{1}{NL}\sum^N\sum^L(M_s - \bar{M}_t)^2 = \frac{1}{NL}\sum^N\sum^L(M_s - \bar{M}_s)^2 + \frac{1}{LK}\sigma^2_{x\cdot c(t)}.$$

(2) $(M_s - \bar{M}_t) = (M_s - M_t) + (M_t - \bar{M}_t)$,

$$\frac{1}{NL}\sum^N\sum^L(M_s - \bar{M}_t)^2 = \frac{1}{K}\sigma^2_{x\cdot c(t)} + \sigma^2_{M(t)}.$$

On equating these two expressions for $(1/NL)\sum^N\sum^L(M_s - \bar{M}_t)^2$,

$$\sigma^2_{M(t)} = \frac{1}{NL}\sum^N\sum^L(M_s - \bar{M}_s)^2 - \frac{L-1}{LK}\sigma^2_{x\cdot c(t)}.$$

An unbiased estimate of the true variance of means from data is thus as follows, on assuming that the classes are exhaustive.

$$E(\sigma^2_M) = \frac{1}{L}\sum^L(M_s - \bar{M}_s)^2 - \frac{L-1}{LK}E(\sigma^2_{x\cdot c}). \tag{12.41}$$

Consider next the case in which the classes may be treated as a random sample. Let $\bar{\bar{M}}_t$ be the theoretical mean of the array of all possible classes. The variances from two ways of analyzing $(M_s - \bar{\bar{M}}_t)$ into independent components may again be equated.

(1) $(M_s - \bar{\bar{M}}_t) = (M_s - \bar{M}_s) + (\bar{M}_s - \bar{\bar{M}}_t)$,

$$\frac{1}{NL}\sum^N\sum^L(M_s - \bar{\bar{M}}_t)^2 = \frac{1}{NL}\sum^N\sum^L(M_s - \bar{M}_s)^2 + \frac{1}{NL}\sum^N\sum^L(\bar{M}_s - \bar{\bar{M}}_t)^2.$$

But

$$\frac{1}{NL}\sum^N\sum^L(\bar{M}_s - \bar{\bar{M}}_t)^2 = \frac{1}{NL^2}\sum^N\sum^L(M_s - \bar{\bar{M}}_t)^2.$$

Thus

$$\frac{1}{NL}\sum^N\sum^L(M_s - \bar{\bar{M}}_t)^2 = \frac{1}{NL}\sum^N\sum^L(M_s - \bar{M}_s)^2\left(\frac{L}{L-1}\right).$$

(2) $(M_s - \bar{\bar{M}}_t) = (M_s - M_t) + (M_t - \bar{\bar{M}}_t)$,

$$\frac{1}{NL}\sum^N\sum^L(M_s - \bar{\bar{M}}_t)^2 = \frac{1}{NL}\sum^N\sum^L(M_s - M_t)^2 + \frac{1}{NL}\sum^N\sum^L(M_t - \bar{\bar{M}}_t)^2$$

$$= \frac{1}{K}\sigma^2_{x\cdot c(t)} + \sigma^2_{M_t}.$$

Thus

$$\sigma^2_{M(t)} = \frac{1}{N(L-1)} \sum^{N} \sum^{L} (M_s - \bar{M}_s)^2 - \frac{1}{K} \sigma^2_{x \cdot c(t)},$$

$$E(\sigma^2_M) = \frac{\sum^{L} (M_s - \bar{M}_s)^2}{L-1} - \frac{1}{K} E(\sigma^2_{x \cdot c}). \tag{12.42}$$

The estimate of the true variance of means from the data on the hypothesis that the classes are a random sample of all possible classes is thus $L/(L-1)$ times the estimate on the hypothesis that the observed classes are exhaustive, as might be expected.

Data are often presented for analysis in which the class frequencies are not the same. This is to be expected if the classes are exhaustive and individuals' records are drawn at random from a total natural population, but it is unfortunate if the classes themselves are intended to be random ones. The estimated variance within classes is in any case

$$E(\sigma^2_{x \cdot c}) = \frac{\sum^{L} \sum^{K} (x - M_s)^2}{\sum^{L} (K-1)}. \tag{12.43}$$

Whether the variance of the theoretical means should be estimated with or without weighting by the class or other frequencies depends on considerations such as those suggested above. If the classes are exhaustive and the values K are representative of them, and the individuals have been drawn at random from the total population, there should be weighting.

$$E\,(\sigma^2_M) = \frac{\sum^{L} K(M_s - \bar{M}_s)^2}{\sum^{L} K} - \frac{L-1}{\sum^{L} K} [E\,(\sigma^2_{x \cdot c})]. \tag{12.44}$$

If the numbers are not representative of the classes, but representative frequencies (w) are known, these may be used.

$$E\,(\sigma^2_M) = \frac{\sum^{L} w(M_s - \bar{M}_s)^2}{\sum^{L} w} - \left(\frac{L-1}{L}\right)\left[\frac{\sum^{L} (w/K)}{\sum^{L} w} E(\sigma^2_{x \cdot c})\right]. \tag{12.45}$$

This becomes the same as above if $w = K$. If the w's are all the same it becomes:

$$E\,(\sigma^2_M) = \frac{\sum^{L} (M_s - \bar{M}_s)^2}{L} - \frac{L-1}{L\hat{K}} E\,(\sigma^2_{x \cdot c}) \tag{12.46}$$

where $\hat{K}$ is the harmonic mean. If some of the means are based on grossly inadequate numbers that are not representative of their classes, the estimate of the variance of means becomes unreliable. It is highly desirable that the frequencies of classes represent closely the actual frequencies in the population if the object is to describe the population as it actually is.

If the classes are treated as a random sample, the means that are used should be given equal weight.

$$E(\sigma_M^2) = \frac{\sum^{L} (M_s - \bar{M}_s)^2}{L - 1} - \frac{1}{\bar{K}} E(\sigma_{x \cdot c}^2). \tag{12.47}$$

The estimate becomes unreliable if means are included that are based on small numbers. It is obviously highly desirable in this case that the collection of data be planned so that all values of K are the same.

In the data on birth weights and year weights of guinea pigs, the means were weighted in calculating the variance of means, which made it possible to calculate the crude variances and the covariance within families merely by subtracting those for the weighted means from the totals. The crude variance of family means was 15.7% of the total for birth weights on this basis. On correcting for the amount expected from accidents of sampling this is reduced to 12.1%. For year weight, the crude figure 32.6% of the total is reduced to 29.6%.

Since weight is a character that is approximately normalized by using a logarithmic transformation, it might have been better to have analyzed variance on this scale but the difference would not be great. In other cases transformation is more important. An extreme case is variability that is specified merely by a dichotomy of the threshold sort. Assume that the percentage below the threshold varies among subpopulations, and it is desired to state the degree to which the postulated underlying normal variability is determined by whatever it is that is the basis for the subdivision of the total population. The appropriate statistic is Pearson's (1910) squared biserial correlation ratio, η^2.

It is assumed that the underlying variance is the same in all subpopulations and may be assigned the value 1. The location of the mean, M, of each subpopulation with respect to the postulated common threshold is given by $M = -\text{pri}^{-1}\, p$ on the unit normal distribution, where p is the proportion below the threshold. In calculating the variance of means the same considerations apply as above. If the subpopulations are treated as exhaustive and their means are weighted by the frequencies, the estimate is given by the following:

$$\sigma_M^2 = \frac{\sum^{L} K(M - \bar{M})^2}{\sum^{L} K} - \frac{L - 1}{\sum^{L} K}. \tag{12.48}$$

If the subpopulations are treated as a random sample and are assigned equal weights, even though their frequencies vary,

$$\sigma_M^2 = \frac{\sum^L (M - \bar{M})^2}{L - 1} - \frac{1}{\hat{K}}. \tag{12.49}$$

The variance of the total population on the scale on which the variances of the subpopulations is 1 is $(\sigma_M^2 + 1)$. Thus the portion of the total variance determined by the factor in question is

$$\eta^2 = \frac{\sigma_M^2}{1 + \sigma_M^2}. \tag{12.50}$$

The variability of an inbred strain (no. 35) with respect to presence or absence of the little toe was analyzed in various respects by this method (Wright 1934*c*). The portion of the strain considered included all animals (1,976) derived from a single mating in the twelfth generation of brother-sister mating and born between 1916 and 1924, inclusively. The total was broken up into 21 groups of closely related individuals to test for heterogeneity. Subdivisions were also made by month of birth and by age of dam from 3 to 27 months. The subdivisions were all treated as exhaustive (Wright 1934*c*).

TABLE 12.4. Estimates of the portions (η^2) of the variability in number of digits in an inbred strain of guinea pigs (no. 35) due to substrain differentiation, month of birth, and age of dam.

	Substrain	Month of Birth	Age of Dam
L	21	12	16
$\sum KM^2/\sum K$	0.5376	0.2938	0.4514
$\bar{M}^2$	0.3000	0.2530	0.2849
$(L - 1)/\sum K$	0.0101	0.0056	0.0076
σ_M^2	0.2275	0.0352	0.1589
$\eta^2 = \sigma_M^2/(1 + \sigma_M^2)$	0.1853	0.0340	0.1371

From data of Wright (1934*c*).

There was a tetrachoric correlation of 0.623 between littermates which may be interpreted as the square of the correlation between either littermate and the array of factors common to them. As this includes 0.185 due to substrain differentiation, 0.034 due to month of birth, and 0.137 due to age of dam, there is 0.267 due to other factors (nongenetic) that are common to the littermates.

In a later branch of this strain (358 animals) derived from a single mating in the twenty-second generation reared in a different laboratory (Chicago) under very different conditions, there was no significant substrain differentiation or relation to month of birth, but a stronger relation to age of dam ($\eta^2 = 0.25$). The tetrachoric correlation between littermates, 0.54, agrees with that expected from the earlier data on elimination of substrain differentiation $(0.623 - 0.185)/(1 - 0.185) = 0.537$. Table 12.5 shows the analysis of variability that could be made.

TABLE 12.5. Analysis of variability in number of digits in an inbred strain of guinea pigs (no. 35).

Factor	Guinea Pigs of Strain 35	
	At Beltsville, Md.	At Chicago
Substrain differentiation	0.18	0
Age of dam	0.14	0.25
Month of birth	0.03	0
Other nongenetic factors common to littermates	0.27	0.29
Individual nongenetic factors	0.38	0.46
Total	1.00	1.00

From data of Wright (1934*c*).

In this case, the best tests of significance were by the χ^2 method. The data on the relatively slight relation to month of birth have been presented earlier (Table 12.2). With 24 classes, 11 degrees of freedom, $\chi^2 = 48$, the probability from random sampling was 0.000 002 with no allowance for grouping of signs. There was no doubt of significance of any of the above components.

For graded data, the best test has been devised by Fisher on the basis of the ratio of independent estimates of the same variance, assuming that the parent populations are normal. The formula for the distribution of this ratio (F) is far from normal, but tables have been published for probability levels 0.20, 0.05, 0.01, and 0.001 and for pairs of degrees of freedom (Fisher and Yates 1938).

The most convenient form for testing whether there is significant differentiation in data from a particular mode of subdivision is given by Fisher (1925). The first row "within classes" relates to deviations of individuals from the class means, the second "between classes" relates to the weighted class means, and the bottom row relates to deviations of individual observations from the grand average of the total data.

(12.51)

	Degrees of Freedom	Sum of Squares	Mean Square
Within classes:	$N - L$	$\sum^{L}\sum^{K}(X - M)^2 = \sum^{L}\sum^{K}X^2 - \sum^{L}(M\sum^{K}X)$	$\sigma^2_{x \cdot c}$
Between classes:	$L - 1$	$\sum^{L}K(M - \bar{M})^2 = \sum^{L}(M\sum^{K}X) - \bar{M}\sum^{N}\sum^{L}X$	$\sigma^2_{x \cdot c} + K\sigma^2_M$
Total:	$N - 1$	$\sum^{L}\sum^{K}(X - \bar{M})^2 = \sum^{L}\sum^{K}X^2 - \bar{M}\sum^{N}\sum^{L}X$	

Note that the sum of squares within classes, that of weighted deviations of the class means from the grand average, and that of individual values from the latter all have the same number of terms and the total sum of squares is the sum of the other two. The degrees of freedom are also additive. Divisions of the sums of squares within classes and between weighted class means by the corresponding degrees of freedom give mean squares that are independent estimates of the same statistic—the variance within classes—on the hypothesis, in the second case, that there are no differences among the true means. The entries under mean squares above are the interpretations of the latter under the hypothesis that the classes are *random samples* of all possible classes. If the classes are exhaustive the mean square between classes is to be interpreted as $\sigma^2_{x \cdot c} + [\bar{K}L/(L - 1)]\sigma^2_M$. The ratio of the larger to the smaller mean square is the quantity F, which is examined for significance on entering the tables with the corresponding degrees of freedom. In the case of birth weights and year weights of guinea pigs, F comes out 4.34 and 11.26, respectively, with 23 and 536 degrees of freedom respectively and probabilities far below 0.001 in both.

If there is significant excess of the mean square between classes, the variance of true means can be estimated by dividing this excess by the factor that is appropriate for the interpretation. If the classes are interpreted as random, this gives an estimate based on the weighted deviations of the means, which may be very poor if a few of the class frequencies are much larger than the others. The great desirability of collecting the data so that the class frequencies are as nearly the same as possible if the classes are to be interpreted as a random sample, is again to be emphasized.

This mode of analysis may be extended to cases in which classes are divided into subclasses. Let K be the number in a subclass, L in a class, and Q the number of classes.

	Degrees of Freedom	Sum of Squares	Mean Square
Within subclasses:	$\sum^{Q}\sum^{L}(K-1)$	$\sum^{Q}\sum^{L}\sum^{K}(X-M)^2 = \sum^{Q}\sum^{L}\sum^{K}X^2 - \sum^{Q}\sum^{L}(M\sum^{K}X)$	$\sigma^2_{x\cdot s}$
Subclasses within classes:	$\sum^{Q}(L-1)$	$\sum^{Q}\sum^{L}K(M-\bar{M})^2 = \sum^{Q}\sum^{L}(M\sum^{K}X) - \sum^{Q}(\bar{M}\sum^{L}\sum^{K}X)$	$\sigma^2_{x\cdot s} + K\sigma^2_{s\cdot c}$
Classes within the total:	$Q-1$	$\sum^{Q}LK(\bar{M}-\bar{\bar{M}})^2 = \sum^{Q}(\bar{M}\sum^{L}\sum^{K}X) - \bar{\bar{M}}\sum^{Q}\sum^{L}\sum^{K}X$	$\sigma^2_{x\cdot s} + \bar{K}\sigma^2_{s\cdot c} + \bar{K}\bar{L}\sigma^2_{c\cdot t}$
Total:	$\sum^{Q}\sum^{L}K-1$	$\sum^{Q}\sum^{L}\sum^{K}(X-\bar{\bar{M}})^2 = \sum^{Q}\sum^{L}\sum^{K}X^2 - \bar{\bar{M}}\sum^{Q}\sum^{L}\sum^{K}X$	

(12.52)

The degrees of freedom and the sums of squares are again additive, and significance at the second or third step is determined from ratio of its mean square to that at the preceding step. The same principles as before apply in estimating the variance components. There may be any number of steps in such a hierarchic pattern.

M_{11}	M_{12}	$\cdots$	M_{1h}	$\sum^{C} M_{1j}$	$M_{1\cdot}$
M_{21}	M_{22}	$\cdots$	M_{2h}	$\sum^{C} M_{2j}$	$M_{2\cdot}$
$\vdots$	$\vdots$	$\ddots$	$\vdots$	$\vdots$	$\vdots$
M_{g1}	M_{g2}	$\cdots$	M_{gh}	$\sum^{C} M_{gj}$	$M_{g\cdot}$
$\sum^{R} M_{i1}$	$\sum^{R} M_{i2}$	$\cdots$	$\sum^{R} M_{ih}$		
$M_{\cdot 1}$	$M_{\cdot 2}$	$\cdots$	$M_{\cdot h}$		$\bar{M}$

$$(M_{ij} - \bar{M}) = (M_{i.} - \bar{M}) + (M_{.j} - \bar{M}) + (M_{ij} - M_{i.} - M_{.j} + \bar{M})$$

Another very important mode of subdivision of data is that in which classes cut across each other instead of being in hierarchic order. The cells must all be given equal weight to obtain independent estimates of the same variance on the hypothesis of homogeneity.

The sums of the first two terms of the analysis of $(M_{ij} - \overline{M})$ give the expected deviations of all means from the grand average on the hypothesis that the deviations of rows and columns are additive. The last term gives the actual deviations from this expectation. The three terms are necessarily uncorrelated in occurrence if the cells are given equal weight.

The significance of the differences among rows (R) is determined from the ratio of the mean square from rows to the residual variance and the significance of differences among columns (C) is analogous.

(12.53)

	Degrees of Freedom	Sum of Squares	Mean Square
Rows:	$R - 1$	$\sum^{R} (M_{i.} - \overline{M})^2 = [\sum^{R} (M_{i.} \sum^{C} M_{ij}) - \overline{M} \sum M_{ij}]$	$\sigma_I^2 + C\sigma_{Mij}^2$
Columns:	$C - 1$	$\sum^{C} (M_{.j} - \overline{M})^2 = [\sum^{C} (M_{.j} \sum^{R} M_{ij}) - \overline{M} \sum M_{ij}]$	$\sigma_I^2 + R\sigma_{Mij}^2$
Residuals:	$(R-1)(C-1)$	$\sum (M_{ij} - M_{i.} - M_{.j} + \overline{M})^2$	σ_I^2
Total:	$RC - 1$	$\sum (M_{ij} - \overline{M})^2 = [\sum^{R} \sum^{C} M_{ij}^2 - \overline{M} \sum M_{ij}]$	

The entries in the cells are often themselves means (as indicated in the table). The variance within cells may be determined from the equation 12.54, letting K represent the number in a cell which may be variable.

$$\sigma_{X \cdot M}^2 = \frac{\sum^{R} \sum^{C} \sum^{K} (X - M_{ij})^2}{\sum^{R} \sum^{C} (K - 1)} = \frac{\sum^{R} \sum^{C} [\sum^{K} X^2 - \overline{M}_{ij} \sum^{K} X]}{\sum^{R} \sum^{C} (K - 1)}. \tag{12.54}$$

This, divided by the harmonic mean $\hat{K}$, is the expected mean square in the above analysis if there is homogeneity. The term $\sigma_{X \cdot M}^2/\hat{K}$, must be added to the expressions interpreting the mean square. If the residual mean square differs significantly from $\sigma_{X.M}^2/\hat{K}$ an interaction between the row and column classes is indicated.

In the two-column case, the test by Fisher's formula for F reduces to "Students'" test of paired comparisons.

(12.55)

	Degrees of Freedom	Sum of Squares	Mean Square
Column:	1	$N \sum^{2} (M_c - \overline{M})^2$	$2N(\overline{\Delta}/2)^2 = N\overline{\Delta}^2/2$
Row:	$N - 1$	$2 \sum^{N} (M_R - \overline{M})^2$	
Residual:	$N - 1$	$\sum^{N} \sum^{2} [(X - M_R) - \overline{(X - M_R]}^2$	$\sum^{2N} \frac{\left(\frac{\Delta}{2} - \frac{\overline{\Delta}}{2}\right)^2}{N-1} = N \frac{\sigma^2_{\overline{\Delta}}}{2}$
Total:	$2N - 1$	$\sum^{2} \sum^{N} (X - \overline{M})^2$	

$$F = \frac{\overline{\Delta}^2}{\sigma^2_{\overline{\Delta}}} = t^2. \tag{12.56}$$

The analysis may readily be extended to three or more interacting modes of classification. There are many other patterns which have been found useful, especially in plant-breeding experiments in which proper allowance for soil heterogeneity is a difficult but all-important consideration.

□ □

CHAPTER 13

Path Analysis: Theory

Basic Equations

Systems of correlation coefficients may be dealt with from two radically different points of view: from that of purely statistical description and from that of interpretation in terms of paths of causation. There are well-established methods for the former. It will be convenient, however, to begin with the method of path analysis (Wright 1921, 1934, 1954, 1960*a*, *b*), which was designed for the purpose of interpretation but which becomes identical with the methods of statistical description (multiple regression, vectorial representation) when the appropriate symmetrical pattern of relations is imposed.

This method presupposes the construction of a qualitative diagram in which every included variable, measured or hypothetical, is represented by arrows as either completely determined by certain others, which may in turn be represented as similarly determined, or as an ultimate factor. Each ultimate factor in the diagram must be connected by lines with arrowheads at both ends, with each of the other ultimate factors to indicate possible correlations through still more remote, unrepresented factors, except in cases in which it can safely be assumed that there is no correlation.

The necessary formal completeness of the diagram requires the introduction of a symbol for the array of unknown residual variables back of each variable that is not represented as one of the ultimate factors, unless there is reason to assume complete determination by the specified factors. Such a residual factor can be assumed by definition to be uncorrelated with any of the other factors immediately back of the same variable but cannot be assumed without adequate reason to be independent of other variables in the system.

Instantaneous reciprocal interaction between two variables may be represented by separate arrows from each to the other. Such a relation tends to complicate the analysis but does not invalidate it, if properly carried through.

It is assumed here that all relations are linear. The approximate results that can be obtained where relations are not linear (or cannot be made linear by systematic transformation) will, however, be considered briefly.

The validity of the system requires that any variable that enters into the system as a common factor back of two or more dependent variables, or as an intermediary in a chain, vary as a whole. If one part of a composite variable (such as a total or average) is more significant in one relation than in another, the treatment of the variable as if it were a unit may lead to grossly erroneous results. Fortunately, the parts of a composite variable are often known to be so strongly correlated in their values or in their action on others, or both, that they may be used to obtain approximate results. It cannot be emphasized too much, however, that the strict validity of the method depends on the properties of formally complete linear systems of unitary variables.

The basic diagram in developing the theory is one in which a variable V_0 (Fig. 13.1) is represented as completely determined by a number of immediate factors $V_1, V_2, \cdots, V_m, V_u$, all of which, except the residual V_u, are represented as intercorrelated.

We consider the correlation of V_0 with any variable, V_q. The latter must be represented as correlated with each of the factors of V_0, including the residual V_u, if there is no reason to the contrary. As all relations are assumed to be linear, we have:

$$V_0 = c_0 + c_{01}V_1 + c_{02}V_2 + \cdots + c_{0m}V_m + c_{0u}V_u. \tag{13.1}$$

The coefficients, c_{0i} are of the type used in estimating V_0 from the other variables by the method of least squares, and are known as partial regression coefficients, except that they include one that pertains to the totality of unknown residual factors. They are defined as path regression coefficients.

The path regression coefficient, c_{01}, measures the concrete contribution

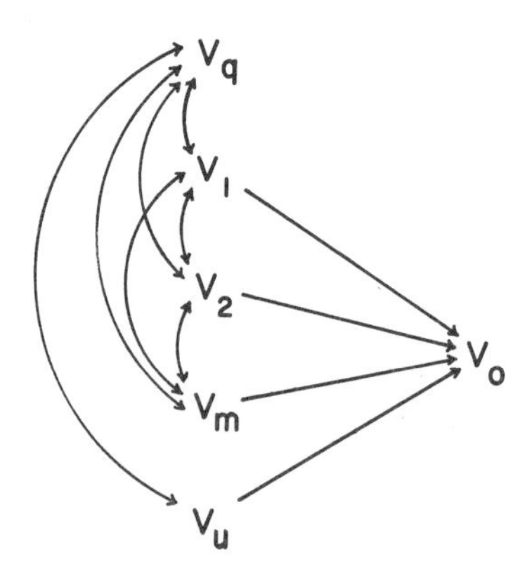

FIG. 13.1. (Redrawn from Fig. 1, Wright 1960*a*.)

that V_1 makes directly to V_0 from the point of view represented in the diagram. If this correctly represents the causal relations, it measures this contribution in an absolute sense, and its value can be used in the analysis of other populations. The standardized path coefficient $p_{01} = c_{01}\, \sigma_1/\sigma_0$ obviously does not have this property, but it has other virtues, including greater convenience in analysis. Let $X_0 = (V_0 - \overline{V}_0)/\sigma_0$, etc.

$$X_0 = p_{01}X_1 + p_{02}X_2 + \cdots + p_{0m}X_m + p_{0u}X_u. \tag{13.2}$$

On scales on which all standard deviations have the value 1, all correlation coefficients are reduced to product moments:

$$\begin{aligned} r_{0q} &= \frac{1}{n}\sum^{n} X_0 X_q \\ &= p_{01}r_{1q} + p_{02}r_{2q} + \cdots + p_{0m}r_{mq} + p_{0u}r_{uq} \\ &= \sum_{i=1}^{u} p_{0i}r_{iq}. \end{aligned} \tag{13.3}$$

This is the basic equation of path analysis. If V_q is one of the immediate factors, such as V_1, $r_{qu} = 0$.

$$r_{01} = p_{01} + p_{02}r_{12} + \cdots + p_{0m}r_{1m} \tag{13.4}$$

If V_q is V_0 itself,

$$r_{00} = p_{01}r_{01} + p_{02}r_{02} + \cdots + p_{0m}r_{0m} + p_{0u}^2 = 1; \tag{13.5}$$

$$r_{00} = \sum_{j=1}^{m} p_{0j}r_{0j} + p_{0u}^2 = 1 \quad \text{(where } V_j \text{ does not include } V_u\text{).} \tag{13.6}$$

Since $p_{0u}^2(=r_{0u}^2)$ measures the degree of determination of V_0 by residual factors, $\sum p_{0j}r_{0j}$ must measure that by the known variables collectively as represented by the best fitting linear function of these.

If $V_E = c_0 + c_{01}V_1 + c_{02}V_2 + \cdots + c_{0m}V_m$ (with no residual),

$$r_{0E}^2 = \sum p_{0j}r_{0j}; \tag{13.7}$$

$$r_{0u}^2 = 1 - r_{0E}^2. \tag{13.8}$$

Returning to (13.3), it should be noted that its derivation does not depend on the assumption of normality of any of the variables or on the nature of the correlations. Thus it is valid if there is reciprocal action of V_0 on any of them. The correlation coefficients, r_{iq}, involved in (13.3) may be analyzed by application of this formula to itself, if any of the immediate factors or V_q are represented as determined by more remote factors in a more extended diagram. In cases in which there are no paths that return on themselves, it is often convenient to make use of the following consequence of such analysis. The correlation between any two variables in a properly constructed diagram of relations, which does not involve reciprocal interaction, is equal to the sum of contributions pertaining to the paths by which one may trace from one to the other in the diagram without going back along any arrow after going forward along one, and without passing through any variable twice in the same path.

Compound Path Coefficients

A coefficient pertaining to a whole path connecting two variables and thus measuring the contribution of that path to the correlation, is known as a compound path coefficient. Its value is the product of the values of the coefficients pertaining to the elementary paths along its course. One, but not more than one, may be a correlation coefficient pertaining to a two-headed arrow without violating the rule against going backward along an arrow after going forward along the preceding arrow in the path.

A unidirectional compound-path coefficient may be indicated by listing the variables in order, from dependent to most remote independent, as subscripts. Thus in Figure 13.2, p_{013} pertains to the path $V_0 \leftarrow V_1 \leftarrow V_3$ and has the value $p_{01}p_{13}$.

In a bidirectional compound path coefficient, it is convenient to list the variables in the subscript in order from either end of the path but to set off the ultimate common factor or pair of factors by parentheses. Thus $p_{01(3)2}$ in Figure 13.2 pertains to the path $V_0 \leftarrow V_1 \leftarrow V_3 \rightarrow V_2$ with value $p_{01}p_{13}p_{23}$, and $p_{01(34)2}$ pertains to the path $V_0 \leftarrow V_1 \leftarrow V_3 \leftrightarrow V_4 \rightarrow V_2$ with value

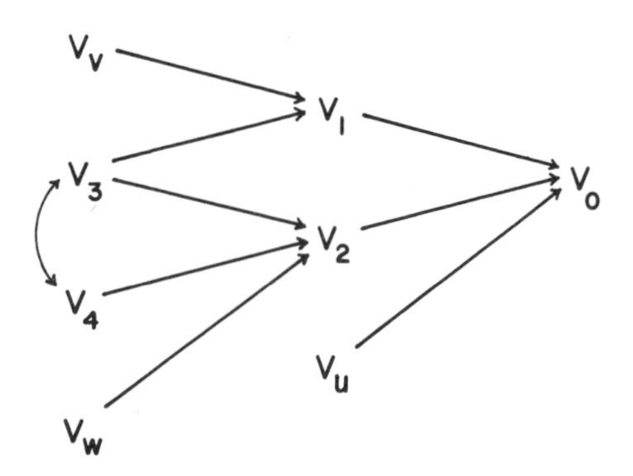

FIG. 13.2. (Redrawn from Fig. 2, Wright 1960*a*.)

$p_{01}p_{13}r_{34}p_{24}$. According to the rule above, $r_{02} = p_{02} + p_{01(3)2} + p_{01(34)2}$. According to this convention one might use $p_{(34)}$ for r_{34}.

Independence of Residual Factor

The residual factor can always be treated as if uncorrelated with the represented factors without loss of mathematical consistency, even though it may be known or suspected that the dependent variable is correlated with the latter, through paths other than those represented. It is instructive to show how a system in which one variable is completely determined by a number of others, all intercorrelated, may be transformed into a system in which one of the determining variables is not correlated with the others.

V_0 is represented as completely determined by V_1, V_2, and V_3 in Figure 13.3*A*. Let V_u be a linear function of V_1, V_2, and V_3 that is to be uncorrelated with V_1 and V_2, but in conjunction with these completely determines V_0 (Figs. 13.3*B* and 13.3*C*). The path coefficients of this new system are given primes to distinguish them from those of Figure 13.3*A*.

By hypothesis

$$p'_{u1}r_{u1} + p'_{u2}r_{u2} + p'_{u3}r_{u3} = 1, \tag{13.9}$$

and

$$\begin{cases} r_{u1} = p'_{u1} + p'_{u2}r_{12} + p'_{u3}r_{13} = 0 \\ r_{u2} = p'_{u1}r_{12} + p'_{u2} + p'_{u3}r_{23} = 0. \end{cases} \tag{13.10}$$

Thus

$$r_{u3} = p'_{u1}r_{13} + p'_{u2}r_{23} + p'_{u3} = \frac{1}{p'_{u3}}. \tag{13.11}$$

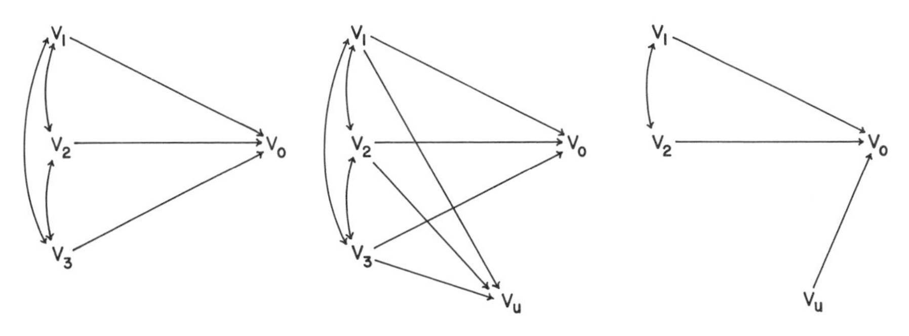

FIG. 13.3. (*C*, redrawn from Fig. 7, Wright 1960*a*.)

Multiply the preceding three equations by p'_{u3} and let $x = p'_{u1}p'_{u3}$, $y = p'_{u2}p'_{u3}$, and $z = p'^2_{u3}$.

$$\text{(13.12)} \quad \begin{cases} x + yr_{12} + zr_{13} = 0 \\ xr_{12} + y + zr_{23} = 0 \\ xr_{13} + yr_{23} + z = 1 . \end{cases}$$

These may be solved for x, y, and z and the values of p'_{u1}, p'_{u2} and p'_{u3} obtained.

$$\text{(13.13)} \quad \begin{cases} p'_{u1} = (r_{12}r_{23} - r_{13})/D \\ p'_{u2} = (r_{12}r_{13} - r_{23})/D \\ p'_{u3} = (1 - r^2_{12})/D . \end{cases}$$

where $D = \sqrt{(1 - r^2_{12})(1 - r^2_{12} - r^2_{23} - r^2_{13} + 2r_{12}r_{13}r_{23})}$.

$$\begin{array}{llll} & \text{Equations} & \text{Solutions} & \\ \text{(13.14)} & \begin{cases} p_{03} = p'_{0u}p'_{u3} \\ p_{01} = p'_{01} + p'_{0u}p'_{u1} \\ p_{02} = p'_{02} + p'_{0u}p'_{u2} \end{cases} & \left. \begin{array}{l} p'_{0u} = p_{03}/p'_{u3} \\ p'_{01} = p_{01} - p'_{0u}p'_{u1} \\ p'_{02} = p_{02} - p'_{0u}p'_{u2} \end{array} \right\} & \text{(13.15)} \end{array}$$

Thus, if the values of V_0, V_1, and V_2 are available in a set of measurements, but those of V_3 are not, Figure 13.3*C* is a valid diagram for mathematical purposes, with V_u the independent residual necessary for completion, even though it may be known or suspected that there are unrepresented factors (V_3) that are correlated with V_1 and V_2. The path coefficients p_{01} and p_{02} are not, however, the same in Figure 13.3*C* as in Figure 13.3*A*, which is more adequate from an interpretative standpoint.

Analysis of Simple Systems

It was noted earlier that the usefulness of the correlation coefficient as a descriptive statistic has been questioned, and it has been suggested that its place had better be taken over wholly by the regression coefficient (or in other connections by the covariance). It has similarly been suggested that only path regressions be used in path analysis (Tukey 1954, Turner and Stevens 1959).

It has seemed to me that the abstract path coefficients have a distinct descriptive value for reasons similar to those in the case of the correlation coefficient. This will be brought out further when interpretative applications are discussed.

Even, however, when the sole objectives of analysis are the concrete coefficients, actual path analysis takes on a simpler, more homogeneous form in terms of the abstract coefficients. Moreover, the use of the abstract coefficients lead naturally to the systematic expression of all the available information in the form of equations to be solved simultaneously—a matter of first importance in path analysis—but this is less likely to be done if the concrete coefficients are used. The application of the method to data usually requires algebraic manipulation of coefficients pertaining to unmeasured variables on the same basis as measured ones. As the former can only be treated in standardized form, homogeneity requires that all be so treated in the course of the algebra. It is such a simple matter to pass from either form to the other (in cases in which the standard deviations are available at all) that the economy of effort in using the concrete coefficients, as far as possible where these are the objectives, is usually outweighed by the loss of economy in the analysis.

This is illustrated below in a number of simple examples (sets 13.16 to 13.20). Numerical subscripts are used here for variables that are supposed to be measured and literal ones for the hypothetical variables that are necessary for completion of the diagrams. All of the equations that can be written from the known correlations and from the cases of complete determination are expressed in terms of path coefficients and correlations. They can all be written from inspection by merely tracing the connecting paths. Of course, they can also be written in terms of concrete coefficients. This is done in some of the cases. Parentheses enclose quantities that are inseparable in analysis restricted to concrete coefficients.

	Standardized Coefficients	Concrete Coefficients
(13.16)	$r_{12} = p_{12}$	$b_{12} = c_{12}$
	$r_{01} = p_{01}$	$b_{01} = c_{01}$
	$r_{02} = p_{01}p_{12}$	$b_{02} = c_{01}c_{12}$
	$r_{00} = 1 = p_{01}^2 + p_{0u}^2$	$\sigma_0^2 = c_{01}^2\sigma_1^2 + (c_{0u}^2\sigma_u^2)$
	$r_{11} = 1 = p_{12}^2 + p_{1v}^2$	$\sigma_1^2 = c_{12}^2\sigma_2^2 + (c_{1v}^2\sigma_v^2)$

In this case (Fig. 13.4), the first three equations are equally simple with concrete and standardized coefficients and overdetermine two paths. One may (*a*) obtain a compromise solution as by the method of least squares, or may (*b*) attribute any inconsistency to unrepresented correlations between V_u and the variables back of V_1 (which must compensate in such a way that

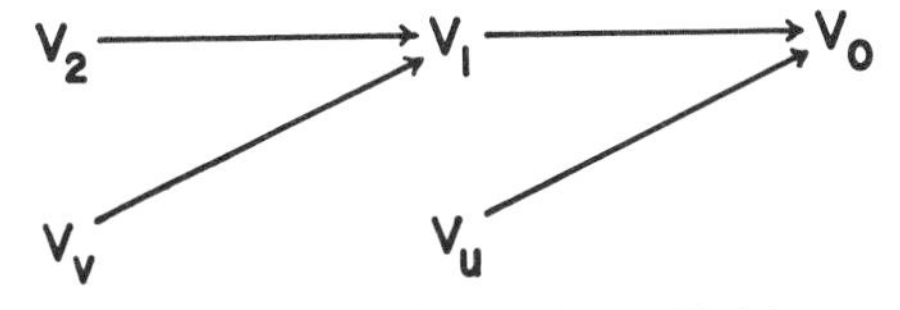

FIG. 13.4. (Redrawn from Fig. 3, Wright 1960*a*.)

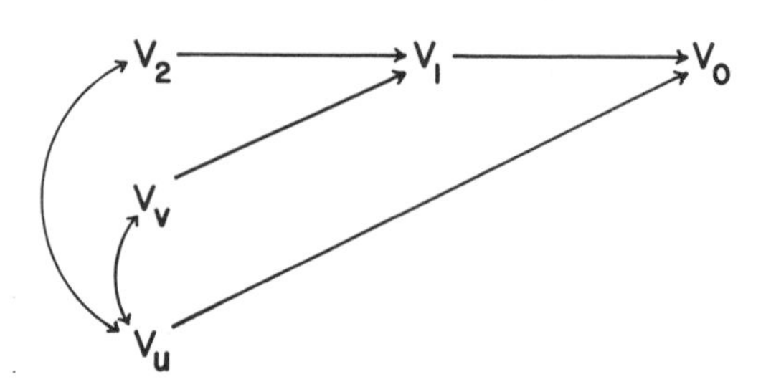

FIG. 13.5. (Redrawn from Fig. 4, Wright 1960*a*.)

$r_{1u} = 0$, as required by the definition of V_u as a residual), or may (*c*) assume that the measurements of V_1 are in error. The hypothesis of errors in either V_0 or V_2 does not resolve inconsistency in the first three equations in 13.16.

Figure 13.5 shows the revision of Figure 13.4 required to represent hypothesis (*b*).

	Standardized Coefficients	Concrete Coefficients
(13.17)	$r_{12} = p_{12}$	$b_{12} = c_{12}$
	$r_{11} = 1 = p_{12}^2 + p_{1v}^2$	$\sigma_1^2 = c_{12}^2\sigma_2^2 + (c_{1v}^2\sigma_v^2)$
	$r_{01} = p_{01}$ (since $r_{1u} = 0$)	$b_{01} = c_{01}$
	$r_{00} = 1 = p_{01}^2 + p_{0u}^2$	$\sigma_0^2 = c_{01}^2\sigma_1^2 + (c_{0u}^2\sigma_u^2)$
	$r_{02} = p_{01}p_{12} + p_{0u}r_{2u}$	$cov_{02} = c_{01}c_{12}\sigma_2^2 + c_{0u}cov_{2u}$
	$r_{1u} = 0 = p_{12}r_{2u} + p_{1v}r_{uv}$	$cov_{1u} = 0 = c_{12}cov_{2u} + c_{1v}cov_{uv}$

The standardized equations can be written at once from inspection of the diagram. Some of the concrete ones are less obvious.

A necessary condition for solution—that there be as many independent equations as paths—is met. This is not, in general, a sufficient condition since a system may be undetermined in one part and overdetermined in another. In the above case, however, the unknown path coefficients and correlation

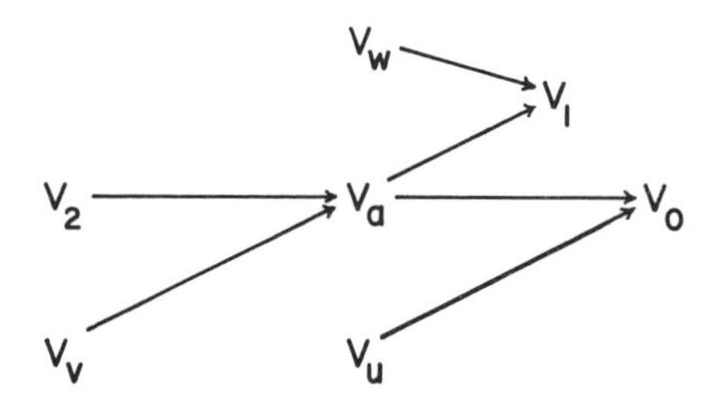

FIG. 13.6. (Redrawn from Fig. 5, Wright 1960*a*.)

coefficients can be obtained in succession from the known correlations on the left.

The hypothesis (*c*) that inconsistency of the first three equations in 13.16 is due to errors of measurement of V_1 represented in Figure 13.6, in which true V_1 is represented by V_a, and observed V_1 is represented as determined by V_a and errors of measurement V_w. Since $V_1 = V_a + V_w$, two of the concrete coefficients are given directly, $c_{1a} = 1$, $c_{1w} = 1$. The corresponding path coefficients are $p_{1a} = \sigma_a/\sigma_1$, $p_{1w} = \sigma_w/\sigma_1$.

	Standardized Coefficients	Concrete Coefficients $(c_{1a} = c_{1w} = 1)$
(13.18)	$r_{01} = p_{0a}p_{1a}$	$b_{01} = c_{0a}c_{1a}\,\sigma_a^2/\sigma_1^2 = c_{0a}\,\sigma_a^2/\sigma_1^2$
	$r_{12} = p_{1a}p_{a2}$	$b_{12} = c_{1a}c_{a2} = c_{a2}$
	$r_{02} = p_{0a}p_{a2}$	$b_{02} = c_{0a}c_{a2}$
	$r_{00} = 1 = p_{0a}^2 + p_{0u}^2$	$\sigma_0^2 = c_{0a}^2\sigma_a^2 + (c_{0u}^2\sigma_u^2)$
	$r_{11} = 1 = p_{1a}^2 + p_{1w}^2$	$\sigma_1^2 = c_{1a}^2\sigma_a^2 + c_{iw}^2\sigma_w^2 = \sigma_a^2 + \sigma_w^2$
	$r_{aa} = 1 = p_{a2}^2 + p_{av}^2$	$\sigma_a^2 = c_{a2}^2\sigma_2^2 + (c_{av}^2\sigma_v^2)$

The six path coefficients can readily be determined from the standardized equations and also $\sigma_a = p_{1a}\sigma_1$ and $\sigma_w = p_{1w}\sigma_1$. The analysis in terms of concrete coefficients is again encumbered with variances, some known or determinable and two undeterminable. There is, of course, indeterminacy if both hypothesis (*b*) and (*c*) are assumed.

Unnecessary encumbrance with variances occurs wherever two variables trace to a third. The simplest case is shown in Figure 13.7.

	Standardized Coefficients	Concrete Coefficients
(13.19)	$r_{13} = p_{13}$	$b_{13} = c_{13}$
	$r_{23} = p_{23}$	$b_{23} = c_{23}$
	$r_{12} = p_{13}p_{23}$	$b_{12} = c_{13}c_{23}\,\sigma_3^2/\sigma_2^2$
	$r_{11} = 1 = p_{13}^2 + p_{1u}^2$	$\sigma_1^2 = c_{13}^2\sigma_3^2 + (c_{1u}^2\sigma_u^2)$
	$r_{22} = 1 = p_{23}^2 + p_{2v}^2$	$\sigma_2^2 = c_{23}^2\sigma_3^2 + (c_{2v}^2\sigma_v^2)$

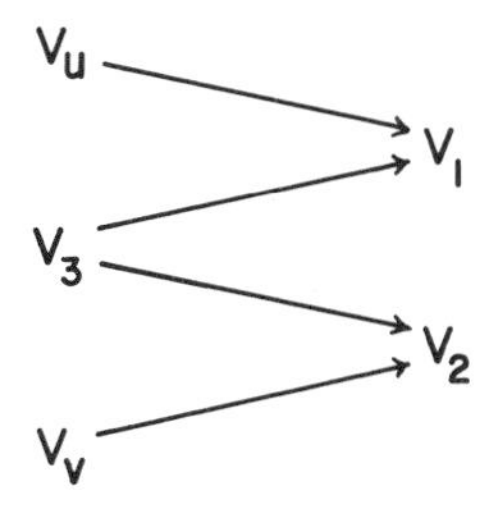

FIG. 13.7. (Redrawn from Fig. 6, Wright 1960*a*.)

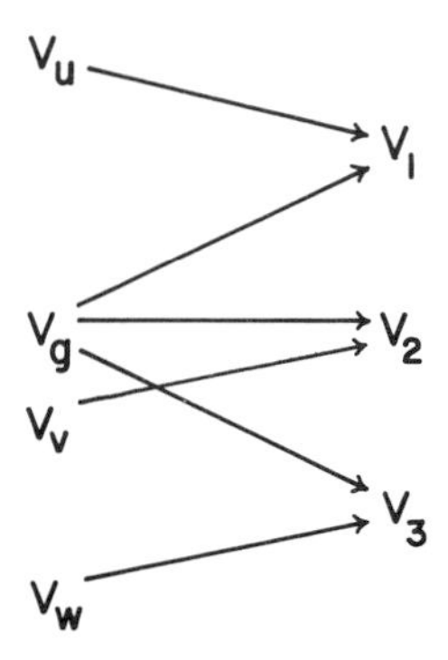

FIG. 13.8. (Redrawn from Fig. 8, Wright 1960*a*.)

There is overdetermination of two of the paths. Again a compromise solution may be obtained or the diagram may be revised to indicate a possible correlation between the two residuals or revised to indicate that V_3, at least as measured, does not fully represent the common factor of V_1 and V_2. Figure 13.8 illustrates a case in which concrete coefficients cannot be used at all because of ignorance of a key variance.

$$(13.20) \quad \begin{cases} r_{12} = p_{1g}p_{2g} \\ r_{13} = p_{1g}p_{3g} \\ r_{23} = p_{2g}p_{3g} \\ r_{11} = 1 = p_{1g}^2 + p_{1u}^2 \\ r_{22} = 1 = p_{2g}^2 + p_{2v}^2 \\ r_{33} = 1 = p_{3g}^2 + p_{3w}^2 \end{cases}$$

The six equations permit solution for the six path coefficients. This is a simple example of the pattern of conventional factor analysis with one general factor, as proposed by Spearman (1904*b*). This subject will be considered later.

Linear Functions

The most direct application of path analysis is to determine the correlation coefficient between two linear functions V_s, V_t of variables, V_i, V_j, etc., of which some or all are in common.

$$(13.21) \quad \begin{cases} V_s = c_s + \sum^m c_{si}V_i \\ V_t = c_t + \sum^m c_{ti}V_i \end{cases}$$

(13.22)
$$\begin{cases} \sigma_s^2 = \sum c_{si}^2\sigma_i^2 + 2\sum^m c_{si}c_{sj}r_{ij}, & j > i \\ \sigma_t^2 = \sum c_{ti}^2\sigma_i^2 + 2\sum^m c_{ti}c_{tj}r_{ij}, & j > i \end{cases}$$

(13.23)
$$\begin{cases} p_{si} = c_{si}\,\sigma_i/\sigma_s \\ p_{ti} = c_{ti}\,\sigma_i/\sigma_s \end{cases}$$

(13.24)
$$r_{st} = \sum p_{si}p_{ti} + 2\sum p_{si}r_{ij}p_{tj}, \qquad j > i$$

This simplifies considerably in the important case in which the components are uncorrelated. There is further simplification if the components are all equally variable ($\sigma_1^2 = \sigma_2^2 = \sigma_m^2 = \sigma^2$). For example, suppose that V_s is the sum of k such components, V_t is the sum of l components, and g of them are in common.

(13.25)
$$\sigma_s^2 = k\sigma_1^2 \qquad \sigma_t^2 = l\sigma_1^2$$

(13.26)
$$p_{si} = \sigma_i/\sigma_s = \sqrt{1/k} \qquad p_{ti} = \sigma_i/\sigma_t = \sqrt{1/l}$$

(13.27)
$$r_{st} = g/\sqrt{kl}$$

If $k = l$, $r_{st} = g/k$, which gives the well-known interpretation of the correlation coefficient as the proportion of common equally important factors.

In dealing with purely mathematical relationships, the representation of certain variables as dependent and others as independent is, of course, wholly arbitrary. If $V_s = V_1 + V_2$, one may equally well represent V_1 as a linear function of V_s and V_2, or V_2 as a linear function of V_s and V_1. Two or more such systems may be combined without leading to inconsistency.

Let $V_s = V_1 + V_2$ and $\sigma_1 = \sigma_2$, $r_{12} = 0$. Then

(13.28)
$$p_{s1} = p_{s2} = r_{s1} = r_{s2} = \sqrt{\tfrac{1}{2}}$$

(13.29)
$$\sigma_s^2 = 2\sigma_1^2$$

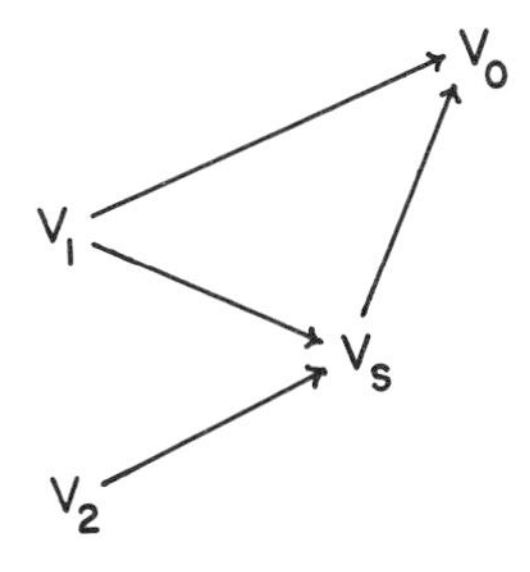

FIG. 13.9. (Redrawn from Fig. 6, Wright 1923.)

Let $V_0 = V_s - V_1$, then $c_{0s} = 1$, $c_{01} = -1$.

$$\sigma_0^2 = \sigma_s^2 + \sigma_1^2 - 2\sigma_s\sigma_1 r_{s1} = \sigma_1^2 = \sigma_2^2 \tag{13.30}$$

$$p_{01} = c_{01}\,\sigma_1/\sigma_0 = -1 \tag{13.31}$$

$$p_{0s} = c_{0s}\,\sigma_s/\sigma_0 = \sqrt{2} \tag{13.32}$$

$$r_{01} = p_{01} + p_{0s}p_{s1} = 0 = r_{12} \tag{13.33}$$

$$r_{0s} = p_{0s} + p_{01}r_{s1} = \sqrt{1/2} = r_{2s} \tag{13.34}$$

$$r_{02} = p_{0s}p_{s2} = 1 = r_{22} \tag{13.35}$$

Thus V_0 comes out identical with V_2 in variance and all of its correlations. It is to be noted, however, that p_{0s} does not equal p_{s2} as these are functions of the relations in Figure 13.9. This case also illustrates the point that a path coefficient may exceed 1.

As another example of the direct use of path coefficients to estimate correlations, consider the estimation of the true correlation between two variables from the correlation between measurements that are subject to error (Spearman's [1904*a*] correction for attenuation) (Fig. 13.10).

$\bar{A}$ = mean of m measurements of A
$\bar{B}$ = mean of n measurements of B
$r_{AA'}$ = average correlation between measurements of A
$r_{BB'}$ = average correlation between measurements of B
r_{AB} = average correlation between measurements of A and B

$$\sum^{m} p_{\bar{A}A}r_{\bar{A}A} = mp_{\bar{A}A}^2\,[1 + (m - 1)r_{AA'}] = 1 \tag{13.36}$$

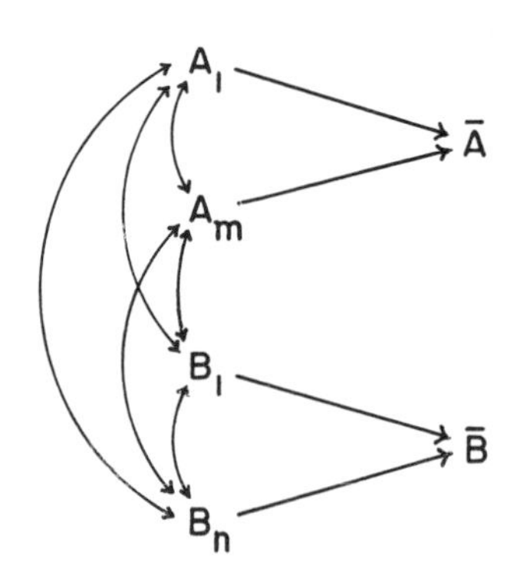

FIG. 13.10. (Redrawn from Fig. 4, Wright 1934.)

$$(13.37) \qquad \sum^{n} p_{\bar{B}B} r_{\bar{B}B} = n p^2_{\bar{B}B} [1 + (n-1) r_{BB'}] = 1$$

$$(13.38) \qquad \begin{cases} p^2_{\bar{A}A} = \dfrac{1}{m[1 + (m-1) r_{AA'}]} \qquad p^2_{\bar{B}B} = \dfrac{1}{n[1 + (n-1) r_{BB'}]} \end{cases}$$

$$(13.39) \qquad r_{\bar{A}\bar{B}} = mn p_{\bar{A}A} r_{AB} p_{B\bar{B}} = r_{AB} \sqrt{\frac{mn}{[1 + (m-1) r_{AA'}][1 + (n-1) r_{BB'}]}}$$

With indefinitely large m and n, $r_{\bar{A}\bar{B}} = r_{AB}/\sqrt{r_{AA'} r_{BB'}}$.

This result can be reached more easily from a system in which the direction of the arrows is the reverse of that in Figure 13.10. Assume that the observed values of A are determined by the true value $\bar{A}$ and by independent errors of observation; and that values of B are similarly determined (Fig. 13.11).

$$(13.40) \qquad r_{AA'} = p^2_{A\bar{A}} \qquad \therefore p_{A\bar{A}} = \sqrt{r_{AA'}}$$

$$(13.41) \qquad r_{BB'} = p^2_{B\bar{B}} \qquad \therefore p_{BB'} = \sqrt{r_{BB'}}$$

$$(13.42) \qquad r_{AB} = p_{A\bar{A}} r_{\bar{A}\bar{B}} p_{B\bar{B}} \qquad \therefore r_{\bar{A}\bar{B}} = r_{AB}/\sqrt{r_{AA'} r_{BB'}}$$

Inter- and Intraclass Correlations

Figure 13.10 can be used to answer a different question, the calculation of the correlation, r_{AB}, between members of L paired classes (one class with m, the other with n members per class) without actually tabulating the Lmn entries.

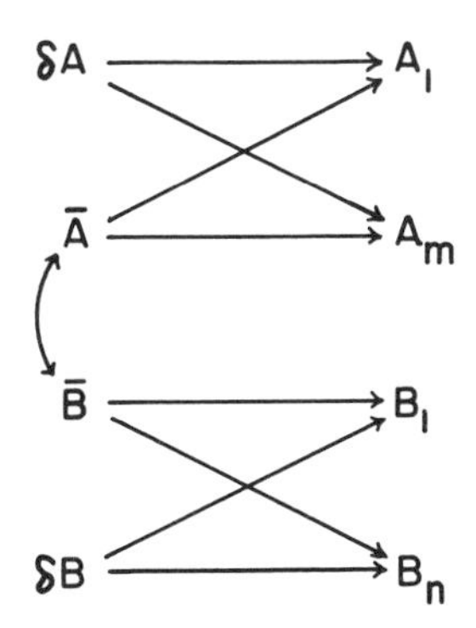

FIG. 13.11. (Redrawn from Fig. 5, Wright 1934.) δA's, δB's all independent.

It is to be noted that

$$\sigma^2_{\bar{A}} = \frac{1}{m^2}\,\sigma^2_{\Sigma A} = \left[\frac{1 + (m-1)r_{AA'}}{m}\right]\sigma^2_A$$

and

$$\sigma^2_{\bar{B}} = \frac{1}{n^2}\,\sigma^2_{\Sigma B} = \left[\frac{1 + (n-1)r_{BB'}}{n}\right]\sigma^2_B.$$

Thus

$$r_{AB} = \frac{r_{\bar{A}\bar{B}}\sigma_{\bar{A}}\sigma_{\bar{B}}}{\sigma_A\sigma_B} \quad \text{from (13.39).} \tag{13.43}$$

The standard error on the hypothesis of no correlation is thus $1/\sqrt{Lmn}$. The average correlation between parents ($m = 2$) and offspring in families of size n is an example of interclass correlation of this sort.

The intraclass correlation such as that among offspring of the same family is closely related (Fig. 13.12).

Assume n members of the class and L classes. There are $\frac{1}{2}n(n-1)$ different pairs of members per class.

$$\text{(13.44)}\quad \begin{cases} np_{\bar{A}A}r_{\bar{A}A} = 1 \qquad p_{\bar{A}A} = 1/(nr_{\bar{A}A}) \\ r_{\bar{A}A} = p_{\bar{A}A}[1 + (n-1)r_{AA'}] = [1 + (n-1)r_{AA'}]/(nr_{\bar{A}A}) \\ r_{AA'} = [nr^2_{\bar{A}A} - 1]/(n-1) \end{cases}$$

These agree with the formulas for inter- and intraclass correlation given by Harris (1913).

The standard error $(1 - r^2)/\sqrt{\frac{1}{2}\sum n(n-1)}$ should be multiplied by

$$\sqrt{\frac{1 + (n-1)r}{1 + r}}$$

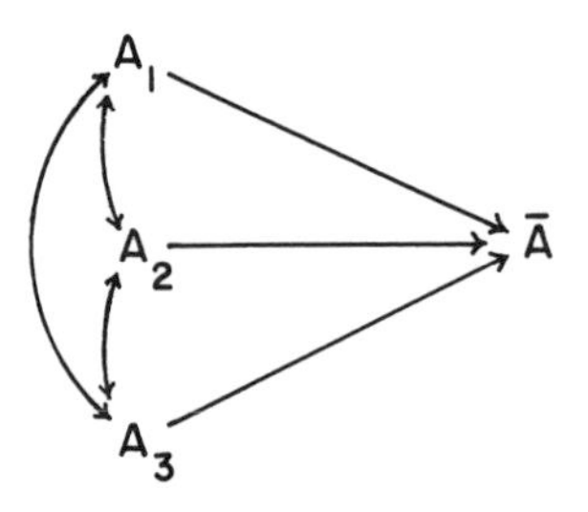

FIG. 13.12

to allow for repetition of individuals where n is greater than 2. The distribution is, however, so far from normal that tests of significance are best made by testing the significance of the variance of means.

Correlations within Subclasses

The interclass correlation above is not to be confused with the correlation between different characters of individuals of a subclass of a population. In Figure 13.13, A and B are the two characters, $\bar{A}$ and $\bar{B}$ are the subgroup means and δ_A and δ_B are the deviations from these means. There can be no correlation between $\bar{A}$ and δA or between $\bar{B}$ and δB. The desired correlation within subclasses is $r_{\delta A\delta B}$.

$$r_{AB} = p_{A\bar{A}} r_{\bar{A}\bar{B}} p_{B\bar{B}} + p_{A(\delta A)} r_{\delta A\delta B} p_{B(\delta B)}.$$

But

$$p^2_{A\bar{A}} = \sigma^2_{\bar{A}}/\sigma^2_A, \qquad p^2_{A(\delta A)} = \sigma^2_{\delta A}/\sigma^2_A,$$
$$p^2_{B\bar{B}} = \sigma^2_{\bar{B}}/\sigma^2_B, \qquad p^2_{B(\delta B)} = \sigma^2_{\delta B}/\sigma^2_B.$$

Thus

$$(13.45) \qquad r_{AB}\sigma_A\sigma_B = r_{\bar{A}\bar{B}}\sigma_{\bar{A}}\sigma_{\bar{B}} + r_{\delta A\delta B}\sigma_{\delta A}\sigma_{\delta B},$$

$$(13.46) \qquad r_{\delta A\delta B} = (r_{AB}\sigma_A\sigma_B - r_{\bar{A}\bar{B}}\sigma_{\bar{A}}\sigma_{\bar{B}})/\sigma_{\delta A}\sigma_{\delta B}.$$

This principle was illustrated in chapter 12 by the correlation between birth weight and year-weight of guinea pigs within inbred strains.

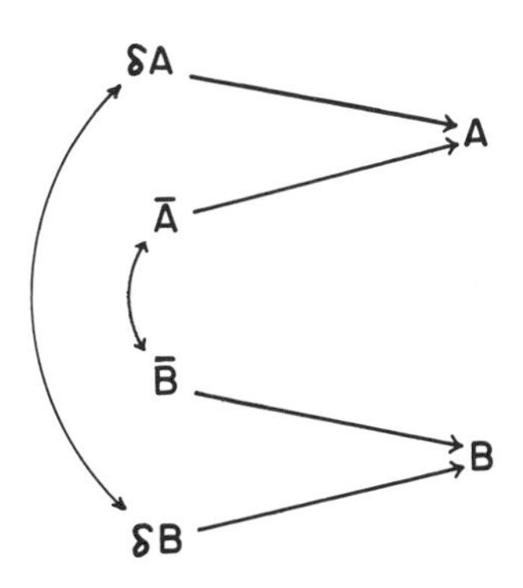

FIG. 13.13

Partial Correlation

A special case of subgroup correlations is that in which the subdivision is on the basis of one or more characters other than those correlated. Assuming linear relations, the correlation $r_{A(C)B(C)}$ between the contributions of given values of a third variable, C, to variables A and B, is 1.

$$r_{AB}\sigma_A\sigma_B = \sigma_{A(C)}\sigma_{B(C)} + r_{AB\cdot C}\sigma_{A\cdot C}\sigma_{B\cdot C}.$$

But

$$\sigma_{A(C)} = \sigma_A r_{AC}, \qquad \sigma_{A\cdot C} = \sigma_A\sqrt{1 - r_{AC}^2},$$
$$\sigma_{B(C)} = \sigma_B r_{BC}, \qquad \sigma_{B\cdot C} = \sigma_B\sqrt{1 - r_{BC}^2}.$$

(13.47)
$$r_{AB\cdot C} = \frac{r_{AB} - r_{AC}r_{BC}}{\sqrt{(1 - r_{AC}^2)(1 - r_{BC}^2)}}.$$

This is Pearson's (1896) coefficient of partial correlation. Additional variables may be held constant by repetition of the process. The standard error is the same as that for the correlation coefficient, $\mathrm{SE}_r = (1 - r^2)/\sqrt{n - 1}$, except that the number of variables held constant should be subtracted from $n - 1$ in the formula.

Imposed Constancy of Variables

It is interesting to determine how the properties of a system of multiple variables change in other respects than that just considered, within subpopulations in which one or more variables are constant. Consider the case of one dependent variable, V_0, that is a linear function of a number of others, $V_1, V_2, \cdots, V_m$, and independent residual variability, V_u.

(13.48)
$$X_0 = p_{01}X_1 + p_{02}X_2 + \cdots + p_{0m}X_m + p_{0u}X_u$$

where $X_0 = (V_0 - \bar{V}_0)/\sigma_0$, etc. The absolute value of p_{01} gives the fraction of the standard deviation of V_0 for which V_1 is directly responsible as the fraction to be expected if V_1 is made to vary as much as in the observed data, although all other independent factors (including the residual) are constant. In the following formulas, all subscripts following a dot are assumed to be constant. The effect of constancy of variables V_2 to V_m and V_u on V_1 through their correlation with the latter is obviated by multiplication by the ratio $\sigma_1/\sigma_{1\cdot 2\cdots m,u}$.

(13.49)
$$|p_{01}| = \frac{\sigma_1\sigma_{0\cdot 2\cdots m,u}}{\sigma_0\sigma_{1\cdot 2\cdots m,u}}.$$

(13.50) $$|c_{01}| = |p_{01}| \frac{\sigma_0}{\sigma_1} = \frac{\sigma_{0\cdot 2\cdots m,u}}{\sigma_{1\cdot 2\cdots m,u}}.$$

(13.51) $$\sigma_{0(1)} = |p_{01}|\,\sigma_0 = |c_{01}|\,\sigma_1 = \sigma_1 \frac{\sigma_{0\cdot 2\cdots m,u}}{\sigma_{1\cdot 2\cdots m,u}} \quad \text{from (13.49).}$$

(13.52) $$\sigma_{0\cdot 1\cdots m} = \sigma_u = \sigma_0\sqrt{1 - r^2_{0(1\cdots m)}} \quad \text{where } r^2_{0(1\cdots m)} = \sum^m p_{0i}r_{0i}.$$

(13.53) $$\sigma_{0(1)\cdot 2\cdots k} = \sigma_{0(1)}\sigma_{1\cdot 2\cdots k}/\sigma_1 \quad k \le m \quad \text{from (13.51)}$$

by replacing σ_1 by $\sigma_{1\cdot 2 \ldots k}$ and fraction by $\sigma_{0(1)}/\sigma_1$.

(13.54) $$p_{01\cdot 2\cdots k} = p_{01}\frac{\sigma_0}{\sigma_1}\frac{\sigma_{1\cdot 2\cdots k}}{\sigma_{0\cdot 2\cdots k}} \quad k \le n \quad \text{from (13.49).}$$
$$= p_{01}\sqrt{\frac{1 - r^2_{1(2\cdots k)}}{1 - r^2_{0(2\cdots k)}}} \quad \text{from (13.52) and (13.54).}$$

(13.55) $$c_{01\cdot 2\cdots k} = c_{01} \quad \text{from (13.50).}$$

The concrete partial regression coefficients are, as expected, unaffected on the average by selecting data in which variables other than that to which the dependent variable is directly related are constant. As shown by (13.54), this is not the case with the abstract path coefficients.

The cases in which only one variable is held constant are especially important.

(13.56) $$\sigma_{0\cdot 1} = \sigma_0\sqrt{(1 - r^2_{01})}.$$

(13.57) $$\sigma_{0(1)\cdot 2} = \sigma_{0(1)}\sqrt{1 - r^2_{12}} \quad \text{from (13.53).}$$

(13.58) $$p_{01\cdot 2} = p_{01}\sqrt{\frac{1 - r^2_{12}}{1 - r^2_{02}}} \quad \text{from (13.54).}$$

In a system in which a variable V_0, is treated as determined by two correlated variables and an independent residual,

(13.59) $$r_{01\cdot 2} = p_{01\cdot 2} = p_{01}\sqrt{\frac{1 - r^2_{12}}{1 - r^2_{02}}} = \frac{r_{01} - r_{02}r_{12}}{\sqrt{(1 - r^2_{02})(1 - r^2_{12})}}.$$

This agrees with the earlier derivation of the partial correlation coefficient (13.47).

Path Coefficients and Probability

If a variable, V_0, consists merely of one of a group of alternatives, with constant values $C_1, C_2, \cdots, C_m$, drawn with probabilities $P_1, P_2, \cdots, P_m$ respectively, these probabilities may be treated as frequencies.

(13.60)
$$\overline{V}_0 = \sum^{m} C_i P_i .$$

(13.61)
$$\sigma_0^2 = \sum^{m} (C_i - \overline{V}_0)^2 P_i .$$

Assume, however, that V_0 is drawn with probabilities $P_1, P_2, \cdots, P_m$ from a group of *variable* alternatives $V_1, V_2, \cdots, V_m$ with momentary deviations from their means, $\delta V_1, \delta V_2, \cdots, \delta V_m$, respectively. These variables may have different variances, σ_i^2, etc., and may be correlated (r_{12}, etc.).

(13.62)
$$\overline{V}_0 = \sum P_i \overline{V}_i , \qquad \overline{\delta V_0} = 0 .$$

The best estimate of the deviation of V_0 is $E(\delta V_0) = \sum^m P_i(\delta V_i)$. The probabilities appear here as path regressions $c_{0i} = P_i$. The corresponding path coefficients may accordingly be written $p_{01} = P_1 \sigma_1/\sigma_0$. The identification of probabilities with path regressions in general, and with path coefficients where all contributing variables have the same variance is useful in analyzing the consequence of mating systems.

Nonlinear Relations

So far it has been assumed that all relations in a system to be analyzed are linear. In some cases nonlinear relations may be removed by systematic transformation of one or more of the variables. If, for example, one variable is a product of two or three others, linearity is obtained by replacing all three by their logarithms, but if transformed in this relation they must be transformed in all others in the system.

Deviation from linearity in relation to an ultimate variable that otherwise has no correlations in the system merely causes the influence of this factor to appear smaller than it actually is. This correlation could indeed be replaced by the correlation ratio.

Deviation from linearity in relation to a variable that is a common factor or an intermediary is more serious but, if slight, does not prevent approximate analysis. In some cases a nonlinear function, $V_0 = f(V_1, V_2, \cdots, V_m)$, may be represented accurately enough by the first-order terms of an expansion by Taylor's theorem. Letting $\delta V_u = (V_0 - \overline{V}_0)$, etc.,

(13.63)
$$\delta V_0 = \frac{\partial V_0}{\partial V_1} \delta V_1 + \frac{\partial V_0}{\partial V_2} \delta V_2 + \cdots + \frac{\partial V_0}{\partial V_m} \delta V_m + R .$$

Thus a product of two uncorrelated variables may be treated as approximately additive, $\delta(XY) = \overline{Y}\,\delta X + \overline{X}\,\delta Y$, if the coefficients of variability, $\sigma_X/\overline{X}$ and $\sigma_Y/\overline{Y}$, are small. Where these are equal, the fraction of the variance of XY that is excluded is less than half the squared coefficient of variability.

There are important cases in genetics of joint determination by uncorrelated variables (deviations due to dominance or interaction) that can be treated exactly. These will be considered in Part II in connection with the correlation between relatives.

Multiple Regression

In its simplest inverse application, the method of path coefficients reduces to that of multiple regression. Variable V_0 is treated as a linear function of a number of others, V_1 to V_m, and residual variability, V_u. The correlations between V_0 and the known variables may be written as a series of simultaneous linear equations, equal in number to the unknown path coefficients.

$$(13.64) \qquad \begin{cases} r_{01} = p_{01} + r_{12}p_{02} + \cdots + r_{1m}p_{0m} \\ r_{02} = r_{12}p_{01} + p_{02} + \cdots + r_{2m}p_{0m} \\ r_{0m} = r_{1m}p_{0m} + r_{2m}p_{0m} + \cdots + p_{0m}. \end{cases}$$

These are the normal equations of the method of least squares with all variables measured in terms of their standard deviations. The p_{0i}'s, found by solution, merely need to be multiplied by σ_0/σ_i to give the concrete partial regression coefficients c_{0i}.

$$(13.65) \qquad \sum^{m} p_{0i}r_{0i} + r_{0u}^2 = 1 \quad \text{same as 13.6}$$

$$(13.66) \qquad \begin{aligned} \sum p_{0i}r_{0i} &= r_{0(12\cdots m)}^2 \text{ squared multiple correlation,} \\ \sigma_0^2(1 - \sum p_{0i}r_{0i}) &= \sigma_0^2 r_{0u}^2 \text{ squared error of estimate.} \end{aligned}$$

The solution was expressed by Yule in terms of the following determinant and its minors.

$$\Delta = \begin{vmatrix} 1 & r_{01} & \cdots & r_{0m} \\ r_{01} & 1 & \cdots & r_{1m} \\ \vdots & \vdots & \vdots\vdots\vdots & \vdots \\ r_{0m} & r_{1m} & \cdots & 1 \end{vmatrix}.$$

The multiple regression equation is

$$(13.67) \qquad \frac{V_{0(E)} - V_0}{\sigma_0} = \frac{\Delta_{01}}{\Delta_{00}}\left(\frac{V_1 - \overline{V}_1}{\sigma_1}\right) - \frac{\Delta_{02}}{\Delta_{00}}\left(\frac{V_2 - \overline{V}_2}{\sigma_2}\right) + \cdots + (-1)^{K+1}\frac{\Delta_{0K}}{\Delta_{00}}\left(\frac{V_K - \overline{V}_K}{\sigma_K}\right),$$

and

(13.68)
$$r_{0(12\cdots m)} = \sqrt{1 - \frac{\Delta}{\Delta_{00}}}$$

is the coefficient of multiple correlation.

The method of path coefficients is used here merely as a convenient way of arriving at these expressions which are usually derived in other ways. Conversely, the method of path coefficients may itself be looked upon as the algebraic use of standardized partial regression coefficients in systems of relations more complicated in general, than the system involved in finding the best linear estimate of one variable in terms of others.

The symmetrical pattern of relations in this case lends itself to geometric representation. Orthogonal axes are assigned each variable, and thus the joint values are located at points in a multidimensional space. The "surfaces" of equal frequency tend to be concentric multidimensional "ellipsoids" in the ideal case in which the distribution can be described adequately as multivariate normal. The multiple regression equation for any given variable corresponds to the hypoplane that includes the estimated mean values of that variable for given values of the others.

Vectorial Representation

A wholly different geometrical representation of a system of multiple correlations was introduced by Pearson (1900). It would be instructive to be able to represent variables by points in a model in such a way that ones that are strongly correlated form clusters while those that are correlated weakly, if at all, are far apart. This can readily be done for all variables, V_i that are linear functions of two reference variables, X_1 and X_2. Assume that all are measured from their means in standardized form.

(13.69)
$$X_i = p_{i1}X_1 + p_{i2}X_2 .$$

(13.70)
$$p_{i1}^2 + p_{i2}^2 + 2p_{i1}p_{i2}r_{12} = 1 .$$

If X_1 and X_2 are independent,

(13.71)
$$X_i = r_{i1}X_1 + r_{i2}X_2 ,$$

(13.72)
$$r_{i1}^2 + r_{i2}^2 = 1 .$$

Thus if X_1 and X_2 are measured along perpendicular axes, and X_i is represented by a vector from the origin, it may be assigned a direction by identifying r_{i1} with the cosine of the angle with X_1 and r_{i2} with sine of this angle (or cosine of the angle with X_2).

If we choose to represent the variable as an entity apart from its particular values, by a vector of unit length, those that are linear functions of X_1 and X_2 all terminate on a circle of unit radius. Ones with zero correlations with each other are 90° apart and those with a correlation of -1 are 180° apart, but can be brought to identity by reversing the sign of one. The arc "distance," (ij), between two variables, X_i and X_j, can be deduced from the equation $r_{ij} = p_{i1}p_{j1} + p_{i2}p_{j2}$.

(13.73) $\cos(ij) = \cos(i1)\cos(j1) + \sin(i1)\sin(i1)$, as expected.

It may be noted that if X_1 and X_2 are not independent, p_{i1} and p_{i2} are still the vector components of X_i. The equation expressing complete determination, 13.70, corresponds to the formula for the length of one side of a triangle in terms of the other sides and the included angle.

Consider next the set of all variables that are linear functions of three independent variables, X_1, X_2, and X_3.

$$X_i = r_{i1}X_1 + r_{i2}X_2 + r_{i3}X_3. \tag{13.74}$$

$$r_{i1}^2 + r_{i2}^2 + r_{i3}^2 = 1. \tag{13.75}$$

The locus of the ends of the unit vectors representing the variables of this set is the surface of a sphere. The distance between any two is still $\cos^{-1} r_{ij}$ and can be deduced from the formula, $r_{ij} = r_{i1}r_{j1} + r_{i2}r_{j2} + r_{i3}r_{j3}$, which translates into the distance formula of spherical trigonometry.

The coefficient of partial correlation,

$$r_{ij\cdot k} = \frac{r_{ij} - r_{ik}r_{jk}}{\sqrt{(1 - r_{ik}^2)(1 - r_{jk}^2)}}$$

translates into the cosine of the dihedral angle at k in the spherical triangle formed by the great circle arcs (ij), (ik), and (jk). The partial correlation is thus the cosine of the angle between the components of X_i and X_j on the plane perpendicular to the constant factor X_k.

The set of variables that are linear functions of four independent variables X_1, X_2, X_3, and X_4 are represented by vectors from the origin to points on a unit hypersphere defined by $r_{i1}^2 + r_{i2}^2 + r_{i3}^2 + r_{i4}^2 = 1$. It is no longer possible to construct a physical model that accurately represents their relations. But just as the relations in the preceding case can be represented in a curved two-dimensional space of which a distorted flat model can be made in the form of two circular areas (such as maps of the Eastern and Western hemispheres), so in this case the relations in a curved three-dimensional space can be represented in distorted form by points in two solid spheres with one-to-one correspondence of points on the surfaces. The true figure is a solid with a definite volume, $2\pi^2$, but no bounding surface (a

Riemann space), since all variables can equally well be the centers of clusters of correlated variables. A variable that is represented at the north pole (of both model spheres) is uncorrelated with any of the variables that are represented in the planes of the two equators, and it has a correlation of -1 with the variable at the south pole of both spheres. The variable in question could equally well be represented as at the center of a model sphere, and thus as uncorrelated with all variables represented as on the surfaces of the spheres, and with a correlation of -1 with the variable at the center of the counter sphere. The correlation between a variable, X_i, related to another, X_k, only through an intermediate, X_j, which in turn is related to X_l only through X_k,($X_l \rightarrow X_k \rightarrow X_j \rightarrow X_i$), is given by $r_{il} = r_{ij}r_{jk}r_{kl}$, which translates into $\cos(il) = \cos(ij)\cos(jk)\cos(kl)$. This defines the distances between X_i and X_l in terms of "distance" in three orthogonal directions (ij), (jk), and (kl). For a variable located at the origin of three orthogonal axes at the center of its model sphere, the "distance" to any other variable is defined by this rule in which the coordinates all have the properties of great circles. If the "distances" are all small (r_{il} close to 1) the squared distance, which can be written $[1 - \sin^2(il)] = [1 - \sin^2(ij)][1 - \sin^2(jk)][1 - \sin^2(kl)]$, becomes approximately $(il)^2 = (ij)^2 + (jk)^2 + (kl)^2$. Thus the law of distance becomes approximately Pythagorean at small distances, and the amount of distortion in representing the variables in a cluster in ordinary Euclidian space becomes negligible.

With functions of more than four independent variables, it no longer becomes possible to represent relations even in a distorted three-dimensional model. It can merely be stated that all variables may be located on the surface of the appropriate unit hypersphere with a law of distance analogous to that above.

It is obvious that, in general, m variables require an m-dimensional hypersphere for location as vectors. The m reference variables, $(X_1, X_2, \cdots, X_m)$ or common factors, returning to the path diagram, may be chosen in an infinite number of ways since any set of m orthogonal axes will do. The path coefficients (or factor loadings or components of factor analysis) are the projections on the axes and are equal to the correlations with the factors in question if, as assumed here, the latter are independent of each other.

It is desirable, where vectorial representation is the object, to have some unique way of choosing a set of factors. Several methods are, however, in use. The most satisfactory convention seems to be Hotelling's, by which one locates first the axis on which the sum of the squared projections is maximum, next the axis at right angles to this on which the squared projections account for as much as possible of the residuum and so on until nothing is left (at most m axes for m variables).

Hotelling (1936) has shown how to derive these projections by a simple iterative process. Let G be the common factor under consideration. The correlation coefficients (including the self-correlations $r_{ii} = 1$) are added in each row (i's), Because of the diagonal symmetry, the sums for the columns (j's) are the same, as is also true for the ratios $R_{i(1)}$ and $R_{j(1)}$ to the sum for a particular variable a.

(13.76) $$R_{i(1)} = \sum_{j=1}^{m} r_{ij} \Big/ \sum_{j=1}^{m} r_{aj}$$

(13.77) $$R_{i(2)} = \sum_{j=1}^{m} R_{j(1)} r_{ij} \Big/ \sum_{j=1}^{m} R_{j(1)} r_{aj}$$

The process is repeated (calculations $R_{i(3)}$, etc.) until stability is reached. When stability is reached, the components are calculated by the formulas.

(13.78) $$p_{aG} = \sqrt{\sum_{j=1}^{m} R_j r_{aj} \Big/ \sum_{j=1}^{m} R_j^2}$$

(13.79) $$p_{iG} = R_i p_{aG}$$

The estimated values, r_{ij}, are given by $p_{iG}p_{jG}$, and the corresponding residuals to be fitted by the next general factor are thus $r_{ij} - p_{iG}p_{jG}$.

Table 13.1 gives the means, standard deviations, coefficients of variability, and all possible correlations of six skeletal measurements in a flock of 276

TABLE 13.1. Mean standard deviation and coefficient of variability of length of skull (L), breadth of skull (B), and lengths of humerus (H), ulna (U), femur (F), and tibia (T), and the correlations in a population of 276 White Leghorn hens.

	Mean, mm	σ, mm	CV, %	Correlation					
				L	B	H	U	F	T
L	38.77	1.26	3.25	1	0.584	0.615	0.601	0.570	0.600
B	29.81	0.93	3.13		1	0.576	0.530	0.526	0.555
H	74.64	2.84	3.80			1	0.940	0.875	0.878
U	68.74	2.73	3.97				1	0.877	0.886
F	77.34	3.20	4.14					1	0.924
T	114.84	5.00	4.35						1

 Basic data from Dunn (1928).

White Leghorn hens (Dunn 1928, Wright 1932), which will be used to illustrate this and other methods. The components by Hotelling's method are shown in Table 13.2.

TABLE 13.2. Components of variables in White Leghorn population by Hotelling's method with complete apportionment of self-correlations to six factors.

	p_{x1}	p_{x2}	p_{x3}	p_{x4}	p_{x5}	p_{x6}	
L	+0.7426	+0.4536	+0.4922	−0.0205	+0.0073	−0.0007	
B	+0.6975	+0.5886	−0.4085	−0.0008	+0.0021	−0.0144	
H	+0.9477	−0.1583	−0.0259	+0.2182	+0.0472	+0.1623	
U	+0.9403	−0.2125	+0.0071	+0.2033	−0.0423	−0.1654	
F	+0.9287	−0.2351	−0.0382	−0.2139	+0.1841	−0.0323	
T	+0.9407	−0.1908	−0.0296	−0.1950	−0.1945	+0.0446	
$\sum p_{xt}^2$	4.5677	0.7141	0.4122	0.1731	0.0758	0.0569	0.59998
%	76.1	11.9	6.9	2.9	1.3	0.09	100.0

The components on the first axis account for 76.1% of the total variance, those on the second for 11.9% those on the third for 6.9%, and those on the remaining three for only 5.1% altogether. There is no simplification, in one sense, in the replacement of 15 observed correlation coefficients by 36 vector components. The analysis does, however, bring out clearly the pattern of relations on focusing on the components with absolute values greater than 0.05.

A variant of this mode of approach usually reduces the number of axes needed to account adequately for the correlations at the expense of an additional special axis for each separate variable. In this, the self-correlations and the correlations with the reference variable are omitted in the sums (Wright 1954).

$$R_{i(1)} = \frac{\sum_{j=1}^{m} r_{ij} - 1 - r_{ia}}{\sum_{j=1}^{m} r_{aj} - 1 - r_{ai}} \tag{13.80}$$

Note that $R_{a(1)} = 1$ as before.

The first estimate of p_{aG} is given by

$$p_{aG(1)} = \sqrt{\frac{\sum^{m(m-1)} r}{(\sum^{m} R)^2 - \sum^{m} R^2}} \tag{13.81}$$

and those for the other variables are given, as before, by

$$p_{iG} = R_{iG(1)} p_{aG(1)}. \tag{13.82}$$

The squared path coefficients as estimates of the self-correlations due to the general factor under this convention are then inserted in Table 13.1 and R's calculated iteratively until stability is reached. At this point the path coefficients are reestimated from the last R's.

$$p_{aG(2)} = \sqrt{\sum_{j=1}^{m} R_j r_{aj} / \sum R_j^2} \tag{13.83}$$

$$p_{xG(2)} = R_i p_{aG(2)}. \tag{13.84}$$

The components of the self-correlations due to the general factor ($p_{iG(2)}^2$) are inserted in place of the self-correlations and the whole process repeated until the path coefficients reach stability. If the changes in all p's continue to decrease without change of sign, the process can be expedited by extrapolation after two trials on the hypothesis of approach to stability in geometric progression. After the path coefficients for the first general factor have been obtained, the residual correlations may be analyzed for a second general factor, and so on until nothing appreciable is left.

Table 13.3 shows the results of such an analysis of the variables in the population of White Leghorn hens analyzed previously by Hotelling's method. It turns out that a common factor space of only three dimensions gives an adequate (although not mathematically complete) determination of the correlations. The average absolute error in calculating the correlations from the 18 coordinates is only 0.003. The portions of the variability determined by the common factors are 73.8%, 6.3%, and 1.8%, respectively, leaving 18.1% to the six special factors.

The geometrical pattern is more easily grasped with only three axes of common space (Fig. 13.14) than with six, but it should be noted that the factors are not of unit length because the special axes for each are omitted. The members of the three pairs of variables, BL, HU, and FT are thus not as close together in the total space as seems indicated in Figure 13.14, because of these omissions. The total nine-dimensional space could be rotated in such a way as to give the components of the previous analysis in six dimensions. There is, of course, no justification for any physiological interpretation of the axes as factors in either case or in any other that may be chosen without bringing in external information.

TABLE 13.3. Components of variables in White Leghorn population by Hotelling's method, except for exclusion of self-correlations, the residuals of which are accounted for by a special factor for each.

	p_{x1}	p_{x2}	p_{x3}	p_{x4} to p_{x5}	
L	+0.685	+0.361	+0.007	+0.632	
B	+0.639	+0.409	−0.030	+0.651	
H	+0.951	−0.083	+0.162	+0.250	
U	+0.946	−0.152	+0.168	+0.232	
F	+0.930	−0.180	−0.166	+0.274	
T	+0.942	−0.124	−0.154	+0.271	
$\sum p_{xi}^2$	4.429	0.375	0.107	1.088	5.999
%	73.8	6.3	1.8	18.1	100.0

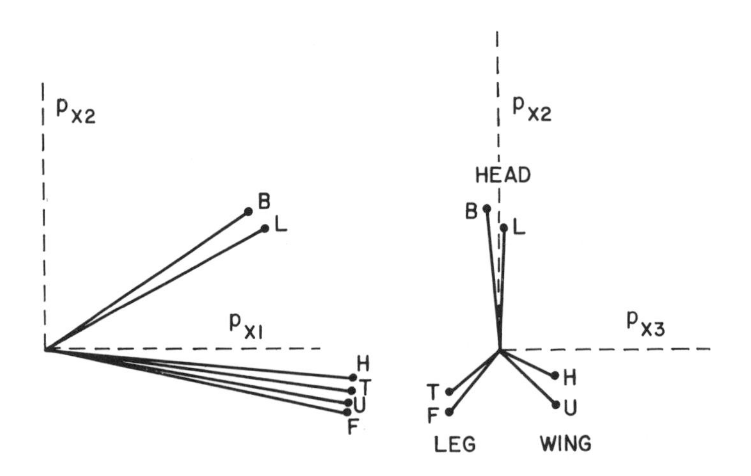

FIG. 13.14. (Redrawn from Fig. 2.1, Wright 1954.)

CHAPTER 14

Interpretation by Path Analysis

Although a correlation coefficient merely describes without explaining, the existence of the association that it describes raises the problem of interpretation. There is usually no difficulty in suggesting a plausible explanation of any single correlation. It is more difficult to interpret a system of correlations and be sure that there is consistency throughout.

The first systematic approach to this problem seems to have been that of Spearman (1904*b*) who interpreted the usually high positive correlations among various mental traits on the hypothesis of one common factor for general intelligence, supplemented by special factors for each, and devised means of testing this hypothesis. It became clear that a single general factor was not enough, and there has been a very extensive development of alternative methods by psychologists (Burt 1937; Thurstone 1935, 1947; Thompson 1939; Holtzinger and Harmon 1941).

I (1918) attempted to interpret the correlations in a set of bone measurements in a rabbit population from a related but somewhat different viewpoint. Factors were postulated for general size, for the size of homologous organs, for single organs as wholes, and for single measurements. The term path coefficients was first used in an analysis of the correlations among related guinea pigs with respect to white spotting (1920). The objective was stated in the first general account of the method (1921):

> The present paper is an attempt to present a method of measuring the direct influence along each separate path in such a system and thus of finding the degree to which variation of a given effect is determined by each particular cause. *The method itself depends on the combination of knowledge of degrees of correlation among the variables in a system with such knowledge as may be possessed of the causal relations.* In cases in which the causal relations are uncertain, the method can be used to find the logical consequences of any particular hypothesis in regard to them.

The sentence italicized above has been overlooked by certain critics of the method who have treated it as an attempt to deduce causal relations wholly from systems of correlations without supplementary information (cf. Niles

1922, and reply, Wright 1923). Actually, attempts at interpretation of extensive systems by the method, usually involves a working back and forth between tentative hypotheses based on external information and the set of correlations until as simple and reasonable a conclusion has been reached as the data permit. The method is not a mill from which interpretations can be arrived at automatically.

Another criticism has been on the use of the words "cause" and "effect" (Niles 1922). In using these words, it was assumed that any event traces back continuously in time and space through successions of previous events and that, statistically, variations in events of a given sort may be traced in principle to variations in previous ones of specified sorts (including joint effects of two or more) with varying degrees of importance, however difficult it may be in practice to disentangle such sequences from correlations through common factors or to deal with almost instantaneous reciprocal interactions. Practically, the method is largely limited to relations that can be treated as approximately linear, although limited extension can be made to joint action. The possibility of dealing with instantaneous reciprocal interaction will be discussed later. Purely mathematical relations such as the determination of a sum by its components may be included in an analysis.

There are various types of supplementary information. Relations in time and space are important. If there is sequence in time, there is at least a possibility of a cause-and-effect relation between earlier and later, while correlations in sets of contemporary measurements can usually only be interpreted in terms of hypothetical common factors. The ordering in space of morphological measurements gives these an advantage in interpretation over unordered ones such as mental tests. In some cases, experimental evidence may be used to designate particular factors and the relation of changes in them to those in other variables.

If such information is sufficiently complete, the interpretative description by path analysis largely gives way to sets of deterministic equations. Even in this case, however, a path analysis may be of value in providing an overall picture which brings out the relative importance of the various chains of causation in a particular population.

With respect to morphological correlations, many cases are known in which pairs of alleles distinguish individuals that are large or small in all of their parts, although not necessarily in the same proportion because of heterauxetic growth (Huxley 1932). Other loci are known that affect largely or exclusively particular parts or groups of homologous parts (such as eyes, wings, or bristles in *Drosophila*, ear size in mice, flower size in many plants, and so forth). All dimensions may be affected more or less alike, or the effect may be largely restricted to one dimension and thus greatly modify form (tail length in mice, number and length of internodes in many plants). In some cases, a gene replacement increases one dimension but reduces another with drastic effect on form. Thus in F_2 of a cross between disk- and sphere-shaped squashes, Sinnott (1935) reported one-factor segregants that differed profoundly in shape but not significantly in weight (Table 14.1).

TABLE 14.1. Means of segregating disks and spheres in F_2 of a cross between squashes with disk- and sphere-shaped fruits.

	No.	Weight (gm)	Length (cm)	Width (cm)
F_2 disks	147	1,036 ± 21	6.71 ± 0.06	16.80 ± 0.13
F_2 spheres	58	1,026 ± 31	13.78 ± 0.21	12.69 ± 0.14

From data of Sinnott (1935).

Occasionally there may be important pleiotropic effects on seemingly unrelated parts.

Nothing approaching a complete analysis of factor effects can be expected from any manipulation of a set of correlations, supplemented by any list of interpretative principles, without isolation of the actual factors. It becomes a matter of choosing a list that seems most generally applicable on the basis of experience in order to arrive at a convention for comparative interpretations.

The list of variables chosen for an analysis is itself important. They should be well balanced. In analyzing a morphological pattern there should not be a disproportionate number of measurements of one aspect. Measurements treated as coordinate in the analysis should be mathematically independent. Thus it is objectionable to include total length and tail length of a mammal as coordinate variables, but unobjectionable to treat body length and tail length as such variables and represent total length as determined by them. Length and width of squashes are independent measurements in the sense intended here. Length and ratio of width to length measure mathematically independent characteristics of size and shape, but are somewhat objectionable as coordinate variables because of the introduction of a tendency toward negative correlation due to factors that act on length alone (including errors). Weight and such a shape index are less objectionable, first because the measure of weight is not a function of the actual measurements from which a width/length ratio is obtained and second, the tendencies toward positive and negative correlations, due to the relation of weight to width and length, respectively, tend to cancel each other. A correlation observed in this case may be interpreted more safely as having a basis in growth physiology than if there were a mathematical functional relation.

The observed prevalence of factors, both genetic and environmental, that affect general size make it desirable, in interpreting sets of correlations among morphological measurements, to assume first that a general factor plays a major role, especially if, as is usually the case, all or nearly all of the correlations are positive. This suggests, as the first step, the determination of the path coefficients that minimize the squared residuals for all correlations except the self-correlations. This, however, results in a balancing of plus and minus residuals and this requires that there be factors that act in opposite senses on different variables like the shape-determining factors in squashes cited earlier. If all the measurements are of one closely integrated part, this may indeed give the best interpretation for purposes of comparison. It is unavoidable where there is a significant negative correlation. Table 14.2 shows the results from a cross between squashes in which one parent was heavier and flatter than the other (1-22) and a cross in which one parent was heavier and more elongated (3-22) (Sinnott 1935).

TABLE 14.2. Comparison of correlations in F_1 and F_2 of crosses between strains of squashes with fruits of different weights and forms.

Experiment		No.	Length vs. Width r	Weight vs. Form Index r
1–22	F_1	32	$+0.41 \pm 0.15$	—
1–23	F_2	392	-0.32 ± 0.05	-0.04 ± 0.06
3–22	F_1	118	$+0.62 \pm 0.05$	—
3–23	F_2	430	-0.41 ± 0.04	-0.05 ± 0.06

From data of Sinnott (1935).

The positive correlations in these and 13 other F_1's (av, + 0.45) reflect largely the differences in general size due to environmental condition, whereas the negative correlations in the corresponding F_2's (av, − 0.37) are due to overbalancing of the positive contributions from both environmental and segregating genetic factors for general size by the segregating shape-determining factors.

The correlations in Table 14.3, reported by Alpatov and Boschko-Stepanenko (1928) relate to the length of the 5 phalanges of digit IV of the domestic fowl (65–75 birds). All correlations are positive. Maximum path coefficients relating to a general factor, p_{iG} were calculated by Hotelling's method,

TABLE 14.3. Correlations between lengths of phalanges of digit IV in a population of domestic fowls, calculated path coefficients, residuals of self-correlations (in parentheses), residuals of other correlations (lower left in group, lower table), and partial correlations (upper right in group, lower table).

	r_{ij}			
	2	3	4	5
1	0.750			
2	0.665	0.711		
3	0.494	0.509	0.625	
4	0.385	0.337	0.502	0.604
5				

	p_{iG}	p_{iH}	p_{iS}	Residuals\Partial Correlations				
1	0.793	−0.194	0.578	(0.333)	0.01	0.01	−0.04	0.03
2	0.860	−0.345	0.378	0.001	(0.141)	−0.02	0.04	−0.02
3	0.840	+0.024	0.542	0.003	−0.003	(0.294)	0.01	−0.01
4	0.729	+0.367	0.578	−0.010	0.009	0.004	(0.334)	0.00
5	0.589	+0.476	0.653	+0.010	−0.005	−0.004	0.000	(0.426)

Upper table from data of Alpatov and Boschko Stepanenko (1928).

omitting self-correlations. The residual correlations were fitted by maximized coefficients, p_{iH}, relating to a form-determining factor. The residuals after allowing for both factors are small. They can be tested by calculating partial correlations with both factors constant. This is given by the residual correlation divided by the product of the path coefficients relating to the special factors:

$$r_{ij \cdot GF} = \frac{r_{ij} - (p_{iG}p_{jG} + p_{iH}p_{jH})}{p_{iS_i}p_{jS_j}}. \tag{14.1}$$

These are shown in Table 14.3 as indicated in the title.

The standard error relative to $r = 0$ may be taken as $1/\sqrt{n-3}$, or about 0.125 in this case. None are significant. The correlation can thus be interpreted satisfactorily as due to a class of factors for general size and to a form-determining class which has no appreciable effect on the middle phalanx but tends to bring about a negative correlation between those on opposite sides of the middle.

It might also be possible, however, to fit the data by assuming a common factor for the first and second phalange and another common factor for the fourth and fifth. This can be done by omitting correlation r_{12} and r_{45} as well as the five self-correlations from the calculations by the modification of Hotelling's method. The ratio R of the sums of correlations remaining in each column to the corresponding ones in one of them are obtained. The first estimate of p_{aG}^2 is given by

$$(14.2) \qquad p_{aG}^2 = \frac{\sum^n r}{(\sum R)^2 - \sum RR}$$

in which $\sum r$ relates to correlations left in the table, and $\sum RR$ is the sum of products of the R's pertaining to each omitted correlation including the self-correlations. The other path coefficients are obtained as before ($p_{iG} = R_i p_{aG}$). Revised estimates of all path coefficients are obtained by iteration as before except that all omitted correlations, and not merely the self-correlations, must be replaced by products of the estimated path coefficients in the process. The results from the above data are shown in Table 14.4.

TABLE 14.4. Second path analysis of the data of Table 14.3. Two correlations r_{12} and r_{45}, in addition to the self-correlations, are excluded in the calculations of p_{iG}.

	Factors				Residuals\Partial Correlations				
	General	1, 2	4, 5	Special					
1	0.719	0.469		0.51	(0.263)	0	−0.09	0.03	0.05
2	0.737	0.469		0.49	(0.220)	(0.237)	0.09	0.04	−0.12
3	0.945			0.33	−0.015	0.015	(0.107)	−0.08	0.08
4	0.676		0.508	0.53	0.008	0.011	−0.014	(0.285)	0
5	0.512		0.508	0.69	0.017	−0.040	0.018	(0.258)	(0.480)

The fit is not as good, but it is good enough. The large residuals for r_{12} and r_{45} are interpreted above in the simplest way: as the squares of path coefficients relating to common factors for phalanges 1 and 2 and phalanges 4 and 5, respectively. In general, there is no way of deciding between this sort of pattern of interpretation and the preceding or an intermediate, in the absence of external information on the actual modes of action of the pertinent developmental factors. In the present case, the interpretation in terms of general size and a single form factor has some advantage because of the more uniform effects attributed to general size. On the other hand, this pattern turns out to give an impossible result (a path coefficient for the form factor greater than unity) when applied to similar data for another bird, *Corvus cornix*, presented by the same authors. In this, the best solution arrived at was the acceptance of a common size factor for phalanges 2 and 3 and a negative relation between phalanges 1 and 5, in addition to the general size factor.

In cases in which some of the variables have little or no apparent relation to each other except as pertaining to the same individuals, the best interpretative procedure seems to be to find as a first trial the coefficients for a general factor, omitting only the self-correlations, and to improve this by omitting also the significant positive residuals. Other factors may be found from the residuals until no significant positive second residuals appear. Significant negative residuals, if any, may then be omitted until the set is whittled down to the correlations that can be accounted for without significant residuals.

In the population of White Leghorn hens, which was described in chapter 13, the coefficients for the general factor (self-correlations omitted in the analysis) leave three positive residuals: for the length and breadth of skull, for the humerus and ulna, and for the femur and tibia. The omission of these and of the self-correlations leaves 12 correlations. On recalculating coefficients for the general factor from these correlations the residuals are small. A negative one, r_{HT}, seems indeed to be significant, but it does not appear that anything would be gained by attempting to push interpretative analysis further.

TABLE 14.5. Path analysis of correlations among bone measurements of White Leghorn fowls (Table 13.1). Three correlations, r_{LB}, r_{HU}, and r_{FT}, in addition to the self-correlations, are excluded in calculating the coefficients relating to general size.

	FACTORS					RESIDUALS					
	General	Head	Wing	Leg	Special	L	B	H	U	F	T
L	0.636	0.461			0.619	(+0.383)					
B	0.583	0.461			0.669	(+0.213)	(+0.448)				
H	0.958		0.182		0.222	+0.006	+0.017	(+0.049)			
U	0.947		0.182		0.265	−0.002	−0.022	(+0.033)	(+0.070)		
F	0.914			0.269	0.304	−0.011	−0.007	0	+0.012	(+0.092)	
T	0.932			0.269	0.243	+0.007	+0.011	−0.015	+0.003	(+0.072)	(+0.059)

Berg (1960) found two sorts of patterns among morphological correlations in plants. The measurements were of vegetative and floral parts. In 19 species that were studied, 12 showed more or less disjunct clusters of positive correlations (correlation "pleiades"); seven did not. She gave the simple examples involving only three characters which appear in Table 14.6.

TABLE 14.6. Contrasting sets of correlations in two plants.

	A	*B*
Anemone nemorosa		
A (stem height)		
B (leaf length)	0.75 ± 0.05	
J (petal length)	0.79 ± 0.05	0.77 ± 0.05
Geranium pratense		
A (stem height)		
B (leaf length)	0.71 ± 0.06	
I (flower diameter)	0.01 ± 0.13	−0.05 ± 0.13

From data of Berg (1960).

Table 14.7 shows four sets of correlation (lower left) and analysis by the method of whittling down to sets which can be accounted for adequately by a single factor. The residuals are in the upper right. Those for the omitted correlations (in parentheses) are analyzed on the simplest basis.

The simplest pattern is in *Delphinium elatum*, in which the set of correlations can be split into two groups fairly satisfactorily, a group of four vegetative characters, including number of florets (*H* not specified), and a group of two flower characters. A closer fit (given in parentheses) can be obtained by assigning to stem height a small component in the floral group, but the correlations on which this is based (0.22 ± 0.13, 0.16 ± 0.13) are not significant.

The pattern is somewhat similar in the case of *Chaminaerium angustifolium*, but there is some overlapping and at least one pair of variables that require common factors. Factor (2) was fitted first with r_{BC} omitted. Factor (1) was then fitted from the residuals omitting r_{AD}. There are some rather large residuals, but further analysis is hardly warranted because of the small number of measurements.

In marked contrast are the sets for the two grains. All measurements in both the vegetative and floral groups are strongly affected by a general factor. In barley (*Hordeum*) the only other common factors are for upper-internode length and leaf length, and for inflorescence length and number of spikelets, neither of which is surprising. In wheat, there is a triad of vegetative character with a common factor and triad of floral characters (including grain length), but the coefficients are not important.

TABLE 14.7. Correlations among various characters of four species of plants (lower left). Residuals after path analysis (incomplete where in parentheses) are in upper right above these. Path coefficients are at right.

Characters	*Delphinium elatum* (53 plants)								
	A	B	G	H	K	M	Vegetative	Floral	Special
A, stem height		−0.03	+0.06	−0.03	(+0.22)	(+0.16)	0.746	(0.297)	0.666 (0.596)
B, leaf length	0.53		−0.03	+0.07	−0.10	−0.10	0.759		0.651
G, number of florets	0.67	0.59		−0.04	+0.12	−0.13	0.815		0.580
H, vegetative	0.45	0.57	0.49		−0.03	−0.08	0.650		0.760
K, corolla tube, length	0.22	−0.10	0.12	−0.03		0.00		0.632 (0.742)	0.775 (0.670)
M, sepal length	0.16	−0.10	−0.13	−0.08	0.40			0.632 (0.539)	0.775 (0.842)

	Chaminaerium angustifolium (50 plants)													
	F	G	A	D	B	C	P	O	I	M	(1)	(2)	(3)	Special
F, inflorescence length		+0.01	+0.01	0.00	−0.07	+0.06	+0.02	−0.13	+0.16	+0.13	0.91			0.42
G, number of plants	0.88		+0.01	−0.02	+0.01	−0.03	+0.01	−0.12	+0.07	+0.08	0.95			0.30
A, stem height	0.85	0.89		0.00	−0.04	0.00	−0.05	−0.09	+0.05	+0.09	0.92		0.31	0.23
D, number of leaves	0.55	0.56	0.78		+0.09	−0.05	−0.10	−0.22	−0.03	−0.10	0.61		0.70	0.38
B, leaf length	0.77	0.90	0.82	0.66		+0.03	+0.05	−0.06	−0.01	+0.05	0.93	0.24		0.28
C, leaf width	0.74	0.69	0.70	0.41	0.84		−0.07	−0.04	+0.03	0.00	0.75	0.48		0.45
P, pistil length	0.29	0.29	0.32	0.08	0.48	0.47		+0.05	0.00	−0.03	0.30	0.66		0.69
O, stamen length	−0.13	−0.12	−0.09	−0.22	0.08	0.24	0.44		−0.01	+0.01		0.59		0.81
I, flower diameter	0.16	0.07	0.05	−0.03	0.20	0.44	0.57	0.50		0.00		0.87		0.50
M, sepal length	0.13	0.08	0.09	−0.10	0.24	0.39	0.50	0.70	0.57			0.81		0.59

(*continued on next page*)

TABLE 14.7—*continued*

Characters	*Triticum aestivum*, variety diamant (100 plants)										
	A	F	G	M	I_1	I_2	L	General	Vegetative	Floral	Special
A, upper internode length		(+0.13)	(+0.14)	0.00	−0.04	+0.03	+0.02	0.731	0.264		0.605
F, inflorescence length	0.73		(+0.26)	+0.02	+0.02	−0.03	−0.01	0.817	0.503		0.282
G, number of spikelets	0.61	0.79		−0.02	0.00	+0.02	−0.01	0.646	0.521		0.558
M, glume length	0.60	0.69	0.51		+0.01	−0.01	0.00	0.821			0.570
I_1, lemma length	0.57	0.70	0.54	0.69		(+ 0.14)	(+0.14)	0.830		0.490	0.266
I_2, palea length	0.67	0.69	0.59	0.71	0.87		(+0.08)	0.880		0.285	0.380
L, grain length	0.43	0.45	0.36	0.47	0.61	0.58		0.567		0.284	0.773

	Hordeum vulgare, variety pallidum (100 plants)											
	A	B	F	G	M	I_1	I_2	L	General	AB	FG	Special
A, upper internode length		(+0.34)	−0.02	−0.03	+0.08	+0.05	−0.09	0.00	0.471	0.587		0.658
B, leaf length	0.65		+0.01	+0.03	+0.02	−0.02	−0.06	+0.01	0.647	0.587		0.486
F, inflorescence length	0.38	0.56		(+0.21)	0.00	0.00	0.00	−0.01	0.846		0.462	0.265
G, number of spikelets	0.37	0.58	0.93		−0.02	−0.02	0.00	+0.02	0.847		0.462	0.264
M, glume length	0.46	0.54	0.68	0.66		−0.04	+0.04	−0.07	0.799			0.602
I_1, lemma length	0.38	0.43	0.60	0.58	0.52		+0.02	+0.02	0.703			0.711
I_2, palea length	0.28	0.44	0.66	0.66	0.66	0.57		+0.03	0.775			0.632
L, grain length	0.34	0.47	0.59	0.62	0.50	0.52	0.58		0.711			0.704

From data of Berg (1960).

Berg finds that the common feature of the 12 species with relative independence of vegetative and floral characters is that they all have specific insect pollinators (butterflies or bees) with adaptations to localize the pollen deposit on some definite part of the insect and corresponding adaptations of the flower. The seven species in which both vegetative and floral measurements are largely determined by general size are pollinated in less specialized ways (wind-pollinated, self-pollinated, or pollinated by unspecialized insects).

Sets of correlations that are available for both F_1 and F_2 of crosses between pure strains are of special interest (as noted in the simple case of Sinnott's cucurbit crosses) because of the presence of a factor, genetic segregation, in F_2 that is absent in F_1. Table 14.8 gives data on a cross made by Castle (1922) between strains of rabbits at opposite extremes in size (Flemish giant, mean weight 3,646 g; Polish, mean weight, 1,404 g). The correlations among 7 measurements of 27 F_1 rabbits are shown in the lower left on the left side and of 112 F_2's in the lower left on the right side of Table 14.9.

On finding the path coefficients that would account best for all these correlations (excluding self-correlations), the largest positive residuals were in both cases r_{FT}, r_{HF}, r_{HT}. The first of these was highly significant in both cases and the others in F_2. On omitting these, as well as the self-correlations from the calculation, all the remaining 18 correlations in each generation could be accounted for by a single factor for general size, of which the length of the skull and the length of the humerus were the best indicators in both F_1 and F_2. The path coefficients are shown in Table 14.10. In F_2 the triad of residuals for the correlations among the leg measurements permit calculation of path coefficients relating to factors that affect the legs. In F_1, calculation on this basis leads to an impossible result (more than 100% determination). Separate factors have been calculated for the legs as a group and for the hind legs on a somewhat arbitrary basis to avoid this. The coefficient relating to all factors other than general size are given in parentheses.

The residual correlations not accounted for by the general factor are shown in the upper right on both sides of Table 14.9. The partial correlation for constant general size can be obtained by dividing these residuals by the products of the coefficients for factors other than general size. Only that for r_{FT} is clearly significant in F_1, but all three residuals for the correlations among leg bones are in F_2.

The squared coefficients of variability are analyzed into components by using the squared path coefficients, in Table 14.11 (combining those for the legs and hind legs in F_1). The entries for F_1 are largely nongenetic since sex differences could be shown to be negligible. There is a great increase in all except breadth of skull in F_2. The differences between the F_2 and F_1 components may be interpreted as measuring the variances due to genetic

TABLE 14.8. Statistics of measurements of F_1 and F_2 rabbits and means of the parental strains.

	F_1 (27)			F_2 (112)			MEAN	
	Mean	σ	CV	Mean	σ	CV	P_1	P_2
W, weight	2506.0 gm.	1.87	7.47	2127.0	255.0	11.99	3646.0	1404.0
L, length of skull	75.6 mm.	1.31	1.74	73.5	3.09	4.21	85.5	65.7
B, breadth of skull	42.2 mm.	1.01	2.40	40.9	1.17	2.86	45.4	38.0
E, ear length	109.4 mm.	3.8	3.44	107.0	5.9	5.47	145	84.0
H, humerus	66.2 mm.	1.18	1.78	64.1	2.93	4.57	75.0	57.7
F, femur	83.4 mm.	1.65	1.97	80.5	3.80	4.72	97.6	72.3
T, tibia	96.2 mm.	2.29	2.38	93.5	4.82	5.15	110.0	83.0

From data of Castle (1922).

TABLE 14.9. Correlations (lower left) and residuals (upper right) in F_1 rabbits (left) and in F_2 rabbits (right)

	W	L	B	E	H	F	T	W	L	B	E	H	F	T
W		+0.037	+0.147	−0.027	−0.065	−0.042	−0.088		−0.019	+0.044	−0.024	+0.015	+0.007	−0.010
L	0.408		−0.051	−0.038	−0.030	+0.045	+0.084	0.731		+0.001	−0.009	−0.019	+0.017	+0.034
B	0.352	0.388		+0.003	−0.010	−0.003	−0.030	0.495	0.491		+0.038	−0.022	−0.026	−0.036
E	0.176	0.396	0.243		+0.119	−0.044	−0.049	0.594	0.663	0.442		+0.022	−0.010	−0.005
H	0.216	0.572	0.323	0.448		(+0.306)	(+0.221)	0.717	0.743	0.437	0.650		(+0.218)	(+0.256)
F	0.172	0.504	0.251	0.206	0.653		(+0.686)	0.676	0.744	0.411	0.589	0.898		(+0.369)
T	0.037	0.351	0.118	0.097	0.423	0.840		0.583	0.679	0.352	0.526	0.859	0.944	

TABLE 14.10 Path coefficients; F_1 rabbits (left), F_2 rabbits (right).

	General	Leg	Hind Leg	Special	General	Leg	Special
W	0.417			0.909	0.831		0.556
L	0.891			0.454	0.903		0.430
B	0.493			0.870	0.543		0.840
E	0.487			0.874	0.744		0.668
H	0.675	0.600		0.429	0.844	0.389	0.368
F	0.515	0.509	0.690	0	0.805	0.561	0.193
T	0.300	0.368	0.676	0.513	0.714	0.658	0.239

segregation. There are squared coefficients of variability of 12% to 14% for all bone measures except breadth of skull (1%) and considerable segregation of factors affecting the legs independently of the other measurements, but not much with respect to special factors except for ear length. In an earlier analysis of these data (Wright 1932), the path coefficients for the general factor were obtained without removing the high positive residuals and thus gave residuals that indicated not only leg and hind leg factors, but ones that affected foreleg and head.

White Spotting

Table 14.12 illustrates a case in which the variabilities of different parts of the body are independent to an extraordinary extent. It concerns the presence or absence of white at various points in the coats of 405 guinea pigs of a strain, no. 35, derived from a single mating in the twenty-second generation of brother-sister mating.

Although there was variation from a mere trace of white to self-white, there was no significant correlation between parent and offspring and thus no indication of genetic variability. The correlations (2 × 2) for all pairs among 24 points in rubber-stamp drawings were calculated by Chase (1939). These points include 11 symmetrical pairs and 2 on the midline (forehead, "rump" spots). The correlations for symmetrical points showed no significant differences and are averaged. Correlations for the same side of the body are shown in the upper right part of the table, those for opposite sides in the lower left.

Table 14.13 gives estimates of the path coefficients relating each unpaired point and each on one side to four paired factors and one unpaired one.

Spotting probably arises from an interaction between two developmental patterns, that of migration of prospective melanocytes from the neural crest

TABLE 14.11. Components of squared coefficients of variability in F_1 and F_2 rabbits and differences, attributable to genetic segregation, in F_2.

	F_1 (Nongenetic)				F_2				$F_2 - F_1$ (Genetic)			
	General	Leg	Special	Total	General	Leg	Special	Total	General	Leg	Special	Total
W	9.7		46.1	55.8	99.2		44.6	143.8	89.5		(−1.5)	88.0
L	2.4		0.6	3.0	14.4		3.3	17.7	12.0		2.7	14.7
B	1.4		4.4	5.8	2.4		5.8	8.2	1.0		1.4	2.4
E	2.8		9.0	11.8	16.5		13.4	29.9	13.7		4.4	18.1
H	1.5	1.1	0.6	3.2	14.9	3.2	2.8	20.9	13.4	2.1	2.2	17.7
F	1.0	2.9	0	3.9	14.5	7.0	0.8	22.3	13.5	4.1	0.8	18.4
T	0.5	3.4	1.4	5.3	13.5	11.5	1.5	26.5	13.0	8.1	0.1	21.2

TABLE 14.12. Correlations with respect to presence or absence of color among 24 points (11 symmetrical pairs, forehead [1] and "rump" spot [13] median) in the coat patterns of 405 guinea pigs of inbred strain no. 35. Those between points on the same side were closely similar and are here averaged (upper right). Corresponding right-left and left-right correlations were also similar and their averages are given in the lower left.

Point	2	3	4	5	6	7	8	9	10	11	12	"rump" 13
1 Forehead	+0.17	+0.16	+0.13	+0.15	+0.13	−0.02	+0.02	−0.01	−0.04	−0.04	−0.04	−0.04
2 Eye	+0.13	+0.23	+0.11	+0.05	+0.07	−0.03	0.00	−0.01	−0.01	−0.01	+0.04	−0.02
3 Ear	+0.06	+0.14	+0.46	+0.43	+0.28	+0.08	+0.04	+0.01	−0.04	−0.04	−0.05	−0.04
4 Throat	−0.01	+0.16	+0.18	+0.64	+0.53	+0.19	+0.05	+0.02	−0.05	−0.06	−0.07	−0.01
5 Shoulder	+0.04	+0.15	+0.21	+0.33	+0.68	+0.25	+0.17	+0.08	−0.02	−0.01	−0.06	−0.02
6 Foreleg	0.00	+0.08	+0.13	+0.23	+0.19	+0.38	+0.29	+0.19	−0.01	+0.02	0.00	−0.02
7 Chest	0.00	+0.04	+0.05	+0.09	+0.04	+0.06	+0.47	+0.55	+0.15	+0.12	+0.04	−0.01
8 Side	−0.05	+0.02	0.00	+0.05	+0.07	+0.16	+0.32	+0.62	+0.29	+0.25	+0.05	−0.01
9 Belly	−0.01	+0.02	−0.02	+0.04	+0.04	+0.07	+0.19	+0.16	+0.37	+0.38	+0.17	+0.09
10 Hip	−0.02	+0.01	+0.04	0.00	+0.05	+0.10	+0.17	+0.16	+0.24	+0.71	+0.58	+0.41
11 Groin	+0.01	0.00	+0.07	+0.05	+0.11	+0.10	+0.16	+0.16	+0.14	+0.11	+0.71	+0.41
12 Hind leg	−0.01	−0.04	+0.04	0.00	+0.04	0.00	+0.06	+0.06	+0.14	+0.06	+0.16	+0.59

From data of Chase (1939*a*).

TABLE 14.13. Array of path coefficients relating four paired and one medial factor to the occurrence of color at points in the coat patterns in an inbred strain of guinea pigs, as deduced from the correlation coefficients of Table 14.12. Symbols *S* and *O* distinguish paths from the paired factors to points on the same or opposite sides, respectively. Points 1 and 13 were medial, the others paired. Points 3–4, 5–7, 8–9, and 10–11 were dorsal-ventral pairs on each side.

	FACTORS								
	Head		Anterior		Middle		Posterior		Rump
Point	*S*	*O*	*S*	*O*	*S*	*O*	*S*	*O*	
1 Forehead	0.20		0.12						
2 Eye	0.66	0.10	0.05	0					
3 Ear	0.31		0.43	0.15					
4 Throat	0.14		0.75	0.11					
5 Shoulder			0.89	0.17	0.12				
6 Foreleg			0.74	0.10	0.26				
7 Chest			0.27		0.60	0.50			
8 Side					0.76	0.22			
9 Belly					0.83	0.10	0.20		
10 Hip					0.32	0.10	0.69	0.13	
11 Groin					0.25	0.10	0.86		
12 Hind leg							0.84	0.06	0.23
13 Rump							0.48		0.48

and that of a differentiation of the skin which slows down the migration (Wagener 1959, Schumann 1960). Normally the melanocytes reach all parts of the skin, but in animals of genotype *ss*, the migration pattern is frozen at a certain stage. The "factors" of the present analysis seem to represent somewhat variable blocks in the skin. The interactions in different blocks seem to be wholly independent in this case.

The absence of any appreciable general factor depends on the fact that this population was essentially isogenic. As brought out in previous discussion of this character, strains with the same major gene (*ss*) but different arrays of modifiers may range from an average percentage of about 10% white to more than 97%. In the branch of strain 35 considered here, the averages were 67.8% in males, 73.1% in females. In a random-bred stock with segregation of modifiers, the array of modifiers of each animal would behave as a general factor that would bring about moderate correlations among all points in the coat. The small sex difference and small effects of age of the mother in strain 35 are insufficient to bring about, when squared, appreciable correlations between remote points.

Systems with Ordering in Time

In the sets of correlations discussed so far, none of the measured variables have been in sequence and thus have been connected only by hypothetical common factors. Where there is ordering in time among measured variables, the patterns become more complicated. Treatment of an intermediary that is changing in value continuously as if it were a point variable introduces an inaccuracy in that the true contributions of compound paths are somewhat less than the product of the components. In our first example this is of relatively little importance.

The vital characters of a population: fertility and fecundity, mortality rate at all ages, and growth rates form a system that is of major importance in its evolution and is one in which there are complicated relations in time. These may be illustrated by data from populations of guinea pigs, inbred and crossbred, studied by the author (Wright 1921, 34*f*, 60*c*).

One of the most important factors affecting perinatal mortality and early growth is size of litter. As shown in Figure 14.1, there are somewhat curvilinear negative relations of fetal growth (V), birth weight (I), and gains to 33 days of age (II) to litter size. In a vigorous stock under favorable conditions, the optimum size of litter for percentage born alive (III, solid line) and for percentage reared of those born alive (III, broken line), is three with four only slightly inferior. Otherwise the optimum tends to be two. The gestation period (IV) shows a negative relation to litter size.

A second important factor is season of birth. Figure 14.2 shows the averages for various characters, corrected for size of litter, for all inbreds born in each of the twelve months in a certain period. The optimum months of birth, for rate of gain and percentage reared of those born alive are naturally earlier than for birth weight and percentage born alive, and much earlier than for litter size (Haines 1931).

The differences between conditions in different years were almost as important as those between different seasons of the same year. Figure 14.3 shows the more or less parallel changes in four of the inbred strains and the control stock (B) in nine successive years in which conditions were on the whole improving, with respect to weaning weights (above) and size of litter (below).

A fourth factor of importance was the condition of the dam. There was a correlation between weight of dam and size of litter of $+0.419 \pm 0.052$ and correlations between weight of dam, and birth weight of 0.428, 0.512, 0.619, and 0.740 (av, 0.575) in litters of 1, 2, 3, and 4, respectively (Eaton 1932).

Heredity is a fifth factor. There were marked differences among the inbred strains in all aspects of vigor, illustrated for weaning weight and size of

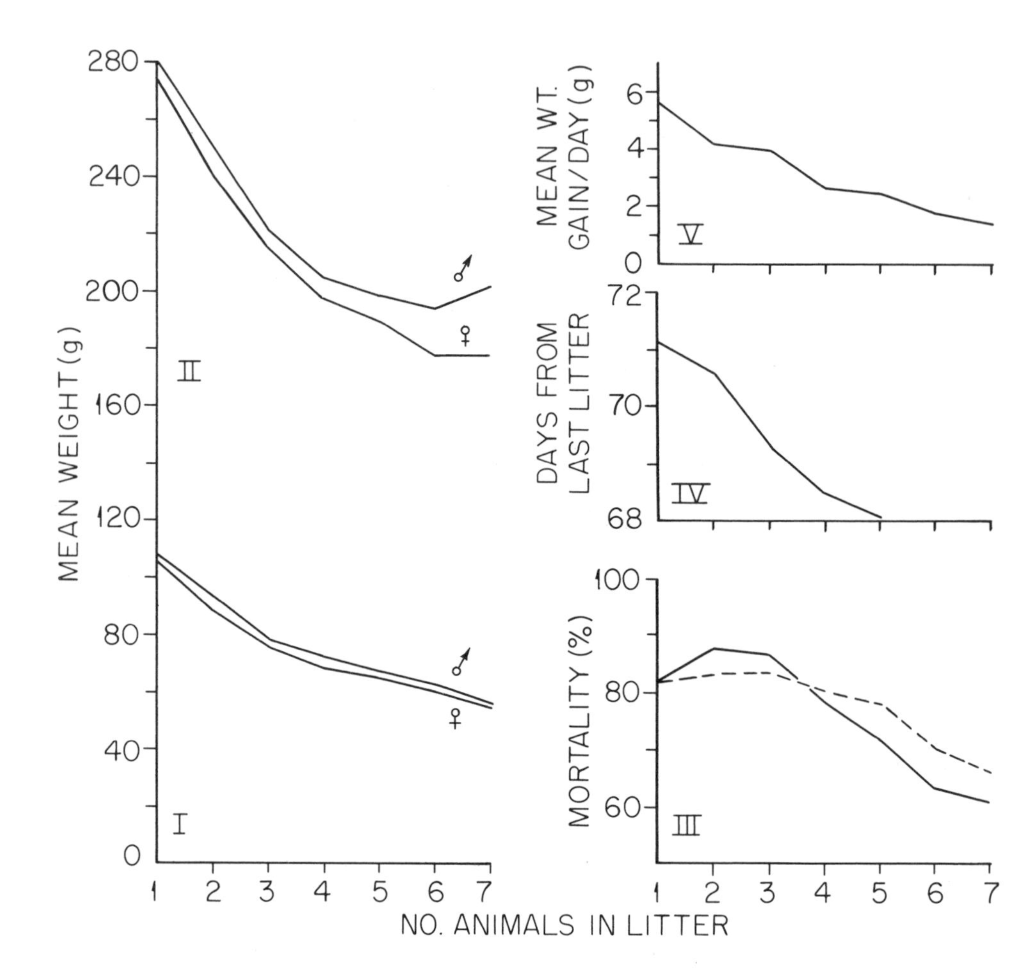

Fig. 14.1. Relation to size of litter in guinea pigs of weight at birth (I) and 33 days (II), mortality percentages (III: born alive [solid line], raised of born alive [broken lines]), mean interval from last litter (65–75 days; IV), and fetal gain from 55th to 65th day (V). (From Fig. 2, Wright 1960*a*.)

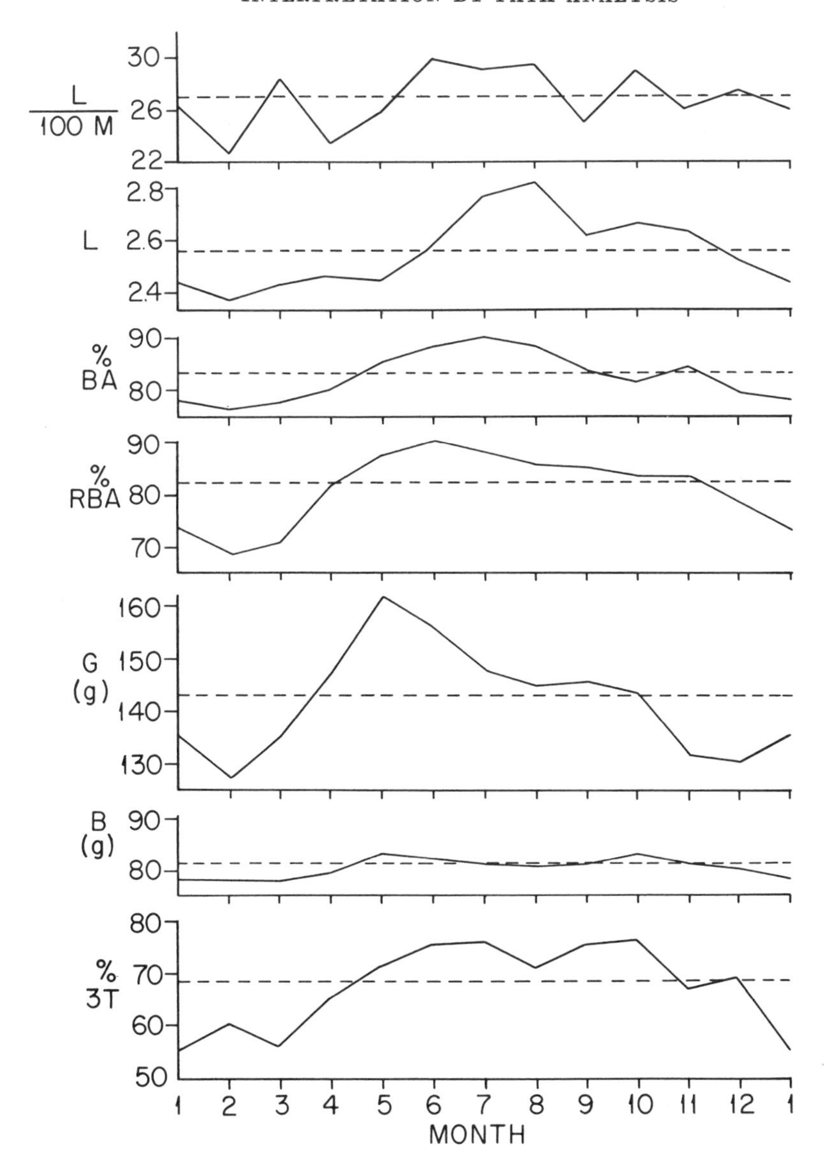

FIG. 14.2. Means for litters born per month per 100 matings (L/100 *M*), size of litter (*L*), percentage born alive (*BA*), percentage raised of those born alive (*RBA*), gain to 33 days (*G*), birth weight (*B*) in array of inbred strains, and percentage with 3 toes on hind feet (instead of 4) in inbred strain no. 35. (From Fig. 3, Wright 1960*a*.)

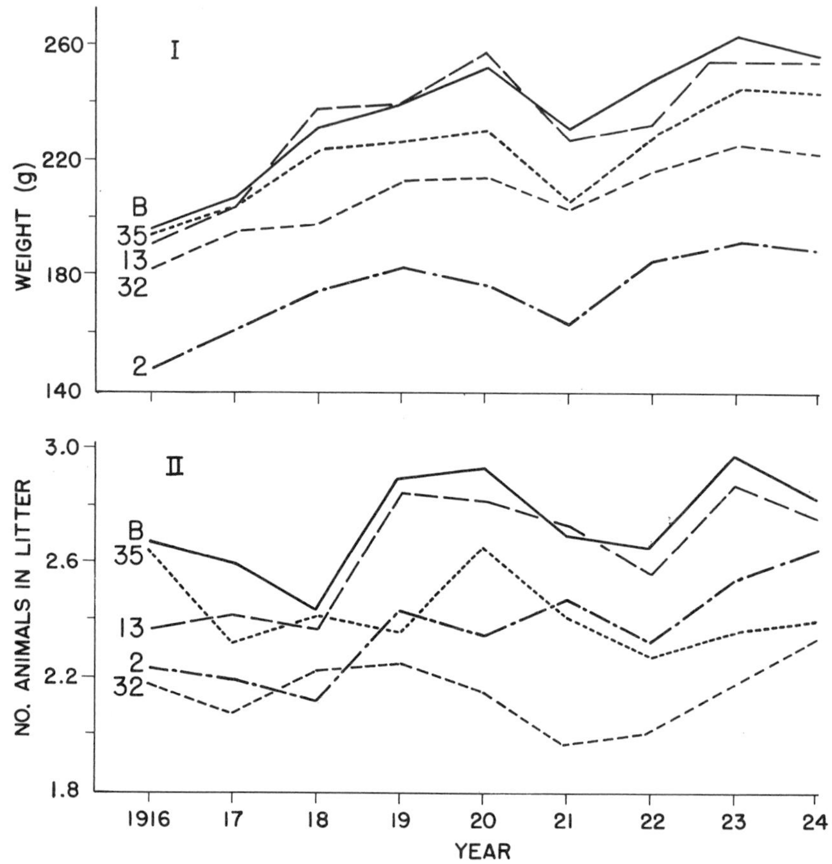

FIG. 14.3. Mean weight at 33 days (above) and mean size of litter (below) in four inbred strains of guinea pigs and controls (*B*) during years 1916–24. (From Fig. 4, Wright 1960*a*.)

litter in Figure 14.3. The ranking of the strains in average weight remained largely the same through life and ranking in size of litter in 23 strains was significantly correlated with this. Otherwise, when ranked, the various aspects of vigor (percentage born alive, percentage reared of those born alive [both corrected for litter size], and number of litters per year) were almost wholly independent of each other and of size of litter and weight (corrected for litter size).

Although genetic differences were clearly manifested in different averages, year after year, the importance of heredity for individual litters is very

TABLE 14.14. Correlation between different litters of the same mating in the control stock with respect to size of litter, mean birth weight, and mean gain to 33 days

	CONSECUTIVE LITTERS		NONCONSECUTIVE LITTERS	
	No.	Correlation	No.	Correlation
Size of litter	833	−0.011	2,213	+0.068
Mean birth weight	833	−0.052	2,312	+0.060
Mean gain to 33 days	601	+0.224	1,663	+0.063

From data of Wright (1960*a*).

slight. Table 14.14 shows correlations in the control stock with respect to size of litter, birth weight, and rate of gain to 33 days in consecutive and nonconsecutive litters. In the latter, individual heredity of the mother is supplemented by any persistent condition due to nongenetic causes. In the former, there is confounding by short-time effects. The only appreciable correlation is between gains of consecutive litters. It is instructive to make a path anaylsis of the principal characteristics of litters.

Minot (1891), on finding that the birth weight of guinea pigs tended to vary inversely with size of litter, pointed out that this might either be an effect of prenatal competition or of a stimulus to early parturition by a large size of litter. He found that the gestation period did in fact tend to vary inversely with size of litter and that birth weight tended to vary directly with gestation period. He concluded from these and other considerations that deviations in the gestation period, induced by the number of fetuses, rather than prenatal competition was the explanation of the inverse relation between birth weight and size of litter.

Data bearing on this question were obtained in 1916 from the records of the control stock, 1910–15, and separately from 11 of the inbred strains, 1906–15. Similar data were obtained later for the control stock in the period 1916–18 in which conditions were much less favorable. The means, standard deviations, and correlations are given in Table 14.15 in which the results for the inbred strains are averaged.

The correlations bear out Minot's observations. Only a rather rough path analysis is possible since the data are not wholly suitable because of differences in numbers of entries. The number of litters in which the young reached weaning age was naturally smaller than that for which average birthweight of litters was available. The gestation period is represented approximately by the interval since the preceding litter, if less than 76 days, since estrus

Table 14.15. Means, standard deviations, and correlation coefficients with respect to litter size, interval since preceding litter (if less than 76 days), birth weight, and gain to 33 days in the control stock (*B*) in 1910–15, 1916–18 and the average in 11 inbred strains (1906–15).

	B (1910–15)				*B* (1916–18)				Inbreds (1906–15)			
	No.	M	σ	r	No.	M	σ	r	No.	M	σ	r
Interval (days)	261	69.3	1.88	−0.500	167	69.1	2.07	−0.476	904	68.8	1.87	−0.450
Litter		3.17	1.50			2.49	1.04			2.91	1.26	
Birth weight (gm.)	587	83.3	18.7	−0.673	459	76.9	20.8	−0.656	2307	83.3	19.1	−0.665
Litter		3.00	1.26			2.58	1.04			2.74	1.14	
Birth weight (gm.)	261	82.2	19.5	+0.507	167	78.3	18.4	+0.485	904	83.0	18.0	+0.569
Interval (days)		69.3	1.88			69.1	2.07			68.8	1.87	
Gain (gm.)	513	152.5	49.6	−0.438	373	137.1	43.1	−0.520	2123	157.5	45.0	−0.347
Litter		3.07	1.20			2.57	0.97			2.77	1.11	
Gain (gm.)	224	147.3	48.3	+0.380	145	136.8	41.3	+0.306	844	156.0	43.8	+0.262
Interval (days)		69.2	1.88			69.0	2.10			68.7	1.74	
Gain (gm.)	513	152.5	49.6	+0.550	373	137.1	42.1	+0.607	2123	157.5	45.0	+0.531
Birth weight (gm.)		84.3	17.3			80.8	18.8			84.3	17.8	

From Wright (1960*a*).

follows immediately after parturition and does not recur for about 17 days. The number of cases in which this interval can be used is much smaller than the number available for the other characters. Restriction of the data to this smaller number might have been desirable but might have distorted other relations. Comparison of the statistics does not, however, indicate any consistent pattern of selection in the data as taken.

The two hypotheses considered by Minot may be represented in a single simple diagram as shown in Figure 14.4. Size of little (L) is assumed to affect the fetal growth curve (F) and the time at which this is interrupted by birth (I). Birth weight (B) is merely the ordinate of the growth curve at this time and is treated as if completely and linearly determined by F and I. The path coefficients can be calculated without difficulty. The following results were derived from the data from the inbred strains (Wright 1934f).

TABLE 14.16. Preliminary oversimplified path analysis of r_{BL} in inbreds.

$p_{FL} = -0.59$	p_{BFL}	$= p_{BF}p_{FL}$	$= -0.51$
$p_{IL} = -0.44$	p_{BIL}	$= p_{BI}p_{IL}$	$= -0.15$
$p_{BF} = +0.86$	r_{BL}	$=$	-0.66
$p_{BI} = +0.33$			

From data of Wright (1934f).

Thus it appeared that the path of influence of litter size through fetal growth was much more than that through the gestation period, contrary to Minot's conclusion (Wright 1921).

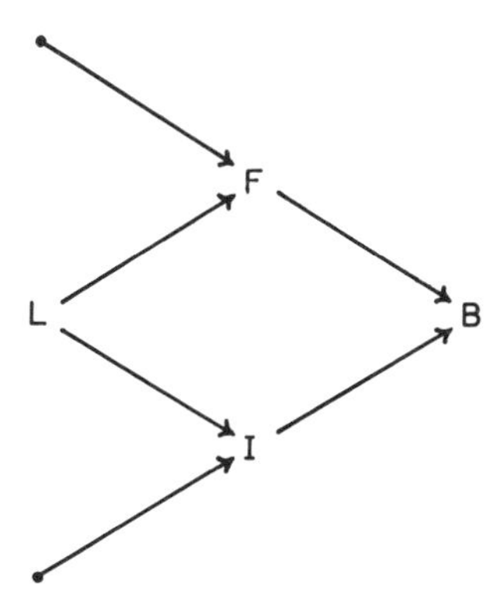

FIG. 14.4. Two hypotheses proposed by Minot on relation of birth weight (B) to size of litter (L). F is fetal growth rate and I is interval since previous litter where it is an index of gestation period.

The diagram was, however, rather seriously oversimplified to get a solution. Prenatal growth might be expected to have a direct (negative) effect on time of birth. The relation of fetal growth to size of litter is, moreover, more complicated than was implied, a point brought out clearly in a study made by Ibsen (1928) of weights of fetuses at successive conception ages. Ibsen found that there was no relation to litter size up to the fiftieth day (mean weight, 36.3 gm.) and not much even at the fifty-fifth day (mean weight, 49.4 gm.). After this there was divergence in negative relation to size of litter.

On calculating the regressions from his data for each size of litter (L) from age 55, weight 49.4 gm. as origin, and smoothing by least squares, the estimated daily rate of fetal gain (F) near the end of gestation comes out $5.739 - 0.664\ L$. The mean birth weight (fetal weight at conception age A) can then be estimated as approximately

$$\bar{B} = 49.4 + (5.739 - 0.664L)(A - 55). \tag{14.3}$$

Since the observed interval between litters in which conception immediately follows birth of the preceding litter is about a day longer than the true gestation period, taking account of errors at both ends, one may write $B = E + F(I - 56)$ in which E is the variable fetal weight at the fifty-fifth day, F is daily rate of gain as a linear function of litter size, and I is the observed interval between litters. This, of course, represents what is really a set of curvilinear growth curves after age 50 by a single line from ages 50 to 55, followed by diverging straight lines for the different litter sizes but should do as an approximation.

The formula involves a nonlinear term, the product of variables F and I, which requires introduction of a residual factor, N, into the diagram back of birth weight.

$$\delta B = \delta E + \bar{F}\,\delta I + (\bar{I} - 56)\delta F + N. \tag{14.4}$$

The correlation between the observed mean birth weight of litters of each size and the best linear estimate of this from size of litter, in the control stock 1910–18, was 0.985. This indicates 3% ($=1 - r^2$) determination of birth weight by deviations from linearity. Applying 0.173 ($=\sqrt{1 - r^2}$) to the observed values of r_{BL} in the three sets of data gives values of 0.116, 0.113, and 0.115, respectively. The value 0.115 will be used for this rather unimportant path coefficient, p_{BN}.

Since there is no possibility of estimating the influence of fetal weight before the effect of litter size begins (p_{BE}) from the set of correlations, it will be well to make a rough estimate of it from Ibsen's direct observations of fetal growth. For this purpose, we will take the ratio of the standard deviation of mean weight in litters at age 50 rather than 55 (at which slight effects of litter size were already apparent) to that at age 65 days three days before average age at parturition. From an analysis of variance, $p_{BE}^2 = 13.4/129.1$, $p_{BE} = 0.322$.

The observed interval (I) between litters is in error by fractions of day at each end, as an estimate of the gestation period. It is estimated that r_{AI} is about 0.95. All observed correlations involving interval (as given in Table 14.15) should be divided by 0.95 to estimate those involving A.

At this point it is instructive to make estimates on the same basis as above, except for introduction of the coefficients $p_{BN} = 0.115$, $p_{BE} = 0.322$, and $p_{AI} = 0.95$. There are six equations to solve for six unknowns. It is convenient to simplify the symbols for the path coefficients as indicated in Table 14.18.

$$(14.5) \quad \begin{cases} r_{IL} = 0.95a_1 \\ r_{BL} = b_1f_1 + b_2a_1 \\ r_{BI} = 0.95(b_1f_1a_1 + b_2) \\ r_{FF} = 1 = f_1^2 + f_2^2 \\ r_{AA} = 1 = a_1^2 + a_3^2 \\ r_{BB} = 1 = b_1^2 + b_2^2 + 2b_1b_2f_1a_1 + (0.115)^2 + (0.322)^2 \end{cases}$$

Solution yields the estimates shown in the first columns in each set in Table 14.18. The results in these three sets do not differ by amounts that can be considered important. The values of the compound path coefficients relating birth weight (B) to litter size (L) by the two routes considered by Minot are shown in Table 14.17.

TABLE 14.17. Revised analysis of r_{BL}.

	Randombreds (B) (1910–15)	Randombreds (B) (1916–18)	Inbreds (1906–15)
$p_{BFL} = b_1f_1 =$	−0.542	−0.534	−0.492
$p_{BAL} = b_2a_1 =$	−0.131	−0.122	−0.173
$r_{BL} =$	−0.673	−0.656	−0.665

From data of Wright (1960a).

The conclusions from the previous analysis are essentially unaltered. It is, however, desirable to get some idea of the error involved in ignoring the possible influence of fetal growth rate on the gestation period. This needs an additional equation, but one can be borrowed from Ibsen's data by making a direct estimate of p_{BA} (or b_2). On making adjustments for average litter size in his data and in the three sets of data here, and for somewhat greater birth

TABLE 14.18. Path coefficients relating to the birth weight, estimated from correlations in Table 14.5 and from assumptions or estimates based on Ibsen's (1928) data (in parentheses).

	B (1910–15)			*B* (1916–18)			Inbreds (1906–15)		
	1	2	3	1	2	3	1	2	3
r_{ES}	(0)	(0)	(+0.500)	(0)	(0)	(+0.500)	(0)	(0)	(+0.500)
$f_1 = p_{FL}$	−0.658	−0.646	−0.749	−0.641	−0.606	−0.731	−0.646	−0.663	−0.744
$f_2 = p_{FS}$	+0.753	+0.763	+0.663	+0.767	+0.796	+0.682	+0.763	+0.749	+0.668
$a_1 = p_{AL}$	−0.526	−0.556	−0.526	−0.501	−0.581	−0.501	−0.474	−0.414	−0.474
$a_2 = p_{AF}$	(0)	−0.046	(0)	(0)	−0.132	(0)	(0)	+0.090	(0)
$a_3 = p_{AU}$	+0.850	+0.850	+0.850	+0.865	+0.859	+0.865	+0.880	+0.892	+0.880
$b_1 = p_{BF}$	+0.824	+0.814	+0.724	+0.833	+0.808	+0.730	+0.761	+0.778	+0.661
$b_2 = p_{BA}$	+0.249	(+0.279)	+0.249	+0.243	(+0.333)	+0.243	+0.366	(+0.314)	+0.366
$b_3 = p_{BE}$	(+0.322)	(+0.322)	(+0.322)	(+0.322)	(+0.322)	(+0.322)	(+0.322)	(+0.322)	(+0.322)
$b_4 = p_{BN}$	(+0.115)	(+0.115)	(+0.115)	(+0.115)	(+0.115)	(+0.115)	(+0.115)	(+0.115)	(+0.115)

From Wright (1960*a*).

weights for given litter sizes in his data, the estimated regression coefficients in grams per day came out 3.05 for the controls (1910–15), 3.11 for the controls (1916–18), and 3.22 for the inbreds. The corresponding path coefficients ($p_{BA} = c_{BA}\, \sigma_B/\sigma_A$, $\sigma_A = 0.95\sigma_I$) are 0.279, 0.332, and 0.318, respectively.

These estimates permit determinate solution on introduction of another path into the diagram. The path relating gestation period to late fetal growth (p_{AF}) seems of most interest. The new set of equations are easily solved.

$$(14.6)\quad \begin{cases} r_{BL} = b_1 f_1 + b_2 r_{AL} & \text{from which } b_1 f_1 = r_{BL} - b_2 r_{AL}. \\ r_{BA} = b_1 r_{AF} + b_2 & \text{from which } b_1 r_{AF} = r_{BA} - b_2. \\ r_{BB} = 1 = b_1^2 + b_2^2 + b_3^2 + b_4^2 + 2b_1 b_2 r_{AF} & \text{to be solved for } b_1. \\ f_1 = (r_{BL} - b_2 r_{AL})/b_1. & \\ \left.\begin{aligned} r_{AF} &= (r_{BA} - b_2)/b_1 = a_1 f_1 + a_2 \\ r_{AL} &= a_1 + a_2 f_1 \end{aligned}\right\} & \text{to be solved for } a_1 \text{ and } a_2. \\ r_{AA} = 1 = a_1^2 + a_2^2 + a_3^2 + 2a_1 a_2 f_1 & \text{to be solved for } a_3. \end{cases}$$

The solutions are given in Table 14.18 in the second columns of each set. The analysis of r_{BL} is given in Table 14.19.

TABLE 14.19. Third analysis of r_{BL}.

		Randombreds (B) (1910–15)	Randombreds (B) (1916–18)	Inbreds (1906–15)
$p_{BFL} = b_1 f_1$	=	−0.526	−0.490	−0.516
$p_{BAL} = b_2 a_1$	=	−0.155	−0.193	−0.130
$p_{BAFL} = b_2 a_2 f_1$	=	+0.008	+0.027	−0.019
r_{BL}	=	−0.673	−0.656	−0.665

From Wright (1960*a*).

The estimates of p_{AF} are small and inconsistent, and the analysis of the correlation between birth weight and size of litter is not modified to an extent that can be considered as of importance.

We have not, however, taken account of the positive correlations that certainly exist to some extent between the ultimate growth factors E and S and probably between these and size of litter (L). Unfortunately, the available data do not permit a solution. We may, however, test the effect of arbitrarily assuming a moderately large correlation (+0.50) between the two growth factors. It will be assumed that r_{LE}, r_{LS}, and r_{AF} are negligible. The solutions are given in the third columns under each set in Table 14.18. There is considerable readjustment, but the components of the correlations between birth weight and litter size are the same as in the earlier solutions in which p_{AF} was ignored.

We may indeed assume any values for r_{ES}, r_{LE}, and r_{LS} without affecting the values of p_{BA}, p_{AL}, and the resulting contributions of litter size to birth weight by way of the gestation period as long as p_{AF} is assumed to be negligible. There is, of course, considerable readjustment among the coefficients by which birth weight is related to litter size in other ways, but the total for the compound residual path $p_{(BL)}$ is necessarily unaffected. Thus if $r_{ES} = 0.50$, $r_{LE} = 0.30$, and $r_{LS} = 0$, we have the analysis of r_{BL} shown in Table 14.20.

TABLE 14.20. Fourth analysis of r_{BL}.

	Randombreds (B) (1910–15)	Randombreds (B) (1916–18)	Inbreds (1906–15)
$p_{BFL} = b_1 f_1 =$	-0.639	-0.631	-0.588
$p_{BEL} = b_3 r_{LE} =$	$+0.097$	$+0.097$	$+0.097$
$p_{BAL} = b_2 a_1 =$	-0.131	-0.122	-0.173
$r_{BL} =$	-0.673	-0.656	-0.665

From Wright (1960*a*).

The gain to 33 days is represented in Figure 14.5 as affected by perinatal mortality (M), size of litter (L), gestation period (A), birth weight (B), and residual factor, (T). The direct effect of deaths of littermates (M) is positive because of relief from competition. Perinatal mortality itself should be affected positively by the direct effect of litter size (large litters, heavy mortality), negatively by the direct effect of A (premature birth, heavy mortality), and positively by the direct effect of B (excessive size at birth, heavy mortality from difficulty in parturition and anoxia). The direct effect of L on G (large litters, severe competition) should be opposite in sign to its indirect effect through M. Similarly the direct effect of A on G (premature birth, slow gain) should be opposite in sign to its indirect effect through M. The effect of L, A, and B on M are probably far from linear (importantly for large L, for small A, and for large B) thereby accounting for the highly nonlinear total effect of L on M (litters of two or three optimal). It thus seems best to combine the effects through M with the direct effects and not include M explicitly in the analysis.

This leaves G a function of L, A, B, and the residual factor T. If T is treated as an independent variate, the path coefficients relating G to its

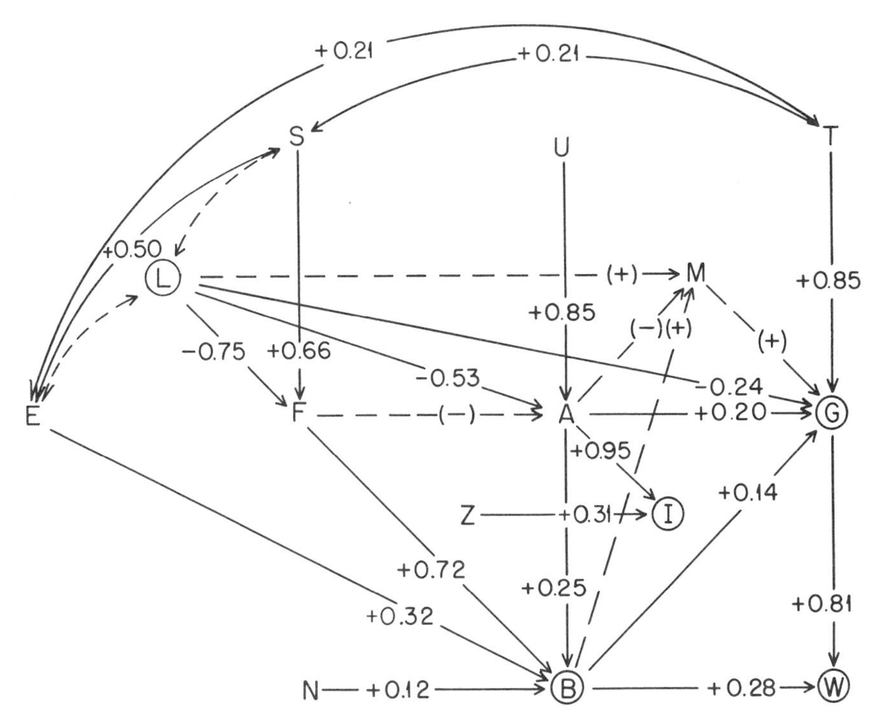

FIG. 14.5. Path diagram showing weight at 33 days (W) as the sum of birth weight (B) and gain (G). B is represented as determined by early embryonic growth (E), fetal growth (F), gestation period (A), and nonlinearity (N). The observed interval since the previous litter, if less than 76 days, is represented as determined by (A) and errors of observation (Z). F is represented as determined by size of litter (L) and residual factors (S). A is represented as determined by L F (negligible), and residual factors (U). G is represented as determined by direct and indirect effects of L, A, B, and residual factors (T). Indirect effects through perinatal mortality (M) are indicated but could not be used in the calculations. Correlations are indicated among some of the ultimate factors, but those with L were not used. The coefficients are estimates from the data of the control stock, 1910–15. (From Fig. 7, Wright 1960*a*.)

factors become ordinary standardized partial regression coefficients. Using the symbols of Figure 14.5 and Table 14.18, we have the following equations:

$$
(14.7) \quad \begin{cases}
r_{GL} = g_1 + r_{AL}g_2 + r_{BL}g_3, \\
r_{GA} = r_{AL}g_1 + g_2 + r_{BA}g_3, \\
r_{GB} = r_{BL}g_1 + r_{BA}g_2 + g_3, \\
r_{GG} = 1 = g_1r_{GL} + g_2r_{GA} + g_3r_{GB} + g_4^2.
\end{cases}
$$

The solutions are given in the first columns under the three sets in Table 14.21.

According to these solutions, gain is affected considerably by birth weight in all cases, but the effects of litter size and of gestation period are small and inconsistent. This is not wholly unreasonable in view of the opposition between the direct effects of L and A and their indirect effects through M. We have, however, taken no cognizance of the positive correlations that undoubtedly exist among the residual factors for all growth processes. Recognition of these tends to subtract from the effect attributed directly to birth weight and to modify those attributed to L and A.

As an extreme hypothesis, assume that $r_{ES} = 0.500$, that r_{ET} and r_{ST} are equal, and that there is no direct effect of birth weight or indirect effect through M,($g_3 = 0$). A term, $g_4 r_{BT}$, is added to the equation for r_{GB} in the preceding equations ($r_{BT} = b_1 f_2 r_{ST} + b_3 r_{ET}$). The new equations yield the results shown in the third columns in Table 14.21 under the various rates. These give fairly consistent and reasonable results as far as the effects of L and A are concerned, but probably are too extreme in allowing no direct influence of birth weight (including here the reinforcement from the effect through M). An intermediate result can be obtained by putting the direct influence, measured by g_3 equal to the indirect one through T,($g_4 r_{BT}$). In this case g_4^2 in the equation for r_{GG} must be replaced by $g_4 r_{GT}$, in which $r_{GT} = g_4 + g_3 r_{BT}$. The solutions are given in the second columns in Table 14.12 for the various sets. These probably give a more satisfactory interpretation than either of the preceding. Those under B (1910–15) are used in Figure 14.5 together with those in the third column of Table 14.18.

Weight at 33 days (W) is merely the sum of birth weight and gain to 33 days. The correlations involving W were calculated from the observed variances of B and G and the correlation r_{BG}.

$$(14.8) \qquad \begin{cases} \sigma_W^2 = \sigma_B^2 + \sigma_G^2 + 2\sigma_B\sigma_G\sigma_{BG}, \\ p_{WB} = \sigma_B/\sigma_W, \qquad p_{WG} = \sigma_G/\sigma_W, \\ r_{WX} = p_{WB} r_{BX} + p_{WG} r_{GX}, \end{cases}$$

where X is any other variable.

The path diagram gives a picture of the network of relations among the perinatal characters in so far as quantitative evaluation has been possible from the data at hand. It is, of course, only a partial picture. Perinatal mortality (M) is introduced only qualitatively. In a more complete qualitative scheme, M should be analyzed at least into mortality up to birth and mortality between birth and weaning since, as noted, those are determined to a considerable extent by different factors. Size of litter (L) could be analyzed

TABLE 14.21. Path coefficients, relating to gain to 33 days, and weight at 33 days, estimated from correlations in Table 14.5 and assumptions (parentheses).

	B (1910–15)			B (1916–18)			Inbreds (1906–15)		
	$r_{BT}=0$	$g_3=g_4r_{BT}$	$g_3=0$	$r_{BT}=0$	$g_3=g_4r_{BT}$	$g_3=0$	$r_{BT}=0$	$g_3=g_4r_{BT}$	$g_3=0$
r_{ES}		(+0.500)	(+0.500)		(+0.500)	(+0.500)		(+0.500)	(+0.500)
r_{ET}	(0)	(+0.209)	(+0.303)	(0)	(+0.248)	(+0.360)	(0)	(+0.264)	(+0.366)
r_{ST}	(0)	(+0.209)	(+0.303)	(0)	(+0.248)	(+0.360)	(0)	(+0.264)	(+0.366)
r_{BT}	(0)	(+0.167)	(+0.243)	(0)	(+0.203)	(+0.295)	(0)	(+0.201)	(+0.279)
$g_1=p_{GL}$	−0.085	−0.238	−0.315	−0.224	−0.391	−0.479	+0.002	−0.191	−0.279
$g_2=p_{GA}$	+0.129	+0.199	+0.234	−0.034	+0.042	+0.082	−0.065	+0.078	+0.144
$g_3=p_{GB}$	+0.424	+0.142	(0)	+0.478	+0.164	(0)	+0.571	+0.178	(0)
$g_4=p_{GT}$	+0.823	+0.847	+0.877	+0.778	+0.810	+0.851	+0.846	+0.886	+0.929
$w_1=p_{WB}$	+0.284	+0.284	+0.284	+0.338	+0.338	+0.338	+0.315	+0.315	+0.315
$w_2=p_{WG}$	+0.815	+0.815	+0.815	+0.758	+0.758	+0.758	+0.796	+0.796	+0.796

From Wright (1960*a*).

into amounts of ovulation, percentage of implantation, and percentage surviving early death and absorption. These are affected by genetic factors of the dam and of the individual and, through the condition of the dam, by external factors. The interval since the preceding litter, according to whether it is more or less than 76 days, is positively related to subsequent and perhaps also preceding litter size. Amount of ovulation is correlated with growth heredity of the dam. The various growth factors (E, S, T) involve the dam's heredity and that of the individual in different degrees, and aspects of the condition of the dam that trace to the external environments at successive times. Because of the network of interactions, the genetic factors may all be expected to have pleiotropic effects on all of the vital characters.

Rate of Gain throughout Life

It is interesting to supplement the analysis of the correlations among perinatal characters by one of growth correlations later in life. In a random-bred stock, there would undoubtedly be positive correlations between rates of growth as long as growth continues because of the common factor of heredity. The ranks of inbred strains with respect to birth weight, rates of gain, and adult weight are essentially the same. We will consider here correlations among 153 males of a particular inbred strain (no. 2) with respect to rates of gain at ages up to about a year when growth is nearly complete.

Table 14.22 shows mean weights, standard deviations, coefficients of variability, and regressions of weight on size of litter at successive ages up to 353 days and also similar data for the rates of gain per day in the intervals. It may be seen that absolute rate of gain is low in the first 3 days after birth but reaches its peak in the next 10 days, falls to a low value at weaning (33 days), rises to another peak at 83 days, and thereafter steadily declines. The negative regression of weight on size of litter continues to increase up to 53 days, but then remains essentially constant. This implies that size of litter continues to have an effect up to about 53 days, even though litter mates were separated at 33 days, and that there is no compensatory recovery from this later.

The correlations among weights at different ages (not shown) were positive. These naturally fell off as the interval of time increased. There was still a correlation of +0.30 between birth weight and weight at 353 days.

Of more interest analytically are the correlations among rates of gain at different periods including birth weight and of these correlations with the important initial factor, size of litter. Those are shown in Table 14.23.

As implied above, size of litter shows negative correlations with gains up to the period 33–53 days. There are positive correlations among all gains up

TABLE 14.22. Data on weight (W) and rate of daily gain (G) at successive ages (days) for 153 male guinea pigs of inbred strain no. 2. The means (M), standard deviations (σ), and coefficients of variability (CV), correlations with size of litter (L), and the corresponding regression coefficients are shown.

Days	Weight					Gain/Day				
	M	σ	CV	r_{WL}	b_{WL}	M	σ	CV	r_{GL}	b_{GL}
0	77	11.5	14.9	−0.683	− 8.8					
3	81	12.6	15.6	−0.644	− 9.1	1.10	1.33	123	−0.031	−0.046
13	127	20.5	16.2	−0.564	−12.9	4.59	1.20	26	−0.299	−0.401
23	165	28.8	17.4	−0.462	−14.9	3.83	1.27	33	−0.183	−0.260
33	192	35.4	18.4	−0.418	−16.5	2.72	1.25	46	−0.099	−0.138
53	266	51.4	19.3	−0.339	−19.5	3.72	1.21	33	−0.100	−0.135
83	381	67.8	17.8	−0.267	−20.2	3.83	0.98	26	−0.007	−0.008
113	474	72.4	15.3	−0.251	−20.3	3.08	0.99	32	+0.047	+0.053
143	540	69.9	13.0	−0.289	−22.6	2.20	0.68	31	−0.135	−0.103
173	585	65.7	11.2	−0.335	−24.6	1.57	0.81	54	−0.117	−0.106
203	621	66.4	10.7	−0.301	−22.3	1.19	0.71	60	+0.023	+0.018
233	648	64.5	10.0	−0.258	−21.5	0.90	0.81	90	+0.046	+0.042
263	667	66.4	9.9	−0.333	−24.7	0.65	0.88	134	−0.093	−0.091
293	689	66.2	9.6	−0.292	−21.6	0.72	0.82	115	+0.017	+0.016
323	700	66.2	9.5	−0.290	−21.5	0.38	0.94	247	+0.055	+0.058
353	709	64.4	9.1	−0.285	−20.5	0.30	0.86	288	+0.030	+0.029

to 83 days. Gain in the period 83–113 days shows no appreciable correlation with any gain earlier (average, −0.009), or later (average, +0.002). Gains in later periods show very little correlation with the early gains, but the average is negative and undoubtedly significant. Average of 42 correlations between intervals from 3–83 days and ones from 113–323 days is −0.11. The correlations among gains after 113 days are all insignificant except for those between adjacent intervals after 143 days. The average for these six correlations is −0.24. The average for the others was +0.007.

The explanation of these negative correlations is obvious. During these later periods in which the average rates of gain are small, the apparent gain or loss from one weighing to the next is largely due to temporary causes—a full or empty digestive tract or a brief period of indisposition. These would tend to give a correlation of −0.50 between successive periods, which would be partially offset by a tendency toward positive correlation due to conditions of somewhat longer duration.

The portion of the table up to 83 days warrants most analysis. Size of litter was treated as a first factor. After removing the correlations among gains

TABLE 14.23. Correlations involving size of litter (L), birth weight (B), and gains at successive intervals up to 353 days in 153 male guinea pigs of inbred strain no. 2.

	B	G_1 0–	G_2 3–	G_3 13–	G_4 23–	G_5 33–	G_6 53–	G_7 83–	G_8 113–	G_9 143–	G_{10} 173–	G_{11} 203–	G_{12} 233–	G_{13} 263–	G_{14} 293–	G_{15} 323–353
L	−0.68	−0.03	−0.30	−0.18	−0.10	−0.10	−0.01	+0.05	−0.13	−0.12	+0.02	+0.05	−0.09	+0.02	+0.05	+0.03
B		+0.10	+0.33	+0.36	+0.19	+0.16	+0.11	−0.03	−0.03	0.00	+0.09	−0.12	−0.01	−0.09	−0.04	−0.13
G_1			+0.22	+0.22	+0.18	+0.24	+0.10	+0.08	−0.10	−0.14	−0.11	+0.06	+0.06	−0.18	−0.06	−0.05
G_2				+0.42	+0.13	+0.36	+0.12	+0.02	−0.10	−0.08	−0.05	−0.06	+0.01	−0.11	−0.16	−0.02
G_3					+0.49	+0.32	+0.19	−0.08	−0.07	−0.33	−0.08	0.00	−0.07	−0.29	−0.23	+0.10
G_4						+0.33	+0.35	−0.01	−0.18	−0.30	−0.04	−0.19	−0.11	−0.08	+0.01	−0.01
G_5							+0.35	−0.01	−0.27	−0.18	−0.21	−0.08	−0.06	−0.11	−0.14	+0.03
G_6								−0.03	−0.15	−0.30	+0.12	−0.24	−0.10	0.00	+0.04	−0.01
G_7									−0.10	+0.07	−0.03	−0.04	−0.02	+0.01	+0.11	+0.01
G_8										+0.08	+0.01	+0.04	+0.03	+0.01	−0.13	+0.06
G_9											−0.19	+0.01	+0.12	−0.01	+0.17	−0.04
G_{10}												−0.20	+0.07	−0.06	+0.06	−0.03
G_{11}													−0.26	+0.11	−0.12	−0.01
G_{12}														−0.23	+0.06	−0.22
G_{13}															−0.24	0.00
G_{14}																−0.34

due to this, the coefficients pertaining to a general factor (E_1) were estimated by the process used in previous cases. The residuals required a factor (E_2) affecting the gains in the two successive intervals 13–23 and 23–33 days, and a factor (E_3) affecting the three intervals in the period from 23–83 days (Table 14.24). Seasonal conditions (E_4) would tend to produce slight effects on the gains in the later intervals which would cause only negligible contributions to the correlations among these intervals (products of two small-path coefficients), but appreciable although small negative correlations with the early gains, about half a year earlier, indicated by negative correlations relating E_1 and E_3 to E_4. Returning to the three factors other than litter size that appear to affect early gains, the general factor (E_1) presumably consists largely of maternal condition and its continuing effects. The factor (E_2) found to operate between 13 and 33 days may relate specifically to persistence of lactation, and the factor found to operate from 23 to 83 days may reflect seasonal and other continuing conditions that affect the success of the young guinea pigs after weaning. No doubt such identification of the factors are more definite than really warranted. The essential point is that there is a succession of factors that persists over two or more of those intervals that bring about positive correlations between rates of gain in different intervals in this period of rapid growth. After 83 days such continuing tendencies almost cease and rate of gain depends almost solely on condition of short duration. Any tendency toward positive correlation due to persistence of condition over two intervals is more than offset (after 113 days) by the negative effect of temporary condition at the time of weighing. There was no permanent differentiation among these animals of the same genotype, but it

TABLE 14.24. Path coefficients relating birth weight (B) and gains at successive periods (G), to size of litter (L) and to factors E_1, E_2, E_3, and E_4.

	L	E_1	E_2	E_3	Special		E_4	Consecutive r_{GG}
B	−0.68	0.28			0.68	G_{83-}	0	0
G_{0-}	−0.03	0.39			0.92	G_{113-}	0.31	−0.10
G_{3-}	−0.30	0.54			0.79	G_{143-}	0.47	−0.07
G_{13-}	−0.18	0.63	0.53		0.53	G_{173-}	0.10	−0.24
G_{23-}	−0.10	0.30	0.53	0.46	0.64	G_{203-}	0.22	−0.22
G_{33-}	−0.10	0.53		0.36	0.76	G_{233-}	0.10	−0.28
G_{53-}		0.27		0.58	0.77	G_{263-}	0.30	−0.26
E_4		−0.72		−0.69	0	G_{293-}	0.20	−0.30
						G_{323-}		−0.34

should be noted that this applies to ones that were vigorous enough to reach a year of age.

Instantaneous Reciprocal Interaction

It is instructive to compare two different methods of dealing with rapid reciprocal adjustment of two variables. One is to treat both as *completely* determined by auxiliary hypothetical variables the nature of which is perhaps most easily grasped by considering the way in which, in classical economic theory, the price, P, and quantity marketed, Q, of some product, are determined by the supply situation, S (quantities that would be brought to market at given prices), and the demand situation, D (quantities that would be purchased at given prices). In Figure 14.6 with prices as abscissas and quantities as ordinates, the supply curves for different times are represented by parallel lines all with the same positive slope, e, the elasticity of supply, and demand curves are represented by another set of parallel lines with constant negative slope, η, the elasticity of demand. It is assumed that the actual quantity marketed and the price at any given time are determined

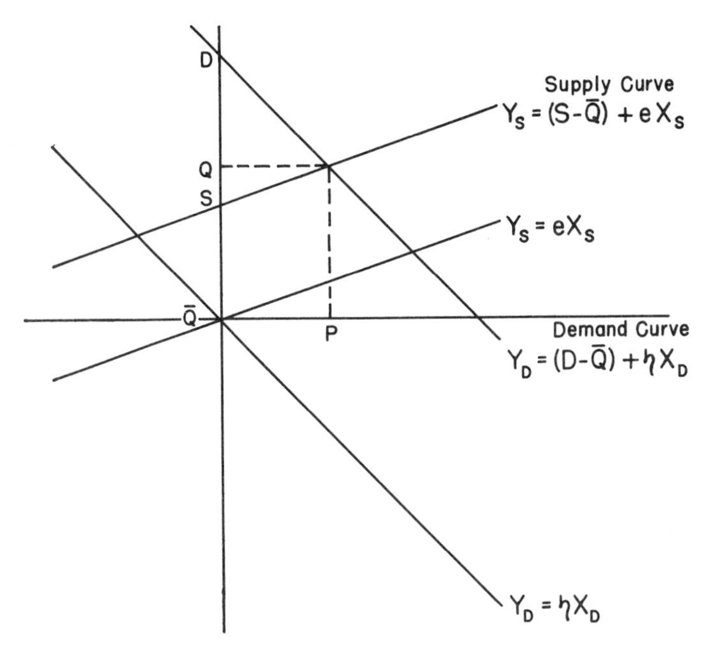

FIG. 14.6. Simplified diagram of relations among quantity (B), price (P), and supply (S) and demand (D) situations. (Redrawn from Fig. 8, Wright 1960*b*.)

by reciprocal interaction (bargaining), as the coordinates of the point of intersection. Supply and demand as hypothetical variables are measured by the intercepts on the vertical axis (quantity marketable or purchasable at the time in question) at the average price for the whole period under consideration.

The path diagram, including a variable A that is known to be correlated with demand, but not supply, and a variable B that is known to be correlated with supply, but not demand, is given in Figure 14.7. It is here assumed that $r_{SD} = 0$.

$$r_{PA} = p_1 r_{DA}. \tag{14.9}$$

$$r_{QA} = q_1 r_{DA}. \tag{14.10}$$

$$r_{PQ} = p_1 q_1 + p_2 q_2. \tag{14.11}$$

$$r_{PP} = 1 = p_1^2 + p_2^2. \tag{14.12}$$

$$r_{QQ} = 1 = q_1^2 + q_2^2. \tag{14.13}$$

These are easily solved.

$$p_2^2 q_2^2 = 1 - p_1^2 - q_1^2 + p_1^2 q_1^2 \quad \text{from (14.12) and (14.13).}$$
$$p_2^2 q_2^2 = r_{PQ}^2 - 2p_1 q_1 r_{PQ} + p_1^2 q_1^2 \quad \text{from (14.11).}$$
$$p_1 = r_{PA}\, q_1 / r_{QA}. \quad \text{from (14.9) and (14.10).}$$

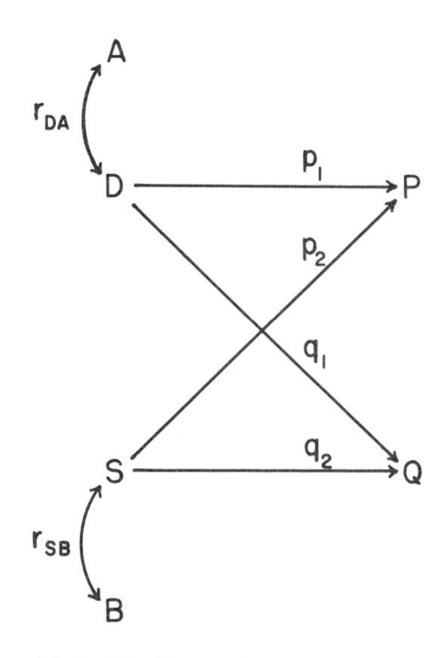

FIG. 14.7. (Redrawn from Fig. 9, Wright 1960b.)

Thus

$$(14.14)\quad \begin{cases} q_1 = \sqrt{\dfrac{1 - r_{PQ}^2}{1 - 2(r_{PA}r_{PQ}/r_{QA}) + (r_{PA}^2/r_{QA}^2)}}, \\ r_{DA} = r_{QA}/q_1, \quad q_2 = \sqrt{1 - q_1^2}, \quad p_2 = \sqrt{1 - p_1^2}. \end{cases}$$

The signs must satisfy equations (14.9), (14.10), and (14.11).

If a variable of type B is known, analogous equations can be written. If there is reason to believe that a correlation exists between the supply and demand situations, it requires knowledge of variables of both types A and B to solve for the seven unknown coefficients.

The elasticities, e and η, are ordinarily of most interest. These can be calculated from the regression equations.

$$(14.15)\quad \begin{cases} Q - \bar{Q} = c_{QD}(D - \bar{Q}) + c_{QS}(S - \bar{Q}). \\ P - \bar{P} = c_{PD}(D - \bar{Q}) + c_{PS}(S - \bar{Q}). \end{cases}$$

$$(14.16)\quad \begin{cases} e = \dfrac{Y_s}{X_s} = \left(\dfrac{Q - \bar{Q}}{P - \bar{P}}\right)_{S=\bar{Q}} = \dfrac{c_{QD}}{c_{PD}} = \dfrac{q_1\sigma_Q}{p_1\sigma_P} = \dfrac{r_{QA}\sigma_Q}{r_{PA}\sigma_P}. \\ \eta = \dfrac{Y_D}{X_D} = \left(\dfrac{Q - \bar{Q}}{P - \bar{P}}\right)_{D=\bar{Q}} = \dfrac{c_{QS}}{c_{PS}} = \dfrac{q_2\sigma_Q}{p_2\sigma_P} = \left(\dfrac{r_{PA} - r_{QA}r_{PQ}}{r_{PA}r_{PQ} - r_{QA}}\right)\dfrac{\sigma_Q}{\sigma_P}. \end{cases}$$

The other mode of approach is to represent reciprocal interaction without lag in a diagram by arrows from each to the other as in Figure 14.8 in which D and S are the residual factors for P and Q from the viewpoint indicated.

This diagram is equivalent to the two diagrams: in the first, determination of P by Q and D, which must be represented as correlated, and in the second determination of Q by P and S, which must be represented as correlated. Thus r_{PQ} in Figure 14.8 can be written in two ways: as $\sum p_{Pi}r_{Qi}$ or as $\sum p_{Qj}r_{Pj}$ where V_i and V_j are the immediate factors of P and Q respectively.

The usual analysis of correlations into compound path coefficients, under the provision that there shall be no passage through the same variable twice in the same path, breaks down here. Valid equations can, nevertheless, be written by analyzing step by step the correlation terms of r_{Qi} or r_{Pj} in the above expressions.

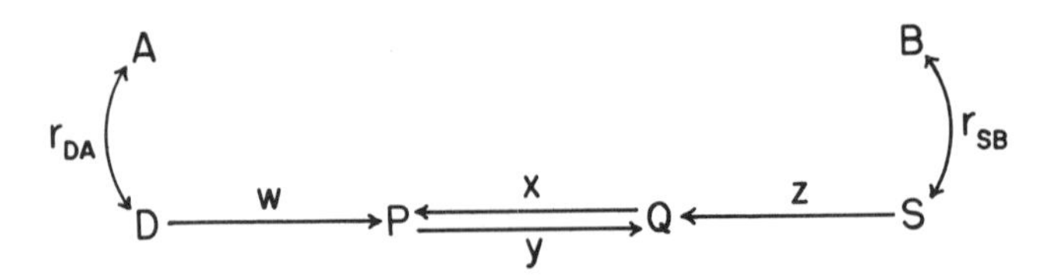

FIG. 14.8. (Redrawn from Fig. 10, Wright 1960b.)

The following equations can be written. Different symbols are used for the path coefficients since even those that connect the same variables (p_{PD}, p_{QS} in Figures 14.7 and 14.8) cannot be expected to have the same values under the different viewpoints.

$$(14.17)\quad \begin{cases} r_{QD} = yr_{PD} = yw/(1 - xy). \\ r_{PD} = w + xr_{QD} = w/(1 - xy). \\ r_{PS} = xr_{QS} = xz/(1 - xy). \\ r_{QS} = z + yr_{PS} = z/(1 - xy). \\ r_{PP} = 1 = xr_{PQ} + wr_{PD}. \\ r_{QQ} = 1 = yr_{PQ} + zr_{QS}. \\ r_{PQ} = x + wr_{QD} = x + ywr_{PD} = x + y(1 - xr_{PQ}) \\ \quad = (x + y)/(1 + xy). \end{cases}$$

As only three known correlations (r_{PP}, r_{QQ}, and r_{PQ}) are here analyzed; solution for four unknowns requires use of a known variable of type A or B. Either of those adds two usable equations (e.g., $r_{QA} = yr_{PA}$; $r_{PA} = wr_{DA} + xr_{QA}$) at the expense of only one additional unknown quantity and thus permits solution.

$$(14.18)\quad \begin{cases} y = r_{QA}/r_{PA}. \\ x = (r_{PA}r_{PQ} - r_{QA})/(r_{PA} - r_{QA}r_{PQ}) \quad \text{from } r_{PQ} \text{ in 14.17}. \\ w = \sqrt{(1 - xy)(1 - xr_{PQ})}. \\ z = \sqrt{(1 - xy)(1 - yr_{PQ})}. \\ r_{DA} = (r_{PA} - xr_{QA})/w = r_{PA}(1 - xy)/w. \end{cases}$$

The regression equations for Q and P are as follows:

$$(14.19)\quad \begin{cases} Q - \bar{Q} = y\dfrac{\sigma_Q}{\sigma_P}(P - \bar{P}) + z\dfrac{\sigma_Q}{\sigma_S}(S - \bar{Q}); \\ P - \bar{P} = x\dfrac{\sigma_P}{\sigma_Q}(Q - \bar{Q}) + w\dfrac{\sigma_P}{\sigma_D}(D - \bar{Q}). \end{cases}$$

By substituting each in the other, these can be expressed in terms of S and D for comparison with the preceding analysis.

$$Q - \bar{Q} = \frac{yw}{1 - xy}\frac{\sigma_Q}{\sigma_D}(D - \bar{Q}) + \frac{z}{1 - xy}\frac{\sigma_Q}{\sigma_S}(S - \bar{Q}).$$

$$P - \bar{P} = \frac{w}{1 - xy}\frac{\sigma_P}{\sigma_D}(D - \bar{Q}) + \frac{xz}{1 - xy}\frac{\sigma_P}{\sigma_S}(S - \bar{Q}).$$

$$(14.20)\quad \begin{cases} e = \left(\dfrac{Q - \bar{Q}}{P - \bar{P}}\right)_{S=\bar{Q}} = y\dfrac{\sigma_Q}{\sigma_P} = \dfrac{r_{QA}}{r_{PA}}\dfrac{\sigma_Q}{\sigma_P}. \\ \eta = \left(\dfrac{Q - \bar{Q}}{P - \bar{P}}\right)_{D=\bar{Q}} = \dfrac{1}{x}\dfrac{\sigma_Q}{\sigma_P} = \left(\dfrac{r_{PA} - r_{QA}r_{PQ}}{r_{PA}r_{PQ} - r_{QA}}\right)\dfrac{\sigma_Q}{\sigma_P}. \end{cases}$$

The concrete elasticities are the same as before. Since *P* and *Q* are both determined by *S* and *D*, as shown in Figure 14.7, there was an arbitrary choice in representing *P* as determined by *Q* and *D*, and *Q* by *P* and *S* in Figure 14.8. Figure 14.9 is an equally legitimate representation. The path coefficients are, of course, all different.

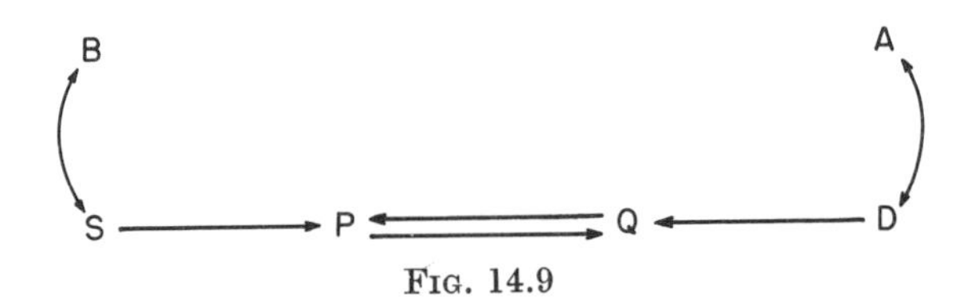

FIG. 14.9

Since the diagrams are formally identical except for exchange of *Q* and *P*, the solution can be written by analogy. The concrete elasticities are again unchanged.

For a somewhat complicated biological example, consider the following data from the classical experiments of J. S. Haldane and J. G. Priestley (1905) on the relations between respired air (*A*), alveolar air in the lungs (*L*), and depth (*D*) and frequency (*F*) of respiration. There were 39 experiments on one subject (J.S.H.) and 67 on another (J.G.P.). Unfortunately for the present purpose only the averages for groupings of similar experiments were published (15 and 21 groups, respectively). The number of experiments in groups varied from 1 to 4 in the former case, 1 to 6 in the latter. The values to be used here for the correlations and regressions are those derived by weighting the reported group averages.

Variables, *A*, *D*, and *F* were direct measurements. Variable *C* was based on the CO_2 content in samples taken after expiration to nearly the greatest possible extent corrected for residual nonalveolar air, on the basis of estimation of the "dead space" in the lungs of the subject. The virtual absence of correlations between *C* and any of the other variables, its high mean, and low standard deviation in the case of J.G.P. suggest an overcorrection for nonalveolar air. The results brought out sufficiently well the close control of alveolar CO_2, in spite of the enormous variability of CO_2 in the respired air, but do not permit satisfactory path analysis. The data from J.S.H. appear to be more satisfactory for this purpose, but the probability that the correction was not exact must be borne in mind. The data will be used to illustrate the possibilities of path analysis in more complicated systems than a single reciprocal interaction, recognizing that they are not as suitable as if recorded with this in mind. Variable *C* is treated as a somewhat imperfect measure of CO_2 in the alveolar air, determined by a hypothetical variable *X* for the CO_2

TABLE 14.25. Statistics from experiments on control of respiration by Haldane and Priestley (1905).

Variables		Subjects			
		J.S.H.		J.G.P.	
		Mean	σ	Mean	σ
% CO_2 in respired air	(A)	3.052	1.350	3.507	1.348
% CO_2 in alveolar air	(C)	5.767	0.478	6.568	0.241
Depth of respiration (cc)	(D)	1,200	358	794	214
Frequency of respiration/sec	(F)	15.21	2.203	17.60	2.161
Correlations	r_{DA}	0.979		0.710	
	r_{FA}	0.501		0.779	
	r_{DF}	0.530		0.416	
	r_{CA}	0.823		0.093	
	r_{CD}	0.772		0.015	
	r_{CF}	0.354		−0.067	

concentration of pulmonary air from which C deviates by random errors of technique (U).

The following pattern, Figure 14.10, was used after it appeared by trial

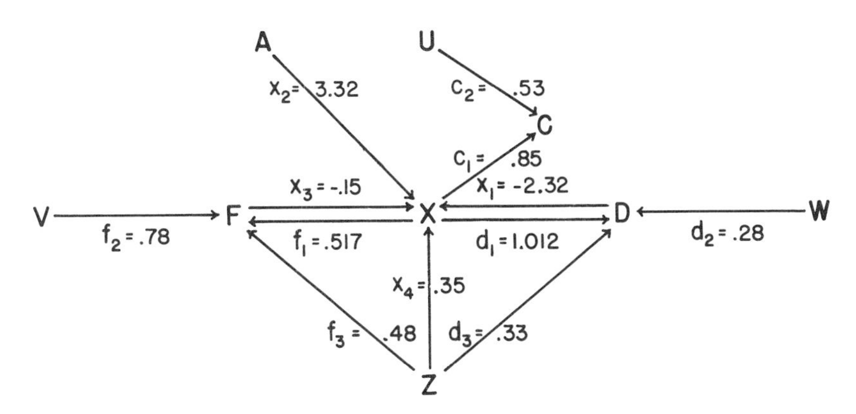

FIG. 14.10. Path diagram of relations of CO_2 concentration of respired air (A), CO_2 concentration of alveolar air—estimated (C) and unknown actual (X), depth of respiration (D), frequency of respiration (F), and residual factors U, V, W, and Z from data from one subject. (Reference from Fig. 13, Wright 1960*b*.)

TABLE 14.26. Formulas for correlations between variables of Figure 14.10 and values calculated from the deduced path coefficients.

No.	Formulas	Value
1	$K = 1 - x_1 d_1 - x_3 f_1$	3.433
2	$r_{XA} = 0.8229/c_1$	0.968
3	$= x_1 r_{DA} + x_2 + x_3 r_{FA} = x_2/(1 - x_1 d_1 - x_3 f_1) = x_2/K$	0.968
4	$r_{DA} = 0.9793 = d_1 r_{XA}$	0.979
5	$r_{FA} = 0.5008 = f_1 r_{XA}$	0.501
6	$r_{XW} = x_1 r_{DW} + x_3 r_{FW} = x_1 d_2/K$	−0.190
7	$r_{DW} = d_1 r_{XW} + d_2$	+0.088
8	$r_{FW} = f_1 r_{XW}$	−0.098
9	$r_{XV} = x_1 r_{DV} + x_3 r_{FV} = x_3 f_2/K$	−0.034
10	$r_{DV} = d_1 r_{XV}$	−0.035
11	$r_{FV} = f_1 r_{XV} + f_2$	+0.762
12	$r_{XZ} = x_1 r_{DZ} + x_3 r_{FZ} + x_4 = (x_1 d_3 + x_3 f_3 + x_4)/K$	−0.143
13	$r_{DZ} = d_1 r_{XZ} + d_3$	+0.186
14	$r_{FZ} = f_1 r_{XZ} + f_3$	+0.406
15	$r_{XD} = 0.7718/c_1$	+0.908
16	$= x_1 + x_2 r_{DA} + x_3 r_{DF} + x_4 r_{DZ}$	+0.912
17	$= d_1 + d_2 r_{XW} + d_3 r_{XZ}$	+0.911
18	$r_{XE} = 0.3541/c_1$	+0.417
19	$= x_1 r_{DF} + x_2 r_{FA} + x_3 + x_4 r_{FZ}$	+0.424
20	$= f_1 + f_2 r_{XV} + f_3 r_{XZ}$	+0.422
21	$r_{DF} = 0.5295$	+0.530
22	$= d_1 r_{XF} + d_2 r_{FW} + d_3 r_{FZ}$	+0.528
23	$= f_1 r_{XD} + f_2 r_{DV} + f_3 r_{DZ}$	+0.532
24	$r_{XX} = 1 = x_1 r_{XD} + x_2 r_{XA} + x_3 r_{XF} + x_4 r_{XZ}$	0.995
25	$r_{DD} = 1 = d_1 r_{XD} + d_2 r_{DW} + d_3 r_{DZ}$	1.004
26	$r_{FF} = 1 = f_1 r_{XF} + f_2 r_{FV} + f_3 r_{FZ}$	1.005

From data of Wright (1960*b*).

that it was necessary to assume that a common factor or factors, Z, contributes positively to correlations among F, X, and D. This residual factor presumably consists primarily of variability in the amount of CO_2 released into the blood by metabolism, throughout the body. If X were a measure of CO_2 in equilibrium with the arterial blood, it might be supposed that the effects of Z on depth and frequency of respiration would be fully taken care of in the paths of interaction of X with D and F. That this is not the case indicates that X does not fully represent CO_2 content of the blood. Simplified symbols are used for the path coefficients.

In solving, it is assumed first that the path regression of X on A is 1, as expected under a steady state with D, F, and Z constant.

$$c_{XA} = x_2 \frac{\sigma_X}{\sigma_A} = x_2 c_1 \frac{\sigma_C}{\sigma_A} = 1. \tag{14.21}$$

From the observed values of σ_C and σ_A, $x_2 c_1 = 2.8249$,

$$\left\{ \begin{aligned} r_{CA} &= c_1 r_{XA} = 0.8229; \\ r_{DA} &= d_1 r_{XA} = 0.9793; \\ r_{FA} &= f_1 r_{XA} = 0.5008; \\ r_{XA} &= x_1 r_{DA} + x_2 + x_3 r_{FA} = x_2/(1 - x_1 d_1 - x_3 f_1) = x_2/K; \\ K &= x_2 c_1 / r_{CA} = 3.433. \end{aligned} \right. \tag{14.22}$$

The equations (excluding those involving C) are given in Table 14.26. Solution requires that path coefficients be found that satisfy the relations arrived at above, that yield the same result from the three equations for r_{XD}, the three for r_{XF}, the three for r_{DF}, and that satisfy the three equations of complete determination. The following approximate values were obtained by a rather tedious process of trial and error. They yield the correlations shown in the last column of Table 14.1, which agree to two decimal places where there is a test.

TABLE 14.27. Estimated path coefficients and path regressions of Figure 14.10.

Path Coefficients		Path Regressions
$c_1 = p_{CX} = 0.85$	$d_1 = p_{DX} = 1.012$	$c_{XD} = -0.00263$
$c_2 = p_{CU} = 0.527$	$d_2 = p_{DW} = 0.28$	$c_{XA} = 1$
$x_1 = p_{XD} = -2.328$	$d_3 = p_{DZ} = 0.33$	$c_{XF} = -0.0280$
$x_2 = p_{XA} = 3.323$	$f_1 = p_{FX} = 0.517$	$c_{DX} = 893$
$x_3 = p_{XF} = -0.15$	$f_2 = p_{FV} = 0.78$	$c_{FX} = 2.81$
$x_4 = p_{XZ} = 0.35$	$f_3 = p_{FZ} = 0.48$	

It may be seen that the relative effect ($x_3 = -0.15$) of rate of respiration on alveolar CO_2 is almost negligible in comparison with that ($x_1 = -2.328$) of depth of respiration.

It is apparent, however, from the value of the coefficients measuring the direct influence of metabolism, Z, on depth and frequency of respiration that X cannot be identified fully with CO_2 content of alveolar air in immediate equilibrium with arterial blood (which will be designated Y). It must be considered to be the function of Z, D, F, and A from which the estimate C may be considered a random deviate. It is instructive to attempt to add Y to the diagram.

It is, however, impossible to entirely eliminate p_{DZ} and p_{FZ} because of the differences between them before Y is added. In defining Y, the assumption will be made that the values of p_{DZ} and p_{FZ} in the new system are to be numerically equal, but the former is negative and the latter positive.

With this assumption, a solution is readily arrived at. Some of the path coefficients in the new diagram (Fig. 14.11), which will be distinguished by primes, are unchanged. Others can be deduced. The rest can all be expressed in terms of y_1' ($=p_{YX}'$) and thus can be obtained by iteration for only this one

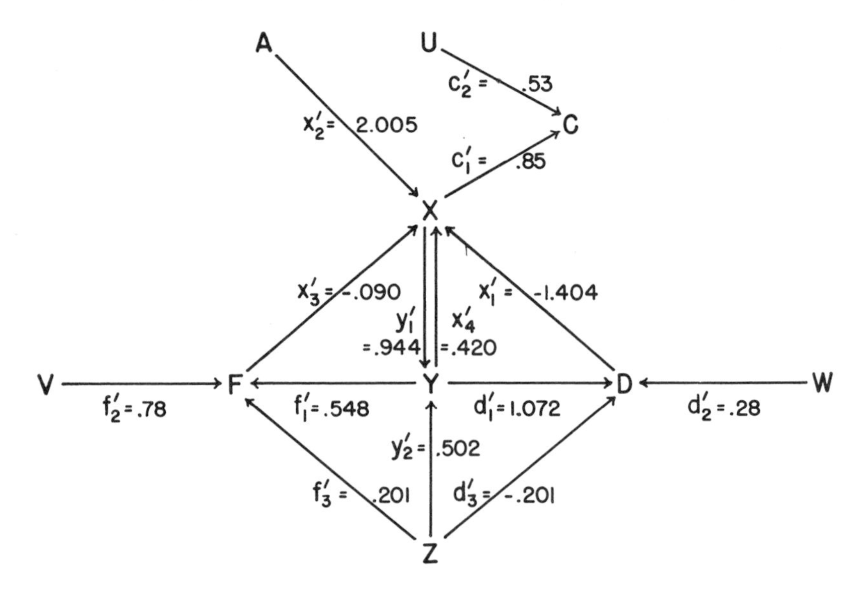

FIG. 14.11. Elaboration of Fig. 13.10, designed to make Z represent more nearly the amount of metabolism as a variable factor. Y here represents the unknown actual CO_2 concentration of alveolar air and X is a statistical intermediary. (Redrawn from Fig. 13, Wright 1960*b*.)

parameter. The details are given by Wright (1960*b*). The path coefficients came out as follows:

TABLE 14.28. Estimated path coefficients of Figure 14.11.

$x_1' = -1.404$	$f_1' = 0.548$	$y_1' = 0.944$	$d_1' = 1.072$
$x_2' = 2.005$	$f_2' = 0.78$	$y_2' = 0.502$	$d_2' = 0.28$
$x_3' = -0.090$	$f_3' = 0.201$		$d_3' = -0.201$
$x_4' = 0.420$			

The rate of increase of CO_2 in Y should be proportional to Z while its rate of decrease in the lungs should be proportional to the difference between the concentrations in Y and X. At flux equilibrium

$$\frac{dY}{dt} = k_1 Z - k_2(Y - X) = 0,$$

$$k_1(\bar{Z} + \delta Z) = k_2(\bar{Y} + \delta Y - \bar{X} - \delta X),$$

$$k_1\,\delta Z = k_2\,\delta Y - k_2\,\delta X,$$

$$\delta Y = \frac{k_1}{k_2}\,\delta Z + \delta X = c_{YZ}\,\delta Z + c_{YX}\,\delta X.$$

Thus

$$c_{YX} = 1, \qquad c_{YZ} = \frac{k_1}{k_2}. \tag{14.23}$$

In the case of X, CO_2 tends to increase at the rate $k_2(Y - X)$ and to decrease jointly according to rate and frequency of respiration and the differences between the concentrations in X and A.

$$\frac{dX}{dt} = k_2(Y - X) - k_3DF(X - A) = 0.$$

Ignoring second order terms,

$$k_2(\delta Y - \delta X) = k_3[\bar{F}(\bar{X} - \bar{A})\,\delta D + \bar{D}\,(\bar{X} - \bar{A})\,\delta F + \bar{D}\bar{F}(\delta X - \delta A)],$$

$$\delta X(k_2 + k_3\bar{D}\bar{F}) = k_2\,\delta Y + (k_3\bar{D}\bar{F})\,\delta A - k_3[\bar{F}(\bar{X} - \bar{A})\,\delta D + \bar{D}(\bar{X} - \bar{A})\,\delta F],$$

$$c_{XY} = \frac{k_2}{k_2 + k_3\bar{D}\bar{F}}, \qquad c_{XA} = \frac{k_3\bar{D}\bar{F}}{k_2 + k_3\bar{D}\bar{F}},$$

$$c_{XD} = -\frac{k_3\bar{F}(\bar{X} - \bar{A})}{k_2 + k_3\bar{D}\bar{F}}, \quad c_{XF} = -\frac{\bar{D}(\bar{X} - \bar{A})}{k_2 + k_3\bar{D}\bar{F}}.$$

Thus $c_{XA} + c_{XY} = 1$,

$$c_{XA} = x_2'\,\frac{\sigma_X}{\sigma_A} = 2.005c_1'\,\frac{\sigma_C}{\sigma_A} = 0.603.$$

It appears from this that $c_{XY} = 0.397$. The value of σ_Y can be estimated from $c_{XY} = x_4' \, \sigma_X/\sigma_Y = 0.397$, giving $\sigma_Y = 0.430$, which while slightly larger is at least close to the estimate for $\sigma_X = 0.406$ as expected.

As variable X has no simple physiological meaning, it may well be removed from the system. The path coefficients in the simplified diagram (Fig. 14.12), to be indicated by double primes, can easily be deduced by comparison of the formulas from Figures 14.11 and 14.12. All but y_1'', y_2'', y_3'', and y_4'' are unchanged. Let $L = (1 - y_1'x_4') = 0.604$.

TABLE 14.29. Estimated path coefficients and path regressions of Figure 14.12.

$y_1'' = y_1'x_1'/L$	$= -2.201$	$y_3'' = y_1x_3'/L$	$= -0.142$
$y_2'' = y_1'x_2'/L$	$= 3.142$	$y_4'' = y_2/L$	$= 0.831$
$c_{YA} = y_2'' \, \sigma_Y/\sigma_A$	$= 1.000$		
$c_{YD} = y_1'' \, \sigma_Y/\sigma_D$	$= -0.00264$	$c_{DY} = d_1'' \, \sigma_D/\sigma_Y$	$= 892$
$c_{YF} = y_3'' \, \sigma_Y/\sigma_F$	$= -0.0277$	$c_{FY} = f_1'' \, \sigma_F/\sigma_Y$	$= 2.81$

The shift from variable X to Y results in no changes in the path regressions. Figure 14.12, however, probably gives a better representation of the physiological situation in this series of experiments. The variations in CO_2 of the outside air (A) tended to produce directly more than three times as great a standard deviation in the CO_2 content of the alveolae (Y) as that actually observed ($p_{YA} = 3.14$), and the variation in the depth of respiration (D) tended to produce more than twice the observed standard deviation

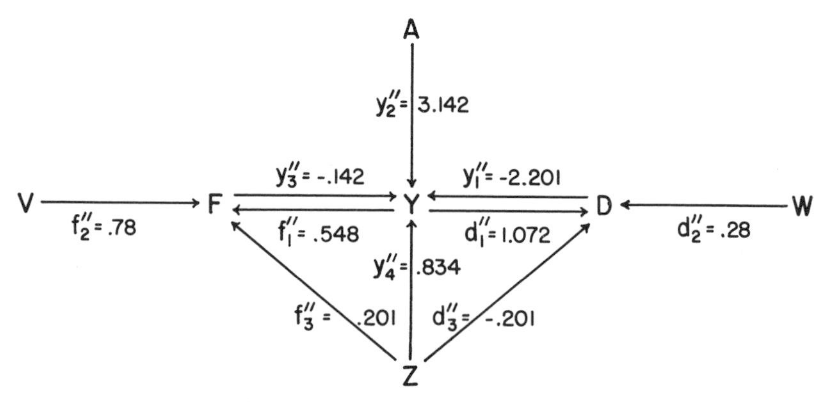

FIG. 14.12. Same as Fig. 14.11, except for removal of the arbitrary variable X. (Redrawn from Fig. 15, Wright 1960*b*.)

($p_{YD} = -2.20$). But these largely cancelled each other because of the almost perfect positive correlation ($r_{DA} = +0.979$) between them, due mainly to stimulatory effect of CO_2 in the arterial blood in equilibrium with that in the alveolar air on the respiratory center and, hence, on depth of respiration ($p_{DY} = 1.07$). There was also a moderately strong stimulatory effect on frequency of respiration ($p_{FY} = +0.55$), but the effect of the latter on CO_2 content of the alveolar air was practically negligible in this subject ($p_{YF} = -0.14$).

A similar analysis of the data from the other subject (J.G.P.) is not practicable for reasons already noted. There appears to have been an overcorrection for dead space in the lung that has left nothing but random variations in the estimates of alveolar CO_2.

A rough comparison may, however, be attempted on the basis of the correlations among A, D, and F and borrowing from the analysis for the first subject. Three statistics must be borrowed to compensate for the three missing correlations.

We might take $c_{YA} = p_{YA}\,\sigma_Y/\sigma_A = 1$ as one of them. This yields $p_{YA} = 5.59$ if σ_Y is equated to observed σ_C. This is considerably larger than that from the first subject (3.14). The observed value of σ_C for J.G.P. is, however, probably much too low because of overcorrection. It seems better for purposes of comparison to borrow the path coefficient 3.14. The other statistics that seem most suitable for borrowing are the coefficients $p_{FZ} = +\,0.201$ and $p_{DZ} = -\,0.201$ because of their smallness. The results are shown in Figure 14.13. They bring out systematically the great difference between the two subjects implied by the correlations r_{DA} and r_{FA}. Alveolar CO_2 content

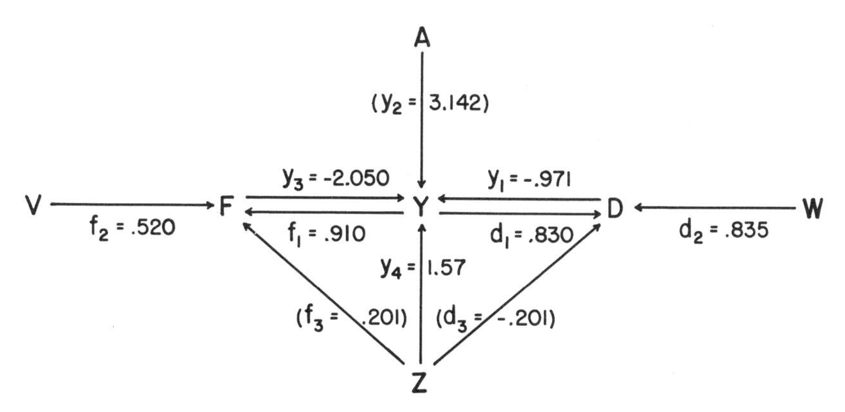

FIG. 14.13. Similar to Fig. 14.12, except based as far as possible on the data from a second subject. (Redrawn from Fig. 16, Wright 1960b.)

(Y) seems to have been controlled about twice as much by frequency as by depth of respiration in the case of J.G.P. in contrast with virtually exclusive control by depth in that of J.S.H.

Other systems involving ordering in time will be treated in Part II. These include correlations in the abundance of species in ecological systems (which involve interactions with lag) and correlations among relatives. The most important applications of path analysis, however, will be in the working out of the genetic consequences of mating systems.

□ □

CHAPTER 15

The Genetics of Quantitative Variability

Early Theories

The prevailing concept of heredity before 1900 was that of the fluid blending implied by such terms as half-blood and quarter-blood. This was probably based on observations of the effects of crossing breeds of livestock that differed in size, conformation, and such characteristics as length, fineness and density of wool fibers in sheep, as well as on observations on crosses between human races. Blending was considered the rule and deviations from this were looked upon as requiring special supplementary explanations such as exceptional "prepotency" of certain individuals or breeds, the transmission of an exceptionally "acquired character," an effect of "telegony" or of "maternal impressions." With respect to plants, the extensive experiments on the production of species hybrids in the eighteenth and nineteenth centuries also indicated that blending of characters was the rule.

Darwin made extensive experiments with many species of plants but did not achieve any substantial advance. His theory of evolution (1859) by a process of gradual natural selection was severely criticized by Fleeming Jenkin (1867) as not allowing for the swamping of variability in each generation that is to be expected under blending heredity unless an amount of new variability, equal to that left in each generation after blending, is supplied from some source. Darwin himself became so impressed with this difficulty that he proposed a Lamarckian interpretation of heredity, "pangenesis."

Galton in 1889 distinguished sharply between the inheritance of continuous and stepwise variability and proposed an attack, especially on the former, by statistical study of individuals related by descent. Pearson developed the method of multiple regression in connection with this problem. The resulting estimation formulas were valid statistically for individuals of the population from which the coefficients had been derived. It was not sufficiently appreciated, however, that this approach gives no indication of the nature of the biological process. Nevertheless the methods developed by the biometric school under Pearson's leadership constitute one of the pillars supporting modern population genetics.

The Multiple-factor Hypothesis

The other main pillar had already been suggested by Mendel in 1865 in his account of a cross between *Phaseolus multiflorus* with colored flowers and *P. nanus* with white flowers that yielded a more or less continuous series of colors from purple red to pale violet and white in F_2. As translated by Bateson 1909:

> Even these enigmatic results, however, might probably be explained by the law governing Pisum if we might assume that the color of the flowers and seeds of *P. multiflorus* is a combination of two or more entirely independent colors which individually act like any other constant characters of the plant.

He noted that $1/16$ of F_2 would be white-flowered if two characters were involved, $1/64$ if three, and that the observed 1 in 31 might be accounted for in one of these ways.

After the rediscovery of Mendelian heredity in 1900, most investigators focused their attention exclusively on sharply segregating characters. A violent controversy ensued between the biometricians, led by Pearson, and the Mendelians, led by Bateson. As early as 1902, however, the biometrician, Yule, pointed the way to the reconciliation that ultimately led to modern population genetics. He showed that there would be no decline in variability in an indefinitely large population derived from a cross, since each pair of alleles would continue indefinitely under random mating to be represented in the three genotypes in the ratio 1:1, and zygotes would continue to be present according to the square of this ratio. Castle (1903) generalized these conclusions for any gene ratio. These principles attracted no attention until pointed out by both Hardy and Weinberg in 1908. This persistence of genetic variability under Mendelian heredity removes Fleeming Jenkin's criticism of Darwin's theory.

There had been much confused thinking on quantitative variability, going back to Darwin's Lamarckian views, from failure to discriminate sharply between the contributions of heredity and environment. Clarification was due especially to Johannsen (1903, 1909) who made experiments with a number of lines of beans, each maintained by self-fertilization. The average seed weights of these lines varied from 35 to 60 cg. but there was wide overlapping of the distributions. Seeds of the same size from different lines developed into plants that produced seeds of very different average weights, the average being in each case the same, within the limits expected from sampling, as had been characteristic of the line of origin. Several generations of selection of small seeds along one subline, large seeds along another from the same pure line, produced no significant deviations from the line average.

Johannsen's pure-line concept was exemplified dramatically in experiments by G. H. Shull (1909) with a variety of maize that was homogeneous as understood at that time. The lines that were obtained from several generations of self-fertilization differed greatly from each other. However, they showed unprecedented uniformity within themselves in numerous plant and ear characters that had not been thought of as varying genetically in the original variety. Shull pointed out what had been noted long before by Mendel that half of the genes that are heterozygous in any plant should become homozygous one way or the other in any offspring plant under selfing. Thus selfing brings about exactly the same decrease in variability in Mendelian characters, along any one line, that is expected in a large random-breeding population under strict blending heredity. There is a strong implication that any variability that persists in large random-breeding populations but is rapidly lost within self-fertilized lines is Mendelian.

Shull noted the profound decrease in plant height, yield, and other aspects of vigor that occurred in maize under self-fertilization and the extraordinary restoration of vigor on crossing weakened inbred lines (both in harmony with extensive observations on various species of plants made by Darwin). He proposed the term "heterosis" for this hybrid vigor, without any implication as to its genetic mechanism. He suggested the enormously useful practical application to hybrid corn.

More specific evidence for the multiple factor hypothesis was supplied by Nilsson-Ehle's (1909) complete Mendelian analysis of quantitatively varying color characters of oats and wheat, most notably of a cross between red-glumed and white-glumed lines of wheat, which gave intermediate red in F_1 and variability but (as it happened) no whites in a rather large F_2 population. Among 78 selfed F_2's, 8 gave ratios of 3 red:1 white, 15 good 15:1 ratios, 5 ratios interpretable as 63:1, while 50 gave only reds, close to the Mendelian expectation of 7:15:10:46 for these classes, on the hypothesis of three independently segregating loci and white a triple homozygote. East (1910) demonstrated similarly that variation ranging from intense yellow to white in maize depended on two independent loci.

The multiple-factor hypothesis for quantitative variability was given substance especially by the extensive experiments of East and Hayes (1911), Emerson and East (1913), Hayes (1913), and East (1916). These consisted of crosses between more or less inbred strains (in various plant species) that differed in various characters, comparison of variability in F_2 (in which segregation should be occurring) with that in the parental strains, and in F_1, followed by comparisons of F_3 progenies from F_2 parents of different grades. Tables 15.1 and 15.2 present a few of their comparisons of parents, F_1 and F_2 in terms of means and coefficients of variability which, the authors

TABLE 15.1. Parents, F_1 and F_2, for various characters of maize. M' and $(\sigma')^2$ are here derived from M and C by the logarithmic transformation. Minimum estimates of the number of loci are from the untransformed (n_1) (above, in parentheses) and transformed data (n_2) (below), basing σ_E^2 on the variances of P_1, F_1, and P_2 in 1:2:1 ratio and correcting for dominance.

	No.	M	σ	C	M'	$(\sigma')^2$	n_1, n_2
Height (cm.)							
P_1 Tom Thumb	15	85	8.84	0.104	1.927	0.00203	
P_2 Missouri Dent	144	229	23.98	0.105	2.357	0.00206	
F_1	20	175	11.18	0.064	2.242	0.00077	(6.1)
F_2	223	159	26.41	0.167	2.195	0.00519	6.8
Nodes (no.)							
P_1 Tom Thumb	15	8.9	0.81	0.091	0.948	0.00156	
P_2 Missouri Dent	145	19.8	1.46	0.074	1.296	0.00103	
F_1	20	12.9	0.86	0.067	1.110	0.00085	(4.3)
F_2	223	13.0	2.05	0.157	1.108	0.00458	4.3
Internode length (cm.)							
P_1 Tom Thumb	15	9.2	0.85	0.093	0.962	0.00162	
P_2 Missouri Dent	142	11.4	1.20	0.105	1.055	0.00206	
F_1	20	13.5	1.22	0.090	1.129	0.00152	(0.2)
F_2	223	12.3	2.27	0.184	1.079	0.00629	0.2
Height (cm.)							
P_1 Watson Flint	97	68	6.48	0.095	1.831	0.00169	
P_2 Leaming Dent	73	101	5.07	0.050	2.004	0.00047	
F_1	56	95	8.21	0.087	1.976	0.00143	(1.8)
F_2	538	83	11.09	0.134	1.915	0.00337	2.2
Ear length (cm.)							
P_1 Tom Thumb	57	6.6	0.81	0.123	0.816	0.00283	
P_2 Black Mexican	101	16.8	1.86	0.111	1.223	0.00231	
F_1	69	12.1	1.51	0.125	1.079	0.00292	(11.9)
F_2	1268	12.7	1.97	0.155	1.099	0.00447	12.4
Weight of seeds (gm./25)							
P_1 Tom Thumb	61	2.7	0.39	0.144	0.427	0.00387	
P_2 Black Mexican	55	8.3	1.19	0.145	0.915	0.00392	
F_1	54	4.6	0.64	0.139	0.658	0.00360	(12.4)
F_2	962	5.6	1.02	0.182	0.741	0.00614	12.4

From data of Emerson and East (1913).

TABLE 15.2. Results of crosses between squashes (*Cucurbito pepo*) differing greatly in length and diameter of fruit (Emerson and East 1913), and of a cross between tobacco varieties (*Nicotiana tabacum*) with the same numbers of leaves (Hayes *et al.* 1913). M', $(\sigma')^2$, n_1, and n_2 as in Table 15.1. Estimates of n_1 and n_2 in second and third cases meaningless because parental strains are not at extremes.

	No.	M	σ	C	M'	$(\sigma')^2$	n_1, n_2
Length (cm.), squash							
P_1 Crookneck		39.6	6.73	0.170	1.592	0.00537	
P_2 Scallop		7.4	1.17	0.158	0.864	0.00465	
F_1		17.5	3.32	0.190	1.235	0.00667	(2.6)
F_2		19.6	8.37	0.427	1.256	0.03159	2.6
Diameter (cm.), squash							
P_1 Crookneck		11.4	1.37	0.120	1.054	0.00270	
P_2 Scallop		17.8	2.24	0.126	1.247	0.00297	
F_1		17.5	2.21	0.126	1.240	0.00297	(0.1)
F_2		13.2	7.76	0.588	1.056	0.00560	0.1
Leaves (no.), tobacco							
P_1 Cuban 1910	150	19.9	1.50	0.075	1.298	0.00106	
1911	124	20.6	1.09	0.053	1.313	0.00053	
P_2 Havana 1910	143	19.8	1.38	0.070	1.296	0.00092	
1911	150	20.3	1.80	0.089	1.307	0.00149	
F_1 (1910)	150	19.8	1.21	0.061	1.296	0.00070	
F_2 (1911)	192	20.9	3.31	0.158	1.318	0.00467	

pointed out, were usually more uniform in parents and in F_1 than standard deviations (also given here). In all cases there is an obvious increase in variability in F_2, although not so great in most cases in which the parents are chosen at opposite extremes as to make possible interpretation in terms of a single pair of alleles.

The results of a cross between two varieties of *Nicotiana longiflora* that differed enormously in corolla length are given more completely in Table 15.3. In his paper, East (1916) listed eight principles that should hold under the multiple factor hypothesis, if all populations after the original cross are obtained by self-fertilization.

1. Crosses between individuals belonging to races which, from long continued self-fertilization or other close inbreeding, approach a homozygous condition, should give F_1 population comparable to the parental races in uniformity.

TABLE 15.3. Results to F_5 from a cross between varieties of *Nicotiana longiflora* that differed in tube length.

Designation	Year	Generation	Parent Size	Class Center in mm.																							No.	M	C
				34	37	40	43	46	49	52	55	58	61	64	67	70	73	76	79	82	85	88	91	94	97	100			
383	1911				13	80	32																				125	40.5	0.043
	1912			1	4	28	16																				49	40.6	0.049
	1913				4	32	1																				37	39.8	0.027
330	1911																					6	22	49	11		88	93.2	0.025
	1912																					2	16	32	6	1	57	93.4	0.024
	1913																					5	7	10	2		24	92.1	0.029
(383×330)	1911	F_1									4	10	41	75	40	3											173	63.5	0.046
1	1912	F_2	61							1	5	16	23	18	62	37	25	16	4	2	2						211	67.5	0.088
2			61							2	4	2	24	37	31	38	35	27	21	5	6	1					233	69.8	0.097
1–1	1912	F_3	72										4	20	25	59	41	19	2								170	73.1	0.052
1–2			46				1	4	26	44	38	22	7	1													143	53.5	0.070
1–3	1913		50				6	20	53	49	15	4															147	50.2	0.063
1–4			60				2	3	9	25	37	70	19	10													175	56.3	0.072
2–1			77						1		1	1	1	2	16	33	43	34	20	6	1						159	73.0	0.068
2–3			81										1	1	8	16	20	32	41	17	3	3	1				143	76.3	0.066
2–4			80										2	8	14	21	39	39	32	10	1						166	74.0	0.065
2–5			50				7	25	55	55	18																160	53.0	0.057
2–6			82												3	5	12	20	40	41	30	9	2				162	80.2	0.059
1–2–1	1914	F_4	44			8	42	95	38	1																	184	45.7	0.052
1–3–1			43			2	23	122	41	1																	189	46.2	0.040
2–6–1			85														4	9	38	75	59	6	3	1			195	82.2	0.040
2–6–2			87												4	5	6	11	21	33	41	29	8	5	1		164	82.9	0.070
1–3–1–1	1915	F_5	41	3	6	48	90	14																			161	42.0	0.055
2–6–2–1			90														2	3	8	14	20	25	25	20	8		125	87.9	0.063

From data of East (1916).

2. In all cases where the parent individuals may reasonably be presumed to approach complete homozygosis, F_2 frequency distributions arising from extreme variants of the F_1 population should be practically identical, since in this case all F_1 variation should be due to external conditions.
3. The variability of the F_2 population from such crosses should be much greater than that of the F_1 population.
4. When a sufficient number of F_2 individuals are available, the grandparental types should be recovered.
5. In certain cases, individuals should be produced in F_2 that show a more extreme deviation than is found in the frequency distribution of either grandparent.
6. Individuals from various points in the frequency curve of an F_2 population should give F_3 populations differing markedly in their modes and means.
7. Individuals from the same or from different points in the frequency curve of an F_2 population should give F_3 populations of diverse variabilities extending from that of the original parent to that of the F_2 generation.
8. In generations succeeding the F_2, the variability of any family may be less but never greater than the variability of the populations from which it came.

East noted that all of these conditions had been met many times in the course of experiments. Most of them are illustrated (with minor exceptions that are not significant) in the data in Table 15.3. With respect to 4, the failure of F_2 to yield the parental types indicates a rather large number of segregating loci. Condition 5 had been predicted before actual cases were discovered. It is well illustrated by Hayes' cross between varieties of tobacco that were almost identical in leaf number and gave a similar F_1 (Table 15.2). The enormous increase in variability in F_2 permitted rapid selection of a variety with a much larger number of leaves than either parent variety.

Higher animals are less suitable than plants for the exhibition of the principles of multifactorial heredity because of the impossibility of self-fertilization and of the greater difficulty in obtaining adequate numbers for statistical analysis. Nevertheless, at least the first three of East's rules were soon found to apply. Table 15.4 gives comparison of mean and variability of F_1 and F_2 for four diverse size characters in a cross between two breeds of rabbits at opposite extremes in size (Castle 1922). Other characters behaved similarly except that genetic variability of measurements of skull width were more obscured by nongenetic variability than in the other cases. The general result was an F_1 intermediate between the parents and an F_2 with somewhat lower mean but significantly more variability, which, however, fell far short of including the parental sizes.

The data in Table 15.5 illustrate a different sort of character. A highly variable strain of rats with the hooding pattern of color on head and midline

TABLE 15.4. Results from cross between Polish Dwarf and Flemish Giant rabbits. Transformation (M', $[\sigma']^2$) and estimations of minimum number of loci as in Table 15.1.

Character and Class	No.	M	σ	C	M'	$(\sigma')^2$	n_1, n_2
Weight (gm.)							
P_1	20	1404	142	0.1011	3.145	0.00192	
P_2	3	3646			3.560		
F_1	27	2512	198	0.0788	3.399	0.00117	(23.4)
F_2	112	2126	257	0.1209	3.324	0.00274	13.6
Ear length (mm.)							
P_1	23	83.5	1.68	0.0212	1.922	0.000085	
P_2	3	145.3			2.162		
F_1	27	109.3	3.76	0.0344	2.038	0.000223	(21.3)
F_2	112	107.0	6.05	0.0565	2.029	0.000601	19.1
Skull length (mm.)							
P_1	21	66.0	1.88	0.0285	1.819	0.000153	
P_2	5	85.5			1.932		
F_1	27	75.5	1.18	0.0156	1.878	0.000046	(5.6)
F_2	125	73.1	3.15	0.0431	1.864	0.000350	5.2
Femur length (mm.)							
P_1	20	72.3	1.71	0.0237	1.859	0.000106	
P_2	5	97.6			1.989		
F_1	27	83.3	1.67	0.0200	1.921	0.000076	(6.7)
F_2	126	80.4	3.83	0.0476	1.905	0.000927	6.0

From data of Castle (1922).

of the back, white elsewhere (genotype *ss*) had been divided into two nonoverlapping substrains by selection in opposite directions (Castle and Phillips 1914). Grade was expressed on an empirical scale in which +6 represented self-color and −4 was self-white. After five or six generations of selection each substrain appeared to be rather homogeneous (as indicated by standard deviations of half a grade or less) while differing in average by nearly 4.5 grades. Nevertheless selection continued to be effective as shown by a mean difference of 5.7 grades in the tenth generation. Variability was considerably reduced but there was still a large enough genetic component to permit extending the difference to 7.4 grades (+4.6 in the plus series, −2.8 in the minus series) by the twentieth generation. Crosses made in the fifth

TABLE 15.5. Results from crosses between strains of hooded rats selected in opposite directions for 5 or 10 generations. In this case M' and σ' are derived from the inverse probability transformation. The minimum number of loci is estimated from the untransformed data (n_1) and from the transformed data (n_2, n_3). In the cases of n_1 and n_2, σ_E^2 is taken as $\sigma_{F_1}^2$, while in n_3 the variances of P_1, F_1, and P_2 are used in 1 : 2 : 1 ratio.

	No.	M	σ	M'	σ'	n_1, n_2, n_3
P_1 minus (gen. 5)	1252	−1.56	0.44	−0.694	0.196	
P_2 plus (gen. 6)	701	+2.90	0.50	+0.496	0.143	4.8
F_1	93	+0.06	0.71	−0.238	0.184	5.2
F_2	305	+0.24	1.01	−0.192	0.261	5.6
P_1 minus (gen. 10)	1451	−2.01	0.24	−0.845	0.086	
P_2 plus (gen. 10)	776	+3.73	0.36	+0.592	0.120	10.4
F_1	14	+1.00	0.60	0.000	0.151	10.1
F_2	73	+0.76	0.87	−0.060	0.220	15.3

From data of Castle and Phillips (1914).

or sixth generation and in the tenth generation gave intermediate F_1's with considerably greater apparent variability than in the parent strains, but much greater variability in F_2. The increase in variability in F_1 may reflect in part a damping of variability toward the extremes but this cannot account for the greater increase in F_2.

Number of Loci

Castle (1922) raised the important question of the number of independently segregating pairs of alleles (or better, of chromosomes) implied by his results from the rabbit-cross and applied a formula (15.8), suggested in correspondence by the present author. Any such formula is, of course, subject to limitations because of the many assumptions which must be made to make any estimate at all.

The first question concerns the specification of the nongenetic variability. Considerable interaction between genotype and environment is usually revealed by differences among parent strains and F_1's, in all of which variability is assumed to be almost wholly nongenetic. The first step in any analysis must be the elimination of this interaction as far as possible by a suitable transformation of scale. As brought out earlier, the primary criterion for such a transformation is the elimination of any systematic relation between means and standard deviations among populations in which genetic variability is absent. A satisfactory transformation requires consideration of many such populations but unfortunately two parental lines and their first cross is often all that are available.

If many lines are available it is likely to be found that there remain significant differences among the standard deviations even on a scale that eliminates all trend. It appears that some lines are inherently more stabilized than others in their responses to environmental differences. The situation is similar with respect to F_1 populations. Moreover, there is sometimes a systematic difference in variability between inbred strains and their F_1's. An obvious theoretical effect of close inbreeding is the fixation, more or less at random, of deleterious genes. We have noted the prevailing lack of vigor in self-fertilized lines from outbreeding species described by Darwin, Shull, and others. Such lines may be expected to be highly sensitive to unfavorable conditions that make little difference in vigorous crossbreds. There is indeed much data that have shown greater variability in inbred lines than in their crosses in characters that relate to vigor in some sense, in contrast with substantially equal variability in more neutral respects.

This consideration raises the question as to what variance, that of inbred strains or of F_1's or some compromise, should be used as a measure of the environmental component of the variance in F_2 or other crossbred populations. As far as any one locus is concerned, F_2 consists 50% of heterozygotes and 50% of homozygotes. This suggests use of the average of the estimates based on the two kinds of populations or, if only the variances of the two parental strains and their F_1 are available, of the average based on weights of 1:1:2. The average of determinations based equally on inbreds and on F_1's is indicated for backcross populations.

There may, however, be reason in certain cases for assigning other weights. Thus, if the only available data for transformation of scale are those from the parent strains and F_1, the usually closer approach of the mean of F_2 to that of F_1 than to either parent strain may indicate that the variance of F_1 had better be used as the estimate of nongenetic variance of F_2 because of the relative unreliability of the transformation at the extremes. In the case of the backcross, however, this same consideration again suggests that equal weight be given F_1 and the pertinent parent strain. Since no universal rule can be stated, it will be convenient to represent the estimate of the environmental variance that is chosen by σ_E^2 and thus, the genetic variance of F_2 by $\sigma_{F_2}^2 - \sigma_E^2$ and similarly with the backcrosses.

It is convenient to begin further analysis with the assumption that the loci are segregating independently of each other and that their effects are additive. Taking the grade of the lower parent as the base, with genotype

$\sum a_i a_i$, it is convenient to represent the differential contribution of a heterozygote Aa by α_1 and the increment contributed by a second dose of A by α_2.

	Contribution	F_2 Frequency
AA	$\alpha_1 + \alpha_2$	$1/4$
Aa	α_1	$2/4$
aa	0	$1/4$

$$\bar{P}_2 - \bar{P}_1 = \sum \alpha_1 + \sum \alpha_2 \tag{15.1}$$

$$\bar{F}_1 = \sum \alpha_1 \tag{15.2}$$

$$\bar{F}_2 = 3/4 \sum \alpha_1 + 1/4 \sum \alpha_2 \tag{15.3}$$

$$\sigma^2_{F_2} - \sigma^2_E = 1/16[3 \sum \alpha_1^2 + 2 \sum \alpha_1\alpha_2 + 3 \sum \alpha_2^2] \tag{15.4}$$

To obtain an estimate of the number of independent loci, it is necessary to make assumptions with respect to the signs and relative magnitudes of the contributions α_1 and α_2 of loci. We assume first that the parents are at opposite extremes (α_1 and α_2 positive at all loci) that there is semi-dominance ($\alpha_2 = \alpha_1$) at all loci and that all loci make equal contributions (all α_1's equal in value). Under these assumptions, the formula for number of loci (n) referred to above may be derived as follows:

$$\bar{P}_2 - \bar{P}_1 = 2n\alpha_1, \tag{15.5}$$

$$\bar{F}_2 = \bar{F}_1 = n\alpha_1, \tag{15.6}$$

$$\sigma_{F_2} - \sigma^2_E = 1/2 n\alpha_1^2. \tag{15.7}$$

Thus

$$n = \frac{(\bar{P}_2 - \bar{P}_1)^2}{8(\sigma^2_{F_2} - \sigma^2_E)}. \tag{15.8}$$

In relaxing these assumptions somewhat, it is convenient to represent the range between the extreme genotypes, whether represented by the parental strains or not, by R and to calculate an index,

$$S = \frac{R^2}{8(\sigma^2_{F_2} - \sigma^2_E)}. \tag{15.9}$$

This has been designated the segregation index (Wright 1952). It must be modified in various ways to give the best estimate of n.

If the lowest genotype is taken as the base, and it is assumed that there is a constant degree of dominance measured by $h = \alpha_1/(\alpha_1 + \alpha_2)$:

$$\sigma^2_{F_2} - \sigma^2_E = \frac{n}{16}(\alpha_1 + \alpha_2)^2[3 - 4h(1 - h)] \tag{15.10}$$

and the number of loci (still assumed equal in effect) can be estimated from the formula

$$n = R^2 \frac{3 - 4h(1 - h)}{16(\sigma^2_{F_2} - \sigma^2_E)} = [1.5 - 2h(1 - h)]S. \quad (15.11)$$

This reduces to $n = S$ if $h = 0.5$. It is increased by less than 12.5% if h is between 0.25 and 0.75. This is usually unimportant in view of the inevitable roughness of any estimate of number of loci, but if dominance is complete ($h = 0$ or 1) the estimate for n becomes 50% greater than the index S. This formula was given by Serebrovsky (1928) for complete dominance. It becomes still greater with overdominance.

In actual cases, the degree of dominance may be expected to vary among loci, but if F_1 is not exactly halfway between the parent strains (on the chosen scale) it is at least known that there is some excess of dominance in the direction indicated by F_1. The minimal effect of dominance on the estimate of n is obtained by assuming its degree to be uniform and given by the ratio $(\bar{F}_1 - \bar{P}_1)/(\bar{P}_2 - \bar{P}_1)$ if the parents are at the extremes.

The assumption that all loci make equal contributions is also, of course, highly improbable. It is desirable to consider the consequences of other assumptions.

At the opposite extreme is the assumption that practically all of the F_2 variance is due to segregation of one pair of alleles with semidominance. A multiplicity of minor factors are assumed to contribute to the range between the parents (at opposite extremes) but to have such small individual effects that their contributions to variance, determined by the squares of their effects, are negligible even when combined. Let 2α be the contribution of the major factor and $2 \sum \beta$ that of the minor ones to R. The F_2 genetic variance is $\frac{1}{2}\alpha^2 + \frac{1}{2} \sum \beta^2$ of which the second term is assumed to be negligible.

$$S = \frac{(2\alpha + 2 \sum \beta)^2}{4\alpha^2} = \left(\frac{R}{2\alpha}\right)^2. \quad (15.12)$$

The portion of R that can at most be attributed to one locus is $2\alpha/R$, which is given by $\sqrt{1/S}$ instead of by $1/S$ as with equal effects (and semidominance in either case).

If the major locus is completely dominant

$$S = \frac{R^2}{1.5\alpha^2}. \quad (15.13)$$

The maximum portion of R attributable to it (α/R) is given by $\sqrt{1/1.5S}$ or $0.815\sqrt{1/S}$ and thus 18% less than with semidominance. There is no mathematical limit to the number of independent minor factors.

There may be a group of equivalent major factors, supplemented by a multiplicity of minor ones that contribute to the range but not appreciably to the F_2 variance. Assuming n equivalent major factors and semidominance

$$S = \frac{R^2}{4n\alpha^2}. \tag{15.14}$$

The formula $2\alpha/R = \sqrt{1/nS}$ here gives the portion of the range attributable to any one of the major factors.

There is also no mathematical limit to the number of segregating factors under the hypothesis that the effects fall off in geometric progression $(\alpha : c\alpha : c^2\alpha : \cdots : c^n\alpha : \cdots)$. Again assuming semidominance, we have $R = 2\alpha/(1 - c)$ and $\sigma^2_{F_2} - \sigma^2_E = \frac{1}{2}\alpha^2/(1 - c^2)$ giving

$$S = \frac{(1 + c)}{(1 - c)}. \tag{15.15}$$

The ratio of successive effects is given by $c = (S - 1)/(S + 1)$ and the portion of the range due to the leading factor is $2\alpha/R = (1 - c) = 2/(S + 1)$.

If the effects of the loci are assumed to fall off in arithmetic progression, a finite number is indicated but it is greater than S (unless $n = 1$). Let $n\alpha : (n - 1)\alpha : (n - 2)\alpha : \cdots : \alpha$ represent the successive effects. Then (again assuming semidominance)

$$R = n(n + 1)\alpha \quad \text{and} \quad \sigma^2_{F_2} - \sigma^2_E = \tfrac{1}{2}[n(n + 1)(2n + 1)/6]\alpha^2.$$

$$S = \frac{3n(n + 1)}{2(2n + 1)} \tag{15.16}$$

$$\begin{aligned} n &= \tfrac{1}{6}[4S - 3 + \sqrt{16S^2 + 9}] \\ &\approx \tfrac{4}{3}S - \tfrac{1}{2} + \frac{3}{16S} \text{ if } S > 2. \end{aligned} \tag{15.17}$$

The number of loci is thus at most one-third greater than with equal effects. The portion of the range attributable to the leading factor is here $2/(n + 1)$.

Table 15.6 shows the implications of observed values of S under various hypotheses with respect to the factor effects, but assuming semidominance throughout. It is important to note that the actual number of loci involved is greater than S under any degree of dominance other than $h = 0.5$ for all. It is also greater if $(\bar{P}_2 - \bar{P}_1)$ is used for R, in a case in which both parental strains supply plus factors, or if the loci have unequal effects or if any of them are linked, or in most cases of nonadditive interaction, assuming that the most suitable transformation has been made for minimizing genotype-environment interaction. The sort of interaction under which the true

TABLE 15.6. Implications of the Segregation Index $S = R^2/[8(\sigma^2_{F_2} - \sigma^2_E)]$ under various hypotheses. Additive effects, semidominance and absence of linkage is assumed in all. R is the range (not necessarily the difference between parental strains). The estimated number of loci is given under the hypotheses of equal genic effects ($n = S$) and effects in arithmetic series ($n = (1/16)[4S - 3 + \sqrt{16S^2 + 9}]$). There is no mathematical limit if the effects are in geometric series or if practically all of the variance is due to one or two major factors. The ratio of successive terms in a geometric series is $(S - 1)/(S + 1)$. The portions of the range due to the leading factor under the various hypotheses are given in the last five columns.

S	No. of Loci		Ratio of Effects	Portion of Range due to Leading Locus				
	Equal Effects	Arith. Series	Geom. Series $(S-1)/(S+1)$	Equal Effects $1/S$	Arith. Series $2/(n+1)$	Geom. Series $2/(S+1)$	2 Major Loci $\sqrt{1/2S}$	1 Major Locus $\sqrt{1/S}$
1	1	1	0	1	1	1		1
2	2	2.26	0.333	0.500	0.614	0.667	0.500	0.707
3	3	3.56	0.500	0.333	0.438	0.500	0.408	0.577
4	4	4.88	0.600	0.250	0.340	0.400	0.353	0.500
6	6	7.53	0.714	0.167	0.234	0.286	0.289	0.408
8	8	10.19	0.778	0.125	0.179	0.222	0.250	0.353
10	10	12.85	0.818	0.100	0.144	0.182	0.224	0.316
15	15	19.51	0.875	0.067	0.098	0.125	0.182	0.258
20	20	26.18	0.905	0.050	0.074	0.095	0.158	0.224
30	30	39.50	0.935	0.033	0.049	0.065	0.129	0.182
100	100	132.83	0.980	0.010	0.015	0.020	0.071	0.100

From data of Wright (1952).

number may be less than S is that in which the effects of factors are less in the middle of the range than near the extremes, causing R^2 to be overestimated relative to $\sigma^2_{F_2} - \sigma^2_E$. There is also, of course, a possibility of overestimating n by overestimating σ^2_E. With reasonably adequate data and a suitable transformation, S corrected for the amount of dominance in F_1 may be taken as a minimal estimate of the number of loci, with only minor qualifications.

In the rabbit measurements in Table 15.4 the numbers in the parental strains and F_1 are too small to give an adequate basis for transformation of scale. On a priori grounds, however, a logarithmic transformation is indicated. An inverse probability transformation has been used in the case of hooding pattern in the rat (Table 15.5). In the latter it has been assumed that unit differences on the scale correspond to 10% differences in the percentage (p)

of pigment with 0% at −4 and 100% at +6. A rough transformation has been made by taking $\text{pri}^{-1}\, p$ for the transformed means (M') and

$$\tfrac{1}{2}[\text{pri}^{-1}(p + \sigma) - \text{pri}^{-1}(p - \sigma)]$$

for the transformed standard deviation (σ'). Calculations have also been made without transformation for comparison.

In all of these cases, F_1 is sufficiently close to halfway between the parental strains on either scale to warrant use of S as a minimal estimate of number of loci. In the case of the rabbit weights, this indicates 23 loci without transformation but only 14 using the logarithmic scale which is undoubtedly better because of the 2.6-fold difference. Since the rabbit has 22 pairs of chromosomes, linkage is certain on the former estimate and is probable on the latter. The only warranted conclusion is that the number of loci is very large. In other cases, transformation makes no great difference; for ear length, 21 on the untransformed scale, 19 on the logarithmic scale; for skull length, 5.6 and 5.2, respectively; for femur length, 6.7 and 6.0, respectively. For the hooded rats the inverse probability transformation slightly increased the estimate for the cross in the fifth or sixth generation (4.8 to 5.2), but decreased it for the cross after the tenth generation (10.4 to 10.1), basing σ_E^2 on the F_1 variance in both cases. The transformation is, however, intended to make the variances of the parental strains comparable with that in F_1. On using all three variances in 1 : 1 : 2 ratio to calculate σ_E^2, the value of S is raised to 5.6 for the earlier cross and 15.3 for the later one. East's data on the inheritance of corolla length in the cross between varieties of *Nicotiana longiflora* (Table 15.3) gives $S = 10.8$ on the basis of the F_2 and F_1 variances.

Among the crosses entered in Table 15.1, it may be seen that all in which the parents were at opposite extremes have values of S that indicate multiple factors. In plant height in the rather extreme cross, Tom Thumb by Missouri Dent, a minimum of seven loci is indicated after making minimal allowance for dominance. The estimate for one of its components, number of nodes, is at least four loci. The parents differ in the same direction in the other component, internode length, but there is clearly transgressive variability. This, of course, implies that there is also transgressive variability with respect to height and thus more than the seven loci estimated above. In the other cross involving height, Watson Flint by Leaming Dent, the parental strains differ much less and the estimate from S (2.2) is worthless. The genetic variance of F_2 is 56% of that from the more extreme classes. If equal effects are assumed, this implies that there are at least four gene differences.

Both ear length and weight of 25 seeds in the wide cross, Tom Thumb by Black Mexican, given minimum estimates of about 12 loci. Since maize has only 10 pairs of chromosomes, there must be linkage and actually more than 12 loci.

The cross between the elongated crookneck squash and the disk-shaped scallop (Table 15.2) yields an estimate of n of only 2.6 with respect to length. That for diameter is obviously meaningless because of transgressive variability. In this case, as already noted, the treatment of length and diameter separately obscures the segregation of two major genes that determine

TABLE 15.7. Data from various crosses between strains of Pearl millet (*Pennisetum glaucum*). M' and $(\sigma')^2$ have been derived from M and C by the logarithmic transformation. Reliable estimates, n_1 and n_2, respectively, cannot be obtained for either M and σ or M' and $(\sigma')^2$ in the first two cases because of the excessive variability. In the other cases, estimate n_1 has been made by basing σ_E^2 on F_1, without transformation. In estimate n_2, σ_F^2 has been based on the transformed values using P_1, F_1 and P_2 in 1:2:1 ratio. Correction has been made for average dominance in both cases.

	No.	M	σ	C	M'	$(\sigma')^2$	n_1, n_2
Plant weight (lb.)							
16	126	4.39	2.35	0.535	0.588	0.0475	
18	115	2.65	1.86	0.702	0.336	0.0756	
F_1	156	6.21	3.59	0.578	0.731	0.0544	
F_2	1172	4.95	2.87	0.580	0.632	0.0547	
Stems with heads (no.)							
16	126	4.07	1.87	0.459	0.568	0.0361	
18	115	6.19	3.48	0.562	0.732	0.0518	
F_1	158	4.75	2.17	0.457	0.631	0.0358	
F_2	1182	5.19	2.75	0.530	0.661	0.0467	
Plant height (5 in.)							
16	126	16.6	1.04	0.063	1.219	0.00075	
18	115	8.6	1.80	0.209	0.925	0.00806	
F_1	156	23.4	1.17	0.050	1.369	0.00047	3.3
F_2	2024	16.8	3.57	0.213	1.216	0.00837	(5.3)
Plant height (5 in.)							
18	169	10.2	1.76	0.173	1.002	0.00556	
782	180	34.9	5.26	0.151	1.538	0.00425	
F_1	145	36.3	4.25	0.117	1.557	0.00256	3.5
F_2	1408	27.3	7.32	0.268	1.421	0.01308	6.0

(*continued*)

TABLE 15.7. *Continued.*

	No.	M	σ	C	M'	$(\sigma')^2$	n_2, n_2
Internode length (in.)							
18	168	5.23	0.84	0.161	0.713	0.00483	
782	178	10.01	1.47	0.147	1.000	0.00403	
F_1	218	11.23	1.39	0.124	1.047	0.00288	2.6
F_2	1523	9.86	2.18	0.221	0.984	0.00899	(3.7)
Leaves/stem							
18	178	12.73	0.99	0.078	1.105	0.00114	
782	187	17.73	2.18	0.123	1.245	0.00282	
F_1	179	14.41	1.24	0.086	1.157	0.00139	(3.7)
F_2	1913	14.51	1.56	0.108	0.336	0.00218	5.2
Leaves/stem							
18	123	10.8	1.10	0.102	1.031	0.00195	
717	129	19.9	2.02	0.102	1.297	0.00195	
F_1	144	15.6	1.35	0.087	1.192	0.00142	(6.9)
F_2	787	14.8	1.84	0.124	1.167	0.00288	7.6

From data of Burton (1951).

shape, discovered later by Sinnott. These were supplemented by minor factors affecting size. Estimation of n is, of course, meaningless for the cross of the two varieties of tobacco with almost the same mean numbers of leaves in both parents, but enormously greater variability in F_2 than in the others.

Burton (1951) has published extensive data from crosses among strains of Pearl Millet (*Pennisetum glaucum*). A few of these are summarized in Table 15.7. The large numbers in F_2 make it interesting to examine the forms of the distribution curves by plotting $\text{pri}^{-1}\, p$ against grade of transformations of the latter.

Plant weight and number of stems with heads in cross (16 × 18) yield graphs that are strongly concave to the right. These are straightened out to considerable extents by using a logarithmic scale (Fig. 15.1), but estimation of the number of loci would require more confidence in the precision of the transformation and in the comparability of the environmental conditions and of the gene-environment interactions of F_2 and F_1 than is possible in view of the enormous coefficients of variability of the pure strains and F_1.

Plant weight depends on number of stems and plant height. The graphs for plant height from the same cross are shown at the bottom of Figure 15.1.

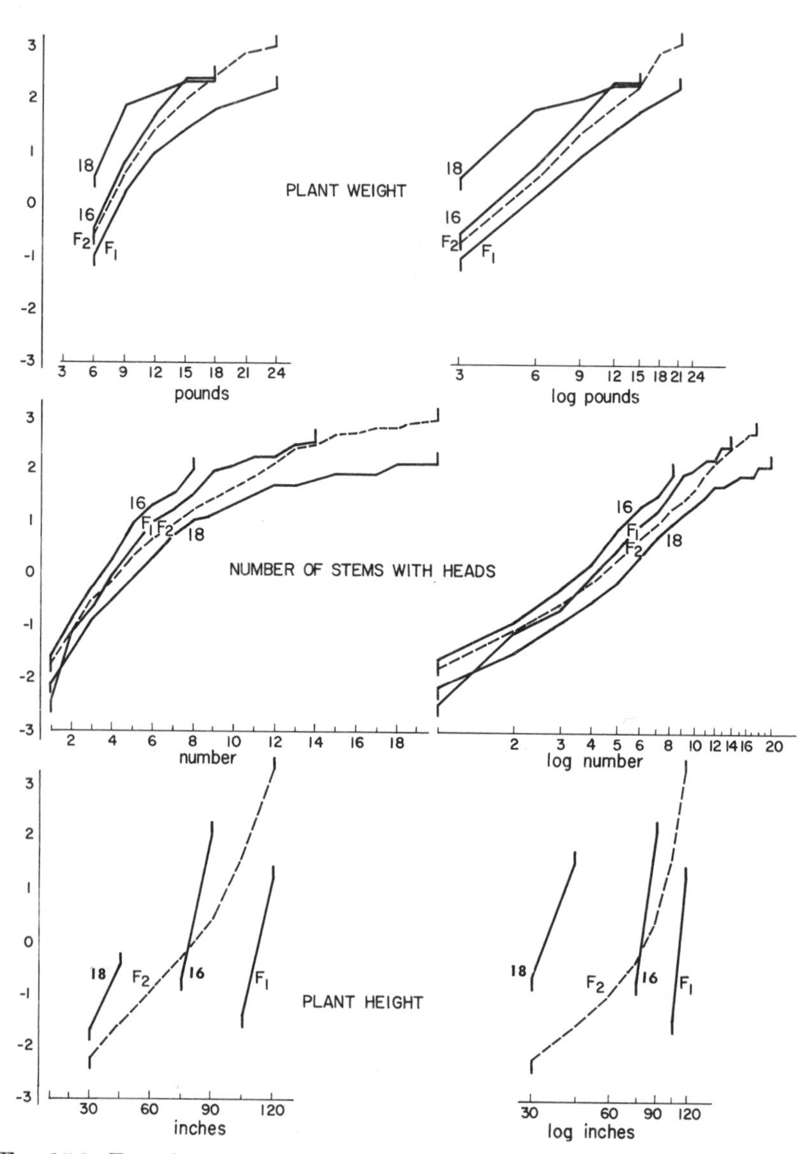

FIG. 15.1. Transformations (pri^{-1}p) of distributions of three characters in strains of Pearl Millet (*Pennisetum glaucum*), and in F_1 and F_2 of crosses. Comparisons are made in each case between the results from natural and logarithmic scales for the characters. (From data of Burton 1951.)

Note that the order of the parents in the components do not agree. The parallelism of the graphs for P_1, P_2, and F_1 is not improved by transformation and neither is the linearity of F_2. At least 4 loci seem to be involved.

In the much greater difference in height in cross 18 × 782 (Fig. 15.2 top) the F_2 graph of $\text{pri}^{-1}\ p$ against height curves in the way that indicates strong platykurtosis. The distribution was in fact bimodal. This suggests that there is an important major factor. If this exists, however, it is not sufficiently important to prevent an estimate of number of loci of 3.5 without the logarithmic transformation on making minimal correction for dominance (overdominance in this case since F_1 is taller than the taller parent). From Table 15.6 it appears that a major factor can determine at most only about 50% of the differences in height.

In the case of internode length from this cross the graphs (Fig. 15.2 middle) are more nearly linear without transformation, although somewhat more parallel on the logarithmic scale. A minimum of three factors seems indicated. There is again a suggestion of platykurtosis in F_2 and thus a major factor presumably the same as for plant height.

With respect to the other component of height, number of internodes (or leaves per stem), there was relatively little difference between the parents in cross 18 × 782 (not shown). In spite of this, a considerable number of loci is indicated, about 5, on the logarithmic scale which showed more parallelism of P_1, F_1, and P_2 than the scale of actual measurements.

The most extreme cross with respect to number of leaves was 18 × 717, the curves for which are shown at the bottom of Figure 15.2. There is good parallelism of P_1, F_1, and P_2 and fair linearity of all curves on the logarithmic scale. The increased variance of F_2 indicates at least 7 loci.

Backcross Data

Data from backcrosses of F_1 to the parent strains considerably extend the inferences that can be made from an extreme cross. As regards any single locus, the backcross population from $Aa \times aa$ consists half of Aa and half of aa. If effects of loci are additive, the mean of the population must be just halfway between the parental strain and F_1, apart from possible differential viability and accidents of sampling, but irrespective of linkage.

$$\bar{B}_1 = \tfrac{1}{2}(\bar{P}_1 + \bar{F}_1). \tag{15.19}$$

$$\bar{B}_2 = \tfrac{1}{2}(\bar{P}_2 + \bar{F}_1). \tag{15.20}$$

$$\bar{F}_2 = \tfrac{1}{2}(M + \bar{F}_1) = \tfrac{1}{2}(\bar{B}_1 + \bar{B}_2). \tag{15.21}$$

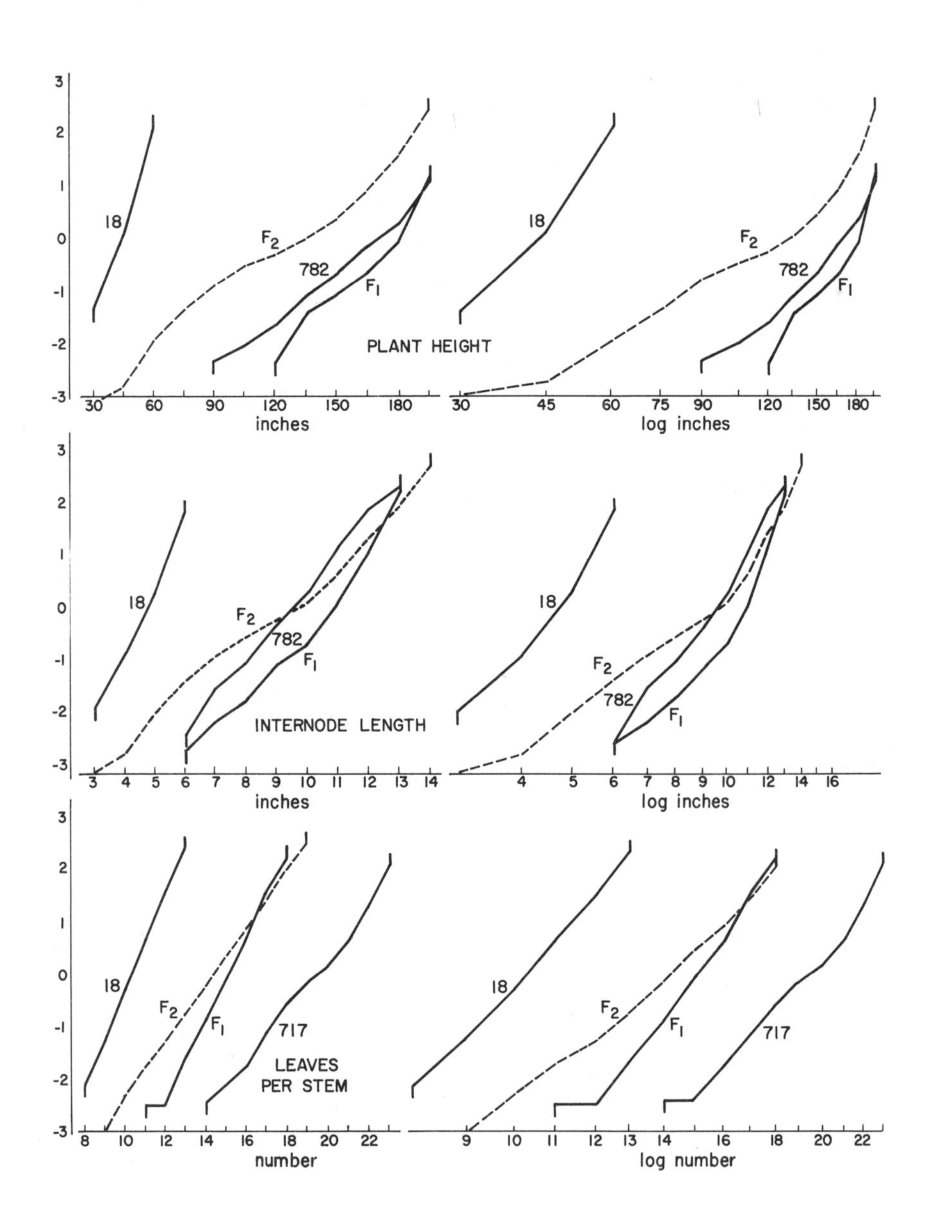

Fig. 15.2. Treatment of three additional characters of Pearl Millet in the same way as in Fig. 15.1.

A significant deviation from any of these relations in the absence of differential viability demonstrates the existence of interaction. The absence of any such deviation does not, however, demonstrate the absence of interaction as may be seen by substituting values of x, y, and z in the following special case of a two-factor array of phenotypes, derivable from $aabb \times AABB$. The above equations necessarily hold, although the effects are not additive unless $x + y = z$.

	bb	Bb	BB
AA	$\alpha_1 + \alpha_2 + z$	$\alpha_1 + \alpha_2 + \beta_1 + y$	$\alpha_1 + \alpha_2 + \beta_1 + \beta_2$
Aa	$\alpha_1 + x$	$\alpha_1 + \beta_1$	$\alpha_1 + \beta_1 + \beta_2 - y$
aa	0	$\beta_1 - x$	$\beta_1 + \beta_2 - z$

Nevertheless, these relations are incompatible with the more familiar patterns of interaction. They are incompatible, for example, with all possible nonadditive variants of the two-factor F_2 ratios with dominance at both loci.

If there is not only no interaction but also semidominance, F_1 is exactly halfway between the parental strains. Deviation demonstrates that there is not semidominance at all loci, but again the absence of deviation does not demonstrate exact semidominance. There may be an exact balancing of the effects of dominant plus and dominant minus factors. Thus a white guinea pig of genotype EEa^da^dffpp and a yellow of genotype $eeCCffpp$ gives an intermediate cream color. In F_2 the ratio is 3 yellows to 10 creams to 3 whites since the double dominants, EC, and the double recessives, eec^dc^d, are both cream in the presence of *ffpp* (Wright 1927). Similarly Ramage and Day (1960) have found a dominant mutation of barley that increases leaf width by 75% and an independent recessive with leaves about two-thirds normal width. Both double dominant and double recessive have normal width and F_2 gave a ratio of 3 wide:10 normal:3 narrow, thoroughly confirmed by F_3 analysis.

Balancing among two or more opposed dominant factors may be distinguished from semidominance at all loci by consideration of the genetic variances. If there is semidominance and no interaction, the genetic variances of the two backcrosses should be the same, and the genetic variance of F_2 just twice as great. Let c_{ij} be the proportion of recombination between loci A_i and A_j:

$$\sigma^2_{B_1} = \sigma^2_{B_2} = \tfrac{1}{4}\{\textstyle\sum \alpha_i^2 + 2\sum[(1 - 2c_{ij})\alpha_i\alpha_j]\} + \sigma^2_E, \qquad j > i; \tag{15.22}$$

$$\sigma^2_{F_2} = \tfrac{1}{2}\{\textstyle\sum \alpha_i^2 + 2\sum[(1 - 2c_{ij})\alpha_i\alpha_j]\} + \sigma^2_E, \qquad j > i. \tag{15.23}$$

If σ^2_E is the same in both cases, $(\sigma^2_{B_1} + \sigma^2_{B_2}) - (\sigma^2_{F_2} + \sigma^2_E) = 0$.

This does not hold in general. It suffices here to consider the simple formulas that hold in the absence of linkage. The phenotypes are taken as 0 for *aa*, α_1 for *Aa*, and $(\alpha_1 + \alpha_2)$ for *AA*, as before.

$$\sigma^2_{B_1} = \tfrac{1}{4} \sum \alpha_1^2 + \sigma_E^2. \tag{15.24}$$

$$\sigma^2_{B_2} = \tfrac{1}{4} \sum \alpha_2^2 + \sigma_E^2. \tag{15.25}$$

$$\sigma^2_{F_2} = \tfrac{1}{16}[3 \sum \alpha_1^2 + 2 \sum \alpha_1\alpha_2 + 3 \sum \alpha_2^2] + \sigma_E^2. \tag{15.26}$$

$$(\sigma^2_{B_1} + \sigma^2_{B_2}) - (\sigma^2_{F_2} + \sigma^2_E) = \tfrac{1}{16} \sum (\alpha_1 - \alpha_2)^2. \tag{15.27}$$

This disappears only if $\alpha_2 = \alpha_1$. It is brought out in Part II that this is the portion of the total variance of F_2 (assuming no interaction or linkage), that may be attributed to dominance deviations. The remainder on subtracting from $\sigma^2_{F_2} - \sigma^2_E$ is the additive variance,

$$2\sigma^2_{F_2} - (\sigma^2_{B_1} + \sigma^2_{B_2}) = \tfrac{1}{8} \sum (\alpha_1 + \alpha_2)^2. \tag{15.28}$$

This formula is equivalent, theoretically, to the estimate of the additive variance from $(\sigma^2_{F_2} - \sigma^2_E)/[1.5 - 2h(1 - h)]$, $h = (F_1 - P_1)/(P_2 - P_1)$, if the degree of dominance is the same at all loci (equation 15.10). Thus the number of loci in an extreme cross might be estimated from

$$n = \frac{R^2}{8[2\sigma^2_{F_2} - (\sigma^2_{B_1} + \sigma^2_{B_2})]}. \tag{15.29}$$

This formula is superior, in cases in which the backcross variances are available, in not assuming that the degree of dominance is uniform. It is, however, likely to be highly unreliable as depending on the difference of large quantities.

If $\sigma^2_{B_1}$ and $\sigma^2_{B_2}$ are available, they may be used without $\sigma^2_{F_2}$ to obtain estimates of the number of loci by which F_1 differs from one or the other parent. There is the advantage over estimates that involve $\sigma^2_{F_2}$ that segregation is not complicated by dominance.

$$F_1 - P_1 = \sum \alpha_1 \qquad \sigma^2_{B_1} = \tfrac{1}{4} \sum \alpha_1^2 + \sigma^2_E$$

$$P_2 - F_1 = \sum \alpha_2 \qquad \sigma^2_{B_2} = \tfrac{1}{4} \sum \alpha_2^2 + \sigma^2_E$$

Let $S_1 = \dfrac{(\bar{F}_1 - \bar{P}_1)^2}{4(\sigma^2_{B_1} - \sigma^2_E)}$ and $S_2 = \dfrac{(\bar{P}_2 - \bar{F}_1)^2}{4(\sigma^2_{B_2} - \sigma^2_E)}$.

Thus S_1 gives the number of loci, n_1, by which F_1 differs from P_1 with assumption that these are all plus factors of equal effect (α_1). S_2 gives the number of loci, n_2, by which P_2 differs from F_1 on the corresponding assumption for the α_2's. If there is semidominance $(\alpha_2 = \alpha_1)$ or the same ratio of α_2 to α_1, these estimates should agree with each other and with the estimate from $n = (\bar{P}_2 - \bar{P}_1)^2[1.5 - 2h(1 - h)]/8(\sigma^2_{F_2} - \sigma^2_E)$. If, on the other hand, it is assumed that there is complete dominance at all loci, the total number of loci involved is given by $S_1 + S_2$, irrespective of the distribution of plus factors, since each contributes to only one of the backcrosses.

S_1 and S_2 can also be interpreted in the other ways suggested in the case of S. The formulas are analogous.

It is instructive to construct profiles of the means and variances of P_1, B_1, (F_1, F_2), B_2, and P_2 in this order. This is done in Figure 15.3 for cases of four pairs of alleles with additive effects that are equivalent except that the direction of dominance may be either positive (A, B, C, D) or negative (W, X, Y, Z). The values of F_2 are indicated by small circles. The odd rows give the means from crosses between opposite extremes with varying numbers of plus and minus dominants. The even rows give the corresponding genetic variances. The effect of one dose of the more dominant allele (if any difference) is assumed to be the same throughout ($\alpha_1 = 1$). The first column gives the case of semidominance ($\alpha_2 = 1$) and thus is the same for all of the crosses. In the second column, $\alpha_2 = 0.5$, in the third, $\alpha_2 = 0$ (complete dominance), in the fourth $\alpha_2 = -0.5$, and in the last, $\alpha_2 = -1$ (pure overdominance, $P_2 = P_1$). These illustrate the rather wide range of patterns that may be expected from extreme crosses with additive effects.

Other patterns for crosses between extremes are found if there is factor interaction. Figure 15.4 shows a number of two-factor cases in which there is symmetry about the diagonal axis from P_1 to P_2 in the array of genotypes, starting from the additive case. There are considerable deviations from additivity among the profiles of means and in some cases among those of the variances. It is obvious, however, that data must be collected under carefully controlled conditions and in such numbers as to make sampling errors negligible for differences to be significant. Moreover, additive and diverse sorts of nonadditive relations are likely to be combined in such a way in multifactorial cases with intermediate F_1 as to resemble closely some additive case.

Figure 15.5 gives the profiles for a number of extreme crosses involving two-factor interactions with dominance for comparison with the linear case (1:6:9 ratio) with which the rows start. There are again considerable differences, but also danger of confusion unless both systematic and random errors are negligible. In both these figures the profiles for means applicable to the reverse ratio of minus to plus (e.g. 9:7 instead of 7:9) are given by turning the figures upside down while the variances remain unchanged.

At this point it will be well to present some results of actual crosses in which backcrosses as well as F_2 were obtained. The numbers are of varying degrees of adequacy.

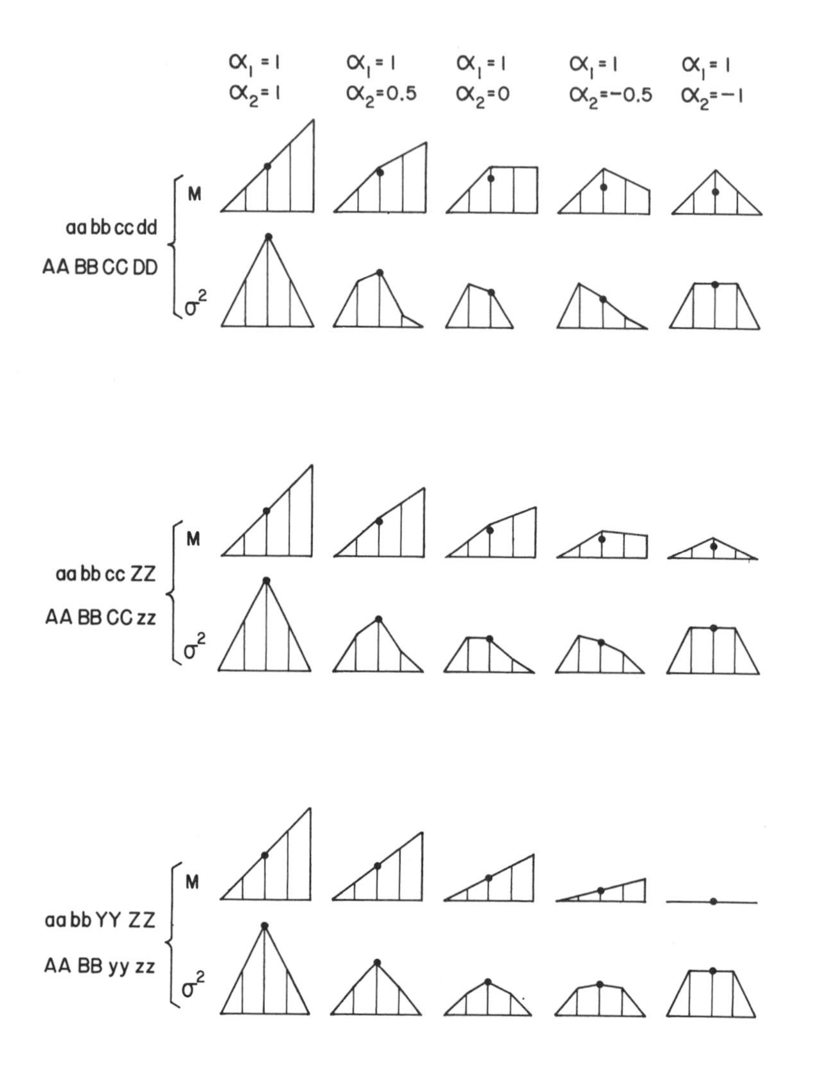

FIG. 15.3. Profiles of theoretical means and variances of P_1, B, F_1, and F_2 (dot), B_2, and P_2 in this order for crosses involving four equivalent additive pairs of alleles with various degrees of dominance ($[Aa] = [aa] + \alpha_1, [AA] = [aa] + \alpha_1 + \alpha_2$).

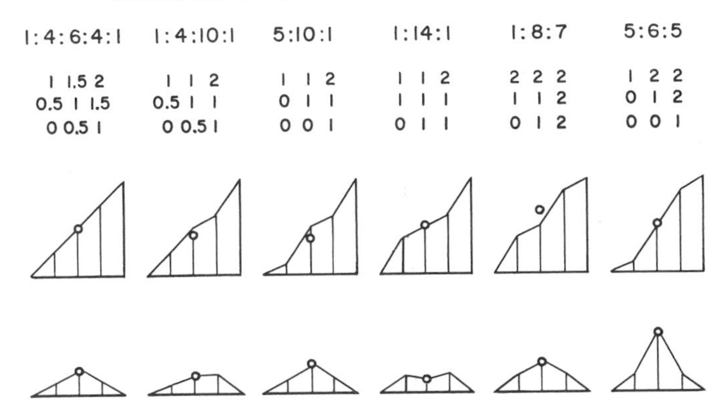

FIG. 15.4. Profiles of theoretical means and variances as in Fig. 15.3 for various two-factor interaction patterns indicated above.

H. H. Smith (1937) crossed small flowered *Nicotiana Langsdorfii* with large flowered *N. Sanderae.* There was full fertility and it was considered doubtful whether these were more than varieties of a single species. The sum of the logarithms of tube length and maximum lobe length was used as an index of flower size. The principal data are shown in Table 15.8 and as profiles in Figure 15.6, *Nicotiana, M* above, σ^2 below. There is no great difference from expectation on the hypothesis of multiple additive semidominant genes but some indication in both means and variances of dominance in excess of 0.50 and a slight suggestion of the type of interaction of Figure 15.4. The slight

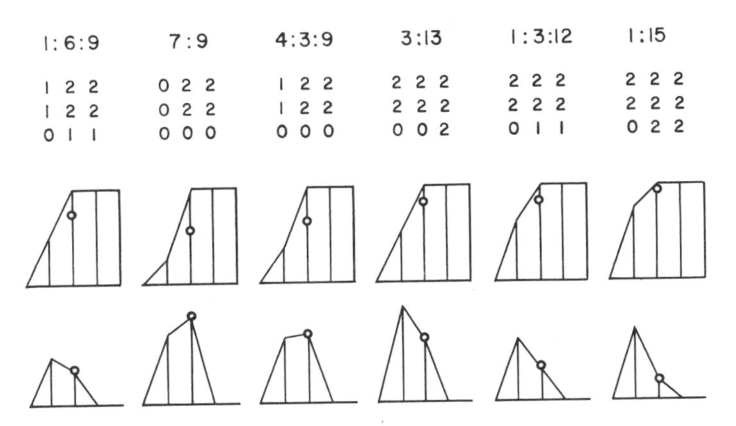

FIG. 15.5. Profiles of theoretical means and variances for two-factor interaction patterns, additional to those of Fig. 15.4.

TABLE 15.8. Data on extreme crosses from various species. M' and $(\sigma')^2$ are derived from M and σ by the logarithmic transformation with modifications described in the text. Estimates of n are derived from B_1, F_2 and B_2, using σ_E^2 from weighted average of P_1, P_2, and F_1. Corrections have been made for average dominance.

	No.	M	σ	C	M'	$(\sigma')^2$	n
log (tube × lobe lengths)							
P_1 *Nicotiana*	62				37	32	
B_1 *Langsdorfii*	279				315	98	$16.2 < B_1$
F_1	78				742	46	
F_2	537				700	140	$11.1 < F_2$
B_2	115				1077	78	$17.6 < B_2$
P_2, *N. Sanderae*	47				1292	48	
Size of fruit, tomato (gm.)							
P_1, Red currant	420	0.915	0.230	0.252	−0.137	0.0165	
B_1	932	2.095	0.862	0.411	0.249	0.0339	$9.7 < B_1$
F_1	475	5.480	1.502	0.274	0.710	0.0144	
F_2	932	5.380	3.107	0.578	0.653	0.0570	$10.0 < F_2$
B_2	931	16.105	7.137	0.443	1.163	0.0344	$12.7 < B_2$
P_2, Danmark	456	51.17	15.39	0.301	1.689	0.0165	
% oil, maize kernels							
P_1, low	22	1.40	0.284	(0.202)	0.513	0.00142	
B_1	68	2.83	0.445	0.157	0.670	0.00169	$20.5 < B_1$
F_1	20	4.70	0.263	0.056	0.817	0.00030	
F_2	146	4.55	0.816	0.179	0.804	0.00303	$18.8 < F_2$
B_2	74	7.54	0.891	0.118	0.972	0.00169	$20.6 < B_2$
P_2, high	19	11.42	0.703	0.062	1.122	0.00053	
Weight (gm.) fowls							
P_1, Leghorn	820	2064	260	0.126	3.311	0.00297	
B_1	128	2578	348	0.135	3.407	0.00340	$11.2 < B_1$
F_1	611	2991	387	0.130	3.472	0.00314	
F_2	560	3111	508	0.163	3.487	0.00496	$4.7 < F_2$
B_2	38	3389	458	0.135	3.526	0.00341	$6.3 < B_2$
P_2, Brahma	124	3940	416	0.106	3.593	0.00209	
Weight, mouse (gm.)							
P_s, small	41	13.61	0.678	0.0498	1.130	0.00044	
B_s $S \times F_1$	95	19.69	2.052	0.1013	1.294	0.00184	$13.0 < B_S$
B_s $F_1 \times S$	59	$21.4 - \frac{1}{2}x - y$					
F_1 $S \times L$	96	24.47	1.131	0.0443	1.397	0.00036	
F_1 $L \times S$	65	$27.16 - x$					
F_2	216	$27.10 - \frac{1}{2}x - y$	2.751	0.1015	1.395	0.00191	$14.1 < F_2$
B_L $F_1 \times L$	114	$32.70 - \frac{1}{2}x - y$	2.588	0.0802	1.471	0.00125	$6.5 < B_L$
B_L $L \times F_1$	50	$31.31 - x$					
P_L Large	65	$37.35 - x$	1.980	0.0530	1.541	0.00049	

(continued)

TABLE 15.8. *Continued.*

Seed size, Lima bean (gm.)							
P_1 small	183	0.4032	0.028	0.069	1.6045	0.00089	
B_1	40	0.4521	0.064	0.142	1.6515	0.00376	2.5
F_1	78	0.5942	0.043	0.073	1.7729	0.00101	
F_2	689	0.5690	0.093	0.164	1.7498	0.00502	(7.3)
B_2	52	0.7046	0.161	0.229	1.8382	0.00962	2.7
P_2, large	95	1.1824	0.109	0.092	2.0712	0.00160	

From data; Smith (1937); Powers (1942); Sprague and Brimhall (1949); Waters (1931); Chai (1956); Ryder (1958).

excess of the F_2 genetic variance over the sum of the genetic backcross variances argues against balancing of strongly dominant plus and minus factors. The estimation of the number of loci comes out 11.1 for F_2, while the less reliable estimates from the backcrosses indicate larger numbers (16.2, 17.6). It is clear that many loci are involved.

There was approximately an 18-fold difference in flower area between the parents in this case. Powers (1942) described a cross between two varieties of tomato (red currant and Danmark) that differed about 56-fold in weight. Table 15.8 shows how far the actual means deviate from additive expectation. Powers used logarithms of the weights. These bring the standard deviations of P_1, P_2 and F_2 to the same order of magnitude (0.114, 0.145, 0.134, in contrast with 0.23, 15.4, and 1.5, respectively). This case will be used to illustrate the application of a transformation of the type $x' = \log(x - a)$ designed to make the modified coefficients of variability, C', of the parent strains the same, relative to a calculated origin.

$$C' = \sigma_{P_1}/(\bar{P}_1 - a) = \sigma_{P_2}/(\bar{P}_2 - a), \qquad a = 0.153.$$

$$\bar{x}'_1 = \log_{10}(\bar{x} - 0.153) - \tfrac{1}{2}\log(1 + C')^2.$$

$$\sigma^2_{x'} = 0.4343 \log_{10}(1 + C'^2).$$

As may be seen from Figure 15.6 (tomato), $\bar{P}_1$, $\bar{B}_1$, $\bar{F}_1$, $\bar{B}_2$, and $\bar{P}_2$ fall approximately in a straight line, $\bar{F}_2$ differs little from $\bar{F}_1$, and the genetic variances of the backcross populations are nearly the same and about half that of F_2. There is no need to go beyond the hypothesis that the effects of loci are additive and that there is semidominance on the transformed scale. The estimated number of loci, assuming equal effects and no linkage is 10.0 from $\sigma^2_{F_2}$, 9.7 from $\sigma^2_{B_1}$, and 12.7 from that of $\sigma^2_{B_2}$, which are in as close agreement as can be expected.

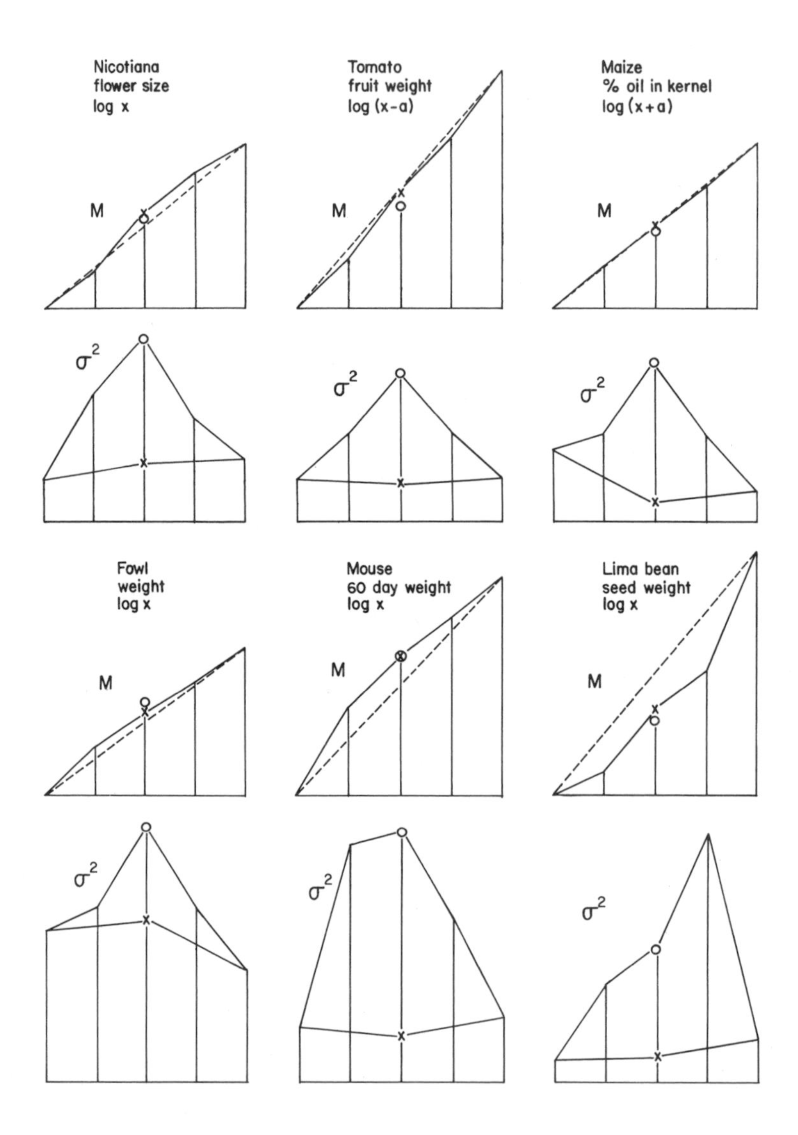

FIG. 15.6. Profiles of observed means and variances (on transformed scales) for six wide crosses discussed in the text. (From data: *A*, Smith 1937; *B*, Powers 1942; *C*, Sprague and Brimhall 1949; *D*, Waters 1931; *E*, Chai 1956; *F*, Ryder 1958.)

One of the most noteworthy selection experiments ever made was begun in 1896 by Hopkins at the Experiment Station of the University of Illinois and continued to the present. It involved changing the protein and the oil contents of maize in both directions. We are concerned here with crosses between the extremes, with respect to oil content, conducted by Sprague and Brimhall (1949). The parent strains, derived from the extremes by two generations of selfing, averaged 1.40% and 11.42% oil (Table 15.8).

It is evident that a transformation should be made but the large coefficients of variability of the low parent and of B_1 indicate that a simple logarithmic transformation is unsuitable. The numbers of individuals of the parental strains and F_1 that were assayed are so small that it seemed best to use the transformation that makes the coefficients of variability of the backcrosses the same above a certain origin. This yielded $x' = \log (x + 1.87)$.

The nongenetic variance was estimated from the transformed variances of P_1, P_2, and F_1 with weights of 1:1:2. As shown in Figure 15.6 (maize), $\bar{P}_1$, $\bar{B}_1$, $\bar{F}_1$, $\bar{B}_2$, and $\bar{P}_2$ fall approximately on a straight line, and $\bar{F}_2$ differs little from $\bar{F}_1$. The genetic variance of F_2 comes out a little greater than twice the equalized genetic variances of the backcrosses. In view of the small numbers there is no warrant for going beyond the simple hypothesis of additive gene effects including semidominance. The number of equivalent factors, based on the transformed F_2 variance is 20 (which agrees with the estimates given by the authors both with no transformation and with a simple logarithmic transformation). In view of the certainty of linkage (10 chromosome pairs) the actual number must be much greater.

Table 15.8 gives data on body weights from a cross made by Waters (1931) between a small breed of fowl (Leghorn) and a large one (Brahma). The approximate equality of the coefficients of variability of the parent strains and F_1 indicate the suitability of a simple logarithmic transformation.

$\bar{P}_1$, $\bar{B}_1$, $\bar{F}_1$, $\bar{B}_2$, and $\bar{P}_2$ (Fig. 15.6, fowl) do not fall quite as well on a straight line on the logarithmic scale as on the scale of actual weights but this may be interpreted as indicating merely a slight degree of excess dominance of plus factors. $\bar{F}_2$ is not far from $\bar{F}_1$. With respect to the genetic variances those of the backcrosses are only about 27% of that of F_2, but the backcross data are much less adequate in numbers than those from F_2 and the large amount of nongenetic variance makes for unreliability. The estimate of equivalent number of loci from the F_2 variance is 4.7. On the whole, the simple hypothesis of additivity and a very slight excess dominance of plus factors (on the logarithmic scale) seems indicated as a first approximation.

Table 15.8 and Figure 15.6 (mouse) give data on body weight of mice at 60 days obtained by Chai (1956) from a cross between two inbred strains with almost a 3-fold difference. The unweighted average of the males and females was used (M), and linear corrections were applied before transformation to eliminate nongenetic influences of crossbreeding (1.42 gm. if F_1 mother) and size of the mother (1.37 gm. if L mother, 0.68 gm. if F_1 mother). The variances are based on those within litters. Chai found that, $\bar{P}_1$, $\bar{B}_1$, $\bar{F}_1$, $\bar{B}_2$, and $\bar{P}_2$ fall remarkably close to a straight line without any transformation, but as he also noted, the much closer similarity among the coefficients of variability of the parental strains and F_1 than that among the standard deviations, indicates that nongenetic factors at least tend to have multiplicative effects. The sex differences are also more nearly constant as ratios than absolutely. The curvilinear graph of means on a logarithmic scale and the pattern of genetic variances indicates a considerable tendency toward excess dominance of factors for large size on this scale. The estimate of equivalent number of loci comes out rather large (about 14 on the basis of $\sigma^2_{F_2} - \sigma^2_E$).

We will consider next a case which deviates to an unusual extent from the expectations on the hypothesis of additive effects of loci. It has to do with a cross between inbred strains of lima beans that differed nearly 3-fold in seed weight, described by Ryder (1958) and interpreted by him as involving very unusual complementary effects. Table 15.8 shows the data both in terms of actual weights and of logarithms. The plants were grown in randomized plots in three replicates, including large numbers of F_3 families not here considered. The standard deviations were not reported as such but are given here as implied by the standard errors of the means of logarithms which are stated to have been based on the grand total variance in each case. The much closer agreement among the standard deviations of the logarithms of P_1, P_2, and F_1 than among the standard deviations of actual weights again indicates that nongenetic factors at least tend to act multiplicatively.

The profiles of the logarithmic means and variances are shown in Figure 15.6 (lima bean). The former show a wide departure from linearity, although not as wide a one as before transformation. $\bar{F}_2$ is somewhat less than $\bar{F}_1$ which is much less than the midparent, but $\bar{F}_2$ is about halfway between the backcross means, both of which are far below the average of F_1 and their inbred parent. The very large genetic variance of B_2, relative to that of F_2, is difficult to interpret but clearly indicates strong interaction. The estimate of n based on the large F_2 population comes out about 7 which must, however, be discounted because of the evidence for strong interaction.

The estimates from the backcross variances indicate a minimum of three loci since the assumption of two major loci implies genetic variances that would be much greater than observed. The assumption of three such loci gives a better fit. With this assumption, it appears that the average excess of the six genotypes of B_1, other than P_1 (*aabbcc*), and F_1 (*AaBbCc*), is only 0.035 over P_1 on the logarithmic scale (M'), in contrast with 0.168 for F_1.

Similarly, the average excess of the six genotypes of B_2, other than F_1 and P_2, is only 0.037 over F_1, in contrast with 0.300 for P_2. The average excess of the 12 genotypes of F_2, other than the 15 included in the two backcross populations (including all homozygous in at least one plus factor and also in at least one minus factor), is 0.168 over P_1 and thus the same as for F_1. Assuming that there is little or no variation among the genotypes within these three categories, the values, derived from the means, also account well for both genetic backcross variances (0.0023 instead of 0.0026 in the case of $\sigma^2_{B_2} - \sigma^2_E$, 0.0081 instead of 0.0085 in the case of $\sigma^2_{B_2} - \sigma^2_E$, but give a calculated value of $\sigma^2_{F_2} - \sigma^2_E$ [0.0065] that is considerably larger than observed [0.0039]). It is not easy to improve essentially on this peculiar three-factor interaction system, although the error could be distributed more equably. The numbers do not warrant an attempt to go further.

Diallel Crosses

Somewhat more can be learned of the genetics of a character from the statistics of all possible crosses made among three strains than merely between two. A really adequate survey of the genetics of a character requires, however, that all possible crosses be made among a considerable number of representative strains and carried to F_2 and the backcrosses according to a design that insures statistical comparability of all populations. This is a very formidable task, but less ambitious systems of diallel crosses have been instructive. We will consider studies of two different characters of maize in which such crosses have been used.

Emerson and Smith (1950) studied the number of rows on the ear of maize. They developed six selfed lines that were on the average 8-rowed, three on the average 10-rowed, and thirteen on the average 12-rowed. Characteristic distributions have been shown in chapter 11. It was brought out that the regression of standard deviation on mean is largely eliminated by a logarithmic transformation. Wherever logarithmic transformations are referred to below, the unit is the double row.

Extensive high and low selection within these lines brought about no significant change. Mutation was thus not an appreciable factor.

Number of rows was systematically a little greater on rich soil than on poor soil, but the difference accounted for only a very small part of the total variance. The average logarithmic variances of the groups of inbred strains and of F_1 from 8-rowed × 12-rowed were as follows. As they differ little, the grand average may be taken as σ^2_E.

	No. of Lines	$\sigma^2_{\log}$
8-rowed	6	0.002547
10-rowed	13	0.002285
12-rowed	13	0.002291
$(8 \times 12)F_1$	4	0.002687
	36	0.002410

There were, however, significant differences among lines. Among the thirteen 12-rowed lines, the variance of the standard deviation (0.000,0401) was 14 times the mean squared standard error (0.000,0028), indicating significantly greater susceptibility to causes of variability in some genotypes than in others.

Diallel crosses were made among the thirteen 12-rowed lines. There was no very striking heterosis in any case (range 12.0 to 14.4), but a small average amount was shown by a rise in mean number from 12.05 in the inbreds to 12.90 in F_1. A similar set of diallel crosses among the six 8-rowed lines gave much less heterosis (inbreds 7.89, F_1 7.98 with range 7.7 to 8.1). The F_1 hybrids between the 8-rowed and 12-rowed lines (77 of possible 78) had a mean row number of 9.82, slightly less than the midparental average (9.97). There were, however, considerable differences (range 8.4 to 11.3).

Some idea of the number and effects of loci may be obtained from crosses of 8-rowed to 16-rowed and 18-rowed strains Table 15.9. The latter two strains were not selfed, but F_2 in each case came from a single ear. The logarithmic scale of number of double rows is used here.

TABLE 15.9. Estimates of segregation index, S, with respect to number of kernel rows, from crosses between strains of maize.

	No.	M	σ		No.	M	σ
16-rowed	109	0.8759	0.0597	18-rowed	210	0.9324	0.0601
8-rowed	111	0.5990	0.0448	8-rowed	111	0.5990	0.0448
F_1	61	0.7127	0.0720	F_1	31	0.7855	0.0402
F_2	163	0.7222	0.0768	F_2	132	0.7589	0.0685
$\sigma^2_{F_2}$	0.005901			$\sigma^2_{F_2}$	0.004737		
σ^2_E	0.002410			σ^2_E	0.002410		
$\sigma^2_{F_2} - \sigma^2_E$	0.003491			$\sigma^2_{F_2} - \sigma^2_E$	0.002327		
$S = 2.7$				$S = 6.0$			

Data from Emerson and Smith (1950).

These are, of course, only minimum estimates and the relatively large F_2 genetic variance of the 16 × 8 cross suggests that the 8-rowed line carried plus factors lacking in the 16-rowed line.

All crosses among twelve of the 12-rowed lines were carried to F_2 and subjected to both high and low selection for a few generations. These brought about highly significant differences in most of the cases, in contrast with the ineffectiveness of selection within the inbred lines. The greatest difference was between averages of 9.3 from low selection, 15.3 from high selection in one case. Four others yielded differences of more than 5 rows while only 7 of the 66 yielded differences of less than 2 rows. Using logarithms of double rows, the average for the 66 high selections was 0.8364 (corresponding to 13.72 rows), and for the 66 low selections was 0.7039 (corresponding to 10.12 rows). The distributions of 13 of the F_2 populations were reported and yielded a mean standard deviation of 0.003008 on the logarithmic scale. Taking the difference achieved after a few generations of high and low selections as a measure of the range due to the more important factors, the value of S comes out 3.7. Thus two 12-rowed lines differed on the average by some four major factors which in combination produced an average difference of about 3.6 rows.

By crossing high-selected strains from two different 12 × 12 crosses and selecting from F_2 to F_5, the average number of rows was raised to about 18 in three of six cases. By crossing the more successful of these, which assembled heredity from seven of the original 12-rowed lines, the average was raised by three generations of selection to about 22 rows in three experiments. In view of the complete failure of selection within the inbred lines, the very limited advance by selection from crosses between two 12-rowed lines but great advance by selection from the array of genes from seven of the lines, it is clear that there are multiple factors (at least a dozen) with more or less additive effects (on the logarithmic scale) and approximate semi-dominance. Thirteen strains with the same phenotype (12-rowed) were all different from each other genetically.

The second character of maize for which the results of diallel crosses will be considered is yield in bushels per acre, reported by Kinman and Sprague (1945). They used 10 selfed lines, all with the characteristically low yields of such lines in maize.

Figure 15.7 shows the results from the crosses of each of the lines with each of the others. The bottom line shows the midparental averages which are necessarily parallel in all of the cases. The top line shows the F_1 yields; the F_2 yields are indicated by circles. The value of F_2 expected from additive effects of loci, $\bar{F}_2 = \frac{1}{2}(\bar{P} + \bar{F}_1)$, is indicated by the dotted line.

The enormous amount of heterosis, shown by the excess of the top line

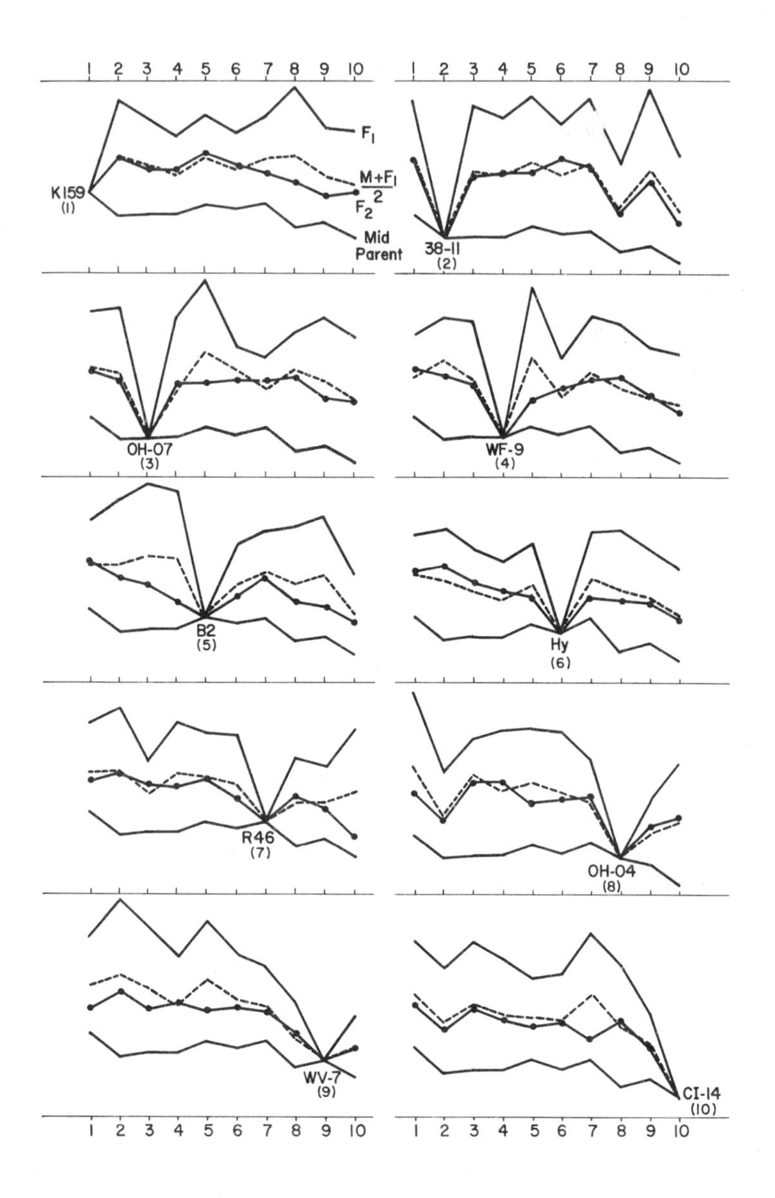

Fig. 15.7. Results of diallel crosses among 10 inbred strains of maize, comparing midparent, F_1, F_2, and expected F_2 (broken line) on the additive hypothesis. (From data of Kinman and Sprague 1942.)

(F_1) (excluding the case of selfing) over the bottom line midparent, is the most striking thing that is brought out. Figure 15.7 also illustrates, by the irregularities in order, the concepts of general and specific combining ability (introduced by Sprague and Tatum [1942]). The former is defined as the average performance of the line in question in hybrid combinations and the latter by the deviations in particular crosses from the expectation on this basis. General combining ability as manifested by the average of F_1 progenies is most important practically, but for our purpose that manifested in the average of the F_2 progenies measures best the additive effects in all combinations. It was used here to arrange the 10 lines in order. These parallel the yields of the inbreds themselves somewhat more closely than do the F_1 averages.

It may be seen that the F_2 averages are for the most part about halfway between F_1 and midparent and thus require no more than that each parent supply more or less dominant factors favorable to yield that are lacking in the other. The difficulty in obtaining inbred lines that have no unfavorable recessives could be accounted for by overdominance at some loci (Hull 1945) but does not require anything more than difficulty in breaking up linkages between favorable and unfavorable genes, suggested by Jones (1918). There are, however, six cases (1×8, 1×9, 3×5, 4×5, 5×9, 7×10) among the 45 crosses in which F_2 is significantly less (at the 5% level) than expected from F_1 and the midparent. There are none in which F_2 is significantly above expectation. These six cases require some mechanism of extreme heterosis based on factor interaction such as disproportionate overdominance in multiple heterozygotes. Comparison of the grand averages ($\bar{P} = 28.90$, $F_1 = 79.89$, $F_2 = 50.84$, $\frac{1}{2}(\bar{P} + \bar{F}_1) = 54.04$), shows that F_2 is 3.2 bushels per acre less than expected on the additive hypothesis.

A very different situation from all those preceding has been reported in Sirk's (1932) very detailed account of the results of 58 crosses, among pure lines of *Vicia faba* (25 lines of the subspecies major and 8 of the subspecies minor). In this case, morphological characters of sorts that are usually highly multifactorial were found to depend on only one or a few loci with virtually no residual genetic variability (though much that was nongenetic). Thus, mean number of internodes was found to depend wholly in both subspecies on three alleles, I_1 (average 5), I_2 (average 8), and I_3 (average 12) with complete dominance of larger over smaller number in each case. Genetic differences in internode length within subspecies major were found to be restricted to four alleles, G_1, G_2, G_3, and G_4, again with complete dominance of greater length in each case. The same loci were involved in crosses between the subspecies, but the cytoplasm from subspecies minor was found to have a strong plus effect which fell off with number of nodes (I-locus). Otherwise, there was no residual genetic variability. The number of leaflets in the leaves varied with phase in the growth cycle but the maximum reached (five, six, or seven) was determined wholly by I_1, I_2, and I_3, respectively. Form and size of leaflets similarly vary with growth phase, but in a given phase depended again (without residual genetic variability) on the G series with respect to general size, and on three loci concerned specifically with form, W, w, with respect to breadth; B, b, with respect to basal length; and T_2, T_1, t with respect to terminal length. These same four loci had somewhat parallel effects on size and form of fruits and seeds.

Summing up, there was much multiple allelism and much pleiotropy, but regular occurrence of complete dominance of all higher alleles over lower ones, and especially an extraordinary absence of any indication of multifactorial residual heredity. One possible explanation for this last phenomenon could be that all strains had descended from a single pure line relatively recently and that all differences that arose depended on the occurrence of few major mutations. This would not, however, explain the absence of residual variability in crosses between the two subspecies, the wide genetic difference between which is indicated by the cytoplasmic effects in crosses, including 25% F_2 zygote sterility in crosses of minor ♀ × major ♂, associated with loss of a particular chromosome. The author did, indeed, find rare intermediates (1% or 2%) in crosses involving extreme alleles which he interpreted as due to some sort of mutational origin of intermediate alleles, but these were not tested.

Evidence from Linkage Tests

Direct evidence that Mendelian genes are concerned with quantitative variability has been provided by linkage experiments. The first demonstration seems to have been from Payne's (1918) test of the genetic basis of the increase in number of scutellar bristles (4 to an average of 9.1) which he brought about in *Drosophila melanogaster* by long-continued selection (referred to in chapter 11). Crosses between the selected line and normals of strains that carried various marker genes gave clear evidence for genes in the first (X) and third chromosomes. A sex-linked inhibiting gene was also demonstrated (1920) to have arisen by mutation in the course of the selection.

Warren (1924) made extensive studies of differences in egg size among wild and mutant strains of the same species by the same method. He found that all four chromosomes were involved. These chromosome effects were not due to pleiotropic effects of the marker genes. Mather (1941) and Mather and Harrison (1949) have found evidence for release of variability by multifactorial recombination within the major chromosomes of the same species in the course of selection for high and low numbers of abdominal chaetae. Sismanides (1942) obtained a similar result for numbers of scutellar bristles, the same character studied by Payne. F. H. Robertson has made use of the absence of crossing over in male *Drosophila* to study systematically the recombination effects on wing length of differences in the three major chromosomes of various strains, including ones selected for large and short wings. All these chromosomes were involved. The effects were not all additive.

Turning to other organisms, Sax (1923, 1926) demonstrated association in F_2 of differences in several quantitatively varying characters (size of seed, date of blooming, leaf type, yield) of the bean, *Phaseolus vulgaris*, with segregants of known Mendelian genes. Lindstrom (1924, 1928) found similar association of fruit size in tomatoes with marker genes of three linkage systems and of row number in maize with markers also of at least three linkage systems. In H. H. Smith's study of the genetics of corolla size in his cross of *Nicotiana Langsdorfii* and *N. Sanderae*, he found association of size differences of more or less similar magnitude with segregation of at least six independent color factors.

Green (1937) found association of differences in body length of mice with segregation of black (*B*) and brown (*b*) in crosses between inbred strains. This was confirmed in extensive studies by Castle and associates (1941) who also found a similar association of large size (especially tail length) with dilution (*dd*) from crosses in which *D* and *d* were segregating, whereas various other color factors showed no association. Since brown segregates were larger than black from several independent crosses, Castle suggested that pleiotropy rather than linkage was involved. So far there seems to be no evidence against this in the case of brown. In the case of dilution, Butler (1954) described a cross in which the *dd* segregants averaged slightly, but significantly, smaller than the *D* segregants instead of larger as in Castle's experiments, indicating linkage rather than pleiotropy. This does not disprove pleiotropy, however, since he may well have been dealing with a different dilution factor (leaden ln).

Dobzhansky (1927) found great differences from a wild stock of *Drosophila melanogaster* in spermatheca shape of 12 mutant stocks. He investigated the extent to which these differences were separable from the mutant gene of each stock by backcrossing to a pure wild strain, extracting the recessive and repeating the process, generation after generation. In 10 of his 12 cases at least part of the difference in spermatheca shape persisted to the end, in one case up to 31 generations, suggesting pleiotropy.

Schwab (1939) made a similar test of 11 mutations of the same species which he carried farther in most cases (15 to 20 backcrosses). There was more shuffling off of differences than in Dobzhansky's experiment and only one gene (dumpy) showed a strong difference at the end, and only three others (including one with no appreciable difference at first) showed small significant differences at the end. Even these do not necessarily indicate pleiotropy since, as brought out by Haldane, the average length of the region of a chromosome that escapes crossing over in long-continued backcrossing is rather large (9.5 units in 20 backcrosses).

Dobzhansky and Holz (1943), however, avoided this difficulty by inducing mutations in an isogenic stock. Four independent white mutations, nine of ten yellow or allelic mutations, and a vermilion mutation all showed small, but significant, differences in spermatheca shape from the unmutated stock. Four other kinds of mutations showed no difference. In this case, at least, it seems clear that a number of color mutations showed pleiotropic effects on a chosen morphological character.

It was brought out earlier that cytoplasmic differences have effects on quantitatively varying characters. It appears, however, from the direct evidence from linkage and pleiotropy, as well as from the less direct inferences from East's rules, that chromosomal heredity is much more important. But, it might be that this is of a fundamentally different sort from that involved in qualitative differences. Such a suggestion was made by Mather (1943, 1944), who proposed a distinction between "oligogenes" concerned with the latter and "polygenes" with the former and further suggested that this distinction might correspond to the cytologic distinction between euchromatin and heterochromatin. In support of this, he noted that the supernumerary chromosomes of maize, sorghum, rye, and other organisms, which are heterochromatic, vary in number, but natural selection for quantitative effects is indicated by persistence of average numbers. He finds effects of this sort in the heterochromatic *Y* chromosome of *Drosophila* and heterochromatic regions of other chromosomes.

Linkage studies such as those of Payne referred to earlier and of Wigan (1949) have, however, shown that polygenic factors need not be associated with recognizable heterochromatin. Moreover, the very nature of multifactorial heredity implies wide distribution. There is indeed an alternative interpretation that heterochromatin consists of numerous replications of the same genetic material and is accordingly highly subject to unequal crossing over with consequent quantitative variability. This, however, clearly does not apply where there is well-demonstrated constancy of pure lines (as in the experiments of Emerson and Smith on row number in maize). The wide distribution of polygenes might also be taken as implying interspersion of heterochromatin with euchromatin but this comes close to begging the question.

There are undoubtedly questions here that need further study. At present, it seems probable that the genes involved in quantitative variability are very heterogeneous, grouped together only by the nature of the measurable character that they are observed to affect.

It is indeed probable that the distinction between qualitative and quantitative is more in the eye of the observer than in the physiological processes underlying the observed character. On the prevailing view that gene mutations represent replacements, losses, or rearrangements of one nucleotide by another in DNA and that the primary effects of these are the imposition of similar changes in the synthesis of protein molecules, all mutations may be considered qualitative. In some cases, the effects carry through as qualitative to the observed character. In others, however, observation is restricted to a scale reading of a rather remote effect and necessarily appears as quantitative. The most qualitative of characters may indeed be reduced to quantitative by this means. In the case of coat color of the guinea pig, for example, the "qualitative" differences in the effects of the epistatic series, *CEB* "black," *CEbb* "brown," *Cee–* "yellow," and $c^a c^a$– – "white," becomes quantitative if measured by reflectionmeter readings (black 41, brown 35, yellow 23, white 0) (on a convenient transformed scale).

From the discussion of the genetics of guinea-pig coat color and hair direction in chapter 5, it may be seen that in addition to major "qualitative" factors as above, characterized by dominance and striking interaction effects, there are minor factors among which dominance is usually lacking and similar effects tend to be additive; and that beyond this, there is always residual genetic variability that becomes difficult to analyze because of the confounding with nongenetic variability, but which is clearly multifactorial in at least some cases. In this last respect, the genetics of these qualitative characters appears more similar to that of most of the cases of quantitative variability that have been considered here than does the genetics of such characters as plant height and leaf form in *Vicia faba* as analyzed by Sirks.

Leading Factors

The ambiguity of formulas for estimating the number of loci involved in a cross has been stressed earlier in this chapter. It is desirable to consider here the possibility of resolving this in so far as it depends on whether there are wide differences among the genes in their contributions. It is probable that a leading factor or factors could be isolated much more often than has been the case.

The most effective method seems to be shuffling off all minor factor differences by repeated backcrossing to one of the parent strains while retaining the leading factor by selection. It is convenient to begin with the assumption that all minus factors by which the parental strains differ are assembled in one (genotypes of P_1 and P_2, $\sum^n a_i a_i$, and $\sum^n A_i A_i$, respectively, without implication of dominance). The average grade of F_1 is $\bar{P}_1 + \sum^n (A_i a_i)$ with variance σ_E^2. The mean of the backcross to P_1 is $\bar{P}_1 + \frac{1}{2} \sum^n (A_i a_i)$ and its variance is $\frac{1}{4} \sum (A_i a_i)^2 + \sigma_E^2$. The mean for a second backcross to P_1 can be written $\bar{P}_1 + \frac{1}{2} \sum^x (A_i a_i)$, where the summation relates to the plus factors that happen to have been present in the backcross parent. The variance is $\frac{1}{4} \sum^x (A_i a_i)^2 + \sigma_E^2$.

If a considerable number of large second (or later) backcross progenies have been obtained, use may be made of the ratio $(\sigma_B^2 - \sigma_{P_1}^2)/(\bar{B} - \bar{P}_1)^2$, the squared coefficient of genetic variability relative to P_1 as zero. This takes the value 1 if there is segregation of a single factor but otherwise is less. Among those in which this index is maximum, the greater the value of $(\bar{B} - \bar{P})$ the more important the factor and the more reliable the indication of a single difference. The quickest way of isolating the leading factor under the above assumption is thus to pick individuals from near the middle of each backcross population for further backcrossing to P_1. About half of these should have the leading factor and should have lost more than half the remaining factors. These may be distinguished from those which have lost the leading factor by the relatively high value of the above index. After repeated backcrossing with this sort of selection, progenies should fall into two sharply distinct classes, if there actually is an important leading factor: those that maintain a mean well above P_1 and high variance, and those that become indistinguishable from P_1 in distribution and breeding behavior.

Unfortunately, the case is more complicated if each parent provides plus factors lacking in the other, since selection of the backcross populations to the low strain (P_1) from near the middle, tends to maintain the minus factors of the plus strain. The best procedure in the earlier backcross seems to be to try to maintain by selection as great a difference from P_1 as possible, irrespective of the index. Such selection tends to maintain not only the leading plus factor but also the more important secondary plus factors of P_2 and to eliminate directly the more important of the minus factors of the latter while permitting its less important factors (whichever their direction of effect) to be shuffled off largely at random in each generation. Only after it seems probable that all minus factors of the plus strains that are capable of contributing appreciably to the variance have been eliminated can the isolation of the leading factor proceed as described first.

Repeated backcrossing of this sort has been used to isolate leading factors from several continuously varying characters of the guinea pig. Distributions of piebald spotting, varying continuously from 0% to 100% white, even within closely inbred lines, have been given in chapter 6. These testify to an extraordinary amount of uncontrollable nongenetic variability, while the different means of different inbred strains testify to wide genetic variability. There are true breeding self-colored strains, but F_1 from crosses to spotted strains usually include both self-colored and spotted, and F_2 is so ambiguous that various interpretations had been made with respect to its leading factors (a dominant factor, a semidominant one, or two leading factors with different sorts of effects on pattern). Backcross tests were decisive. Table 15.10 shows the results of F_1 from the cross between the whitest of the spotted strains (no. 13) and self-colored animals ($^7/_8$ blood strain D), and of repeated backcrosses to strain 13, and finally of the results of inbreeding within the apprently segregating classes. F_1 included 19% self-colored but none appeared in the backcrosses to strain 13. The backcross progeny, while varying continuously from a trace of white (X) to self-white, showed two modes. Selection of parents from the lower half of the distribution (grades X to 9) but near its middle on the average gave, in generation after generation of backcrossing to strain 13, a similar bimodal distribution. There was some shifting of the distributions in the white direction from the first to the second backcross, but no significant change from the second to the sixth backcross. The two apparently segregating classes of the third backcross generation were inbred. Those from the lower half of the distribution (grades X to 9) gave a bimodal distribution in which some 75% instead of about 50% were clustered about the lower mode and which included self-colored. As the latter constituted only 12.5% instead of 25% of the total it appeared that even SS showed some white after fixation of minus factors from strain 13. Progeny of parents that both were from the whiter half of the backcross population bred much like strain 13, except that the average amount of white was a little lower.

Another inbred strain (no. 2) had on the average slightly less white than strain 13 but in crosses with the latter gave F_1 and F_2 populations that were intermediate in character. F_1 from strain 2 and the self-colored strain (D) was largely self-colored (lower part of Table 15.10); and in repeated backcrosses to D (made in the study of another character) white spotting almost disappeared. On testing $^7/_8D$, by crossing with strain 13, most of the former were found to be SS but one male and two females were shown to be Ss (fifth line in the lower part of Table 15.10). Inbreeding of these yielded the expected proportion of recessives with white spotting, but the grades were very low. On inbreeding these, they were shown to be in fact ss by producing a progeny with no self-colored young, but most of them had only from a trace to 25% white in marked contrast with the distribution of strain 2 from which they had derived gene s.

Crosses have been made between strain 13 and an inbred strain (39) which

Table 15.10. Results of repeated backcrossing of guinea pigs with low-grade piebald from a cross between self (SS) and the inbred strain (no. 13), with the highest grade, to the latter; tests of low and high grades that are $^{15}/_{16}$ of strain 13, results of backcrossing of a self strain (D), to the next to whitest strain (no. 2), and extraction of low-grade piebald from this.

	Parental Genotypes	Grade of S-parent	Blood of 13,2	Grade									No.
				0	X–2	3–5	6–8	9–11	12–14	15–17	18–20	W	
13 × 13	$ss \times ss$		1					0.1	1.2	12.5	62.4	23.8	1,688
$^7/_8D \times 13$	$SS \times ss$	0	$^1/_2$	19.3	72.3	6.0	2.4						83
$^1/_2(13) \times 13$	$Ss \times ss$	all X	$^3/_4$		43.6	5.1	2.6	2.6	6.4	11.5	24.4	3.8	78
$^3/_4(13) \times 13$		X to 6	$^7/_8$		17.9	14.9	14.9		6.0	9.0	25.4	11.9	67
$^7/_8(13) \times 13$		X to 6	$^{15}/_{16}$		18.9	16.8	6.3	3.2	2.1	12.6	38.9	1.1	95
$^{15}/_{16}(13) \times 13$		3	$^{31}/_{32}$		16.7	25.0	8.3			8.3	25.0	16.7	12
$^{31}/_{32}(13) \times 13$		2 to 6	$^{63}/_{64}$		5.9	11.8	21.6	2.0	7.8	11.8	29.4	9.8	51
$^{63}/_{64}(13) \times 13$		8 to 9	$^{127}/_{128}$		22.7	9.1	4.5		9.0	18.2	18.2	18.2	22
Tests		Sire Dam											
$^{15}/_{16}(13) \times {}^{15}/_{16}(13)$	$Ss \times Ss$	1–9 X–8	$^{15}/_{16}$	12.5	31.6	11.8	10.3	2.2	2.9	11.0	13.2	4.4	136
	$ss \times ss$	16–20 16–20						0.8	2.3	29.0	56.5	11.5	131
2 × 2	$ss \times ss$		1				0.2	2.0	5.8	20.8	68.2	3.0	1,650
$2 \times D$	$ss \times SS$		$^1/_2$	91.5	8.5								59
$^1/_2(D) \times D$	$Ss \times SS$		$^1/_4$	95.7	4.3								207
$^3/_4(D) \times D$	$S- \times SS$		$^1/_8$	99.0	1.0								305
$^7/_8$, $^{15}/_{16}D \times D$	$S- \times SS$		$^1/_{16}$, $^1/_{32}$	100.0	0.0								182
$^7/_8D \times 13$	$Ss \times ss$		$^9/_{16}$		46.2		7.7	7.7	7.7	23.1	7.7		13
$^7/_8D \times {}^7/_8D$	$Ss \times Ss$		$^1/_8$	71.0	22.6	6.4							31
$^7/_8D \times {}^7/_8D$	$ss \times ss$	X–5	$^1/_8$	0	69.7	20.7	6.9	0.7	1.4	0.7			145

From data of Wright and Chase (1936).

was spotted, but near the opposite extreme. F_1 and F_2 were spotted, ranging from a trace to near white (grade 20). There was clearly segregation of more than one factor, but the enormous amount of nongenetic variability makes any estimate beyond this of no value.

One factor that somewhat enhanced the amount of white in the spotting pattern of strain 13 (but not of strain 2) was isolated by a second effect—scattered white hairs (silvering), especially on the belly. Crosses of a silvered strain with self give both selfs and low-grade silvers in F_1, and more variability but no clear-cut segregation in F_2. Again, repeated backcrossing to the silvered strain indicated one leading semidominant factor; there was much less bimodality in these backcrosses than in the case of piebald. But, the persistence of the variability from self to high-grade silvering through five successive backcrosses and the sharply different behavior on inbreeding from the lower and upper halves of the backcross progenies made the segregation of a leading pair of alleles (*Si*, *si*) unmistakable (Wright 1959*b*).

In marked contrast with these cases were backcross tests of another character (see chapter 5) which superficially had given more indication of unifactorial segregation. This was presence and grade of development of the little toe. As stated in chapter 5, on crossing the 3-toed strain (no. 2) with the true-breeding 4-toed strain (*D*) (the crosses referred to above in connection with spotting), F_1 was wholly 3-toed, F_2 did not differ significantly from 75% 3-toed to 25% 4-toed, and the backcross of F_1 to *D* did not differ significantly from 50% of each. All suggested segregation of a recessive gene for presence of the little toe, even though residual genetic variability was indicated by imperfect development in many individuals in contrast with invariably perfect development in strain *D*. This interpretation was completely destroyed by a second backcross to strain *D*. The supposed 3-toed and 4-toed segregants bred almost alike, the former producing only 23% instead of 50% 3-toed and the latter 16% instead of 0%. Another equally 3-toed strain (no. 13) crossed with strain *D*, produced 33% 4-toed young in F_1 (both reciprocals). In F_2, 3-toed × 3-toed and 4-toed × 4-toed gave substantially the same results; this was also true of the backcrosses of the two types to strain *D*. There was no simulation of unifactorial ratios in this case (46% 3-toed in F_2, 14% in the backcrosses). A third inbred strain (no. 35) included both good and poor 4-toed as well as 3-toed. There were average differences in percentage among different substrains but 3-toed × 3-toed and 4-toed × 4-toed bred alike within substrains. There was clearly much nongenetic variability. By assuming a normal distribution of tendencies on an underlying physiological scale and division into three phenotypes (3-toed, poor 4-toed, good 4-toed) by two thresholds at a unit interval apart, the standard deviation ($\sigma = 0.86$) due to nongenetic factors could be established for strain 35 and the position of the mean relating to the lower threshold as 0 (-0.42) could be determined by the formulas:

$$\sigma' = 1/(\text{pri}^{-1}\, p_2 - \text{pri}^{-1}\, p_1) = 1/(1.655 - 0.493) = 0.861,$$
$$m' = -\sigma' \, \text{pri}^{-1}\, p_0 = -0.861 \times 0.493 = -0.424.$$

On allowing for substrain differences, the value $\sigma = 0.80$ was used for rough estimates. Strain D, true breeding, but produced by selection, was assumed to have a mean at 2.5σ above the upper threshold (Figs. 5.14, 5.15). Strain 13 was assumed to have a mean at such a point below the lower threshold that no 4-toed could be produced, but F_1 with mean halfway between those of the parents and the standard deviation 0.80 would overlap the lower threshold to give the observed 33% 4-toed. The calculation of the mean and standard deviation of F_2 from the proportions of the three phenotypes indicated a mean close to that of F_1 but much greater variability. In strain 2, the absence of 4-toed in F_1 indicated a much lower mean than in strain 13. The results in crosses of D with another strain (32) were very similar to those from strain 2, except that one 4-toed individual appeared among 26 in F_1. Thus the mean in both $2 \times D$ and $32 \times D$ was assumed to be sufficiently far below the lower threshold to permit only very rare overlap (2.5σ). Again the mean of F_2 (calculated from the proportions of the three phenotypes) came out close to that of F_1, but the variance was, of course, greatly increased. The mean of the backcross came out about halfway between that of F_1 and D, and the variance was less than that of F_2. Segregation indexes from either F_2 or the backcross indicated the equivalent of about four equal factors, assuming only plus factors in D and minus in the extreme strains 2 and 32. Later backcrosses suggested that there might be a leading factor, which if assumed to be responsible for substantially all the variance accounted for about half the difference between the parental strains on the physiological scale. All data from all crosses fell into order on the hypothesis of multifactorial heredity, semidominance, and a continuous physiological scale with a threshold below which no little toe developed and a second threshold above which development of the little toe was complete (Wright 1934*d*).

Another case in which repeated backcrossing has demonstrated the absence of any important leading factor for a character with continuous variability is given by Chai's studies of size inheritance in mice. His results in P_1, P_2, F_1, F_2, B_1, and B_2 have already been discussed. A large number (at least 14) of factor equivalents was indicated. Backcrossing to the large strain was carried to the fifth generation and to the small strain to the seventh generation. After the second backcross each series was divided into a line in which the crossbred parent was selected for large size and one in which it was selected for small size. The two backcrosses to the large line became essentially indistinguishable from each other or the large line. In the backcrosses to the small line, a difference amounting to about 10% of that between the small line and F_1 was maintained by selection to the end, indicating that the leading factor for large size had this much effect.

Next to the question of equal or unequal factor effects, in the interpretation of the genetics of quantitative variability are those of the prevalence of dominance and of factor interaction. These are difficult questions on which only tentative conclusions can be arrived at here since there are important lines of evidence which must be deferred to volume 3.

We may note here that most of the profiles of parental first and second generation means and variances that have been presented are in harmony with the interpretation that effects of loci are predominantly additive and that there is approximate semidominance. Where there is evidence of interaction (as in Ryder's study of weight of lima beans and a few of Kinman and Sprague's crosses with respect to yield in maize), it is quite possible that this pertains to only two or three leading factors. With respect to extreme heterosis in general growth and especially in yield in maize, it is possible that this depends on relatively small numbers of loci at which recessive inactivations of major genes are so closely linked with the active dominant phase of other major genes that strains assembling only the favorable genes are almost impossible to arrive at, as suggested long ago by Jones (1917). Closely similar in effect but even more irremediable are consequences of overdominance at a restricted number of important loci (Hull 1945). It is possible, in short, that additivity of loci and semidominance is the rule for most of the loci that are concerned and that deviations from these depend on only a few major loci.

Other Information on the Genetics of Quantitative Variability

Only the simpler and more direct modes of analyzing the genetics of quantitative variability have been considered in this chapter. A large part of the data of population genetics, to be considered in later volumes, bear on this matter.

The consequences to be expected from multifactorial theory in F_3 progenies were touched on in connection with East's pioneer results. Fisher, Immer, and Tedin (1932) listed a considerable number of statistics that could be derived from large, randomly selected arrays of third generation progenies of various sorts that would in principle carry the analysis farther than possible from F_2 and first backcross data. Mather (1949) and others have carried out extensive experiments from this point of view. These have given abundant verification of East's rules.

The results have been rather disappointing from the standpoint of determination of the actual genetic differences between strains that have been crossed or even of the components of the F_2 variance (environmental, additive genetic, dominance, and, in the more ambitious efforts, kinds of interaction effects). With respect to the number of loci involved, there is an advantage over F_2 data in that the parental strains need not be selected as at opposite extremes, but differences in the effects at the loci cannot be determined and dominance, interaction and linkage can only be allowed for very imperfectly.

Mather (1949) compared the theoretical effects of linkage in F_2 and F_3 progenies in considerable detail. He showed that the interpretation depends not only on whether there is predominant coupling or repulsion of plus factors, but also on whether dominance goes with factors that act in the same direction (reinforcement) or in an opposite one (opposition). There is greatest excess variance due to linkage if one homolog carries only plus factors, all dominant or all recessive. In the limiting case of complete linkage, the results are, of course, as if there were only one pair of alleles with effects equal to the sum of those of the actual genes. There is greatest reduction of the variance expected under random assortment if in each chromosome region, plus and minus factors tend to alternate, and dominance and recessiveness are distributed in such a way as to balance the expression of these in heterozygotes. With complete balancing in both respects and close linkage there could be no apparent segregation except a little by rare crossing over.

Statistics of second and third generation progenies reveal linkage most effectively where the recombination percentage is between 20 and 30. Loose linkage is naturally difficult to detect, but this is not very important. Unfortunately, close linkage, which is the most important from the standpoint of ultimate release of variability in balanced systems, is also not likely to be revealed statistically by changes that occur in a single generation.

Even if linkage and interaction are treated as negligible in effect, the difficulty of obtaining sufficiently reliable data for estimation of the variance due to dominance deviations is great. A large number of individuals in each of a large number of randomly selected progenies is required. In 9 of 15 cases, cited by Mather (1949) from his own and earlier data, the dominance component came out negative and in 7 of these the absolute value was more than half that of the additive genetic variance. The results tend to confirm the view that dominance is not a general property of the minor gene differences involved in quantitative variability, but there was great uncertainty in each individual case.

Extensive attempts have been made to determine the average degree of dominance in a variety of quantitatively varying characters by Robinson, Comstock, and Harvey (1949) and others, summarized by Gardner (1963). The statistic they used was $\sqrt{2\sigma_D^2/\sigma_G^2}$ where σ_D^2 and σ_G^2 are the dominance and additive components, respectively. This was derived from the analysis of variance of data from maize in which F_2 individuals were backcrossed to the parental strain. The amount of dominance exhibited by most of the plant characters studied was slight, but there was much apparent overdominance with respect to yield. This was dissipated to such an extent in repeated backcrossing (8 generations) as to indicate that it was largely due to repulsion linkage of plus factors which did not have more than moderate degrees of dominance when isolated.

Hayman (1958) and others have explored in various ways the possibilities of deriving statistics from large arrays of diallel crosses carried to F_2, designed to bring out the overall importance of dominance and interaction relative to the additive variance.

Among the most important lines of evidence on the nature of the genetic differences present in a population are inbreeding experiments and selection experiments. These have been touched on here, but largely only as sources of more or less homallelic material for crosses. Such experiments have brought out the practically universal occurrence of genetic variability in seemingly homogeneous random breeding populations, capable of yielding wide differences under selection. The multifactorial theory is abundantly confirmed, but the inadequacy of the simple additive concept of gene effect and the importance of pleiotropy have usually been manifest after a moderate number of generations of selection. The detailed discussion of these matters must be deferred until after development of the theories of inbreeding, selection and the correlations among relatives.

Selective Value

The character of most importance in evolution has only been touched on. This is selective value in nature, a character that is a function of all others and one that is practically impossible to measure in individuals in such a way that distributions and profiles of means and variances of populations, involved in a cross, can be determined. The grade of any measurable character such as size, length of life, or fecundity may constitute selective value under artificial selection, but the more rigorously this is done, the more certainly will the strain become extinct because of selection that runs counter to natural selective value.

One thing that can be inferred confidently about natural selective value is that it is maximum in a combination of intermediate grades of most of its measurable components in any population that has been subject to natural selection in relation to essentially the same array of conditions for a long time. For most such characters the optimum must be close to the mean. This insures that all genes that approach additivity in their effects on quantitatively varying characters will be favorable in some combinations and unfavorable in others in terms of natural selective value and, thus, exhibit interaction effects of the most extreme sort in the latter respect.

It has been noted that pleiotropy may be assumed to be practically universal property of genes in relation to the superficial characters with which organisms encounter their environments, even though their primary physiological effects are single. This again insures that natural selective value is a function of the system of genes as a whole rather than something that can be assigned individual genes.

CHAPTER 16

Conclusions

Since this volume has been concerned merely with the foundations of population genetics, the only conclusions that are warranted at this point are with respect to these foundations themselves and to the restrictions they impose on the structure of theory that is to be erected upon them. This structure is to consist primarily of a working out of the statistical consequences of the well-established genetic mechanisms in populations of diverse sorts under diverse conditions. Its grand problem is the interpretation of the process of organic evolution in collaboration with other disciplines: physiological genetics, ecology, taxonomy, and paleontology. Since firm postulates can only be derived from the genetics of organisms living today, it cannot make much of a contribution to the problems of the origin of life and of the first grand phase of evolution in which the Mendelian mechanism itself evolved. Neither can it contribute much, except as a foundation, to theories on the actual courses taken in the evolution of the higher categories. Its contribution is to the relatively modest but nevertheless crucial problems of the modes of transformation of characters during the lifetimes of species and of the relation of this to their splitting into permanently isolated phyletic lines. It focuses on microevolution rather than on macroevolution.

The introductory chapter gave a brief account of the ideas on the origin of species and their evolution that had been proposed up to the rediscovery of the Mendelian mechanism in 1900. There was little or no evaluation at that juncture, but a tentative evaluation on the basis merely of present knowledge of the genetic mechanism will be appropriate at the end of this discussion.

Chapters 2 to 5 reviewed briefly the principles of genetics with regard to their importance as postulates for population genetics. The overwhelming importance of the Mendelian mechanism among currently existing organisms, from the simplest to the most complexed, was stressed although not, of course, fully documented, a project that would require many volumes.

Chapter 2 dealt with the parallelism between laws of heredity and the behavior of chromosomes and their constituents in mitosis and meiosis that has firmly established the latter as the physical basis. The diversity of ways in which the processes of reproduction involving only equational mitosis (clonogeny), biparental heredity by syngamy or otherwise, and the segregation and assortment of genes in meiosis, combine in the reproductive cycles of different organisms was discussed as prerequisite for understanding the diverse forms taken by population genetics. The primary postulates of equal segregation of alleles and random assortment of nonalleles were emphasized as the basis for first approximations, but the possibilities of complication by occasional unequal segregation (meiotic drive) and much more importantly, by linkage, were also emphasized as matters that must be taken into account. The latter indeed becomes of primary importance in organisms such as the *Drosophila*, in which the number of chromosomes is small and recombination within chromosomes is severely restricted.

Primary attention should be given to the cycle of diploid organisms and monoploid gametes, or much reduced gametophytes, characteristic of most higher animals and plants, respectively, but some attention should also be given to cases in which the characters of interest are those of the reduced generation. The complication of sex linkage is common enough to require systematic attention. So also is the pattern of segregation of autopolyploids and polysomics. Other mechanisms such as self-incompatibility will be dealt with less systematically. The combination of clonogeny with occasional crossing will require special attention in the synthetic discussion of evolutionary processes.

Chapter 3 dealt with properties of the gene that are of major significance for population genetics. It touched only briefly on the recent revolutionary discoveries on its chemical constitution and mode of duplication. While these require primary consideration in speculations on the origin of life and the first grand phase of evolution, they are of relatively little importance for consideration of evolution on the basis of orderly mitosis and meiosis.

The chemical nature of the gene is, however, of immediate importance in connection with the postulates to be made on mutation. The evidence that genes consist of very long successions of nucleotides and that point mutation is typically an accidental replacement of one nucleotide by another implies the probability of the occurrence of enormously more alleles at each locus of a species than has been apparent from observation. A gene consisting of only 1,000 nucleotides has the potentiality for 3,000 different mutations at a single step, granting that the degeneracy of the genetic code relating to amino acids would considerably reduce this as far as immediate consequences are concerned. There is the potentiality for some 4.5 million different two-step mutations and so on to virtually infinite numbers of nucleotide replacement alone.

It is convenient in many problems of population genetics to postulate merely pairs of alleles, a type allele and a mutant in each case. It must be recognized, however, that the type allele probably includes numerous iso-alleles which can be treated collectively to a first approximation but that it may be expected to undergo a gradual evolutionary change in its average primary and pleiotropic effects and in the rate and quality of its mutations as its composition changes. The collective mutant allele is likely to include a highly heterogeneous group of alleles at all times. The theory is thus decidedly inadequate unless it can be extended to multiple alleles.

The likelihood that observed mutations include not only nucleotide replacements but changes involving multiple nucleotides up to conversions, duplications and deficiencies of whole genes, and rare crossing over within pseudoallelic complexes treated as single loci, must be borne in mind and allowed for where practicable in the development of theory.

Some of the chromosome aberrations, especially inversions and duplications and deficiencies of groups of genes, may be treated for some purposes as if single genes. This may even be the case with translocations, although usually with serious qualifications with respect to equality of segregation. Polysomy and polyploidy, on the other hand, involve wholly different patterns of heredity that require consideration as already noted.

It has been apparent almost from the first that there are genetic phenomena, associated especially with chlorophyl deficiencies, male sterility, and nucleoplasmic incompatibility, that are not Mendelian and not chromosomal (chap. 4), although the overwhelming predominance of Mendelian heredity with respect to observable variations justifies us in putting major emphasis in population genetics on the latter. Nevertheless, considerable account must be taken of non-Mendelian heredity.

One difficulty with the development of any general population theory for non-Mendelian heredity is the extraordinary diversity of the phenomena. There is no general pattern even in the class of chlorophyl variegations. Many supposed cases of non-Mendelian heredity have turned out to be infectious and interpretable as due to a virus or other parasite. This does not wholly rule them out from consideration in population genetics, since an infection to which tolerance has been acquired may develop into a mutually beneficial symbiosis and even give rise to a system that may be treated as a single organism, as in the groups of lichens. Some, and perhaps all, cell organelles contain nucleic acid and are to some extent autonomous, even though most of their variations (in chloroplasts, for example) are under nuclear control. It is quite possible that symbiosis was involved in the ultimate orgin of the cell.

In one-celled organisms there may be a merging of physiological maintenance and heredity in such phenomena as cytoplasmic lag and in repertoires of more or less self-regulatory cytoplasmic states. These are too evanescent to be of much concern in connection with evolutionary transformation.

Such physiological phenomena as well as effects of partially autonomous particles in the cells of higher organisms may, however, be involved in nucleoplasmic incompatibility. The building up of this under isolation may be of first-order importance in clinching genetically a splitting of a species that was initially merely geographic.

Chapter 5 was concerned with the relation of genes to characters, a subject of fundamental importance in population genetics. It is not the mode of primary gene action that is of immediate concern here. Of most concern is the array of effects of allelic differences on the ultimate characters that are directly related to selective value. These characters are in general the consequences of complicated chains and networks of interacting processes, to which numerous primary gene products contribute. This usually applies even to physiological adaptations and practically always to adaptations to the outside world. Seven broad generalizations were listed: (1) characters are in general affected by a great many loci; (2) each locus in general affects many characters (pleiotropy); (3) each of the innumerable possible alleles at a locus has a unique array of effects; (4) dominance is not an attribute of genes or pairs of genes and is often easily modified; (5) the effect of genes on a character is rarely fully additive and may be highly nonadditive; (6) homology, whether ontogenetic or phylogenetic depends on the calling into play of similar gene complexes by similar local conditions during the developmental process; (7) selective values of characters with respect to natural selection usually fall off from an intermediate optimum which imposes interaction effects of the most extreme sort, and implies the existence of numerous selective peaks.

A number of illustrations of the complexity of the patterns of gene interaction were discussed, including ones concerned in relatively superficial characters such as the colors of animals and plants as well as ones affecting morphological characters that may be supposed to be more deeply imbedded in the interaction system of the organism as a whole. Where extensive analysis of a character is practicable, it tends to reveal a few essential genes, inactivation of which leads to drastic change. These tend to be dominant. They are supplemented by allelic differences at less essential loci, at which there tends to be semidominance. Beyond these are a host of quantitatively varying series, dependent on genetic modifiers, environmental effects and accidents of development.

While all characters must be involved to some extent in one great interaction system, the organism as a whole, it is clear that many subsystems are sufficiently independent that their effects may be treated as additive with little error. This does not imply that the effects of individual genes may be treated as additive.

The ultimate character with which population genetics is concerned is selective value itself. Under conditions in which selection is directed toward an extreme grade of some character, the corresponding component of total selective value may often be treated as proportional to the grade of the latter. It is maintained, however, that natural selection is usually directed toward an optimum grade that is not far from the average. In this case, genes that tend at a given time to increase the grade are favorable if the total effect of factors tends toward a grade below the optimum but are unfavorable if the opposite is true. Thus selective value, as a character, usually superimposes interaction effects of the most extreme sort, upon whatever interaction effects there may be among genes with respect to the underlying characters. Mathematical treatment of gene interaction must obviously be a primary concern in the theory of population genetics.

Chapters 6 to 14 dealt with statistical methods with special reference to the classification, description, and interpretation of biological variability, subjects that are necessarily fundamental for population genetics.

The classification of observed frequency distributions was the subject of chapter 6. It was brought out that the central type in homogeneous data is that in which there is dense clustering about a single mode and more or less symmetrical tapering off of frequencies in both directions to give a bell-shaped form. Actual distributions, of which many examples were given, may, however, have more than one mode or, if a single one, may exhibit skewness in one sense or the other, or an unduly broad flat peak (platykurtosis), or an unduly high narrow one (leptokurtosis). There may be continuous variability, or a succession of discrete steps (meristic variability). The general form may differ little in these cases, but not infrequently, a particular number of elements in a meristic distribution may be unduly abundant. There may be a threshold below which, or a ceiling above which, there is no variability, sometimes from mathematical necessity (at 0% and 100% in a distribution of percentages) but often for obscure physiological reasons. In many cases only two or three alternative categories can be recognized. A point distribution that reflects discreteness of underlying factors is to be distinguished from a meristic distribution that reflects a succession of all-or-none developmental reactions.

The mathematical theory was introduced in chapter 7 by consideration of the relation between the concepts of frequency and probability. The two basic methods of describing frequency distributions mathematically, that of percentiles and that of moments, were introduced and compared.

In chapter 8, the method of moments provided the basis for consideration of properties of compound (heterogeneous) distributions and those of the compound variables, typically involved in apparently homogeneous ones. Of special importance is Laplace's principle that if there are many factors or factor groups, with distributions of any sort within themselves, that contribute not too unequally to the variability of a character and that are largely independent in occurrence and additive in their effects, the resultant distribution approaches a certain bell-shaped curve, the normal probability curve. This can explain the central position occupied by this type of distribution in biological variability. The mathematical properties of the normal curve and related distributions (binomial, Poisson, Chi and Chi square, gamma) were discussed.

It was brought out in chapter 9, that second order statistics (standard errors) that indicate the degree of reliability of the first order parameters of frequency distributions, are an essential part of the description. This chapter also introduced some of the more important methods of testing the significance of observed differences between populations with respect both to parameters (especially where numbers are small), and frequency distributions as wholes (Chi square).

Chapter 10 took up from chapter 8 the interpretation of observed deviations from normality. The effects of heterogeneity of the means or of the variances of components of compound distributions was considered. So also were the effects of skewness and kurtosis in the distributions of components of compound variables. The consideration of the effects of correlations in occurrence or effect of factors led to various additional kinds of distributions (Beta, hypergeometric, negative binomial).

Special attention was given to methods of eliminating the class of interactions in which the contribution of each factor is a function of the grade determined by the whole array. Various logarithmic transformations of scale, with a rational basis in the exponential nature of growth, and of damping near an upper limit, took the leading place here but were not the only transformations that were considered. The attempt to normalize aberrant frequency distributions on the hypothesis of additivity on an appropriate scale could be reconciled with the emphasis on the complexity of gene interaction in chapter 5 by noting that additivity of the effects of largely independent gene systems is all that is required for approximate normality, not additivity of the effects of the genes.

A wide variety of actual frequency distributions, including many of those used in illustrating types in chapter 6 were taken up in chapter 11 for interpretation on the principles discussed in chapters 8 and 10. Systematic use was made of the inverse probability of the running sums of the percentage frequencies up to each class limit ($\text{pri}^{-1} p$), which is a straight line when applied to a normal distribution. This brings clearly to the eye the nature of the deviations from normality. There may be a strong suggestion of excessive heterogeneity that was not clearly apparent in the actual frequency distribution. If this is not the case, the general nature of the transformation that is required to normalize adequately a whole family of distributions may be clearly indicated. This applies to both continuous and meristic distributions, although in the latter there is, as already noted, a considerable likelihood, especially if the number of elements is small, that a standardization of one of them will be revealed. It was again emphasized that while some approach to normalization is usually possible for either physiological or morphological variability of a quantitative nature, this would not be possible for the contributions of such variability to selective value if the optimum is intermediate. It may, of course, be possible for total selective value as the result of many more or less independent contributions.

The discussion up to this point dealt only with single variables, abstracted from the actual complex variability of the organisms. Chapter 12 was concerned with the mathematical description of systems of correlated variables. It began with a brief account of Gauss's method of least squares and continued with a discussion of the product-moment correlation and related statistics. Finally, the analysis of variability into components was taken up briefly together with the appropriate tests for the significance of apparent heterogeneity.

Path analysis, a general method for dealing with linear systems of correlated variables, was the subject of chapter 13. The main concern here was with the general principles for dealing with any postulated network of linear relations. It was brought out that in a symmetrical pattern in which one variable is represented as directly determined by a number of correlated variables, the method is identical with multiple regression. In another sort of symmetrical system, that in which a number of measurable dependent variables are all represented as determined by a number of hypothetical factors, the method becomes essentially identical with conventional "factor analysis." The extent to which nonlinear relations can be treated by the method was considered.

The quantitative evaluation by means of path analysis of the irregular systems of variables, measurable or hypothetical, that result from attempts to interpret causal relations was considered in chapter 14. Examples include systems of morphological correlations, systems that involve temporal relations and systems of reciprocal interactions. The application to important problems of population genetics is deferred to the later volumes.

In chapter 15, the biometric principles of the preceding chapters are used in the extension of the genetic postulates to the inheritance of quantitative variability. The massive evidence that the genetic component consists in general of numerous, more or less independent, Mendelian differences was brought out.

Unfortunately, the attempt to deduce in detail the genetics of the differences between pure strains that are crossed, from analysis of F_1, F_2, backcross, and even of F_3 progenies, turns out to be frustrated by the large number of possible parameters: the numbers and effects of allelic pairs, linkage, degrees of dominance, and possible kinds of interaction effects. The profile of parental grades, of F_1, F_2, and backcross averages gives important information, however, and the segregation index based on the difference between F_2 or backcross variances and that of F_1 and parental strains, in relation to the difference between the parental means, adds considerably more in the case of an extreme cross. Analysis of F_3 data on means and variances within and among progenies not discussed here in detail should add still more in principle, but it requires such large numbers within each of a large number of progenies from wholly random matings to obtain statistically significant results that the conclusions are usually disappointing, especially since at best there are always more parameters to be determined than there are independent equations.

Considerable independent evidence has often been obtained from the use of linked markers and has been especially valuable in furnishing direct evidence for chromosomal heredity. It was noted that attempts to isolate at least the leading factor difference between strains by a suitable program of repeated backcrossing could probably be made to advantage more frequently than has been the case.

F_2 distributions differ notably from natural variability because all frequencies of allelic genes from a cross between pure strains are 0.50. Yule pointed out within two years after the rediscovery of Mendelian heredity that this gene frequency tends to persist indefinitely under random mating in the absence of disturbing factors. A year later, Castle showed that this applies to any gene frequency and that zygotic frequencies thus tend to persist in frequencies given by the square of the array of gametic frequencies. This is indeed an immediate consequence of Mendel's demonstration of the symmetry of segregation, and it is only slightly more difficult to show that in the long run all pairs of alleles may be expected to be combined at random in spite of linkage. These provide conditions for an approach to the normal distribution on a scale in which effects are approximately additive if there are a sufficient number of largely independent interaction systems with not too unequal effects. This ties the Mendelian theory of quantitative variability to the prevailing nature of actual biological variability.

Finally, it is appropriate to return to the introductory chapter on ideas of evolution proposed before 1900 to consider how the near universality of the Mendelian mechanism of heredity bears on these. It is merely necessary to substitute gene mutation, segregation, and recombination for Darwin's random variability to bring his conception of natural selection up-to-date as the major process. The most serious contemporary criticism of Darwin's theory, the rapid dissipation of new variability under the blending heredity that he assumed, wholly disappears under Mendelian heredity with its implied tendency toward persistence of all gene frequencies in the absence of disturbing factors.

De Vries' mutation theory led rapidly to recognition of the role of chromosome aberrations in the splitting off of new genetically isolated species. Darwin's objection to "sports" (abrupt genetic changes with major effects) as the basis of adaptive change still holds, however, since the likelihood that any large change will be adaptive is enormously less than that a succession of minute changes, guided at each step by selection, may lead to a similarly large adaptive change. This invalidates de Vries' mutation theory, in so far as it relates to unbalanced chromosome aberrations as the major basis for species transformation. Balanced aberrations do not, indeed, have large effects on characters as a rule, but they become established too infrequently to contribute much to transformation. They may, on the other hand, be of first importance in the building up of reproductive isolation in geographically isolated lines. The relative importance of this process, of gradually built up nucleoplasmic incompatibility, and of accumulations of gene differences, remains to be evaluated.

As these statements imply, the early theories that attributed great importance to spatial isolation are still valid, not as alternatives to natural selection, but as accessories. Darwin was greatly influenced toward acceptance of evolution by the striking differences between the faunas of South America and the Old World and by the endemic groups of species on the Galápagos Islands which he saw could best be accounted for by divergent evolution under long continued isolation. The necessity for isolation in accounting for speciation was emphasized even more emphatically by Moritz Wagner. This aspect of his views may be accepted without accepting his theory that divergence is brought about by the direct influence of different environments on the heredities. Gulick postulated a role of fine-scaled geographical isolation in giving an opportunity for accidental differences to arise and accumulate.

This is, indeed, alternative to natural selection as a cause of divergence but applied only to very small colonies. The possibility will be explored later that its occurrence among many local populations may give a basis for selection among these that is even more important for transformation of the species as a whole, than selection among individuals. This possible role of isolation in species transformation must be distinguished from its undoubted role in the splitting of species.

All the earlier theories that implied specific guidance of the course of evolution by control of the direction of elementary genetic change are incompatible with the accidental nature of mutations of genes as DNA molecules, and with the randomness of segregation and recombination. We are not concerned here with possible control over rates. The theory of direct control by environmental conditions proposed by Buffon, Wagner, and others is ruled out in so far as evolution depends on Mendelian genes. Similarly the guidance of change in the DNA by physiological adaptations of the organism in such a way as to give such adaptation in the absence of the conditions that invoked them in the parents is wholly untenable. The doctrine of the inheritance of acquired characters advocated by Erasmus Darwin, Lamarck, by Charles Darwin in his later writings, and by Cope and others can have no application to evolution based on Mendelian heredity.

The case is not so clear with non-Mendelian heredity, especially in one-celled organisms. There may be environmental effects on cytoplasmic lag or on self-regulatory states of the cell that might be considered to come under the head of the inheritance of acquired characters but probably more for short-term processes than for long-term evolution. An infection by a parasite, to which the host becomes adapted and ultimately dependent, is certainly an acquired character in the most literal sense. Such possible qualifications are far from justifying the general theory, however. The latter may be dismissed without detailed consideration of the experiments that have, from time to time, been supposed to demonstrate it, but that have in no case stood up under critical examination.

The theories under which the course of evolution is supposed to be inherent in the first created form of life (Lamarck's scale of being, leading to man) or under which there is a perfecting principle in life that guides this course over long periods (Nägeli, Osborn) are merely expressions of a belief that the process is one that transcends the possibility of scientific explanation. There seems no necessity for taking this viewpoint in the face of the possibilities of rational explanation on the basis of what is now known of heredity and the action of selection.

The hypothesis of a sort of organic momentum, the orthogenesis of Eimer and others, invoked to account for evolutionary trends that lead to exaggerated development of characters, that may even be injurious to the species, is again contrary to what is known of genetic processes, but the supposed cases need careful examination. Even if rejected as an explanatory term, orthogenesis may still be useful descriptively for long-continued trends.

The treatment of the opposite sort of evolutionary course by Lloyd Morgan as due to a wholly inexplicable "emergence" is again essentially an abandonment of scientific investigation. The term itself, like orthogenesis, may still be useful descriptively, and the supposed cases should be examined from the scientific standpoint.

The basic facts of genetics, discovered since 1900, thus largely dispose of several of the earlier theories. Natural selection of small, favorable genetic variations, the occurrence of major mutations—now identified with chromosome aberrations—and isolation remain as valid factors, the roles of which in evolution are to be interpreted in the light of the new data. It will require much detailed examination of the statistical consequences of the Mendelian mechanism under diverse conditions in conjunction with other lines of evidence to develop a fully rounded theory.

REFERENCES

Allen, C. E. 1930. Gametophytic inheritance in Sphaerocarpos. IV. Further studies of tuftedness and polyclady. *Genetics* 15:150–88.

Allison, A. C. 1955. Aspects of polymorphism in man. *Cold Spring Harbor Symp. Quant. Biol.* 20:239–55.

Alpatov, W. W. and Boschko-Stepanenko, A. M. 1928. Variation and correlation in serially situated organs in insects, fishes and birds. *Amer. Nat.* 62:409–24.

Anderson, Edgar. 1928. The problem of species in the northern blue flags. Iris versicolor L. and Iris virginica L. *Ann. Missouri Bot. Garden* 15:241–332.

———. 1949. *Introgressive hybridization.* New York: John Wiley & Sons.

Anderson, E. G. 1923. Maternal inheritance of chlorophyl in maize. *Bot. Gaz.* 76:411—18.

Andersson-Kottö, I. 1923. The genetics of variegation in a fern. *Jour. Genet.* 13:1–12.

———. 1930. Variegation in three species of ferns (Polystichum angulare, Lastraea atrata, and Scolopendrium vulgare). *Zeit. ind. Abst. Vererb.* 56:115–201.

Auerbach, C. 1949. Chemical mutagenesis. *Biol. Rev.* 24:355–91.

Avery, O. T., MacLeod, C. M., and McCarty, M. 1944. Studies on the chemical nature of the substance inducing transformation of Pneumococcal types. Induction of transformation by a desoxyribonucleic acid fraction isolated from Pneumococcus type III. J. *Exp. Med.* 76:137–58.

Baltzer, F. 1925. Untersuchungen über die Entwicklung und Geschlechtsbestimmung der Bonellia. *Publ. Staz. Zool. Napoli* 6:223–85.

Bateson, W. 1909. *Mendel's principles of heredity.* Cambridge: At the University Press, 1930.

Bateson, W. and Brindley, H. H. 1892. On some cases of variation in secondary sexual characters statistically examined. *Proc. Zool. Soc.* 1892:585–94.

Bateson, W. and Punnett, R. C. 1905. Experimental studies in the physiology of heredity. *Rept. to the Evolution Com. of Roy. Soc. II.*

Baur, E. 1909. Das Wesen und die Erblichkeitsverhältnisse der "Variatates albomarginatae hort." von Pelargonium zonale. *Zeit. ind. Abst. Vererb.* 1:330–51.

Beadle, G. W. and Ephrussi, B. 1936. The differentiation of eye pigments in Drosophila as studied by transplantation. *Genetics* 21:225–47.

Beadle, G. W. and Tatum, E. L. 1941. Genetic control of biochemical reactions in Neurospora. *Proc. Nat. Acad. Sci.* 27:499–506.

Beale, G. H. 1952. Antigen variation in Paramecium aurelia variety 1. *Genetics* 37: 62–74.

Beermann, W. 1957. Nuclear differentiation and functional morphology of chromosomes. *Cold Spring Harbor Symp. Quant. Biol.* (1956) 21: 217–32.

Belling, J. 1927. A working hypothesis for segmental interchange between homologous chromosomes. *Proc. Nat. Acad. Sci.* 13: 717–18.

———, 1928. The ultimate chromomeres of Lilium and Aloë with regard to the number of genes. *Univ. Calif. Publ. in Botany* 14: 307–18.

———. 1933. Crossingover and gene rearrangement in flowering plants. *Genetics* 18: 388–413.

Benzer, S. 1957. The elementary units of heredity. In *Symposium on the chemical basis of heredity,* eds. W. D. McElroy and B. Glass, pp. 70–93. Baltimore: The Johns Hopkins Press.

Berg, R. L. 1960. The ecological significance of correlation pleiades. *Evolution* 14: 171–80.

Bergson, H. 1911. *Creative evolution.* Translated by Arthur Mitchell. New York: Henry Holt & Co.

Bernstein, F. 1924. "Ergebnisse einer biostatistischen zusammenfassenden Betrachtung die erblichen Blutstruskturen des Menschen. *Klin. Wochenschr.* 33: 1495–97.

Bittner, J. J. 1937. Mammary tumors in mice in relation to nursing. *Amer. J. Cancer* 30: 530–38.

Blakeslee, A. F. 1921. An apparent case of non-Mendelian inheritance in Datura due to disease. *Proc. Nat. Acad. Sci.* 7: 116–18.

———. 1928. Genetics of Datura. *Zeit. ind. Abst. Vererb.* Suppl. 1:117–30.

Blakeslee, A. F. and Avery, A. G. 1937. Methods of inducing doubling of chromosomes in plants by treatment with Colchecine. *J. Hered.* 28:392–411.

Bliss, C. I. 1935. The calculation of the dosage mortality curve. *Ann. Applied Biol.* 22: 134–67.

———. 1937. The calculation of the time mortality curve. *Ibid.* 24: 815–52.

Bock, F. C. 1950. White spotting in the guinea pig due to a gene (Star) which alters hair direction. Unpublished thesis, University of Chicago.

Boivin, A., Vandrely, R., and Vandrely, C. 1948. L'acide désoxyribonucleique in noyau cellulaire dépositaire des caractères héréditaires; arguments d'ordre analytique. *C. R. Acad. Sci.* 226: 1061–63.

Boycott, A. E. and Diver C. 1923. On the inheritance of sinistrality in Limnaea peregra. *Proc. Roy. Soc.* 95B: 207–13.

Boycott, A. E., Diver, C., Garstang, S. L., and Turner, F. M. 1930. The inheritance of sinistrality in Limnaea peregra (Mollusca, Pulmonata) *Phil. Trans. Roy. Soc.* B219: 51–130.

Braden, A. W. H. 1958. Influence of time of mating on the segregation ratio of alleles at the T-locus in the house mouse. *Nature* 181: 786–87.

Bridges, C. B. 1916. Non-disjunction as proof of the chromosome theory of heredity. *Genetics* 1: 107–63.

———. 1927. The relation of age of the female to crossingover in the third chromosome of Drosophila melanogaster. *Jour. Gen. Physiol.* 8:689–700.

———. 1935. Salivary chromosome maps with a key to the banding of the chromosomes. *Jour. Hered.* 26:60–64.

———. 1936. The Bar gene a duplication. *Science* 83:210–11.

Bridges, C. B. and Brehme, K. S. 1944. The mutants of Drosophila melanogaster. *Carnegie Inst. Washington*, Publ. No. 522.

Bridges, C. B. and Morgan, T. H. 1923. The third chromosome group of mutant characters of Drosophila melanogaster. *Carnegie Inst. Washington*, Publ. No. 327.

Brink, R. A. 1958. Paramutation at the R locus in maize. *Cold Spring Harbor Symp. Quant. Biol.* 22:379–91.

Brink, R. A. and Nilan, R. A. 1952. The relation between light variegated and medium variegated pericarp in maize. *Genetics* 37:519–44.

Browne, E. T. 1895. On the variation of the tentaculocysts of Aurelia aurita. *Quart. Jour. Micr. Sci.* 37: 255–51.

Burks, Barbara S. 1928. The relative influence of nature and nurture upon mental development; a comparative study of foster-parent—foster-child resemblance and true-parent—true-child resemblance. In *Nature and nurture, their influence on intelligence*, Part I, pp. 219–316. Nat. Soc. Study of Education 1928, 27th Yearbook.

Burton, G. W. 1951. Quantitative inheritance in pearl millet (Penisetum glaucum). *Agron. Jour.* 43:409–17.

Burt, Cyril. 1941. *The factors of mind.* New York: MacMillan Co.

Butler, L. 1954. The effect of the coat colour dilution gene on body size in the mouse. *Heredity* 8:275–278.

Carpenter, J. R. 1939. Recent Russian work on community ecology. *Jour. Animal Ecol.* 8:354–86.

Caspari, Ernst. 1933. Über die Wirkung einer pleiotropischen Gens bei der Mehlmotte, Ephestia kühniella, Zeller. *Arch. Entw.-Mech. der Org.* 130: 353–81.

———. 1948. Cytoplasmic inheritance. *Advances in Genetics* 2:1–66.

Caspersson, T. 1941. Studien über den Eiweissumsatz der Zellc. Naturwissenschaften 29:33–43.

Castle, W. E. 1903. The laws of Galton and Mendel and some laws governing race improvement by selection. *Proc. Amer. Acad. Arts. Sci.* 35:233–42.

———, 1906. The origin of a polydactylous race of guinea pigs. *Carnegie Inst. Washington*, Publ. No. 49.

———. 1911. *Heredity in relation to evolution and animal breeding.* New York: D. Appleton and Co.

———. 1916*a*. An expedition to the home of the guinea pig and some breeding experiments with the material there obtained. *Carnegie Inst. Washington*, Publ. No. 241:3–55.

———. 1916*b*. *Genetics and eugenics.* Cambridge: Harvard University Press.

———. 1921. An improved method of estimating the number of genetic factors concerned in cases of blending inheritance. *Science* 54:223.

———. 1922. Genetic studies of rabbits and rats. *Carnegie Inst. Washington*, Publ. No. 320:1–55.

———. 1929. A mosaic (intense-dilute) coat pattern in the rabbit. *Jour. Exp. Zool.* 52:471–80.

———. 1941. Influence of certain color mutations on body size in mice, rats, and rabbits. *Genetics* 26:177–91.

Castle, W. E. and Phillips, J. C. 1914. Piebald rats and selection. *Carnegie Inst. Washington*, Publ. No. 195.

Chai, C. K. 1956. Analysis of quantitative inheritance of body size in mice. II. Gene action and segregation. *Genetics* 41:165–78.

———. 1961. Analysis of quantitative inheritance of body size in mice. IV. An attempt to isolate polygenes. *Gen. Res.* 2:25–32.

Chargaff, E. 1950. Chemical specificity of nucleic acids and mechanisms of their enzymatic degradation. *Experientia* 6:201–09.

Chase, H. B. 1939. Studies on the tricolor pattern of the guinea pig.
I. The relation between different areas of the coat in respect to the presence of color. *Genetics* 24:610–21.
II. The distribution of black and yellow as affected by white spotting and by imperfect dominance in the tortoiseshell series of alleles. *Ibid.*: 622–43.

Child, C. M. 1911. Experimental control of morphogenesis in the regulation of Planaria. *Biol. Bull.* 20:309–33.

Clancy, C. W. and Beadle, G. W. 1937. Ovary transplants in Drosophila: studies of the characters singed, fused, and female sterile. *Biol. Bull.* 72:47–56.

Cole, L. C. 1946. A study of the cryptozoa of an Illinois woodland. *Ecol. Monographs* 16:49–86.

Collins, G. N. and Kempton, J. H. 1916. Patrogenesis. *J. Hered.* 7:106–18.

Cooper, K. W. 1956. Phenotypic effects of Y chromosome hyperploids in Drosophila melanogaster and their relation of variegation. *Genetics* 41: 242–64.

Correns, C. 1900. G. Mendels Regel über das Verhalten der Nachkommenschaft Rassenbastarde. *Ber. deutsch. bot. Ges.* 18:158–68.

———. 1908. Die Rolle der männlichen keimzellen bei der Geschlechtsebestimmung der Gynodiöcischen Pflanzen. *Ibid.* 26:686–701.

———. 1909. Verebungsversuche mit blass (gelb) grünen und buntblättrigen Sippen bei Mirabilis, Urtica and Lunaria. *Zeit. ind. Abst. Vererb.* 1:291–329.

———. 1937. Nichtmendelnde Vererbung. *Handbuch der Vererb. Wiss.* 22:1–159; ed. by F. von Wettstein, Berlin: Gebrüder Borntraeger.

Cramér, Harold. 1946. *Mathematical methods of statistics*. Princeton: Princeton University Press.

Cuénot, L. 1904. L'hérédité de la pigmentation chez les souris. *Arch. Zool. Exp. et Gen.* (Ser. 4) 2:45–56.

Darlington, C. D. 1929. Meiosis in polyploids. II. Aneuploid hyacinths. *Jour. Genet.* 21:17–56.

Darwin, Charles. 1859. *The origin of species by means of natural selection.* 6th ed. New York and London: D. Appleton & Co., 1910.

———. 1868. *The variation of animals and plants under domestication.* 2d ed. 2 vols. New York: D. Appleton & Co., 1883.

Davenport, C. B. 1904. *Statistical methods with special reference to biological variation.* New York: John Wiley & Sons.

———. 1913. Heredity of skin color in negro-white crosses. *Carnegie Inst. Washington,* Publ. No. 188.

Davidson, F. A. 1928. Growth and senescence in purebred Jersey cows. *Univ. Ill. Agr. Exp. Sta.* Bull No. 302:183–235.

De Garis, C. F. 1935. Heritable effects of conjugation between free individuals and double monsters in diverse races of Paramecium candatum. *Jour. Exp. Zool.* 71:209.

DeHaan, H. 1933. Inheritance of chlorophyl deficiencies. *Bibl. Genet.* 10:357–416; The Hague: M. Nijhoff.

Demerec, M. 1927. A second case of maternal inheritance of chlorophyl in maize. *Bot. Gaz.* 84:139–55.

———. 1928. Mutable characters of Drosophila virilis. I. Reddish-alpha body character. *Genetics* 13:359–88.

———. 1929. Genetic factors stimulating mutability of miniature-gamma wing character of Drosophila virilis. *Proc. Nat. Acad. Sci.* 15:834–38.

———. 1937. Frequency of spontaneous mutation in certain stocks of Drosophila melanogaster. *Genetics* 22:469–78.

———. 1955. What is a gene? Twenty years later. *Amer. Nat.* 89:15–20.

———. 1957. A comparative study of certain gene loci in Salmonella. *Cold Spring Harbor Symp. Quant. Biol.* (1956) 21:113–21.

Deol, M. S. 1958. Genetical studies on the skeleton of the mouse. XXIV. Further data on skeletal variation in wild populations. *Jour. Embr. Exp. Morph.* 6:569–74.

Deol, M. S. and Truslove, C. M. 1957. Genetical studies on the skeleton of the mouse. XX. Maternal physiology and variation in the skeleton of C57B1 mice. *Jour. Genetics* 55:288–312.

Detlefsen, J. A. 1914. Genetic studies on a cavy species cross. *Carnegie Inst. Washington,* Publ. No. 205.

———. 1918. Fluctuations of sampling in a Mendelian population. *Genetics* 3:599–607.

Diakanov, D. M. 1925. Experimental and biometric investigations on dimorphic variability of Forficula. *Jour. Genet.* 15:201–32.

Diver, C. and Andersson-Kottö, I. 1938. Sinistrality in Limnaea peregra (Mollusca, Pulmonata). The problem of mixed broods. *Jour. Genet.* 35: 447–525.

Dobzhansky, T. 1927. Studies on the manifold effects of certain genes in Drosophila melanogaster. *Zeit. ind. Abst. Vererb.* 43:330–88.

Dobzhansky, T. and Holz, A. M. 1943. A re-examination of the problem of manifold effects of genes in Drosophila melanogaster. *Genetics* 28:295–303.

Dobzhansky, T. and Queal, M. L. 1938. Genetics of natural populations. II. Genic variation in populations of Drosophila pseudoobscura inhabiting isolated mountain ranges. *Genetics* 23:463–84.

Dobzhansky, T. and Sturtevant, A. H. 1938. Inversions in the chromosomes of Drosophila pseudoobscura. *Genetics* 23:28–64.

Dobzhansky, T. and Wright, S. 1943. Genetics of natural populations. X Dispersion rate in Drosophila pseudoobscura. *Genetics* 28:304–40.

Dubinin, N. P. 1929. Allelomorphentreppen bei Drosophila melanogaster. *Biol. Zentr.* 49:328–39.

Dunn, L. C. 1928. The effect of inbreeding on the bones of the fowl. *Storrs Agr. Exp. Sta. Bull.* 52:1–112.

———. 1934. Analysis of a case of mosaicism in the house mouse. *Jour. Genet.* 29:317–26.

——. 1957. Analysis of a complex gene in the house mouse. *Cold Spring Harbor Symp. Quant. Biol.* (1956) 21:187–95.

East, E. M. 1910. A mendelian interpretation of variation that is apparently continuous. *Amer. Nat.* 44:65–82.

———. 1916. Studies on size inheritance in Nicotiana. *Genetics* 1:164–76.

East, E. M. and Hayes, H. K. 1911. Inheritance in maize. *Conn. Agr. Exp. Sta. Bull.* 167:142 pp.

Eaton, O. N. 1932. Correlation of hereditary and other factors affecting growth in guinea pigs. *U.S. Dept. Agr. Tech. Bull.* No. 279.

Edgar, R. S. and Epstein, R. H. 1965. Conditional lethal mutations in bacteriophage T4. *Proc. XI Int. Congr. Genetics (1963)* 2:2–16, Oxford: Pergamon Press.

Edgeworth, F. Y. 1908. On the probable errors of frequency constants. *Jour. Roy. Stat. Soc.* 11:651–78.

Elderton, W. P. 1906. *Frequency curves and correlation.* 2d ed. London: C. & E. Layton, 1924.

Emerson, R. A. 1914. The inheritance of a recurring somatic variation in variegated ears in maize. *Amer. Nat.* 48:87–115.

Emerson, R. A., Beadle, G. W., and Fraser, A. C. 1935. A summary of linkage studies in maize. *Cornell Univ. Agr. Exp. Sta. Mem.* 80.

Emerson, R. A. and East, E. M. 1913. The inheritance of quantitative characters in maize. *Bull. Agr. Exp. Sta. Nebraska* No. 2:118 pp.

Emerson, R. A. and Smith, H. H. 1950. Inheritance of number of kernel rows in maize. *Cornell Univ. Agr. Exp. Sta. Mem.* 298:130 pp.

Emerson, S. 1938. The genetics of self-incompatibility in Oenothera organensis. *Genetics* 23:190–202.

———. 1939. A preliminary survey of the Oenothera organensis population. *Genetics* 24:524–37.

Ephrussi, B. 1953. Nucleocytoplasmic relations in microorganisms. Oxford: Clarendon Press.

Fisher, R. A. 1918. The correlation between relatives on the supposition of Mendelian inheritance. *Trans. Roy. Soc. Edinburgh* 52:399–433.

———. 1921*a*. On the probable error of a coefficient of correlation deduced from a small sample. *Metron* 1 (no. 4):1–32.

———. 1921*b*. On the mathematical foundations of theoretical statistics. *Phil. Trans. Roy. Soc.* A222:309–68.

———. 1922*a*. On the dominance ratio. *Proc. Roy. Soc. Edinburgh* 42:321–41.

———. 1922*b*. On the interpretation of χ^2 from contingency tables and the calculation of *P*. *Jour. Roy. Stat. Soc.* 85:87–94.

———. 1924. On a distribution yielding the error functions of several well-known statistics. *Proc. Int. Math. Congress* (*Toronto*) 1924:805–13.

———. 1925. *Statistical methods for research workers.* 7th ed. Edinburgh: Oliver and Boyd, 1938.

———. 1936. Has Mendel's work been rediscovered? *Annals of Science* 1: 115–37.

Fisher, R. A., Immer, F. R., and Tedin, O. 1932. The genetical interpretation of statistics of the third degree in the study of quantitative inheritance. *Genetics* 17:107–24.

Fisher, R. A. and Yates, F. 1938. *Statistical tables for biological, agricultural, and medical research.* Edinburgh: Oliver and Boyd.

Foster, M. 1956. Enzymatic studies of the physiological genetics of guinea pig coat coloration. I. Oxygen consumption studies. *Genetics* 41:396–409.

Fox, A. S. 1949. Immunogenetic studies of Drosophila melanogaster. II. Interaction between rb and v loci in production of antigens. *Genetics* 34:647–64.

Føyn, B. and Gjøen, I. 1954. Studies on the Serpulid, Pomatoceros triqueta L. II. The color pattern of the branchial crown and its inheritance. *Nytt. Mag. Zoologi* 2:85–90.

Freese, E. 1963. Molecular mechanisms of mutation. In *Molecular genetics,* ed. J. H. Taylor, part 1, chap. 5. New York: Academic Press.

Gairdner, A. E. 1929. Male sterility in flax II. A case of reciprocal crosses differing in the F_2. *Jour. Gen.* 21:117–24.

Galton, F. 1879. The geometric mean in vital and social statistics. *Proc. Roy. Soc.* 29:365.

———. 1888. Co-relations and their measurements, chiefly from anthropometric data. *Proc. Roy. Soc.* 45:135–45.

———. 1889. *Natural inheritance.* London: MacMillan Co.

———. 1897–98. An examination of the registered speeds of American trotting horses with remarks on their value as hereditary data. *Proc. Roy. Soc.* 62:310–15.

Gardner, C. O. 1963. Estimates of genetic parameters in cross-fertilizing plants and their implications in plant breeding. *Statistical Genetics and Plant Breeding:* 225–52. Eds. W. D. Hanson and H. F. Robinson. *Nat. Acad Sci.—Nat. Res. Council,* publ. 982.

Garrod, A. E. 1902. The incidence of alkaptonuria, a study of chemical individuality. *Lancet* Dec. 13.

———. 1908. The inborn errors of metabolism. *Lancet.* Jan. 4, 11, 18, 25.

Gates, R. R. 1907. Pollen development in hybrids of Oenothera lata × O. Lamarckiana and its relation to mutation. *Bot. Gaz.* 43:81–115.

Gershenson, S. 1928. A new sex ratio abnormality in Drosophila obscura. *Genetics* 13:488–507.

Gieres, A. and Mundy, K. W. 1958. Production of mutants of tobacco mosaic virus by chemical alteration of its ribonucleic acid *in vitro*. *Nature* 182:1457–58.

Giles, N. H. 1965. Genetic fine structure in relation to function in Neurospora. *Proc. XI Congr. Genet. (1963)*. 2:17–30. Oxford: Pergamon Press.

Ginsburg, B. 1944. The effects of the major genes controlling coat color in the guinea pig on the dopa oxidase activity of skin extracts. *Genetics* 29: 176–98.

Glass, Bentley. 1959. Maupertuis, pioneer of genetics and evolution. In *Forerunners of Darwin 1745-1859*, eds. B. Glass, O. Temkin, and W. L. Strauss, Jr., pp. 51–83. Baltimore: The Johns Hopkins Press.

Goldschmidt, R. 1933. Untersuchungen zur Genetik der Geographischen Variation VII. *Arch Entw. Mech. Org.* 130:562–615.

———. 1934*a*. The influence of the cytoplasm upon gene controlled heredity. *Amer. Nat.* 68:5–23.

———. 1934*b*. Lymantria. *Bibl. Genet.* 11:1–186.

———. 1946. Position effect and the theory of the corpuscular gene. *Experientia* 2:197–203, 250–56.

Goldschmidt, R., Hannah, A., and Piternick, L. K. 1951. The podoptera effect in Drosophila melanogaster. *Univ. Calif. Publ. in Zoology* 55:67–294.

Gordon, H. and Gordon, M. 1950. Colour patterns and gene frequencies in natural populations of the platyfish. *Heredity* 4:61–73.

Gowen, J. W. and Gay, Helen. 1934. Chromosome constitution and behavior in ever-sporting and mottling in Drosophila melanogaster. *Genetics* 19:189–208.

Green, C. V. 1931. On the nature of size factors in mice. *Amer. Nat.* 65:406–16.

Green, E. L. 1954. Quantitative genetics of skeletal variation in the mouse. I. Crosses between three short-ear strains. *J. Nat. Cancer. Inst.* 15:609–24.

Green, M. M. 1965. Genetic fine structure in Drosophila. *Proc. XI Congr. Genet. (1963)* 2:37–49. Oxford: Pergamon Press.

Green, M. M. and Green, K. C. 1956. A cytogenetic analysis of the lozenge pseudoalleles in Drosophila. *Zeit. ind. Abst. Vererb.* 87:708–21.

Gregory, P. W. and Castle, W. E. 1931. Further studies on the embryological basis of size inheritance in the rabbit. *Jour. Exp. Zool.* 59:199–211.

Gregory, R. P. 1915. On variegation in Primula sinensis. *Jour. Genet.* 4: 305–21.

Grüneberg, H. 1938. Some new data on the grey lethal mouse. *J. Genet.* 36:153–70.

———. 1947. *Animal genetics and medicine.* London: Paul B. Hueber, Inc.

———. 1951. The genetics of a tooth defect in the mouse. *Proc. Roy. Soc. B* 138: 437–51.

———. 1952. Genetical studies on the skeleton of the mouse. IV. Quasi-continuous variation. *J. Gen.* 51:95–114.

Gulick, J. T. 1905. Evolution, racial and habitudinal. *Carnegie Inst., Washington,* Publ. No. 25.

Hagemann, R. 1965. Advances in the field of plastid inheritance in higher plants. *Proc. XI Int. Congr. Genetics (1963)* 3:613–25. Oxford: Pergamon Press.

Haines, G. 1931. A statistical study of the relation between various expressions of fertility and vigor in the guinea pig. *Jour. Agr. Res.* 42:123–64.

Haldane, J. B. S. 1930. Theoretical genetics of autopolyploids. *Jour. Genet.* 22:359–72.

———. 1931. The cytological basis of genetical interference. *Cytologia* 3:54–65.

———. 1932. The time of action of genes and its bearing on some evolutionary problems. *Amer. Nat.* 66:5–24.

———. 1935. The rate of spontaneous mutation of a human gene. *Jour. Genetics* 31:317–26.

Haldane, J. S. and Priestley, J. G. 1905. The regulation of the lung ventilation. *Jour. Physiol.* 32:225–66.

Hannah, Aloha. 1951. Localization and function of heterochromatin in Drosophila melanogaster. *Advances in Genetics* 4:87–125.

Hardy, G. H. 1908. Mendelian proportions in a mixed population. *Science* 28:49–50.

Harris, J. A. 1912. Biometric data on the inflorescence and fruit of Crinum longifolium. *Missouri Bot. Garden 23, Annual Rept.* pp. 75–99.

———. 1913. On the calculation of inter- and intraclass coefficients of correlation from class moments when the number of possible combinations is large. *Biometrika* 9:446–72.

Harris, J. A. and Benedict, F. G. 1919. Biometric standards for energy requirements in human nutrition. *Scientific Monthly* 1919:385–402.

Hayes, H. K., East, E. M., and Beinhart, E. G. 1913. Tobacco breeding in Connecticut. *Conn. Agr. Exp. Sta. Bull.* 176:5–68.

Hayman, B. I. 1954. The theory and analysis of diallele crossing. *Genetics* 39:789–809.

Heidenthal, G. 1940. A colorimetric study of genic effect on guinea pig coat color. *Genetics* 25:197–214.

Heitz, E. 1928. Das Heterochromatin der Moose. *Jahrb. wiss. Bot.* 69:762–818.

———. 1934. Über α- and β- heterochromatin sowie Konstanz and Bau der Chromomeren bei Drosophila. *Biol. Zbl.* 54:588–609.

Hickl, A. 1913. Die Gruppierung der Haaranlagen ("Wildzeichnung") in der Entwickelung des Hausschweines. *Anat. Anz.* 44:393–402.

Holley, R. W., Apgar, J., Everett, G. A., Madison, J. T., Marquisee, M., Merrill, S. H., Penswick, J. R., and Zamir, A. 1965. Structure of ribonucleic acid. *Science* 147:1462–65.

Holtzinger, K. J. and Harmon, H. H. 1941. *Factor analysis*. Chicago: University of Chicago Press.

Hoshino, Y. 1940. Genetical studies of the pattern types of the Lady-bird beetle, Harmonia axyridis. *J. Genet.* 40:215–28.

Hotelling, H. 1936. Simplified calculation of principal components. *Psychometrica* 1:27–35.

Hubbs, C. L. 1922. Variation in the number of vertebrae and other meristic characters of fishes, correlated with the temperature of water during development. *Amer. Nat.* 56:360–72.

Hubbs, C. L. and Ramey, E. C. 1946. Endemic fish fauna of Lake Waccaman, N.C. *Misc. Publ. Mus. Zool. U. of Mich.* No. 5:1–30.

Hull, F. H. 1945. Recurrent selection for specific combining ability in corn. *Jour. Amer. Soc. Agron.* 37:134–45.

Huxley, J. S. 1927*a*. Discontinuous variation and heterogeny in Forficula. *Jour. Genet.* 17:309–53.

———. 1927*b*. Studies on heterogonic growth. IV. The bimodal cephalic horn of Xylotrupes gideon. *Jour. Gen.* 18:45–53.

———. 1932. Problems of relative growth. London: Methuen & Co.

Ibsen, H. L. 1928. Prenatal growth in guinea pigs with special reference to environmental factors affecting weight. *Jour. Exp. Zool.* 51:51–91.

Ibsen, H. L. and Goertzen, B. L. 1951. Whitish, a modifier of chocolate and black hair in guinea pigs. *Jour. Hered.* 42:231–36.

Ikeno, S. 1930. Studien über einen eigentümlichen Fall der infektiösen Buntblätterigkeit bei Capsicum annuum. *Planta* 11:359.

Imai, Y. 1928. A consideration of variegation. *Genetics* 13:544–62.

———. 1936. Recurrent auto- and exomutation of plastids resulting in tricolored variegation of Hordeum vulgare. *Genetics* 21:752–57.

Ingram, V. M. 1956. A specific chemical difference between the globins of normal human and sickle-cell anaemia haemoglobin. *Nature* 178:792–94.

Irwin, M. R. 1947. Immunogenetics. *Adv. in Genetics* 1:133–59.

Jacob, F. and Monod, J. 1961. Genetic regulatory mechanisms in the synthesis of proteins. *Jour. Mol. Biol.* 3:318–56.

Jenkin, Fleeming. 1867. Origin of species. *North British Review* 46:277–318.

Jennings, H. S. 1916. Heredity, variation and the results of selection in the uniparental reproduction of Difflugia corona. *Genetics* 1:407–534.

———. 1940. The cell and cytoplasm in protozoa. The Cell and Protoplasm: *AAAS Publ.* No. 14:44–55.

Johannsen, W. 1903. *Über Erblichkeit in Populationen und in Reinen Linien.* Jena: Gustav Fischer.

———. 1909. *Elemente der exakten Erblichkeitslehre.* Jena: Gustav Fischer.

Jollos, V. 1934. Dauermodifikationen und Mutationen bei Protozoen. *Arch. Protistenkunde* 83:197–219.

———. 1934. Inherited changes produced by heat treatment in Drosophila melanogaster. *Genetics* 16:476–94.

Jones, D. F. 1917. Dominance of linked factors as a means of accounting for heterosis. *Genetics* 2:466–79.

Kaufmann, B. P. and MacDonald, M. R. 1957. Organization of the chromosome. *Cold Spring Harbor Symp. Quant. Biol.* (*1956*) 21:233–46.

Kellogg, V. L. 1907. *Darwinism today.* New York: Henry Holt & Co.

Kihara, H. 1959. Fertility and morphological variation in the substitution and restoration backcrossing of the hybrids, *Triticum vulgare* and *Aegilops candatum. Proc. X Inter. Congr. Genet.* 1:142–71.

Kinman, M. L. and Sprague, G. E. 1945. Relation between number of parental lines and the theoretical performance of synthetic varieties of corn. *Jour. Amer. Soc. Agron.* 37:341–51.

Kitzmiller, J. B. and Levan, H. 1959. Speciation in mosquitoes. *Cold Spring Harbor Symp. Quant. Biol.* 24:161–76.

Klauber, L. M. 1936. A statistical study of the rattlesnakes. Occasional Papers: San Diego Soc. Nat. Hist. No. 1:2–24.

———. 1937. A statistical study of the rattlesnakes. IV. The growth of the rattlesnake. Occasional Papers: San Diego Soc. Nat. Hist. No. 3:1–56.

———. 1938. A statistical study of the rattlesnakes. V. Head dimensions. Occasional Papers: San Diego Soc. Nat. Hist. No. 4:1–53.

———. 1941. The frequency distributions of certain herpetological variables. *Bull. Zool. Soc. San Diego* No. 17:5–31.

———. 1946. A new gopher snake (Pituophis) from Santa Cruz Island, California. *Trans. San Diego Soc. Nat. Hist.* 11:4–48.

Kornberg, A. 1962. *Enzymatic synthesis of DNA.* New York: John Wiley & Sons.

Kröning, F. 1930*a*. Über die Eichung von Fellfarben auf Grund der Ostwaldschen Methode, nach Untersuchungen am Meerschweinchen. *Zeit. ind. Abst. Vererb.* 53:355–67.

———. 1930*b*. Die Dopareaktionb ei verschiedene Farbenrassen des Meerschweinchens und der Kaninchens. *Arch. Ent. Mech.* 121:470–84.

Kühn, A., 1927. Die Pigmentierung von Habrobacon juglandis, Ashmead: ihré Prädetermination und ihre Vererbung durch Gen und Plasmen. *Ges. Wiss. Göttingen Math. Phys. Klasse,* 1927:407–21.

Lack, D. 1948. Natural selection and family size in the starling. *Evolution* 2:95–110.

Landsteiner, K. and van der Scheer, J. 1936. On cross-reactions of immune sera to azo proteins. *Jour. Exp. Med.* 63:325–39.

Laughnan, J. R. 1955. Structural and functional basis for the action of A alleles in maize. *Amer. Nat.* 89:91–103.

Laven, H. 1960. Speciation by cytoplasmic isolation in the *Culex pipiens* complex. *Cold Spring Harbor, Symp. Quant. Biol.,* 34:166–73.

Lawrence, W. J. C. and Scott-Moncrieff, R. 1935. The genetics and chemistry of flower colour in Dahlia: a new theory of specific pigmentation. *Jour. Genet.* 30:155–226.

Lederberg, J. and Tatum, E. L. 1947. Novel genotypes in mixed cultures of biochemical mutants of bacteria. *Cold Spring Harbor Symp. Quant. Biol.* (*1946*) 11:113–14.

Lewis, D. 1948. Structure of the incompatibility gene. I. Spontaneous mutation rate. *Heredity* 2:219–36.

———. 1949. Incompatibility in flowering plants. *Biol. Rev.* 24:472–96.

Lewis, E. B. 1942. The star and asteroid loci in *Drosophila melanogaster*. *Genetics* 27:153–54.

———, 1950. The phenomenon of position effect. *Adv. in Genetics* 3:73–112.

L'Héritier, P. and Hugon DeScoeux, F. 1947. Transmission par greffe et injection de la sensibilité héréditaire au gaz carbonique chez le Dorsophile. *Bull. Biol.* 81:70–91.

L'Héritier, P. and Teissier, G. 1947. Transmission héréditaire de la sensibilité au gaz, carbonique chez Drosophila melanogaster. *Publ. Lab. de L'Ecole Normale Supérieure Biologie* 1:35–76.

Lindegren, C. C. 1933. The genetics of *Neurospora* III. Purebred stocks and crossingover in *N. crassa*. *Bull. Torrey Bot. Club* 59:85–102.

———, 1942. The use of the fungi in modern genetical analysis. *Iowa State Coll. Jour. Sci.* 16:271–90.

———. 1953. Gene conversion in Saccharomyces. *Jour. Genet.* 51:625–37.

Lindegren, C. C. and Lindegren, G. 1942. Locally specific patterns of chromatid and chromosome interference in Neurospora. *Genetics* 27:1–24.

Lindstrom, E. W. 1924. A genetic linkage between size and color factors in the tomato. *Science* 60:182–83.

———. 1928. Linkage of size, shape, and color genes in Lycopersicon. *Verh. V. Int. Kongr. Vererb., Zeit. ind. Abst. Vererb.* Suppl. 2:1031–57.

Loeb, J. 1916. The organism as a whole from the physico-chemical viewpoint. New York: G. P. Putnam's Sons.

Lovejoy, A. O. 1959. Buffon and the problem of species. In *Forerunners of Darwin, 1745-1859*, eds. B. Glass, O. Temkin, and W. L. Straus, Jr., pp. 84–113. Baltimore: The Johns Hopkins Press.

Luria, S. E. and Delbrück, M. 1943. Mutations of bacteria from virus sensitivity to virus resistance. *Genetics* 28:491–511.

Lutz, Anne M. 1907. A preliminary note on the chromosomes of Oenothera Lamarckiana and one of its mutants, O. gigas. *Science* 26:151–52.

Lwoff, A. 1950. *Problems of morphogenesis in ciliates: the kinetosomes in development, reproduction, and evolution.* New York: John Wiley & Sons.

MacDowell, E. C. 1917. Bristle inheritance in Drosophila. II. Selection. *Jour. Exper. Zool.* 23:109–46.

Mather, K. 1941. Variation and selection of polygenic characters. *Jour. Gen.* 41:159–93.

———. 1943. Polygenic inheritance and natural selection. *Biol. Res.* 18:32–64

———. 1944. The genetical activity of heterochromatin. *Proc. Roy. Soc. B.* 132: 308–32.

———. 1949. *Biometric genetics*. New York: Dover Publications.

Mather, K. and Harrison, B. S. 1949. The manifold effects of selection. *Heredity* 3:1–52.

McAlister, D. 1879. The law of the geometric mean. *Proc. Roy. Soc.* 29:367.

McClintock, Barbara. 1952. Chromosome organization and genic expression. *Cold Spring Harbor Symp. Quant. Biol.* (*1951*) 16:15–47.

——— 1956. Controlling elements and the gene. *Cold Spring Harbor Symp. Quant. Biol.* 21:197–216.

Mendel, Gregor. 1866. Versuche über pflanzenhybriden. *Verh. Naturforsch. Ver. Brünn* 4:3–17. Translated by W. Bateson in *Mendel's Principles of Heredity* 1909. Cambridge: at the University Press.

Metz, C. W. 1938. Chromosome behavior, inheritance, and sex determination in Sciara. *Amer. Nat.* 72:485–520.

———. 1947. Duplication of chromosome parts as a factor in evolution. *Amer. Nat.* 81:81–103.

Michaelis, P. 1933. Entwicklungsgeschichtliche-genetische Untersuchungen an Epilobium II. Die Bedeutung des Plasmas fur die Pollen-fertilität der Epilobium luteum-hirsutum-Bastard. *Zeit. ind. Abst. Vererb.* 65:1–71, 353–411.

———. 1958. Cytoplasmic inheritance and the segregation of plasma genes. *Proc. X Int. Congr. Genetics* 1:375–85.

Minot, C. S. 1891. Senescence and rejuvenescence. *Jour. Physiol.* 12:97-153.

Mirsky, A. L. and Ris, H. 1949. Variable and constant components of chromosomes. *Nature* 163:666–67.

Mitchell, M. B. 1955. Aberrant recombination of pyridoxine mutants of Neurospora. *Proc. Nat. Acad. Sci.* 41:215–20.

Mohr, O. L. 1923. A genetic and cytologic analysis of a section deficiency involving four units of the X-chromosome in *Drosophila melanogaster*. *Zeit. Abst. Vererb.* 32:108–232.

Morgan, C. Lloyd. 1933. *The emergence of novelty*. London: Williams and Norgate.

Morgan, T. H. 1922. On the mechanism of heredity. *Proc. Roy. Soc. B* 94: 162–97.

Morgan, T. H., Sturtevant, A. H., Muller, H. J., and Bridges, C. B. 1915. *The mechanism of mendelian heredity*. New York: Henry Holt & Co.

Mozley, Alan. 1935. The variation of two species of Limnaea. *Genetics* 20:452–65.

Muller, H. J. 1928. The problem of genic modification. *Zeit. Abst. Vererb.* Suppl. 1:234–60.

———. 1929. The gene as the basis of life. *Proc. Bot. Congr. Plant Sciences* 1:897–921.

———. 1932. Further studies on the nature and causes of gene mutation. *Proc. 6th Int. Congress Genetics* 1:213–55.

———. 1936. Bar duplication. *Science* 83:528–30.

———. 1950. Our load of mutations. *Amer. Jour. Hum. Genetics* 2:111–76.

———. 1956. On the relation between chromosome changes and gene mutation. *Brookhaven Symp. Biol.* 8:126–47.

Muller, H. J. and Mott-Smith, L. M. 1930. Evidence that natural radioactivity is inadequate to explain the frequency of "natural" mutations. *Proc. Nat. Acad. Sci.* 16:277–85.

Muller, H. J. and Settles, F. 1925. The nonfunctioning of genes in spermatozoa. *Zeit. Abst. Vererb.* 43:285–312.

Muller, H. J., Valencia, J. J., and Valencia, R. N. 1949. The frequency of spontaneous mutations of individual loci in *Drosophila*. *Genetics* 35: 125–26.

Murray, W. S. and Little, C. C. 1935. The genetics of mammary tumor incidence in mice. *Genetics* 20:466–96.

Nabours, R. K. 1930. Mutation and allelomorphism in the grouse locust (Tettigidae, Orthoptera). *Proc. Nat. Acad. Sci.* 16:350–53.

Neel, J. V. and Schull, W. J. 1954. *Human Heredity*. Chicago: The Univ. of Chicago Press.

———, 1956. The effect of exposure to the atomic bombs on pregnancy termination in Hiroshima and Nagasaki. Washington, D.C.: Nat. Acad. Sci. Nat. Res. Council, Publ. No. 461.

Neyman, J. 1939. On a new class of contagious distributions, applicable in entomology and bacteriology. *Ann. Math. Stat.* 10:35–57.

———. 1941. Fiducial argument and the theory of confidence intervals. *Biometrika* 32:128–50.

Niles, H. E. 1922. Correlation, causation, and Wright's theory of "path coefficients." *Genetics* 7:258–73.

Nilsson-Ehle, H. 1909. Kreuzungsuntersuchungen an Hafer und Weizen. *Lunds. Univ. Aarskr. NF* 5:2:1–122.

Nirenberg, M. W. and Matthaei, J. H. 1961. The dependence of cell-free protein synthesis in E. coli upon naturally occurring or synthetic polyribonucleotides. *Proc. Nat. Acad. Sci.* 47:1588–1602.

Nordenskiöld, E. 1928. *The history of biology*. Translated by L. B. Eyre. New York: Tudor Publishing Co.

Novick, A. and Szilard, Leo. 1952. Genetic mechanisms in bacteria and bacterial viruses. *Cold Spring Harbor Symp. Quant. Biol.* (*1951*) 16:337–43.

Offermann, C. A. 1935. The position effect and its bearing on genetics. *Bull. Acad. Sci. URSS Cl. Sci. Math. Nat.* 1:129–52.

Oliver, C. P. 1940. A reversion to wild type, associated with crossing over in Drosophila melanogaster. *Proc. Nat. Acad. Sci.* 26:452–54.

Osborn, H. F. 1925. The origin of species. II. Distinctions between rectigradations and allometrons. *Proc. Nat. Acad. Sci.* 11:749–52.

———. 1934. Aristogenesis, the creative principle in the origin of species. *Amer. Nat.* 68:193–235.

Owen, Ray D. 1959. Immunogenetics. *Proc. X Int. Congr. Genetics,* (*1958*) 1:364–74.

Painter, T. S. 1933. A new method for the study of chromosome rearrangements and the plotting of chromosome maps. *Science* 78:585–86.

———. 1934. A new method for the study of chromosome aberrations and the plotting of chromosome maps in Drosophila melanogaster. *Genetica* 19:175–88.

Park, T. 1933. Studies in population physiology. IV. Factors regulating initial growth of Tribolium confusum populations. *Jour. Exp. Zool.* 65: 17–42.

Parker, G. H. and Bullard, C. 1913. On the size of litters and the number of nipples in swine. *Proc. Amer. Acad. Arts Sci.* 49:399–426.

Pascher, A. 1918. Über die Beziehung der Reduktionsteilung zur Mendelschen Spaltung. *Ber. Dentsch. Bot. Ges.* 30:163–68.

Patterson, J. T., Stone, W., Bedichek, S., and Suche, M. 1934. The production of translocations in Drosophila. *Amer. Nat.* 68:359–69.

Pavan, C., Cordeiro, A. R., Dobzhansky, N., Dobzhansky, T., Malogolowkin, C., Spassky, B., and Wedel, M. 1951. Concealed genic variability in Brazilian populations of Drosophila willistoni. *Genetics* 36:13–30.

Payne, F. 1918*a*. The effect of artificial selection on bristle number in Drosophila ampelophila and its interpretation. *Proc. Nat. Acad. Sci.* 4:55–58.

———. 1918*b*. An experiment to test the nature of the variation on which selection acts. *Indiana Univ. Studies,* 5 No. 36.

———. 1920. Selection for high and low bristle number in the mutant strain "reduced." *Genetics* 5:501–42.

Pearl, R. 1928. *The rate of living.* New York: Alfred A. Knopf.

Pearson, Karl. 1894. Contributions to the mathematical theory of evolution. I. On the dissection of asymmetrical frequency curves. *Phil. Trans. Roy. Soc. A* 185:71–110.

———. 1895. Contributions to the mathematical theory of evolution. II. Skew variation in homogeneous material. *Phil. Trans. Roy. Soc. A.* 186:343–414.

———. 1896. Contributions to the mathematical theory of evolution. III. Regression, heredity and panmixia. *Phil. Trans. Roy. Soc. A* 187: 253–318.

———. 1900*a*. On the criterion that a given system of deviations from the probable in the case of a correlated system of variables is such that it can be reasonably supposed to have arisen from random sampling. Phil. Mag. Series 5, 50:157–175.

———. 1900*b*. Mathematical contributions to the theory of evolution. VII. On the correlation of characters not quantitatively measurable. *Phil. Trans. Roy. Soc. A* 190:1–47.

———. 1902. On the systematic fitting of curves to observations and measurements. Part I. *Biometrica* 1:265–303; Part II. *Ibid.* 2:1–23.

———. 1904. On the generalized theory of alternative inheritance with special reference to Mendel's laws. *Phil. Trans. Roy. Soc. A* 203:53–86.

———. 1905. Mathematical contributions to the theory of evolution. XIV. On the general theory of skew correlation and nonlinear regression. Drapers Co. Res. Mem. Biom. Series 2.

———. 1909*a*. On the ancestral genetic correlations of a Mendelian population mating at random. *Proc. Roy. Soc.* 81:219–25.

———. 1909*b*. On a new method of determining correlation between measured characters A and a character B, of which only the percentage of cases wherein B exceeds (or falls short of) a given intensity is recorded for each grade of A. *Biometrika* 7:66–105.

———. 1910. On a new method of determining correlations where one variable is given by alternative and the other by multiple categories. *Biometrika* 7:248–57.

———. 1913. On the measurement of the influence of "broad" categories on a correlation. *Biometrika* 9:116–39.

———. 1914. *Tables for statisticians and biometricians.* Cambridge: At the University Press.

Peffley, R. L. 1953. Crossing and sexual isolation of Egyptian forms of Musca domestica (Diptera, Muscidae). *Evolution* 7:65–75.

Philip, Ursula and Haldane, J. B. S. 1939. Relative sexuality in unicellular algae. *Nature* 143:334.

Plough, H. H. 1917. The effect of temperature on crossing over in Drosophila. *Jour. Exp. Zool.* 24:147–208.

Plough, H. H. and Ives, R. J. 1935. Induction of mutations by high temperature in Drosophila. *Genetics* 20:42–69.

Pontecorvo, G. 1944. Structure of heterochromatin. *Nature* 153: 365–67.

———. 1956. The parasexual cycle in fungi. *Ann. Rev. Microbiol.* 10:393–400.

Pontecorvo, G. and Roper, J. A. 1952. Genetic analysis without sexual reproduction by means of polyploidy in *Aspergillus nidulans*. *J. Gen. Microbiol.* 6:7.

———, 1956. Resolving power of genetic analysis. *Nature* **178:**83–84.

Powers, Leroy. 1942. The nature of the series of environmental variances and the estimates of the genetic variances and the geometric means in crosses involving species of Lycopersicon. *Genetics* 27:561–75.

Preer, J. P. 1950. Microscopically visible bodies in the cytoplasm of the "killer" strains of Paramecium aurelia. *Genetics* 35:344–62.

Punnett, R. C. 1923. Linkage in the sweet pea. *Jour. Gen.* 13:101–23.

Race, R. R. 1944. An "incomplete" antibody in human serum. *Nature* 153: 771–72.

Ramage, R. T. and Day, A. D. 1960. A 10:3:3 ratio for leaf width in barley. *Agron. Jour.* 52:241.

Rasmussen, J. M. 1933. A contribution to the theory of quantitative character inheritance. *Hereditas* 18:245–61.

Reed, S. C. 1937. The inheritance and expression of fused, a new mutation in the mouse. *Genetics* 22:1–13.

Renner, O. 1925. Untersuchungen über die faktorielle Konstitution einiger komplexheterozygotischen Oenotheren. *Bibl. Genet.* 9:1–168.

Rhoades, M. M. 1933. The cytoplasmic inheritance of male sterility in Zea mays. *Jour. Gen.* 27:71–93.

———. 1938. Effect of the Dt gene on the mutability of the a_1 allele in maize. *Genetics* 23:377–97.

———. 1950. Gene-induced mutation of a heritable cytoplasmic factor producing male sterility in maize. *Proc. Nat. Acad. Sci.* 36:634–35.

Ris, H. 1957. Chromosome structure. In *Symposium on chemical basis of heredity*, eds. W. D. McElroy and B. Glass, pp. 23–69. Baltimore: The Johns Hopkins Press.

Robertson, F. H. 1954. Studies in quantitative inheritance. V. Chromosome analysis of crosses between selected and unselected lines of different body size in Drosophila melanogaster. *Jour. Genet.* 52:494–520.

Robinson, H. F., Comstock, R. E., and Harvey, P. H. 1949. Estimates of heritability and the degree of dominance in corn. *Agron. J.* 41:353–59.

———. 1955. Genetic variances in open pollinated varieties of corn. *Genetics* 40:45–60.

Russell, E. S. 1939. A quantitative study of genic effects on guinea pig coat colors. *Genetics* 24:332–55.

———. 1949. A quantitative histologic study of the pigment found in the coat color mutants of the house mouse. IV. The nature of the effect of genic substitutions in five major allelic series. *Genetics* 39:146–66.

Russell, W. L. 1939. Investigation of the physiological genetics of hair and skin color in the guinea pig by means of the dopa reaction. *Genetics* 24:645–67.

———. 1962. An augmenting effect of dose fractionation on radiation induced mutation rate in mice. *Proc. Nat. Acad. Sci.* 48:1724–27.

Russell, W. L. and Green, E. L. 1943. A skeletal difference between reciprocal F_1 hybrids of a cross between two inbred strains of mice. *Genetics* 28:87.

Ryder, E. J. 1958. The effect of complementary epistasis on the inheritance of a quantitative character, seed size in lima beans. *Agron. Jour.* 50: 298–301.

Sager, Ruth. 1965. Non-chromosomal genes in Chlamydomonas. Genetics today. *Proc. XI Int. Congr. Genetics (1963)* 3:579–89. Oxford: Pergamon Press.

Sandler, L., Hiraizami, Y., and Sandler, Iris. 1959. Meiotic drive in natural populations of Drosophila melanogaster. The cytologic basis of segregation distortion. *Genetics* 44:234–50.

Sandler, L. and Novitski, E. 1957. Meiotic drive as an evolutionary force. *Amer. Nat.* 91:105–10.

Sax, Karl. 1923. The association of size differences with seed coat pattern and pigmentation in Phaseolus vulgaris. *Genetics* 8:552–60.

———. 1926. Quantitative inheritance in Phaseolus. *Jour. Agr. Res.* 33: 349–54.

———. 1941. Types and frequencies of chromosomal aberrations induced by X-rays. *Cold Spring Harbor Symp. Quant. Biol.* 9:93–101.

Schultz, J. 1936. Variegation in Drosophila and the inert chromosomal regions. *Proc. Nat. Acad. Sci.* 22:27–33.

Schumann, Helga. 1960. Die Entstehung der Scheckung bei Mäuser mit weissen Blesse. *Developmental Biology* 2:501–15.

Schwab, J. J. 1940. A study of the effects of random group of genes on shape of spermatheca in Drosophila melanogaster. *Genetics* 25:157–77.

Schwemmle, J., Haustein, E., Sturm, J., and Binder, M. 1938. Genetische und Zytologische Untersuchungen an Eu-Oenothera. *Zeit. Abst. Vererb.* 75:358–800.

Scott, J. P. 1937. The embryology of the guinea pig. III. The development of the polydactylous monster. A case of growth acceleration at a particular period by a semidominant gene. *Jour. Exp. Zool.* 77:123–57.

———. 1938. The embryology of the guinea pig. II. The polydactylous monster. A new teras produced by the genes PxPx. *Jour. Morph.* 62:299–321.

Serebrovsky, A. S. 1928. An analysis of the inheritance of quantitative transgressive characters. *Zeit. Abst. Vererb.* 48:229–43.

———. 1930. Untersuchungen über Treppen-allelomorphismus. IV. Transgenation Scute 6 und ein Fall des "Nicht-allelomorphismus von Gliedern einer Allelomorphenreihe bei Drosophila melanogaster. *Arch. Entw. mech. Org.* 122:88–104.

Sexton, E. W. and Pantin, C. F. A. 1927. Inheritance in Gammarus chevreuxi, Sexton. *Nature* 1927:119.

Sheppard, W. E. 1898. On the calculation of the most probable values of frequency constants from data arranged according to equidistant divisions of a scale. *Proc. London Math. Soc.* 29:353–80.

Shull, A. F. 1932. Clonal differences and clonal changes in the aphid. Macrosiphum solanifolii. *Amer. Nat.* 66:385–419.

Shull, G. H. 1908. The composition of a field of maize. *Rept. of Amer. Breed Assoc.* 4:296–301.

Silrers, W. K. 1958. An experimental approach to action of genes at the agouti locus in the mouse. *Jour. Exp. Zool.* 137:189–96.

Sinnott, E. W. 1931. The independence of genetic factors governing size and shape. *Jour. Hered.* 22:381–87.

———. 1935. Evidence for the existence of genes controlling shape. *Genetics* 20:12–21.

Sirks, M. J. 1931. Plasmatic influences upon the inheritance of Vicia faba. *Proc. Kon. Akad. Wet. Amsterdam* 34:1057–62, 1164–72, 1340–46.

———. 1932. Beitrage zur einer Genotypischen Analyze der Ackerbohne, Vicia faba L. *Genetica* 13:209–631.

Sismanides, A. 1942. Selection for an almost invariable character in Drosophila. *Jour. Genet.* 44:204–15.

Slifer, E. H. 1942. A mutant stock of Drosophila with extra sex-combs. *J. Exp. Zool.* 90:31–40.

Smith, H. H. 1937. The relation between genes affecting size and colors in certain species of Nicotiana. *Genetics* 22:361–75.

Smith, P. E. and MacDowell, E. C. 1930. A hereditary anterior-pituitary deficiency in the mouse. *Anat. Rec.* 46:249–57.

Snell, G. D. 1929. Dwarf, a new Mendelian recessive character of the house mouse. *Proc. Nat. Acad. Sci.* 15:733–34.

Sô, M. 1921. Inheritance of variegation in barley. *Jap. Jour. Genet.* 1:21–36.

Sonneborn, T. M. 1943. Gene and cytoplasm. I. The determination and inheritance of the killer character in variety 4 of Paramecium aurelia. *Proc. Nat. Acad. Sci.* 29:329–47.

———. 1948. The determination of hereditary antigenic differences in genetically identical Paramecium cells. *Proc. Nat. Acad. Sci.* 34: 413–18.

Sonneborn, T. M. and Lynch, R. S. 1934. Hybridization and segregation in Paramecium aurelia. *Jour. Exp. Zool.* 67:1–72.

Spearman, C. 1904*a*. The proof and measurement of association between two things. *Amer. Jour. Psychol.* 15:72–101.

———. 1904*b*. General intelligence objectively determined and measured. *Amer. Jour. Psychol.* 15:201–92.

Sprague, G. F. and Brimhall, B. 1949. Quantitative inheritance of oil in the corn kernel. *Agron. Jour.* 41:30–33.

Sprague, G. F. and Tatum, L. A. 1942. General *vs.* specific combining ability in single crosses of corn. *Jour. Amer. Soc. Agron.* 34:923–32.

Srb, A. M. and Horowitz, N. H. 1944. The ornithine cycle in Neurospora and its genetic control. *Jour. Biol. Chem.* 154:129–39.

Stadler, L. J. 1928. Mutation in barley induced by X-rays and radium. *Science* 68:186–87.

———. 1930. The frequency of mutations of specific loci in maize. *Anat. Rec.* 47:381.

Stebbins, G. L. 1950. *Variation and evolution in plants.* New York: Columbia University Press.

Steffensen, J. F. 1930. *Some recent researches in the theory of statistics and actuarial science.* Cambridge: At the University Press.

Stern, C. 1960. O. Vogt and the terms "Penetrance" and "Expressivity." *Amer. Jour. Human Genetics* 12:141.

Stone, W. S., Wyss, D., and Haas, F. 1947. The production of mutations in Staphylococcus aureus by irradiation of the substrate. *Proc. Nat. Acad. Sci.* 33:59–66.

Stormont, G. 1959. On the application of blood groups in animal breeding. *Proc. X Inter. Congr. Genetics* 1:206–24.

Stormont, C., Owen, R. D., and Irwin, M. R. 1951. The B and C systems of bovine blood groups. *Genetics* 36:134–61.

Stresinger, G. and Franklin, N. C. 1957. Mutation and recombination at the host range genetic region of phage T2. *Cold Spring Harbor Symp. Quant. Biol.* (*1956*) 21:103–11.

"Student" (W. S. Gosset). 1908. The probable error of a mean. *Biometrika* 6:1–25.

Sturtevant, A. H. 1913. The linear arrangement of six sex-linked factors in Drosophila as shown by their mode of association. *Jour. Exp. Zool.* 14:43–59.

———. 1921. A case of rearrangement of genes in Drosophila. *Proc. Nat. Acad. Sci.* 7:235–37.

———. 1923. Inheritance of direction of coiling in Limnaea. *Science* 58: 269–70.

———. 1925. The effects of unequal crossing over at the Bar Locus in Drosophila. *Genetics* 10:117–47.

Sturtevant, A. H. and Beadle, G. W. 1936. The relations of inversions in the X chromosome of Drosophila melanogaster to crossing over and nondisjunction. *Genetics* 21:554–664.

Sturtevant, A. H. and Dobzhansky. 1936*a*. Inversions in the third chromosome of wild races of Drosophila pseudoobscura and their use in the study of the history of the species. *Proc. Nat. Acad. Sci.* 22:448–50.

———. 1936*b*. Geographical distribution and cytology of "sex ratio" in Drosophila pseudoobscura and related species. *Genetics* 21:473–90.

Sumner, F. B. 1920. Geographical variation and Mendelian inheritance. *Jour. Exp. Zool.* 30:369–402.

Sutton, W. S. 1903. The chromosomes in heredity. *Biol. Bull.* 4:231–51.

Swanson, C. P. 1957. *Cytology and cytogenetics.* Englewood Cliffs, N.J.: Prentice-Hall Inc.

Swift, H. F. 1950*a*. The constancy of desoxyribose nucleic acid in plant nuclei. *Proc. Nat. Acad. Sci.* 36:643–54.

———. 1950*b*. The desoxyribose nucleic content of animal nuclei. *Physiol. Zool.* 23:169–98.

Tan, C. C. 1946. Mosaic dominance in the inheritance of color patterns in the lady-bird beetle. Harmonia axyridis. *Genetics* 31:105–210.

Tanaka, G. 1924. Maternal inheritance in Bombyx mori. *Genetics* 9:479–86.

Thiele, T. N. 1889. *Almindelig Iagttagelseslaere*, 1903. *Theory of observations.* London: C. & E. Layton, (reprinted in *Ann. Math. Statistics* (1931) 2:165–308).

Thompson, G. H. 1939. *The factorial analysis of human ability.* New York: Houghton Mifflin Co.

Thurston, L. L. 1942. *Multiple-factor analysis.* Chicago: The University of Chicago Press.

Timofeeff-Ressovsky, N. W. 1934. The experimental production of mutations. *Biol. Rev.* 9:411–56.

———. 1935. Über "mutterliche Vererbung" bei Drosophila. *Die Naturwiss.* 1935:493–96.

Toldt, K. 1912. Beitrage zur Kenntnis der Behaarung der Säugetiere. *Zool. Anz.* 44:393–402.

Toyama, K. 1913. Maternal inheritance and Mendelism. *Jour. Genet.* 2:351–404.

Tschermak, E. 1900. Über künstliche Kreuzung bei Pisum sativum. *Ber. deutsch. bot. Gesells.* 18:232–39.

Tukey, J. W. 1954. Causation, regression and path analysis. In *Statistics and mathematics in biology*, eds. O. Kempthorne, T. A. Bancroft, J. W. Gowen, and J. L. Lush, pp. 35–66. Ames: Iowa State University Press.

Turner, M. E. and Stevens, C. E. 1959. The regression analysis of causal paths. *Biometrics* 15:236–58.

Visconti, N. and Delbrück, M. 1953. The mechanism of genetic recombination in phage. *Genetics* 38:5–33.

Vries, H. De 1901, 1909–10. *The mutation theory: I. The origin of species by mutation; II. The origin of varieties by mutation.* Translated by J. B. Farmer and K. D. Darbishire. Chicago: The Open Court Publishing Co.

Waddington, C. H. 1943. The development of some "leg genes" in Drosophila. *Jour. Genet.* 45:29–43.

———. 1955. On a case of quantitative variation on either side of the wild type. *Zeit. Abst. Vererb.* 87:208–28.

Wagener, Gertrud. 1959. Die Entstehung der Scheckung bei der Haubenratte. *Biol. Ztrbl.* 78:451–60.

Wald, A. 1947. *Sequential analysis.* New York: John Wiley & Sons.

Wang, H. 1943. The morphogenetic functions of the epidermal and larval components of the papilla in feather regeneration. *Physiol. Zool.* 16:325–50.

Warren, D. C. 1924. Inheritance of egg size in Drosophila melanogaster. *Genetics* 9:41–69.

Waters, N. F. 1931. Inheritance of body weight in domestic fowls. *Rhode Island Agr. Exp. Sta. Bull.* 228:7–103.

Watson, J. D. 1965. *Molecular biology of the gene.* New York: W. A. Benjamin, Inc.

Watson, J. D. and Crick, F. H. C. 1953. Genetic implications of the structure of deoxyribonucleic acid. *Nature* 17:964–69.

Weinberg, W. 1908. Über den Nachweis der Vererbung beim Menschen. *Jahreshelfts Ver. vaterl. Naturf. Württemberg* 64:369–82.

Weinstein, A. 1936. The theory of multiple strand crossing over. *Genetics* 21:155–99.

Wenrich, D. H. 1916. The spermatogenesis of Phrynotettix magnus with special reference to synapsis and the individuality of the chromosomes. *Bull. Museum Compar. Zool. Harvard* 60:57–133.

Werbitzki, F. W. 1910. Über blepharoblastlose Trypanosomen. *Zeit. f. Bakt.* 53:303–15.

Wettstein, F. von. 1924. Morphologie und Physiologie des Formwechsels der Moose auf genetische Grundlege. *Zeit. Abs. Vererb.* 33:1–226.

Wettstein, F. von. 1928. Morphologie und Physiologie des Formwechsels der Moose auf Genetisches Grundlage. *Bibl. Genetica* 10:1–216.

Wettstein, D. von, and Eriksson, Gösta, 1965. The genetics of chloroplasts. *Proc. XI. Int. Congr. Genetics (1963)* 3:591–610.

Whitehouse, H. L. K. 1942. Crossing over in Neurospora. *The New Phytogolist* 41:23–62.

Wheldale, Muriel. 1916. The anthocyan pigments of plants. Cambridge: At the University Press; 2d ed., M. W. Onslow, 1925.

Whiting, P. W. and Gilmore, K. A. 1932. Genetic analysis of synapsis and maturation in eggs of Habrobracon. *Proc. 6th Int. Congress Genetics* 2:210–11.

Wiener, A. S. and Wexler, I. B. 1952. The mosaic structure of red blood cell agglutininogens. *Bact. Rev.* 16:69–87.

Wigan, L. G. 1949. The distribution of polygenic activity on the X chromosome of Drosophila melanogaster. *Heredity* 3:53–66.

Williams, W. 1947. Genetics of red clover (Trifolium pratense L.) compatibility. III. The frequencies of incompatibility S alleles in two nonpedigree populations of red clover. *Jour. Genet.* 48:69–79.

Wilson, E. B. 1925. *The cell in heredity and inheritance.* 3d ed. New York: The Macmillan Co.

Winge, Ö. 1917. The chromosomes, their number and general importance. *C. R. Trav. Labor. Carlsberg* 13:131–275.

———. 1919. On the non-Mendelian inheritance in variegated plants. *C. R. Trav. Labor. Carlsberg* 14:1–20.

———. 1937. Goldschmidt's theory of sex determination. *Jour. Genet.* 34:81–89.

Winge, Ö and Laustsen, O. 1937. On two types of spore germination and on genetic segregation in Saccharomyces, demonstrated through single-spore cultures. *C. R. Trav. Labor. Carlsberg* 22:99–119.

Winge, Ö. and Roberts, Catherine. 1954. On tetrad analysis apparently inconsistent with Mendelian law. *Hereditas* 8:295–304.

Winkler, H. 1930. *Die Konversion der Gene.* Jena: G. Fischer.

Wolff, G. L. 1954. A sex difference in the coat color change of a specific guinea pig genotype. *Amer. Nat.* 88:381–83.

———. 1955. The effects of environmental temperature on coat color in diverse genotypes of the guinea pig. *Genetics* 40:96–106.

Woods, M. W. and Du Buy, H. G. 1951. Hereditary and pathogenic nature of mutant mitochondria in *Napata. J. Nat. Cancer Inst.* 11:1105–51.

Wright, S. 1916. An intensive study of the inheritance of color and other coat characters in guinea pigs with special reference to graded variation. *Carnegie Inst. Washington*, Publ. 241:59–160.

———. 1917. The average correlation within subgroups of a population. *Jour. Washington Acad. Sci.* 7:532–35.

———. 1918. On the nature of size factors. *Genetics* 3:367–74.

———. 1920. The relative importance of heredity and environment in determining the piebald pattern of guinea pigs. *Proc. Nat. Acad. Sci.* 6:320–32.

———. 1921. Correlation and causation. *Jour. Agric. Res.* 20:557–85.

———. 1923*a*. The theory of path coefficients—a reply to Niles' criticism. *Genetics* 8:239–55.

———. 1923*b*. The relation between piebald and tortoiseshell color patterns in guinea pigs. *Anat. Rec.* 23:393.

———. 1925. The factors of the albino series of guinea pigs and their effects on black and yellow pigmentation. *Genetics* 10:223–60.

———. 1926*a*. A frequency curve adapted to variation in percentage occurrence. *Jour. Amer. Stat. Assoc.* 21:162–78.

———. 1926*b*. Effects of age of parents on characteristics of the guinea pig. *Amer. Nat.* 60:552–59.

———. 1927. The effects in combination of the major color factors of the guinea pig. *Genetics* 12:530–69.

———. 1928. An eight factor cross in the guinea pig. *Genetics* 26:650–69.

———. 1929. Fisher's theory of dominance. *Amer. Nat.* 63:274–79.

———. 1931. Evolution in Mendelian populations. *Genetics* 16:97–159.

———. 1932. General, group and special size factors. *Genetics* 17:603–19.

———. 1934*a*. Physiological and evolutionary theories of dominance. *Amer. Nat.* 68:25–53.

———. 1934*b*. On the genetics of subnormal development of the head (otocephaly) in the guinea pig. *Genetics* 19:471–505.

———. 1934*c*. An analysis of variability in number of digits in an inbred strain of guinea pigs. *Genetics* 19:506–36.

———. 1934*d*. The results of crosses between inbred strains of guinea pigs differing in number of digits. *Genetics* 19:537–51.

———. 1934*e*. Genetics of abnormal growth in the guinea pig. *Cold Spring Harbor Symp. Quant. Biol.* 2:137–47.

———. 1934*f*. The method of path coefficients. *Ann. Math. Stat.* 5:161–215.

———. 1935*a*. A mutation of the guinea pig tending to restore the pentadactyl foot when heterozygous, producing a monstrosity when homozygous. *Genetics* 20:84–107.

——. 1935*b*. The emergence of novelty. A review of Lloyd Morgan's "Emergent Theory of Evolution." *Jour. Hered.* 26:369–73.

———. 1935*c*. On the genetics of rosette pattern in guinea pigs. *Genetica* 17:547–60.

———. 1941*a*. A quantitative study of the interactions of the major colour factors of the guinea pig. *Proc. 7th Int. Genetics Congr.* (1939):319–29.

———. 1941*b*. The "age and area" concept extended. (A review of book by J. C. Willis). *Ecology* 22:345–47.

———. 1941*c*. The physiology of the gene. *Physiological Reviews* 21:487–527.

———. 1945. Genes as physiological agents. General considerations. *Amer. Nat.* 79:289–303.

———. 1947. On the genetics of several types of silvering in the guinea pig. *Genetics* 32:115–41.

———. 1929. The persistence of differentiation among inbred families of guinea pigs. Washington, D.C.: U.S. Dept. of Agriculture, Tech. Bull. No. 103.

Wright, S. and Wagner, K. 1934. Types of subnormal development of the head from inbred strains of guinea pigs and their bearing on the classification and interpretation of vertebrate monsters. *Amer. Jour. Anat.* 54:383–447.

Wyss, O., Stone, W. S., and Clark, J. B. 1947. The production of mutations in Staphylococcus aureus by chemical treatment of the substrate. *Jour. Bact.* 54:767–72.

Yates, F. 1934. Contingency tables involving small numbers and the χ^2 test. *Suppl. to Jour. Roy. Stat. Soc.* 1:217–35.

Yule, G. U. 1902. Mendel's laws and their probable relation to intra-racial heredity. *New Phytol.* 1:193–207, 222–38.

———. 1907. On the theory of correlation for any number of variables treated by a new system of notation. *Proc. Roy. Soc.* A79:182–93.

Zeleny, C. 1922. The effect of selection for eye facet numbers in the white-Bar eye race of Drosophila melanogaster. *Genetics* 7:1–115.

Zinder, N. D. and Lederberg, J. 1952. Genetic exchange in Salmonella. *Jour. Bact.* 64:679–99.

———. 1949*a*. Estimates of amounts of melanin in the hair of diverse genoty of the guinea pig, from transformation of empirical grades. *Gene* 34:245–71.

———. 1949*b*. On the genetics of hair direction in the guinea pig. I. Variabili in the patterns found in combinations of R and M loci. *Jour. Exp. Zo* 112:303–24; II. Evidences for a new dominant gene, Star, and tests f linkage with eleven other loci. *Jour. Exp. Zool.* 112:325–40.

———. 1950. On the genetics of hair direction in the guinea pig. III. Interactio between the processes due to the loci R and St. *Jour. Exp. Zool.* 113:33–64

———. 1952. The genetics of quantitative variability. In *Quantitative inheritance* p. 5–41. Agric. Res. Council, London: Her Majesty's Stationery Office.

———. 1954. The interpretation of multivariate systems. In *Statistics and mathematics in biology*, eds. O. Kempthorne, T. A. Bancroft, J. W. Gowen, and J. L. Lush, pp. 11–33. Ames: Iowa State University Press.

———. 1959*a*. Genetics, the gene, and the hierarchy of biological sciences. *Proc. X Int. Congr. Gen.* (*1958*) 1:475–89.

———. 1959*b*. On the genetics of silvering in the guinea pig with especial reference to interaction and linkage. *Genetics* 44:383–405.

———. 1959*c*. Qualitative differences among diverse colors of the guinea pig due to diverse genotypes. *Jour. Exp. Zool.* 142:75–114.

———. 1959*d*. Silvering (si) and diminution (dm) of coat color of the guinea pig, and male sterility of the white or near white combination of these. *Genetics* 44:563–90.

———. 1960*a*. Path coefficients and path regression: alternative or complementary concepts? *Biometrics* 16:189–202.

———. 1960*b*. The treatment of reciprocal interaction, with or without lag, by path analysis. *Biometrics* 16:423–45.

———. 1960*c*. Residual variability of intensity of coat color in the guinea pig. *Genetics* 45:583–612.

———. 1960*d*. Postnatal changes in the intensity of coat color in diverse genotypes of the guinea pig. *Genetics* 45:1503–29.

——— 1960*e*. The genetics of vital characters of the guinea pig. *Jour. Cell Comp. Physiol.* 56 (Suppl.) 1:123–51.

———. 1963. Genic interaction. In *Methodology in mammalian genetics*, ed. W. J. Burdette, pp. 159–92. San Francisco: Holden-Day.

Wright, S. and Braddock, Zora I. 1949. Colorimetric determinations of the amounts of melanin in the hair of diverse genotypes of the guinea pig. *Genetics* 34:223–44.

Wright, S. and Chase, H. B. 1936. On the genetics of the spotted pattern of the guinea pig. *Genetics* 21:758–87.

Wright, S. and Eaton, O. N. 1923. Factors which determine otocephaly in the guinea pig. *Jour. Agric. Res.* 26:161–82.

———. 1926. Mutational mosaic coat patterns of the guinea pig. Genetics 11:333–51.

AUTHOR INDEX

SUBJECT INDEX

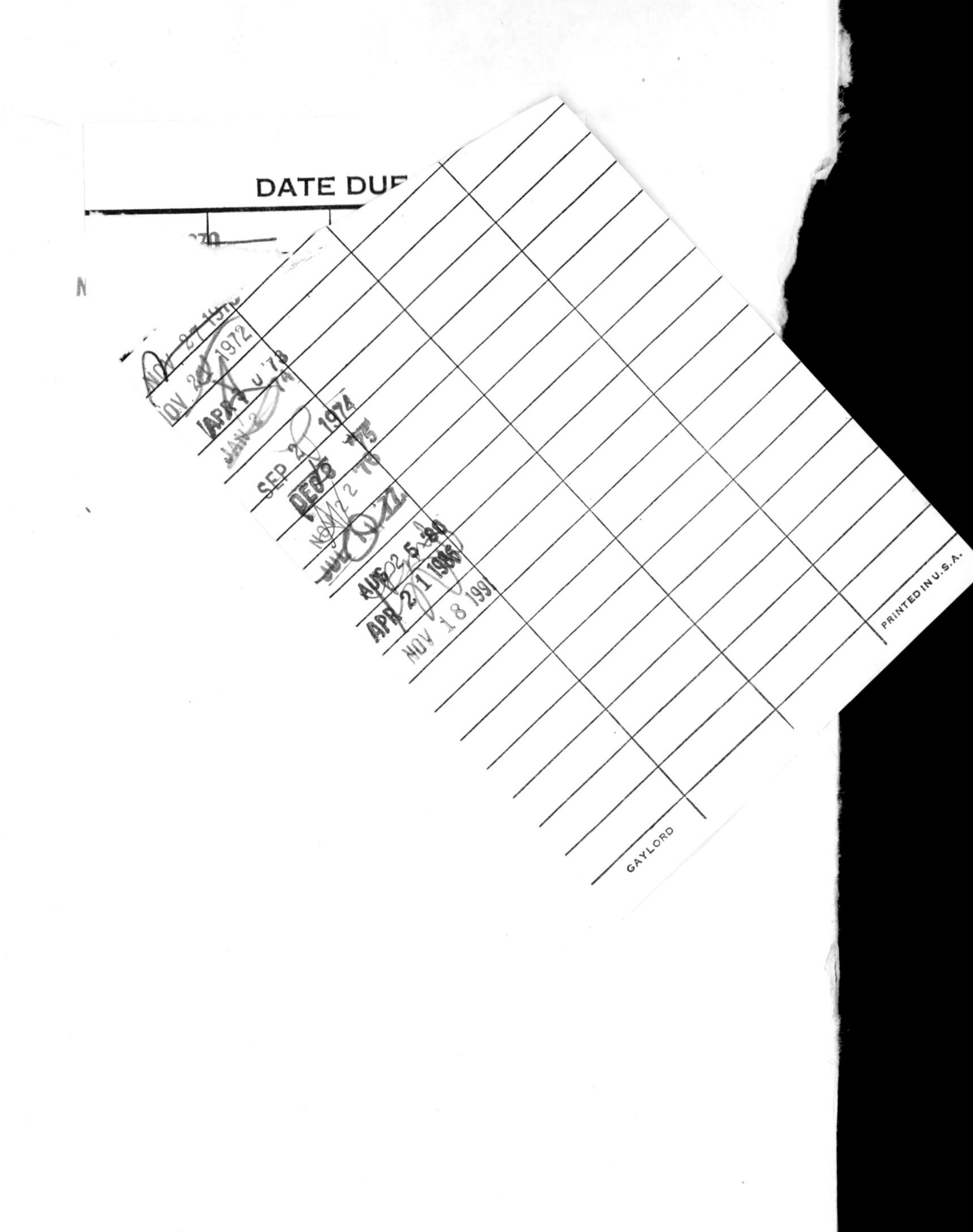
DATE DUE
GAYLORD
PRINTED IN U.S.A.